S0-BDP-556

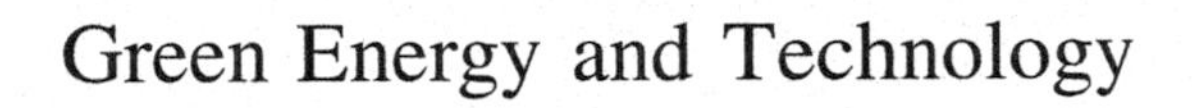

For further volumes:
http://www.springer.com/series/8059

Efstathios E. (Stathis) Michaelides

Alternative Energy Sources

Efstathios E. (Stathis) Michaelides
Department of Engineering
TCU
Fort Worth, TX
USA
e-mail: e.michaelides@tcu.edu

ISSN 1865-3529
e-ISSN 1865-3537
ISBN 978-3-642-20950-5
e-ISBN 978-3-642-20951-2
DOI 10.1007/978-3-642-20951-2
Springer Heidelberg Dordrecht London New York

Library of Congress Control Number: 2011942743

Printed on acid-free paper

Springer is part of Springer Science+Business Media (www.springer.com)

To the memory of my parents
Emmanuel and Eleni

Preface

In the beginning of the twenty-first century, our society is faced with an *energy challenge*: as highly populous, developing countries become more affluent and as the developed nations continue to increase their energy consumption, the energy demand in the entire world has reached levels that cannot be sustained in the future. At the same time, fossil fuels, which are currently providing more than 85% of the total global energy supply, are limited and, in addition, their widespread use has significant adverse environmental consequences. The combustion of fossil fuels produces carbon dioxide, which is one of the causes of global warming as well as of other environmental effects, such as acid rain; higher ozone concentration in urban areas; particulates; and aerosols that are detrimental to air quality. The limited supply of the fossil fuels and their effects on the global environment indicate the only long-term solution of the *energy challenge*: a significant increase in the use of alternative energy sources for the production of electricity as well as for meeting other energy needs of the industrial and post-industrial human society.

This book on *Alternative Energy Sources* is designed to give the reader, a clear view of the role each form of alternative energy may play in supplying the energy needs of the human society in the near and intermediate future (20–50 years). The book is aimed at two types of audience:

a. The student of science and engineering who may take an elective course on one of the subjects of "alternative energy," "renewable energy," "sustainability," etc. For this purpose, the students will review and expand on several concepts taught in the traditional disciplines of Thermodynamics, Fluid Dynamics and Heat Transfer. If "repetition is the mother of learning," students in the engineering disciplines will learn a great deal of the material taught in the Thermal Sciences courses by studying this book.
b. The educated reader, who has a basic knowledge of mathematics and science (e.g. algebra, elementary physics and elementary chemistry). The book assumes minimum prior knowledge on behalf of the reader and imparts some of the

pre-requisite knowledge by including chapters on basic thermodynamics, and elementary financial accounting (investment appraisal methods).

A unique aspect of this book is that it includes two chapters on nuclear fission and one on fusion energy. The reason for this is that nuclear energy does not contribute to the greenhouse effect and is currently viewed by many decision makers as an excellent alternative option for the production of electricity in the twenty-first century. As a result, and despite the recent accident at Fukushima Daiichi, the licensing process for additional nuclear power plants has accelerated in several countries and the debate for the future of nuclear energy has been recently renewed. Nuclear energy, a perennial pariah of environmental groups, may actually become one of the solutions to the global climate change.

The two first chapters on *energy demand and supply* and *environmental effects*, set the tone as to why the widespread use of alternative energy is essential for the future. The third chapter exposes the reader to the laws of energy conversion processes, as well as the limitations of converting one energy form to another. The sections on *exergy* give a succinct, quantitative background on the capability/potential of each energy source to produce power on a global scale. The fourth, fifth and sixth chapters are expositions of *fission and fusion nuclear energy*. The following five chapters (seventh to eleventh) include detailed descriptions of the most common renewable energy sources—*wind*, *solar*, *geothermal*, *biomass*, *hydroelectric*—and some of the less common sources, such as *tidal* and *wave* energy. The emphasis of these chapters is on the global potential of each source; the engineering/technical systems that are currently used in harnessing the potential of each one of these energy sources; the technological developments that will contribute to wider utilization of the sources; and the environmental effects associated with their current and their wider use. The last three chapters are: *energy storage*, which is the main limitation of the wider use of solar and wind power and will become an important issue if renewable energy sources are to be used widely; *energy conservation*, which appears to be everyone's favorite issue, but by itself is not a solution to our *energy challenge*; and *energy economics*, a necessary consideration in market-driven economies.

A number of individuals have helped in the writing of this book: first among them are the students who took my course on *Alternative Energy*. I have learned from them and their questions more than they have learned in my classes. Two of these students contributed significantly to the writing of the book: Maria Andersson reviewed several chapters and gave me valuable suggestions. Eric Stewart drew some of the figures. I am very thankful to my colleagues at the University of Texas at San Antonio and at Texas Christian University, for several fruitful discussions on energy and the great challenge our society is facing. I am also very indebted to my own family, not only for their constant support, but also for lending a hand whenever it was needed. My wife, Laura, has been a constant source of inspiration and help. My father-in-law, Dionisio Garcia proofread some of the chapters and gave me valuable comments. My son Dimitri, who decided to become a student of nuclear energy, devoted a good part of his vacation time to proof-read the entire manuscript and gave me many excellent suggestions. Emmanuel and Eleni were always there and ready to help. I owe to all my sincere gratitude.

Fort Worth, TX, September 2011 Efstathios E. (Stathis) Michaelides

Contents

Chapter 1
Energy Demand and Supply

Abstract The use of energy defines the beginning of human civilization: when the prehistoric human mastered the use of fire for domestic comfort and cooking, human civilization began and evolved to reach the age of the locomotive, the nuclear power plant, the automobile, the airplane, the personal computer and the wireless internet. Throughout the centuries, the human society has evolved by increasingly using energy to the point where the consumption of energy is necessary for the functioning of the contemporary society, the prosperity of the nations and the survival of our civilization. Energy is produced and is being used in different forms: airplanes and automobiles use liquid hydrocarbon fuels; electric power plants convert primarily the energy in coal, natural gas, nuclear and hydroelectric into electricity; and a contemporary household uses electricity and natural gas for domestic comfort, entertainment and the preparation of meals. Because most functions of our society are based on the use of energy, elaborate networks of energy supply have been developed in the last three centuries: electricity is fed into communities by the transmission lines of the electric grid at high voltage; natural gas by a complex system of pipelines, which transcend national boundaries; and tanker ships crisscross the oceans daily to supply crude oil to refineries. The economic impact of the energy supply and the energy trade is of paramount importance to all nations. The geopolitical activities of most modern nations are significantly influenced by their need for a constant and secure energy supply. Most of modern wars (those after 1950) have been fought for the control and security of energy supplies and many treaties and international agreements have been cemented with energy resources as the primary issue. This is a general chapter on energy, not just alternative energy, which explains in a quantitative way what is the quantity we call "energy," whence it comes and where it goes. The forms that energy is produced and consumed are, first, explained. The several units that are commonly used for different energy forms and quantities are listed and their equivalencies are explained. Secondly, the importance of energy in the economic activities of the contemporary society is described qualitatively and the

E. E. (Stathis) Michaelides, *Alternative Energy Sources*,
Green Energy and Technology, DOI: 10.1007/978-3-642-20951-2_1,

primary energy resources are identified. The current energy trade between groups of nations is also described briefly and the main flow of primary energy resources is identified. Thirdly, historical data on energy production and consumption in several groups of nations are offered as well as some acceptable predictions for the future demand and supply of energy.

1.1 Forms and Units of Work, Heat and Energy

From the theory of Newtonian Mechanics, work has been defined as the scalar product of a force and the distance its point of application moves during a period of time. Power is the work per unit time. In the case of a force that is variable, the work is given by an integral, which describes the motion of the force between two end points, *a* and *b*. The work and its time derivative, the instantaneous power, are defined as follows:

$$W = \int_a^b \vec{F} \bullet d\vec{x} \quad and \quad \dot{W} = \vec{F} \bullet \frac{d\vec{x}}{dt} = \vec{F} \bullet \vec{V}. \tag{1.1}$$

Power is the scalar product of the force and of the velocity its point of application moves. For an engine to produce or consume power there must be forceful motion of some of its components. Usually, this motion is circular and is converted by gears to linear motion.

All the engines that are currently used for the performance of the several tasks desired by the human society produce work and consume energy resources. For example, the internal combustion engine consumes gasoline or diesel and produces work, which is used in propulsion. "Energy" is a very broad concept and is defined as the potential of materials and systems to perform useful work. Energy is measured in the same units as work and comes in several forms. Heat or thermal energy is a form of energy that is transferred from materials at higher temperatures to materials at lower temperatures. It is usually produced by the combustion of fuels, which are regarded as "energy sources." There are several engineering systems, such as the power plants and the internal combustion engines, which convert heat into work. Other forms of energy, which are commonly met in engineering practice and everyday applications, are:

1. The potential energy (mgz) of matter, at a higher level.
2. The kinetic energy ($1/2\ m\ |\vec{V}|^2$) of matter that moves with velocity $\vec{V}$.
3. The chemical energy (ΔG) of substances, such as coal and hydrocarbons.
4. The elastic energy ($Vm\sigma\varepsilon$) of a stressed solid with volume V and strain ε.
5. The electric energy (qV) of a charge q in a voltage difference V.
6. The magnetic energy (HM) of a magnetic quantity M in a magnetic field with intensity H.
7. The wave energy ($1/2A\rho g\alpha^2$) of waves on an area A with amplitude α.

8. The nuclear energy (mc^2) which is equivalent to the mass, m.
9. The surface energy (γA) of fluids with surface tension γ and area A.

Our society uses the energy content of several substances to produce work and to accomplish several tasks and activities. Different forms of energy undergo conversions and produce work during these activities. The energy conversions are subjected to the Laws of Thermodynamics. Details of these laws, a general overview of the subject of Thermodynamics, and examples on the conversions of energy forms are given in Chap. 3.

1.1.1 Units of Energy

From the kinetic energy of a single electron that may have been created during the "electron–hole" pair process in a photovoltaic cell, to the net power produced in a large nuclear power station, the units of energy span a very large range. Although they are different concepts and are not interchangeable, work, heat and energy are measured by the same units. The unit of energy in the *Systeme Internationale* (S.I.) is the Joule (J) and is defined as the work done when a constant force of 1 Newton (N) moves its point of application by 1 meter (m). One Joule is also performed when a charge of 1 Coulomb (1 Cb) moves through an electric potential difference of 1 Volt (1 V). Multiples of the Joule—as well as of all the other S.I. units—have been defined by international convention and are as follows:

- 1 yocto-Joule (yJ) = 10^{-24} J
- 1 zepto-Joule (zJ) = 10^{-21} J
- 1 atto-Joule (aJ) = 10^{-18} J
- 1 femto-Joule (fJ) = 10^{-15} J.
- 1 pico-Joule (pJ) = 10^{-12} J.
- 1 nano-Joule (nJ) = 10^{-9} J.
- 1 micro-Joule (μJ) = 10^{-6} J.
- 1 milli-Joule (mJ) = 10^{-3} J.
- 1 kilo-Joule (kJ) = 10^{3} J.
- 1 mega-Joule (MJ) = 10^{6} J.
- 1 giga-Joule (GJ) = 10^{9} J.
- 1 tera-Joule (TJ) = 10^{12} J.
- 1 peta-Joule (PJ) = 10^{15} J.
- 1 exa-Joule (EJ) = 10^{18} J
- 1 zetta-Joule (ZJ) = 10^{21} J
- 1 yotta-Joule (YJ) = 10^{24} J

It must be noted that all these prefixes may be applied to all other units of the S.I. Also that the S.I. units and their multiples have very precise notation and that differences in notation between lower case letters and capital letters denote differences of several orders of magnitude. For example, while mJ/mJ = 1 by definition, MJ/mJ = 10^9, and PJ/pJ = 10^{27}. All science and engineering students have to be very careful and precise with the notation of units.

The unit of power in the S.I. is the Watt (W). When an engine performs one Joule of work per second, it produces a power of one Watt (1 W). The multiples of the Watt as well as for all other units in the *Systeme Internationale* are the same as those of the unit Joule, as listed above.

Because energy is such a ubiquitous and practical subject of significant economic importance and everyday use, several other units of energy have been defined in the past and are still in common use:

- In the c.g.s. system of units the erg is the main energy unit, with 1 erg being equal to 10^{-7} J.
- In the British system of units, the British thermal unit (1 Btu) is very commonly used in the heating and air-conditioning industries and is equal to 1.055 kJ. One Btu is defined as the amount of energy needed to increase the temperature of 1 lb of water from 14.5 to 15.5°F.
- A larger unit in the British system is the Therm (1 therm), which is equal to 10^5 Btu or $1.055*10^8$ J.
- The calorie (cal) is equal to 4.184 J.
- An extremely large amount of energy is the Quad (1 Q), which is equal to 10^{15} Btu or, approximately, 10^{18} J. The total energy consumed in the USA in 2009 was approximately 100 Q and in the entire world approximately 490 Q.
- For extremely small amounts of energy, usually related to atoms or nuclei, the electron-volt (eV) has been defined as the potential energy gained by an electron when it moves through an electric potential difference of 1 Volt. Given that the charge of an electron is $1.6*10^{-19}$ Cb, 1 eV is equal to $1.6*10^{-19}$ J.

A third type of energy units is related to the chemical energy content of fuels, such as coal, crude oil and natural gas. Because the chemical composition and energy content of actual fuels, e.g. sweet East Texas crude or Saudi Arabian crude, depends largely on the location the fuels are extracted from, these units have been fixed according to international convention. Among these units their S.I. equivalents are the following:

- 1 ton of coal equivalent (1 tce) which is equal to $2.931 * 10^{10}$ J.
- 1 barrel of oil (1 bbl) which is equal to $6.119 * 10^9$ J.
- 1 cubic foot of natural gas (1 scf), measured at standard pressure and temperature (1 atm and 25°C), which is equal to $1.072 * 10^6$ J. 1 scf is often approximated as 10^6 J.

Table 1.1 shows the conversion factors for several energy units that are commonly used. For example, 1 cal = 4.18 J, 1 ton of coal is equivalent to $2.78 * 10^7$ Btu, etc. The diagonal terms of this table are equal to 1 by definition, and each term is equal to the reciprocal of its diagonally mirror term, that is $c_{ij} = 1/c_{ji}$.

Table 1.1 Conversion factors for energy units

	eV	calorie	erg	J	Btu	gallon of gasoline	barrel of oil
eV	1	3.82E − 20	1.60E − 12	1.60E − 19	1.52E − 22	1.21E − 27	2.61E − 29
calorie	2.62E + 19	1	4.19E + 07	4.18	3.97E − 03	3.17E − 08	6.84E − 10
erg	6.25E + 11	2.39E − 08	1	1.00E − 07	9.48E − 11	7.58E − 16	1.63E − 17
J	6.25E + 18	0.24	1.00E + 07	1	9.48E − 04	7.58E − 09	1.63E − 10
Btu	6.59E + 21	252	1.06E + 10	1,055	1	8.00E − 06	1.72E − 07
gallon of gasoline	8.25E + 26	3.15E + 07	1.32E + 15	1.32E + 08	1.25E + 05	1	2.16E − 02
barrel of oil	3.82E + 28	1.46E + 09	6.12E + 16	6.12E + 09	5.80E + 06	46.4	1
kw-h	2.25E + 25	8.60E + 05	3.6E + 13	3.60E + 06	3,412	2.73E − 02	5.88E − 04
ft-lb	8.48E + 18	0.32	1.36E + 07	1.36	1.29E − 03	1.03E − 08	2.22E − 10
therm	2.62E + 28	2.52E + 07	1.06E + 15	1.06E + 08	1.00E + 05	0.8	1.72E − 02
ton-TNT	2.62E + 28	1.00E + 09	4.18E + 16	4.18E + 09	3.97E + 06	31.73	0.68
quad	6.59E + 36	2.5E + 17	1.06E + 25	1.06E + 18	1E + 15	8.00E + 09	1.72E + 08
ton of coal	1.83E + 29	7.00E + 09	2.93E + 17	2.93E + 10	2.78E + 07	2.22E + 02	4.79E + 00
scf of nat. gas	1.01E + 25	2.56E + 05	1.07E + 13	1.07E + 06	1,016	8.12E − 03	1.75E − 04

	kw-h	ft-lb	therm	ton-TNT	quad	ton of coal	scf naturalgas
eV	4.44E − 26	1.18E − 19	3.82E − 29	3.82E − 29	1.52E − 37	5.46E − 30	9.89E − 26
calorie	1.16E − 06	3.09	3.97E − 08	1.00E − 09	3.97E − 18	1.43E − 10	3.90E − 06
erg	2.78E − 14	7.37E − 08	9.48E − 16	2.39E − 17	9.48E − 26	3.41E − 18	9.33E − 14
J	2.78E − 07	0.74	9.48E − 09	2.39E − 10	9.48E − 19	3.41E − 11	9.33E − 07
Btu	2.93E − 04	7.78E + 02	1.00E − 05	2.52E − 07	1.00E − 15	3.60E − 08	9.84E − 04
gallon of gasoline	3.66E + 01	9.73E + 07	1.25	3.15E − 02	1.25E − 10	4.50E − 03	123.1
barrel of oil	1.70E + 03	4.51E + 09	5.80E + 01	1.46	5.80E − 09	2.09E − 01	5,708
kw-h	1	2.66E + 06	3.41E − 02	8.61E − 04	3.41E − 12	1.23E − 04	3.358
ft-lb	3.77E − 07	1	1.29E − 08	3.24E − 10	1.29E − 18	4.63E − 11	1.27E − 06
therm	29.31	7.78E + 07	1	2.52E − 02	1.00E − 10	3.60E − 03	98.42
ton-TNT	1162	3.09E + 09	39.66	1	3.97E − 09	1.43E − 01	3,899
quad	2.93E + 11	7.78E + 17	1.00E + 10	2.52E + 08	1	3.60E + 07	9.84E + 11
ton of coal	8.14E + 03	2.16E + 10	2.78E + 02	7.00	2.78E − 08	1	2.73E + 04
scf of nat. gas	0.2978	7.88E + 05	1.02E − 02	2.56E − 04	1.02E − 12	3.66E − 05	1

1.2 Energy Demand and Supply

The beginning of human civilization is marked by the use of energy. When the prehistoric humans started using fire for cooking, heating, and protection from wild animals, the human civilization began. A few millennia later, the human society evolved and used the various forms of energy for heating during the winter, air-conditioning for summer comfort, cooking, transportation, and other activities that define the contemporary civilization. In the twenty-first century, the era of factories, individual houses, airplanes, automobiles and personal computers, energy is being increasingly used for the production of consumer goods, living comfort, transportation and entertainment.

It is hard to envision the human society of the twenty-first century without the use of energy in its several forms: on a typical working day, a man in Paris, France will wake up to the sound of his electric alarm; will make a cup of coffee and breakfast, using gas or electricity; will take the electric elevator to the ground floor of his apartment; will walk to the closest *metropolitaine* (metro) station and, after descending in the electric escalator will take one or more of the electrically powered metro lines towards his working place. If he does not use the metro system, he will use one of the many diesel-powered buses. When he reaches his destination will ascend again in an electric escalator; will walk to his place of work, which is heated by burning natural gas and lit by a system of fluorescent lights. Most of the tasks he will perform during a typical working day, for example, telephone calls, writing, e-mails, power-point presentations, involve the use of significant amounts of electric energy. The day of a typical woman in Los Angeles, California, has a very similar energy-use pattern. One difference may be that, instead of using public transportation, she will use her own automobile to drive to work and that her office is typically air-conditioned instead of heated. Men or women in developing countries may use fewer energy consuming devices and engines in their everyday lives, but, nevertheless, still use a significant amount of energy for living comfort, cooking, transportation and the production of goods.

The contemporary human society depends to a large extent on the use of energy in its various forms. The importance of energy in the human society is accentuated by cases when the supply of sufficient energy has been interrupted: the coal miners' strike in Britain in the winter of 1973–1974 resulted in the disruption of the transportation system, the reduction of the working week to a three-day week, economic recession and several deaths, which were due to lack of heating, electricity and lack of transportation. During the oil embargo of the mid 1970s, shortage of gasoline in Europe and North America resulted in the rationing of oil products, industrial production disruption, long waiting lines in the gas stations and, in a few instances in USA, shootings and murders. The interruption of the electric power supply in New York City on July 13, 1977 and in the entire eastern part of the USA on August 14, 2003 resulted in two infamous blackouts: in both cases the loss of electric power in several American cities resulted in the disruption of civic power, lootings, and large-scale municipal unrest. During these two blackouts, the interruption of electric power supply clearly signaled the disruption of the rule of law and brought social disorder. On such occasions there is always clear evidence that the temporary interruption of electric power and energy supply is accompanied by a disruption of civilization as we know it. The use of energy in its different forms clearly has defined the contemporary human civilization and the uninterrupted and secure supply of energy is essential for the preservation of human civilization.

1.2.1 Energy Demand

Humans use energy in several different forms in order to run machinery and accomplish different tasks: An air-conditioner unit uses electric energy to drive its compressor and provide cool air in a building in Miami, Florida; a burner demands heating oil or coal to heat up the cold air in a factory in Nanjing, China; a stove uses natural gas to prepare food in Germany; a car assembly line in Japan uses electricity to weld parts of cars and trucks; cars, trucks and buses use gasoline or diesel fuel in their internal combustion engines to transport people and goods; airplanes use kerosene to also transport people and goods.

The modern human society needs the accomplishment of such tasks, since the machinery to perform these tasks only runs on energy. The air-conditioner, the kiln of a cement factory, the welding equipment of a shipyard, the train locomotive that brings food to a city and the ship that transports goods between continents, all operate because they use various forms of energy. In the absence of energy supply, these equipment would not operate and the societal tasks would not be performed: when ships do not have fuel, food is not shipped and rots in ports; when the factory does not have electricity, cars are not manufactured to be sold later; with a scarcity of heating fuel, homes in Berlin are very cold in January and the inhabitants do not have the expected domestic comfort; without air-conditioning, working is very difficult during the summer months of Fort Worth, Texas and productivity suffers; without kerosene for airplanes or gasoline for cars, families do not re-unite for holidays.

Energy is demanded in different forms, the most important of which are: electricity, gaseous fuels, liquid fuels and solid fuels. Global energy demand has been constantly growing since the industrial revolution and is affected by two factors:

a) The increase of the global population.
b) The increasing energy demand per capita as agrarian and less affluent societies transform to industrial and more affluent societies.

In general, the more affluent a society is the more energy it consumes per capita. Citizens of the more affluent nations, as a group, buy more energy-intensive consumer goods; use more frequently private means of transportation and travel more; and have higher levels of energy-intensive comfort in their homes. All these activities require higher amounts of energy per person. Figure 1.1 shows the relationship between affluence and energy use in the plot of the consumption of electricity, measured in MWh/per year per capita, versus the Gross Domestic Product

(GDP) per capita for several countries.[1] The GDP values in the figure have been adjusted to take into account the different purchasing power of the US dollar in the respective countries.

It is apparent in this figure that there is a direct relationship between the affluence of a nation and the average energy its citizens consume. Nations in the Organization for Economic Cooperation and Development group[2] (OECD), which encompasses most of the industrialized and more affluent nations have significantly higher GDP per capita and consume significantly more electricity per capita than developing nations. The United States of America, Canada, Switzerland, most European Union countries, Japan, Australia and New Zealand appear at the top of the figure. At the bottom are several developing countries, primarily in the African continent, where societies are in general less affluent and the energy consumption is significantly less. In the middle of the diagram, there are several countries in intermediate stages of economic development, which are commensurate with their energy consumption. Among these, the economies of several countries including China, India and Brazil have improved significantly in the first ten years of the twenty-first century. Citizens of these countries have become more affluent and, as a result, both the total energy consumption and the electric energy consumption per capita have increased significantly.

A correlation of the variables in Fig. 1.1 yields the following relationship between GDP and the electric mega-watt-hours per capita consumed:

$$MWh = 0.06*(GDP)^{1.184}. \tag{1.2}$$

The correlation coefficient of the data is 0.8279, which signifies that the correlation between the MWh consumed per capita and GDP (adjusted for purchasing power) is very significant.

Figure 1.2 presents an interesting consequence of affluence and energy consumption for all nations. This figure depicts the average life expectancy in the chosen countries versus the electric energy consumption per capita. The outlier countries in this figure are essentially countries with high incidence of cases of the AIDS epidemic (several sub-Saharan countries), war-torn countries (Afghanistan) or countries with exceptionally good preventive health care (Cuba, Costa Rica, Sri Lanka and Nicaragua). It is obvious from this figure that a significant and positive correlation exists between the energy consumption and the average longevity of the citizens of a nation. Since the latter is connected to the "quality of life," one may reasonably correlate the energy consumption per capita with the quality of the life of the citizens in a nation.

Regarding Fig. 1.2 and the contemporary use of energy, it is important to note that energy consumption is not the *cause* of affluence and longevity in the nations,

---

[1] The ordinate of Figs. 1.1 and 1.2 is the electric energy consumption per capita. The same trends are observed and the same conclusions may be drawn with the total primary energy consumption per capita being the ordinate of the figures.

[2] The OECD group of countries comprises most of the European Union countries, Switzerland, Canada, the USA, Australia and New Zealand. The countries in this group are sometimes referred to as the "developed" countries.

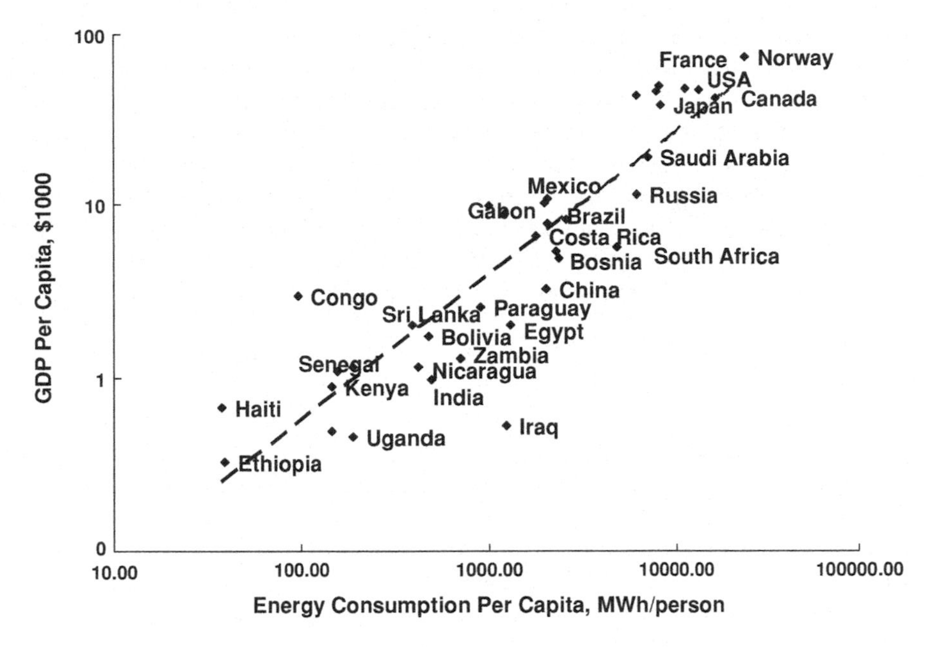

Fig. 1.1 Relationship between electric energy demand and GDP adjusted according to purchasing power. (Data from *Key World Energy Statistics*, IEA, 2009)

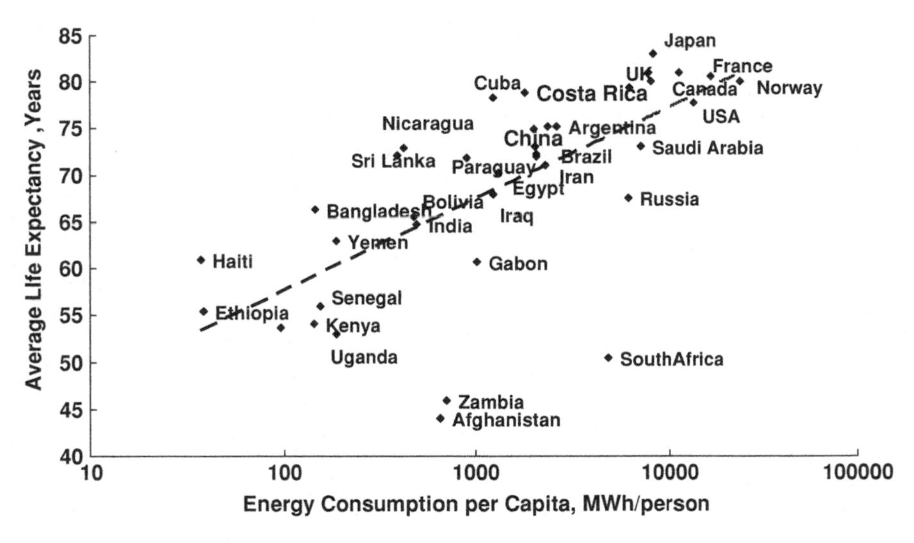

Fig. 1.2 The relationship between electric energy consumption and average life expectancy for several countries. (Data from *Key World Energy Statistics*, IEA, 2009 and the World Bank)

but the *effect* of this affluence: A more affluent society demands more goods and services, travels more, and typically has better public or private health care than a poorer society. Hence, citizens of the affluent nations consume more energy and live longer as the last two figures demonstrate.[3] One may also establish a cause and effect relationship by looking at the use of energy since the beginnings of humans: the prehistoric humans had to perform all the tasks by themselves. As a result the fossils of their bones show a great deal of "wear and tear" and their life expectancy was approximately 20 years. The early historic-period humans domesticated certain animals that performed most heavy tasks for them, but worked hard in agriculture. The life expectancy of these humans increased to 30–35 years. The humans of the early twenty-first century are employing mechanical engines to do most of their work, use machinery in agriculture, do not physically stress their bodies and, as a result, their life expectancy has increased to more than 70 years.

From the public administration point of view, governments look after the welfare of their citizens and, through long-term strategic planning, aim to increase the economic prosperity of their people. By doing so, they also aim to increase the total consumption of energy as well as of electric energy in particular. For this reason, national, regional and local governments adopt long-term plans to secure the adequate supply of energy products and the production of sufficient electric power, in order to sustain their growing economies. Three notable examples of rapid economic development in the recent past accompanied by significantly increased energy demand are: the Peoples Republic of China (PRC), India and Brazil. Figure 1.3 shows the GDP per capita (adjusted for the purchasing power of the US dollar) and electric energy consumption per capita in these three countries in the thirty years from 1980 to 2010.

It is apparent in Fig. 1.3 that the road to affluence for these three emerging economies was accompanied by an immediate increase of the electric energy consumption.[4] The trend is particularly evident in the case of the Peoples Republic of China, where the rapid industrialization and rise in GDP in the first decade of the twenty-first century was accompanied by an equally rapid increase of the use of energy. It must also be noted that these three countries account for more than 40% of the entire population of the Earth and that their further industrialization, increased affluence and increased energy use are expected to continue in the near future.

Similar developmental patterns were followed in the past by the countries of the OECD group when they were in their early development stages and very likely will be followed by most developing countries. The desire and tendency of all nations to become more affluent and the dramatic increase of energy consumption that follows have had very significant implications on the total world energy consumption growth. This trend is depicted in Fig. 1.4, which shows the total energy demand in the entire world from 1970 to 2010 according to the form of energy demanded and used. The values for years 2009 and 2010 are projections.

---

[3] While it may be true that "money does not buy happiness," Fig. 1.2 and this cause-effect relationship demonstrate that, at least, money can buy a longer life.

[4] The fluctuations of the GDP of Brazil are due to the devaluation of the *Real,* the national currency of the country in 1998 and the short-term disruption of the economic activities that followed.

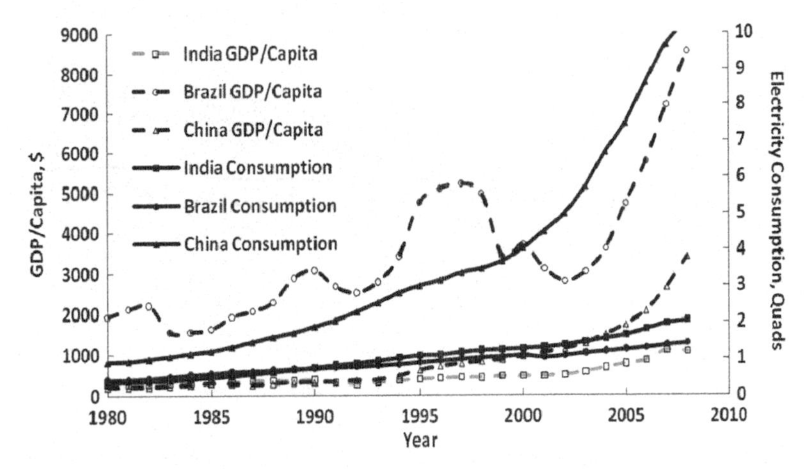

Fig. 1.3 Electric energy demand and GDP for Brazil, China and India. (Data from *Key World Energy Statistics,* IEA)

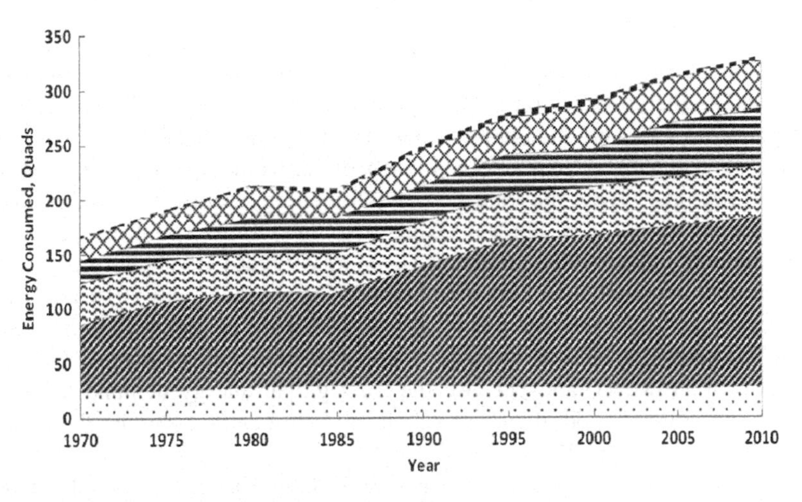

Fig. 1.4 Total global energy demand by energy form from 1970 to 2010. The areas (from *bottom* to *top*) represent coal, petroleum, natural gas, wood and waste, electricity and other energy forms. (Data from *Key World Energy Statistics,* IEA, 2009)

It is apparent that the global energy demand more than doubled in these 40 years. This growth is accounted for by two factors:

a) The improved economic prosperity in the entire world, and
b) The increase of the world's population.

These two factors are expected to continue their upward trends in the foreseeable future, thus driving the global energy demand continuously to higher levels. Energy conservation and improved efficiency in engineering processes

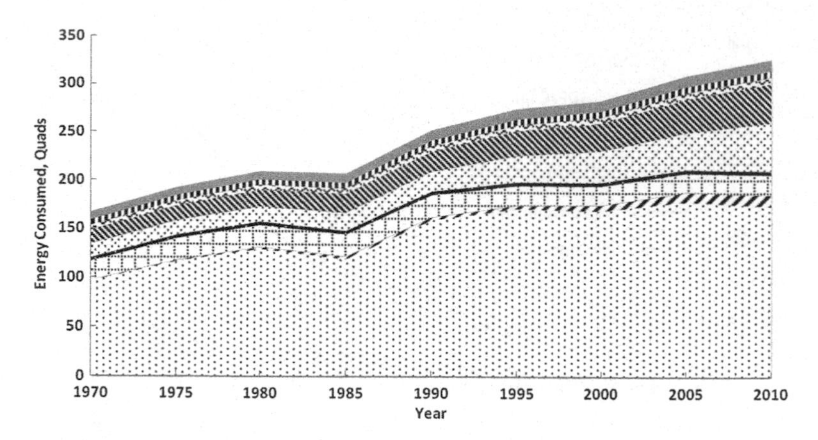

Fig. 1.5 The evolution of the global energy demand by region and groups of countries (data from *Key World Energy Statistics,* IEA, 2009). The areas (from *bottom* to *top*) represent OECD, Middle East, Former Soviet Union, non-OECD Europe, China, Rest of Asia, Latin America and Bunker fuel (the latter is used in transportation internationally)

simply cannot keep up with the growing world population and the increasing affluence. For the continuation of the "global economic progress" the world will need more energy to produce and to consume in the near future.

The rise in the energy demand has not been uniform among nations and geographical regions. Given the characteristics of energy demand according to the affluence of nations, it is not surprising that the OECD countries consume a significantly larger proportion of the world energy. Figure 1.5 shows the evolution of the energy demand from 1970 to 2010 (values for 2009 and 2010 are projections) in several regions of the planet. It is apparent in this figure that the relative energy demand in the OECD countries has fallen from approximately 60% of the total to 45%. However, and because the total global energy demand more than doubled during this period, the absolute energy consumed by the OECD countries increased by 44%, from 101 to 144 Q.

While the energy consumed in the OECD countries increased, the rate of growth in these countries was less than the average rate of growth of the global energy demand. The Peoples Republic of China[5] and the rest of Asia almost doubled their share of the total energy demand, while the share of Latin American and African nations also increased by 50 and 60% respectively. As the economies of the countries in these developing regions grow and the populations become more affluent, the consumption of energy increases according to the trends shown in Fig. 1.1. The only geographical region that has experienced a decrease of energy demand is the former Soviet Union during the years that followed the split

[5] Although the P.R. of China is an Asian country, it merits separate mention because of the size of its population—in 2009 the country had 12% more inhabitants than the entire OECD group—and because of the tremendous economic and energy demand growth it has shown since 1990.

of the Union into several countries. The lower energy consumption in these new countries was a consequence of the economic recession that followed the break-up of the Soviet Union. The improved economic activity in these countries during the first decade of the twenty-first century was immediately followed by increased energy consumption. Once again, this demonstrates the cause-effect relationship between economic prosperity, reflected in the higher GDP, and energy use.

A rather somber conclusion about the energy demand trends in the OECD countries is that, despite the energy conservation and improved energy efficiency measures that were adopted since 1970, the total energy consumed in these nations increased significantly. The global energy demand and the environmental effects associated with energy consumption have continued growing at an accelerated pace in the first decade of the twenty-first century. The current trends indicate that the global energy demand is not going to decrease in any geographic region, during the foreseeable future. On the contrary, the energy demand will grow significantly in several regions. The growth of energy consumption will occur primarily in Asia, Africa and Latin America, where most of the Earth's population lives and where most of the economic growth occurred in the beginning of the twenty-first century. The combined factors of increasing affluence and population growth will be the driving forces for the growth of energy demand in these continents and the entire planet.

1.2.2 Energy Supply

The national energy demand as a total must be met by the global energy supply. Since the natural laws dictate that energy may neither be created nor destroyed, all the energy that is consumed must be produced from energy sources, which exist in the natural environment. These natural energy sources are:

1. Fossil fuels, such as the various forms of coal, crude oil and natural gas.
2. Nuclear fuels, such as uranium and thorium.
3. Renewable energy forms such as solar, wind, biomass, geothermal, wave, and hydroelectric energy.

The first two forms of energy are essentially minerals that were formed several millennia ago. Their formation processes take place over very long periods of time (geological periods). Because these minerals may not be reproduced naturally in the foreseeable future, they will be exhausted at some point in the future. The category in the third item represents renewable energy forms that are reproduced and are inexhaustible. Humankind may expect that these energy forms will continue to supply energy to the humans in the foreseeable and far future.

The human society demands energy in different forms, such as electricity, transportation fuels, natural gas etc. Depending on the form of energy they supply, we distinguish the sources of energy supply as:

1. *Primary* energy encompasses the forms of energy that are directly consumed as they are found in nature, without any processing. The various forms of coal,

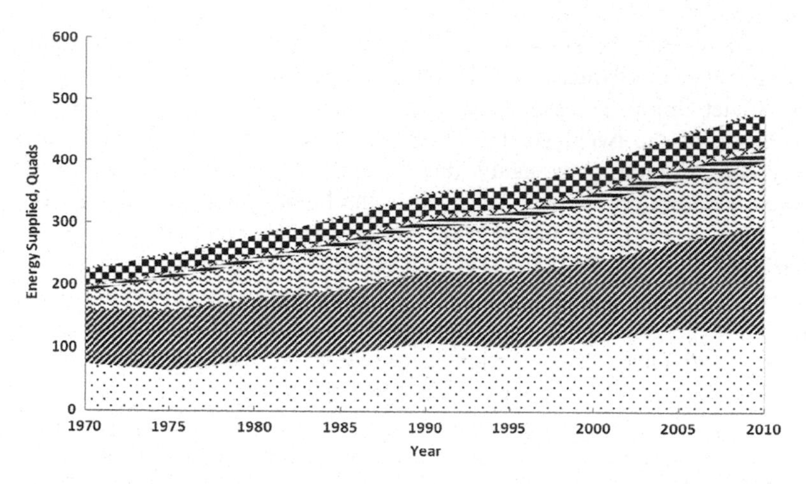

Fig. 1.6 The primary energy sources that supplied the global demand between 1970 and 2010. The areas (from *bottom* to *top*) represent coal, petroleum, natural gas, wood and waste, electricity and other energy forms (data from *Key World Energy Statistics,* IEA, 2009)

crude oil, natural gas, hydraulic energy and passive solar energy are among these forms.

2. *Secondary* energy forms are used by the consumers in a refined, processed form. The liquid petroleum products derived from crude oil, such as gasoline, diesel and kerosene; fuel from biomass; biodiesel; solar collector energy; and thermal geothermal are among the secondary energy forms.
3. *Tertiary* energy forms involve one or more transformations of energy. Electric energy, in any way it is produced, is a tertiary form of energy. Nuclear energy, wind power and most of the other renewable energy sources when they are used to produce electricity contribute to the supply of tertiary energy. In general, the transformations leading to the production of tertiary energy involve a significant percentage of dissipation, which is governed by the 2nd Law of Thermodynamics as explained in Chap. 3.

Secondary and tertiary forms of energy must be produced from primary sources. The main primary energy sources that have satisfied the global energy demand since 1970 are shown in Fig. 1.6 (the values for 2009 and 2010 are estimates). These primary sources are classified as:

a) The various forms of coal (anthracite, bituminous, lignite, peat)
b) Crude oil/petroleum
c) Natural gas
d) Nuclear
e) Hydroelectric energy or water energy
f) Biomass and waste, which primarily comprises trees, and
g) Other renewable forms such as solar, wind and geothermal.

Table 1.2 Evolution of the world primary energy supply between 1973 and 2007

	1973 (total 243 Quads)	2007 (total 475 Quads)
Coal and coal products	24.5	26.5
Crude oil	46.1	34.0
Natural gas	16.0	20.9
Nuclear	0.9	5.9
Hydraulic	1.8	2.2
Wood, biomass and wastes	10.6	9.8
Solar, wind, geothermal	0.1	0.7

All values are in percentages (data from *Key World Energy Statistics,* IEA, 2009)

It is apparent from this figure that the energy supply more than doubled between 1970 and 2010, following the increased energy demand. It is of interest to know the contributions of the several primary energy sources to the total global energy demand during the period covered in Fig. 1.6. Table 1.2, which was produced from the same data shows two "snapshots" of the total global energy supply in the years 1973 and 2007 and elucidates the changes that occurred in the consumption of the several primary energy forms.

A number of conclusions may be drawn from Table 1.2 regarding the trends of the use of primary energy sources in the recent past:

1. The relative amount of coal and its products that have been used primarily for the production of electricity has remained almost the same. Because the total energy consumption almost doubled, the coal consumption also doubled.
2. While the relative amount of crude oil consumption decreased significantly between 1973 and 2007, from 46.1 to 34% of the total, the absolute amount of crude oil consumed actually increased from 112 Quads to 162 Quads, or from 2,819 million tons of crude oil (19,264 million bbl) to 4,090 million tons (27,864 million bbl). This implies that, despite all the campaigns for the reduction of oil consumption and policy measures for energy independence in the OECD countries, the demand for crude oil is still very strong and its use will likely grow significantly in the near future.
3. Most of the global energy supply, 81.4%, comes from fossil fuels (coal, oil and natural gas) which are finite and will be exhausted at some point in the future. For a sustainable future energy supply, the supply and consumption of non-fossil forms of energy must increase significantly.
4. Both the relative and the actual amount of nuclear energy used have increased significantly. However, most of the increase occurred during the 1970s and early 1980s. Only a handful of nuclear reactors have been built in the OECD countries since 1986, when the Chernobyl accident occurred.

5. A great deal of the world energy supply (10%) stems from the burning of trees and other biomass products. This practice occurs primarily in the developing countries, where biomass is the main source of fuel for cooking and heating.
6. The production of hydroelectric energy, in kWh, has more than doubled.
7. Even though the relative use of solar, wind and geothermal energy have increased by a factor of 7 and the energy supplied by them increased by a factor of 15, these renewable energy sources, still contribute less than 1% of the total energy supply. The world has to make significant technological progress and great strides in order to achieve a sustainable energy future that is based on renewable energy sources.

An important conclusion that may be deduced from the data of Table 1.2 and Fig. 1.6 is that the global primary energy demand is satisfied principally by exhaustible energy sources, such as coal, oil, natural gas, and nuclear. The time-scale of replenishment of these fuels is of the order of thousands of years, and the rate of their consumption by far exceeds the rate of their replenishment. Because of this, it is apparent that these fuels will become scarce and will be exhausted in the future. Since the entire contemporary human civilization is based on the consumption of energy, the continuation of our civilization demands that humans must ensure the adequate supply of renewable energy in the near future. Technologies for the utilization of solar, wind, geothermal, and hydrogen fusion energy[6] must become widely available in the future to satisfy the global energy demand and to ensure the continuation of our civilization.

All of the world's countries utilize and promote domestic energy sources to satisfy their energy demand to the extent possible. For the production of electricity, most countries use their domestic coal supply, supplemented by nuclear fuels and a small percentage of fluid hydrocarbons and renewable sources. The Peoples Republic of China has launched a very ambitious electrification program, based on domestic coal supply and the USA has relied on domestic coal to provide more than 70% of its electric consumption. On the other hand, countries with smaller coal reserves, such as France and Japan, have aggressively promoted nuclear programs that supply most of their electric energy. Unlike their neighbor France, Germany and the United Kingdom, two countries with significant coal deposits, rely more on their domestic coal to produce electricity than their nuclear reactors.

The situation is different in the transportation sector, which demands liquid fuels: crude oil and natural gas exist in only a few regions of the world and are in very high demand in Europe and North America. There is a significant trade in fluid hydrocarbons, both crude oil and natural gas, which primarily flow towards the OECD countries. The Middle East, the countries of the former Soviet Union and some countries of Latin America supply most of the crude oil and the natural

[6] Hydrogen is not a renewable energy source. However, the amount of hydrogen on Earth is very high and the energy released by fusion is large enough to make this energy source virtually inexhaustible (Chap. 6).

Table 1.3 World population, GDP, and primary energy production, imports and total supply

Region	Population, (Million)	GDP (billion, $US)	Production, (Quads)	Imports, (Quads)	TPES, (Quads)
OECD and rest of Europe	1,238	32,870	155	74	229
Middle East	193	1,552	61	−38	23
Former Soviet Union	284	2,472	65	−24	41
P. R. China	1,327	10,156	72	8	80
Rest of Asia	2,148	8,292	49	8	57
Latin America	461	3,714	28	−5	23
Africa	958	2,372	45	−19	26
Total world	6,609	61,424	475	0	475

Data of the year 2007 from *Key World Energy Statistics,* IEA, 2009

gas consumed in Europe and the USA. The flow of primary energy forms in the several regions of the world is shown in Table 1.3 together with the population of the regions and the average Gross Domestic Product adjusted for the purchasing power of the US dollar in the respective region. The Total Primary Energy Supply (TPES) is the sum of the domestic production and all imports. In national statistics, it is the TPES that is commonly referred to as "energy."

It is apparent from Table 1.3 that energy imports are primarily directed to the OECD countries and the rest of Europe. It is also of interest to note that these countries also have the highest energy consumption per capita. The energy consumption in the OECD countries is 0.185 Quads per million inhabitants, while the average for the world is 0.072 Quads per million inhabitants. If the entire world had the same level of affluence and consumed energy at the same rate as the OECD countries, in 2007 the total world primary energy demand would have been 1,223 Quads or 2.57 times higher than it actually was. In the absence of serious efforts in conservation and energy efficiency as well as of internationally mandated measures for the curtailment of energy consumption globally, it is likely that the future global energy consumption will reach these levels.

1.2.3 Energy Prices, OPEC and Politics

According to microeconomic theory, price is what connects the supply and demand of a commodity. The demand curve of a commodity is inversely related to its price, because consumers buy less of the commodity if the price increases. Supply curves are positively correlated with price because suppliers produce more of the commodity when the commodity price rises. At the intersection of the supply and demand curves, the supply and demand of a given commodity are at equilibrium. The intersection point of the two curves determines the price of the commodity at equilibrium. Figure 1.7 depicts typical demand and supply curves for a relatively inelastic commodity such as petroleum. The equilibrium price, P,

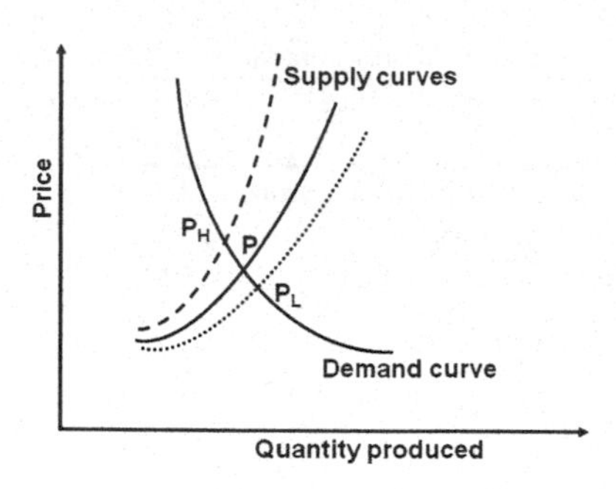

Fig. 1.7 Typical demand and supply curves of a commodity such as petroleum

is also shown at the intersection of the demand and supply curves, which are represented by the solid curves.

Now let us assume that a political event happens that affects one of the oil producing regions—a war, an embargo, or a political unrest—that disrupts the petroleum production in the region and decreases the produced quantity.[7] Other suppliers or other methods may enter the market to supply the demand, but the supply curve shifts to the left as denoted by the dashed line in Fig. 1.7. A new equilibrium is reached with a higher price, P_H. It must be noted that, because of the higher price, the quantity demanded and produced has slightly decreased at this equilibrium state. On the other hand if, again as result of an international political event, the suppliers decide to increase the quantity of the commodity produced, the supply curve shifts to the right and is represented by the dotted curve in Fig. 1.7. The new equilibrium is reached at a lower price, P_L. At this equilibrium, the consumption of the commodity is higher than that at the original equilibrium position because consumers respond to the lower price and use more of this commodity.

The price of crude oil and other hydrocarbons fluctuates almost daily, responding to demand and supply conditions. Absolute crude oil prices are historically much more volatile than prices of other commodities for two reasons:

(a) They are significantly influenced by international politics and events; and
(b) A group of relatively few countries, which effectively control the petroleum production, supply most of the imports to the OECD countries.

The prices of the rest of the energy forms also fluctuate, following the petroleum prices, but with lower variability. Political events in the Middle East—the major supply region of crude oil to the OECD countries—have significant influence on production and the prices, not only of petroleum, but also of all liquid hydrocarbons. Between 1960 and 2010 the price of crude oil jumped by more than

[7] The political upheavals of the first three months of 2011 in Tunisia, Egypt, Libya and other Arab nations caused a temporary disruption to the petroleum supply from North Africa. The immediate result was a jump in the oil price from approximately $80/bbl in January 2011 to more than $110/bbl in April 2011. When the political situation in these countries became more stable, in June 2011 the price of crude dropped to approximately $95/bbl.

100% within a few months in the aftermath of the following political events: the Arab–Israeli war of 1973 and the following oil embargo; the aftermath of the Iranian revolution in 1979; the Iraqi invasion of Kuwait and the Persian Gulf war of 1990–1991; the production reduction of the mid 1990s by the Organization of Petroleum Exporting Countries (OPEC). In the twenty-first century the higher oil prices have been principally driven by the tight control of crude oil supply by OPEC, the war in Iraq, the increased demand in China and India and the domestic/ national strifes in several Arab countries for more political freedom.

The OPEC is an economic cartel composed of Algeria, Angola, Ecuador, Iran, Iraq, Kuwait, Libya, Nigeria, Qatar, Saudi Arabia, the United Arab Emirates, and Venezuela. The cartel has quarterly meetings in Vienna, Austria, where they try to regulate the crude oil supply by issuing quotas among the member countries. In 2007 the OPEC accounted for the production of 29,600,000 bbl of crude oil per day, most of which was exported to the OECD countries. This is equal to 62.8 Quads and, according to Table 1.3, it represents 85% of the oil imports of the OECD countries and the rest of Europe. By regulating the crude oil supply, the actions of OPEC have a very important effect on the world production of crude oil and shift the supply curves. By the supply–demand mechanism shown in Fig. 1.7, the equilibrium position of the market and the price of crude oil are also affected significantly. Consequently, the OPEC's actions affect the price of all the other energy commodities in the world.

Regarding the price of all energy commodities, we distinguish between *real* and *nominal* prices. The nominal prices are those quoted in everyday transactions and frequently appear on the gasoline station billboards. For example the nominal price of one liter of gasoline in Oxford, England was £0.21 in 1976 and £1.56 in 2008. The real price of gasoline and other petroleum products accounts for the inflation that has occurred during a time interval and differs from the nominal price. The real price is often quoted in constant currency of a particular year, such as constant 1990 US dollars or constant 2000 British pounds.[8] The official inflation rate in a country, which is often given by the respective national Treasury Department, determines the annual rate of inflation of the currency for the corrections to be made.

The fluctuations of the liquid hydrocarbon prices from the end of the 2nd World War to 2010 are shown in Fig. 1.8, which depicts the nominal and real crude oil prices, the former measured in constant 2010 US dollars.[9] Several of the important political events that influenced the price of liquid hydrocarbons are also noted in the figure. The following conclusions may be drawn from Fig. 1.8:

---

[8] The conversion to Euro in many European nations, which occurred between 1995 and 2005, makes this computation more cumbersome. Most of the energy resources are quoted in constant US dollars or British pounds.

[9] The price of crude oil (petroleum) is a well-publicized economic parameter. This price is followed daily and appears in most news outlets of the globe. This price is usually quoted per barrel of oil, \$/bbl (1 bbl = 42.0 gallons = 159 l).

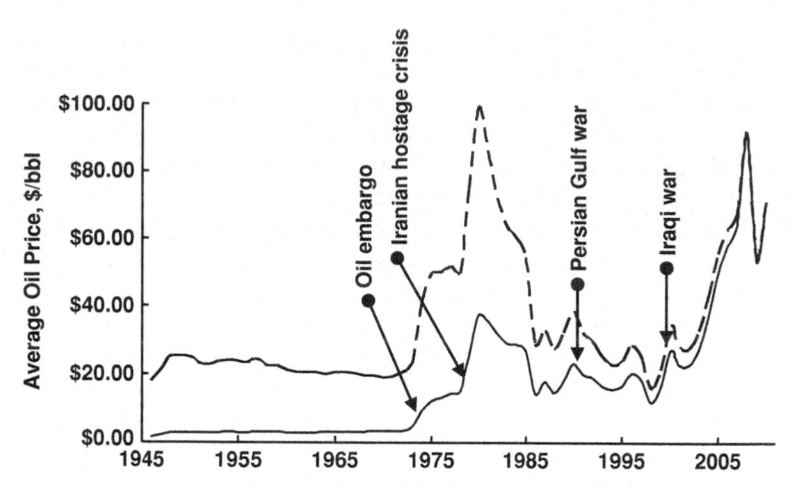

Fig. 1.8 Nominal price (*solid line*) and real price in 2010 $US (*dashed line*) of crude oil. Data obtained from the International Energy Agency and the US Bureau of Labor Statistics

- The effect of major international events in the price of crude oil is significant, but, typically, of short duration. Prices are restored close to their previous level once the "political crisis" is over.
- Despite the significant fluctuations in its nominal price, the real price of crude oil, measured in constant 2010 US dollars, has been historically almost constant in the range $20–$40 per barrel. The exception to this was the decade 1975–1985, when the OPEC yielded significant political influence, and the first decade of the twenty-first century, when the demand from several developing nations became significant enough to affect the global crude oil prices.

The prices of the other fuels and other energy commodities have a high, positive correlation to the price of crude oil. Usually, the price of crude oil leads the trends in the prices of other fuels. However, the prices of the other fuels have higher inertia and do not exhibit the wild fluctuations of crude oil prices. The economic phenomenon of *substitution* becomes important and drives the consumption of alternative energy sources, energy efficiency and conservation: When energy prices rise, the society strives to satisfy its energy needs with other sources of energy, such as wind, solar, geothermal or reduces the energy consumption using conservation and higher efficiency measures. For example, ordinary citizens switch to cars with higher mileage, drive less and use less energy at home; electricity corporations invest more in higher efficiency power plants or in power plants that use alternative energy sources; and nations initiate and promote policy measures that favor alternative energy sources, more energy conservation and higher efficiency. When crude oil and the other conventional energy sources become expensive, alternative energy sources become viable, economic alternatives to satisfy the energy demand of the society.

The experience of the years 2006–2010 demonstrates how the significant energy price increase resulted in more conservation measures and increased use of alternative energy sources at the individual, the corporate, the national and the global levels. On the other hand, when there is an "oil glut" and the energy prices drop to a low level, energy conservation, alternative energy and higher efficiency projects become relatively more costly and less important to the general population. The period 1984–2000 is an example of such low energy prices era: during this period there was very little investment in solar and wind energy projects; conservation and higher efficiency efforts were relaxed or abandoned; and the automobile companies throughout the world introduced vehicles with very low mileage, such as the family caravan, the SUV and the "Hummer," which became very popular with affluent consumers in several countries.

1.3 Reserves, Resources and Future Demand for Energy

It is apparent that the global TPES consumption has been continuously increasing during the last two centuries and that this increase is expected to continue at least in the near future. The growth of the world's population and the improving economic conditions are expected to accelerate the global energy consumption, while the nominal rise in the prices of fuels is expected to promote conservation and to slow the growth of TPES use. One of the factors that contribute to this growth is the increasing energy demand in developing countries, and especially in China and India, which together account for almost 40% of the Earth's population.

The globally-averaged compounded annual rate of the TPES growth between 1970 and 2009 was 1.91%. During this period, the TPES growth rate in the OECD countries was a mere 0.85%. These numbers signify that the developing countries account for most of this TPES growth rate. Given that the developing countries have most of the world's population and, in addition, have the highest GDP growth rates, it is expected that the global energy consumption will continue to increase in the near future. The global energy consumption is expected to grow, with a rate that is similar to the rate of the last forty years despite any reasonable rise in the energy prices and energy conservation measures, which may be adopted by individual nations or the global community.

One of the reasons that the TPES consumption growth rate may become lower in the future is international regulations and international agreements designed to curtail the global greenhouse gas emissions. As it will be seen in more detail in Chap. 2, the combustion of fossil fuels produces large quantities of carbon dioxide, which cause the warming of the atmosphere and global environmental change. Several countries have already adopted measures for the curtailment of carbon dioxide emissions. It is possible that other countries will follow with such regulations and the world will have meaningful restrictions on the production of carbon dioxide. If this occurs, the consumption of fossil fuels, such as coal, liquid and gaseous hydrocarbons, will be reduced significantly and the total TPES will also be

Table 1.4 a Expected primary energy demand in 2030

	RS-2030 Total TPES	Percentage	450 PS-2030 Total TPES	Percentage
a In Quads (data from *Key World Energy Statistics,* IEA, 2009)				
OECD countries	250.4	37.1	222.3	39.0
OME countries	308.5	45.7	240.5	42.2
Rest of world	108.0	16.0	98.6	17.3
Transportation fuel	8.1	1.2	8.6	1.5
Total	675.0		570.0	
b By energy sources in Quads (data from *Key World Energy Statistics,* IEA, 2009)				
Coal	194.4	28.8	94.6	16.6
Petroleum	203.2	30.1	171.0	30
Gas	145.8	21.6	116.9	20.5
Nuclear	35.8	5.3	54.2	9.5
Hydroelectric	16.2	2.4	22.2	3.9
Renewables	79.7	11.8	111.2	19.5
Total	675.0		570.0	

reduced significantly. The *International Energy Authority* (IEA), an agency of the United Nations, has prepared two scenarios for the future energy consumption that extend to the year 2030: The first scenario assumes that there will not be a significant regulatory intervention by the international community. The second scenario assumes that an international treaty will reduce carbon dioxide emissions significantly to stabilize the concentration of the gas at 450 parts per million (450 ppm). The effect of this action will be a reduction of the TPES growth, especially that of fossil fuels and the substitution of fossil fuels by alternative energy and conservation measures. The first scenario is often called Reference Scenario (RS-2030) and the latter is often called the Scenario 450 PS (450 PS-2030).

Tables 1.4a and b show the expected consumption of TPES in the year 2030 under these two IAE scenarios. The first table depicts the expected energy demand in 2030 in the various regions of the globe and the second table gives the supply of primary energy forms that would satisfy this demand. In Table 1.4a the designation of OECD countries also include the rest of Europe. OME denotes the Other Major Economies, which comprise Brazil, the P.R. of China, India, Indonesia, the Russian Federation and the countries of the Middle East. These economies are expected to play an important role in the future energy consumption and the growth of TPES between 2010 and 2030.

It must be noted that the term "renewables" in these Tables comprise: human and animal wastes; solar energy; wind energy; geothermal energy; and biomass, which includes the combustion of timber from live trees. By analyzing the numbers in the two tables, one will draw the following conclusions:

1. Despite all the conservation measures that were adopted in the recent past, the TPES consumption is continuously increasing and is not expected to decrease

in the near future. Under the reference scenario, the expected compound annual growth rate of TPES consumption is 1.75% and under the more restrictive 450 policy scenario this annual growth rate is 0.9%.
2. Given that the 2010 TPES consumption in the OECD countries is approximately 230 Quads, it is apparent that the expected TPES consumption in these countries will remain almost constant between now and 2030. The consumption will slightly increase to 250 Quads under the reference scenario or will slightly decrease to 222 quads under the more restrictive 450-PS scenario. Most of the increase of the TPES consumption is expected to occur in the Other Major Economies (OME) and, to a lesser extent, in the rest of the world. The OME are expected to surpass the OECD countries in TPES consumption by 2030.
3. Comparing the 2030-RS and the more restrictive 450-PS scenario, the majority of relative TPES savings are expected to come from the OME countries (22%). This is expected to occur from the reduction of the rate of growth of TPES consumption in the OME countries. The OECD reduction will be only 11% and the reduction in the rest of the world will be 7%.[10] Given this disparity in the burden of efficiency and conservation, the current rates of economic and population growth in the OME countries, and the current political realities, it is rather unlikely that the OME countries will consent to adopt such an international treaty or scenario.
4. The fossil fuels, coal, petroleum and gas, are still expected to provide the bulk of the energy demand, even under the 450-PS.
5. If the 450-PS restrictions or similar restrictions are adopted, the use of nuclear energy and renewable energy is expected to increase significantly. The fractional contributions of both are expected to double in comparison to the reference scenario contributions. The use of hydroelectric energy is not expected to double, primarily because the best and most significant hydroelectric resources have already been utilized or are expected to have been utilized by 2030 under the reference scenario.

1.3.1 Energy Reserves and Resources

The fossil fuels, coal, natural gas and petroleum are mineral resources, which have been formed eons ago and are currently extracted from the Earth's interior. The current rate of extraction and combustion of the fossil fuels exceeds by far the rate of the replenishment of these resources. Because the mass of the Earth's crust is finite, it is reasonable to deduce that the total amount of the fossil fuels that exist in the crust and may be recoverable must be finite too. With the continuous and

[10] Such assumptions do not appear to be acceptable to the OEM countries and may become reasons for not reaching an international agreement on CO_2 stabilization in the future. The OEM countries have repeatedly denounced international accords that do not include significant TPES reductions in the OECD countries.

increasing mining of these finite resources, it is also reasonable to deduce that these resources will be depleted in the near or far future.

Of the total amount of the existing fossil resources, a small fraction is located in known areas and may be extracted economically with the existing technology. The fraction, which may be economically extracted under present conditions, is called the *proven reserves*. Another part of the resources is known to exist, but may not be recovered economically at present. These are the *potential reserves*. If the price of the fuel rises or if a technological breakthrough that lowers the cost of extraction is achieved, potential reserves become proven reserves and are mined economically.

In addition to the reserves, there are also large quantities of the mineral/fuel resources, which are located in the vicinity of the proven and the potential reserves. Such resources have not been fully explored and quantified at present, but, from past experience with mining, it may be reasonably assumed that when they are explored and quantified they will be recovered economically with the currently known technology. These are the *inferred resources*. Finally, there are additional resources that might exist in the crust of the Earth, but have not been discovered yet and may not be extracted economically or with known technology at present. These resources may become available at a future date and they are the *undiscovered* or *hypothetical resources*.

The concepts of *reserves* and *resources* are strongly tied to the technology and economics of mineral extraction. For this reason, their numbers are not fixed by a scientific method. Reports of their quantities fluctuate significantly by year and by nation. When the price of a fuel rises, and technology improves, potential reserves become proven reserves. Inferred or hypothetical resources eventually become proven resources with additional exploration and technological improvements. For example, during the 1960s, most of the oilfields in the North Sea were classified as potential reserves or inferred resources. With the technological advances in offshore exploration, drilling and oil production as well as with the significant increase of crude oil prices during the 1970s, these reserves became economical to be exploited and were promoted to proven reserves. By 2000, most of these potential reserves have yielded oil-producing fields. Similarly, the technological advances of the 1990s that allowed deep sea drilling operations resulted in the promotion of several offshore oilfields in the Gulf of Mexico from resources to proven reserves.

Because of the strong dependence of the reserves and resources on the current technology and prices, and because the quantities of the reserves suffer from a great deal of uncertainty, reports on them fluctuate significantly and are not scientifically reliable. An additional source of uncertainty stems from inaccuracies of national governments in reporting their energy reserves and resources: Some governments or governmental agencies deliberately under-report or underestimate their national resources, while others over-estimate them.[11]

---

[11] Over-reporting or reporting very optimistic estimates was prevalent until the 1990 s in the former Soviet Union and the Eastern European nations. This gave an inaccurate picture of the global reserves in the 1970 and 1980 s.

1.3.2 The Finite Life of a Resource

All mineral resources including fossil energy resources exist in the crust of the Earth in finite amounts. The molecules and the total quantities of coal, petroleum and natural gas are finite. It is self-evident that, if we continue to mine them and use them for the production of energy, these fossil fuels will be exhausted at some point in the future. Therefore, the question for a society, which overwhelmingly relies on the fossil fuel resources for the production of its energy, is not *if the resources will be exhausted* but *when the resources will be exhausted.* One of the mitigating and rather fortunate factors for the human society is that, since 1950, with the advances of technology and the intensified exploration in all the Continents, new mineral energy resources have been discovered and many of these resources have been mined and exploited. Among these are the near-surface coal deposits in Wyoming, USA, the oil deposits of Alaska and the North Sea. The discovery, development and exploitation of these natural resources has contributed to the lower nominal and real prices of energy resources during the 1980 and 1990s and to the dramatic rise in the consumption of fossil fuels in the first decade of the twenty-first century, which is apparent in Table 1.2. However, it is also apparent that, since the fossil fuel resources are finite, they will be exhausted in the future.

The increasing rate of fossil fuel consumption, if continued unchecked, will accelerate the depletion of these energy resources. Figure 1.9 depicts schematically this depletion scenario. The figure pertains to a hypothetical resource of finite total amount, equal to 32,000 units. This resource was consumed at an annual rate of 10 units from the beginning of the industrial revolution to 1850; at a rate of 20 units from 1850 to 1900; and at a rate of 30 units from 1900 to 1950. From that point in time onwards, the consumption of the hypothetical resource doubles every 20 years (the global crude oil consumption has followed similar trends in the past). The calculations and the figure show that, because of the continuing increase in the consumption, this finite resource is going to be exhausted by 2110. Although this resource and the consumption scenario are hypothetical, it becomes apparent from Fig. 1.9 that:

(a) The most significant drop in the proven reserves of this resource occurs in the last 20 years of its use; and
(b) In 2090, 20 years before the depletion of this resource, one may observe that there is still as much of this resource available for consumption (16,080 units) as it was consumed in the past 240 years. Thus, twenty years before the depletion of the resource, a society may be lulled into a false sense of security by statements such as "...we have as much of this resource left as we have consumed in the last two-and-a-half centuries." When the potential of an energy resource is to be evaluated, it is the current and future consumption of the resource that matter and not the past history of utilization of the resource.

The *lifetime* of a resource is often used as a measure of the time in the future when the resource will be exhausted. This lifetime depends on the total amount of

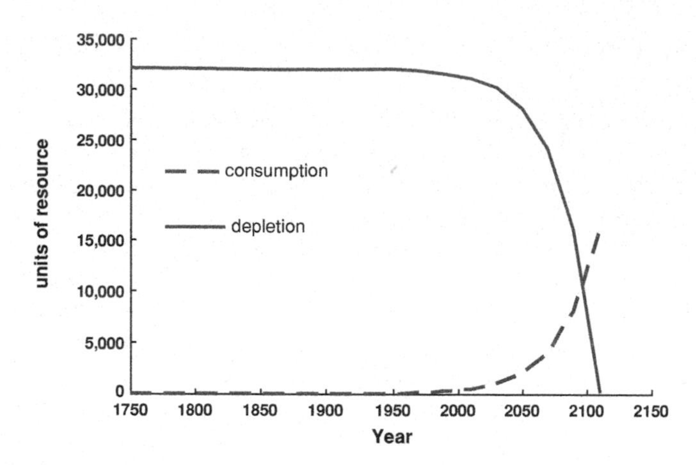

Fig. 1.9 The life cycle of a hypothetical energy resource, whose consumption doubles every 20 years since 1950

the reserves of the resource, M_T, as well as the annual consumption of the resource, M_A. In the simple case when the consumption of the resource is constant, the lifetime of the reserves, T, is simply equal to the ratio M_T/M_A. In most cases though, the consumption of the resources grows at an annual rate, r. In this case of growing consumption, the lifetime of the resource, in years, may be calculated from the following expression:

$$T = \frac{1}{r}\ln\left[r\frac{M_T}{M_A} + 1\right]. \tag{1.3}$$

Predictions for the lifetime of several fossil fuel resources have been made using this expression. A glance in the predictions of the past fifty years, however, proves that these predictions have been inaccurate.[12] The main reason for this is that the total reserves, M_T, is actually a variable that is adjusted upwards as potential reserves and inferred or hypothetical resources are promoted to reserves.

1.3.3 The Hubbert Curve and the Hubbert Peak

The life cycle and the consumption of actual energy resources may somehow differ from the trends shown in Fig. 1.9 because of price adjustments, new discoveries and technological breakthroughs. According to a theory, which is frequently

[12] For example, during the energy crisis of the 1970 s there was an almost unanimous belief by the "experts," which was promulgated in the popular press, that the lifetime of the petroleum reserves was 30–35 years and that, hence, petroleum would have been almost exhausted by 2015. The same "experts" were predicting $12 per gallon of gasoline in the USA by 2010.

referenced in the popular press, the life cycle of a resource should have the shape of a curve that approximates the *bell-shaped curve.* This curve is sometimes called the *Hubbert curve* [1, 2] and its functional form is:

$$y = \frac{e^{-t}}{(1 + e^{-t})^2} = \frac{1}{2(1 + \cosh t))}, \tag{1.4}$$

where t is the time and y the normalized fraction of the resource that is consumed. The shape of this curve is bell-shaped. It is apparent that the Hubbert curve is not a Gaussian, but its shape has a similar form and approaches zero more slowly than a Gaussian curve. According to Hubbert's theory, during the first stages of the utilization of any resource, its consumption rises exponentially because it is accelerated by two factors:

1. Relatively low nominal price.
2. New technologies and new uses are developed for this resource to perform several societal functions and tasks that were accomplished using another resource before. This increases significantly the demand and consumption of the resource.

At the initial stages of the utilization of the resource, there is an increasing growth in its consumption. During the first stages of its consumption the resource is defined as a *new* or *emerging resource.*

It is well known that, given the finite amount of all materials and resources, an exponential growth in the consumption of any commodity cannot be indefinitely sustained. This ushers the second stage of the growth in the consumption of the resource. The rate of growth of the consumption slows down because of two other factors:

1. There are no more new applications that make use of this resource.
2. The price of the resource increases significantly because of profit-taking and because of the realization that the resource may be depleted and may become more valuable in the future.

Following the slowing of the rate of growth, the resource becomes a *mature resource* and its consumption is governed by its price, which is determined by demand and supply. The total production and consumption of the resource starts decreasing during this stage, depending on the price level. Thus, the production of all resources passes through a maximum, which is often called *Hubbert peak.*

The third stage in the theoretical life-cycle of the energy resource occurs when there is a realization that there are few reserves of the resource left. At this point, the resource is classified as a *rare* or *depleting resource.* After this stage, the real and nominal prices of the resource rise continuously and significantly. The widespread use of the resource becomes uneconomical and new technologies are developed to substitute this resource with other, more abundant, longer lasting, more secure and more affordable resources. The immediate consequence of this *substitution effect,* which is driven by the commodity's price increase, is that the consumption of the resource slows down and finally diminishes when the resource is about to be depleted.

An example of the life cycle of an energy resource in the United States is the production and consumption of natural gas. The production and consumption of this energy resource was very low at the beginning of the twentieth century. Discoveries of several large deposits of natural gas were made in the beginning of that century and several applications of this energy resource were invented and adopted. The development of a national distribution network of pipelines for natural gas brought this resource into the major urban centers, thus, reaching new consumers and creating high demand. Natural gas substituted coal and wood for domestic heating and cooking in most of the USA cities during the first half of the twenty first century. Gas turbines were invented and natural gas started being used in gas turbines for the production of electricity, again substituting coal and petroleum products. The driving force behind this substitution was the relatively low price of the resource, the relative safety, the wide availability and the significant convenience of its use. The further development and optimization of the distribution network across the country made this energy resource widely available to the American consumer. As a result, the use of natural gas increased almost exponentially in the first seventy years of the twenty-first century, from 23 billion scf in 1900 to more than 20 trillion scf in 1970 (1 scf of gas provides approximately 10^6 J).

Following this initial growth period, during the energy crisis of the 1970s, the real price of natural gas started increasing continuously, even though several new discoveries of natural gas were made. In addition, no new domestic applications for the use of natural gas were adopted, because most of the households in the USA already were using gas for heating, hot water supply and cooking. As a result, the rate of growth of production and consumption of this resource diminished and actually became slightly negative. The consumption of natural gas slightly decreased, between 1978 and 1990, when energy conservation and the use of relatively cheap electricity substituted for the use of natural gas in some households. Between 1990 and 2010, the price of the natural gas fluctuated significantly and its consumption also fluctuated, largely following the price and the increase of the population in the USA. One reason behind these fluctuations is the relatively ease of substitution of natural gas by other energy forms, especially electricity. Thus, between 1970 and 2010, the demand of natural gas in the USA has followed its price in an inverse manner. It may be said that the natural gas consumption in the USA follows the trend of a *mature resource.*

Figure 1.10 depicts the consumption of natural gas in the USA and shows the trends that were described above. The figure also depicts a *Hubbert curve* that best fits the scenario of gas production and consumption. It is observed that in the early stages of the development of this energy resource, the production followed rather faithfully the shape of the *Hubbert curve.* However, after the peak production was reached in the 1970s, there has been a significant diversion from this analytical

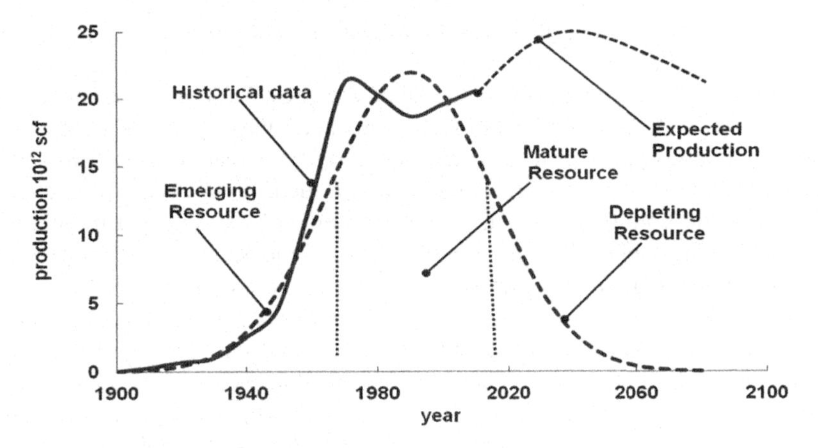

Fig. 1.10 The Hubbert curve life cycle of an energy resource (*bell*-shaped curve) and the production of natural gas in the USA

shape. The actual natural gas consumption in the USA decreased temporarily and then increased again, fueled by new domestic discoveries, such as the Barnett field in Texas, and the significant increase of new family house construction between 1995 and 2008. In addition, liquefaction of natural gas, which is abundant abroad, has made possible the transoceanic transportation of this resource. These imports have also provided additional, albeit limited, supplies of natural gas in the USA market. The figure shows that a local maximum was reached in the 1970s followed by a local minimum in the early 1990s and a continuous but slow consumption increase since then. The more recent discoveries of vast natural gas resources in several regions from the East Coast to the Tennessee Valley in the USA are expected to keep prices low and to further increase the consumption of natural gas in the United States. The expected production part of the curve from 2010 to 2100 reflects these recent resource discoveries and the reaction of the consumers to the expected stable price of natural gas in the USA.

It is also observed in Fig. 1.10 that, neither the curve obtained by the actual data until 2010 nor its extension with the expected production fit well with the entire *Hubbert curve* and, especially, with the rather sharp and well defined *Hubbert peak.* The actual consumption of natural gas in the USA seems to have reached a plateau close to $20*10^{12}$ scf per year. What is more important for the comparison of the two curves is that the demand for natural gas does not show any signs of sharp reduction as the *Hubbert curve* predicts. The main reason for this trend is the new exploration and drilling activities that resulted in the development of new natural gas fields. The more recent discovery of additional resources in the middle of the country indicates that the price of the natural gas will not rise dramatically so as to discourage demand in the foreseeable future. The trends indicate that the natural gas production will stay at the approximately same level in the near future, rather than show a sharp maximum and a significant, sharp reduction as the *Hubbert curve* predicts. Therefore, based on the recent natural gas production data, there is no clear evidence that a well-defined *Hubbert peak* exists in the consumption of natural gas.

The crude oil consumption in the Unites States follows a similar pattern. The domestic production of oil in the USA shows that a maximum was reached in the 1980s and that this production has been decreasing since that decade. However, the consumption of petroleum and liquid hydrocarbons in the USA continued to increase as the domestic production and the increased demand was satisfied by relatively cheap petroleum imports from the Middle East and the South American countries. Even though the domestic production of petroleum in the USA has decreased since the 1980s, this decrease has been gradual and does not fit well to a *Hubbert curve*. The offshore oil production in the Gulf of Mexico and the opening of the Alaska oilfields has contributed to the flattening of the domestic oil production curve. In addition, imports have contributed to an ever-increasing demand for petroleum in the USA. Actually, Hubbert [1] predicted that the domestic petroleum production will peak in the 1970s and that petroleum will be substituted by nuclear power. While the first part may be partly correct, the second prediction was proven to be wrong.

It must be emphasized that the *Hubbert curve* and the *Hubbert peak* are not part of a physical law, which has universal validity, but only a hypothesis that was promulgated in the 1960s and is based on the limited data at the time. While the early part in the consumption of energy resources may be well approximated by this curve, all the empirical data we have collected since the 1980s show that the consumption of natural resources does not follow this curve. In particular, the *Hubbert peak* does not exist as a sharp and well defined maximum, beyond which a substitute must be found for the resource. Instead, the production and consumption curves of the resource reach a rather flat level and fluctuate close to this level. Price adjustments and technological advances in exploration and production make possible the augmentation of the resources and the addition to the proven reserves. In addition, the ease of transportation and the worldwide trade of primary energy sources have made possible the almost continuous supply of energy resources in the world economies. For this reason all the predictions for future production, consumption and pricing that have been made in the past, based on the existence of a *Hubbert peak* have proven to be unreliable. As with all predictions about the future, future energy production and, especially, future energy prices may not be forecasted with any degree of certainty. Based on the existence of a finite quantity of fossil fuels, what is certain is that *eventually* the fossil fuels will be exhausted. However, at present, science cannot tell us with any degree of certainty, when in the future and in what way the depletion of the fossil fuels will occur.

1.4 Concluding Remarks

The supply of sufficient energy has become very important to the affluence of nations and the well-being of the global society. Securing energy supplies for the present and the future has become a prime consideration of all the governments and plays an important role in regional and global political considerations. Energy sources have become prime commodities, on which economic growth and national prosperity are based. The control of vital energy supplies becomes the principal reason for regional conflicts and wars. Because of this, the subject of alternative energy sources is not simply technological, but also extends into areas of economics, history, political science and public policy. As a result, a great deal of complexity and a certain degree of mystery surround the subject of energy. Other factors that make this subject complex and multifaceted are: the multitude of the scientific units that have been used; the significant price fluctuations of the primary and secondary energy sources; the discovery of new resources; and the unpredictability of future supplies.

While simple predictions on the energy production and prices, such as the *Hubbert curve,* are not reliable, it is certain that at some point in the future—very likely the far future—the fossil fuels, which now satisfy more than 80% of the world primary energy demand, will be exhausted. Alternative energy sources, which include nuclear energy, will have to substitute the fossil fuel consumption and provide the human society with sufficient energy to maintain the desired standards of living. In addition, societal concerns about global climate change may impose regulatory limitations on the consumption of fossil fuels in the near future. For this reason it is very important to know: what are the alternative energy sources that may be harnessed with the current technology; what are the engineering systems that will harness this energy; how the alternative energy sources may satisfy the energy demand of the human society; and what is the potential of these sources to supply energy on a large scale.

This book provides the means to understand the scientific principles of energy conversion and the operation of the technical systems that are employed for the harnessing of all the currently known alternative energy sources. The students of the book will become familiar with the governing principles for the harnessing of alternative energy sources and will be able to apply this knowledge to conduct feasibility studies and to design engineering systems in order to make the best use of these energy resources. At the same time, the students will become familiar with the political, social and economic issues that surround the use of alternative energy sources and the global climatic change and will be able to participate as experts in the ongoing societal debates related to these issues.

Problems

1. Convert to Btu the following quantities: 300 kWh; 450 bbl of oil; 25 scf of natural gas.
2. Convert to MJ the following: 500 MeV; 15 Therm; 15 tce.
3. What was the energy demand in your country in 2010 and which primary sources supplied the demand?
4. How much was the electricity production in your country in 2009 and which primary energy resources produced it?

5. The GDP of the Peoples Republic of China is expected to increase by an average rate of 9.7% in the next ten years. What is expected to be the corresponding increase of the rate of electricity consumption?
6. How many kWh of energy were produced in 2007 from solar, wind and geothermal energy? How do these compare to the energy from coal?
7. Oil sands abound in several provinces of Canada. Oil sands contain bitumen, trapped in the sandstone. The best oil sand resources contain close to 0.7 bbl of oil per ton of sand. One method of extracting this resource is to inject high temperature steam or carbon dioxide into the oil reservoir to liquefy the bitumen and bring it to the surface in a pipeline. The heat capacity of the sandstone is 1.1 kJ/kgK. The sand and bitumen are heated from 45 to 250°C for the bitumen to melt and flow into the pipeline. Assuming that all the heat supplied to the extracted bitumen part is recovered at the wellhead, how much heat must be supplied to the sandstone for this process? What percentage of the heating value of the extracted oil does this represent?
8. What is the projected demand for oil, in bbl, in 2030 according to the RS-2030 and 450-PS-2030? Based on your knowledge of oil consumption, suggest four systems or methods that will result in decreasing the future oil energy demand.
9. "Based on the average oil demand of the last hundred years, we have enough petroleum reserves to last us for another three centuries." Comment in an essay of 300–350 words.
10. The world total coal reserves (adjusted for the heating value of the various types of coal) are estimated to be approximately 10^{12} tons. The current annual consumption rate is $6.62*10^9$ tons and the consumption of coal increases by 0.85% per year. Estimate the lifetime of this resource.
11. The world's petroleum reserves are approximately $1.7*10^{12}$ bbl. From outside sources obtain the current global consumption of petroleum and its average rate of growth in the last five years. Based on these, calculate the lifetime of petroleum reserves.

References

1. Hubbert MK (1956) Nuclear energy and the fossil fuels. Proceedings of the spring meeting of the Southern District, American Petroleum Institute, San Antonio Texas
2. Hubbert MK (1971) Energy sources of the earth. Sci Am 224:60–70

Chapter 2
Environmental and Ecological Effects of Energy Production and Consumption

Abstract The exponential increase of energy consumption, since the beginning of the industrial revolution, has produced significant changes in the global environment, chief among which is the increase of the average concentration of carbon dioxide in the atmosphere from 280 ppm in 1750 to more than 390 ppm in 2011. Climatologists predict that this change will cause an increase of the average temperature of the planet as well as regional and global and climatic changes. Other significant environmental effects of energy consumption are: the several ecological problems caused by acid rain, which has threatened in the past the ecosystems of several lakes and rivers; lead contamination of the atmosphere; nuclear waste, which is produced by the more than 430 nuclear power plants in continuous operation worldwide; and, the waste heat rejection by all thermal power plants, which is accompanied by fresh-water consumption. Environmental threats are neutralized by public policy, either national policy or concerted international efforts and protocols that are ratified by several countries. Despite the efforts of the environmental community, there is not yet a global agreement for the mitigation of the effects of high carbon dioxide concentration, which poses the principal environmental threat of the twenty-first century. The problem of nuclear waste is being addressed at several national and regional levels and it appears that solutions for the long term storage of radionuclides will become available in the near future. National public policies and international collaboration has almost solved the acid rain and lead contamination problems. The two are viewed as success stories stemming from international collaboration and successful public policy. This chapter starts with a short section on the environment and ecosystems, continues with descriptions of the most significant environmental problems that are caused by energy consumption and delineates adopted and proposed suggestions on the mitigation of environmental threats.

E. E. (Stathis) Michaelides, *Alternative Energy Sources*,
Green Energy and Technology, DOI: 10.1007/978-3-642-20951-2_2,

2.1 Environment, Ecology and Ecosystems

The *environment* is everything that surrounds the humans and where all the economic activity occurs. The lithosphere, the atmosphere and the hydrosphere are the three distinct components of the environment. Processes and events interact in different ways with the environment. For example, a hurricane is formed in the atmosphere and encompasses water that comes from the hydrosphere. When the hurricane washes over land, it dumps to the ground very high quantities of water as rain, which causes local flooding, erodes the soil and caries it into the sea. These types of interactions produce *environmental changes,* most of which are undesirable.

Ecology is the study of the relationships of organisms with one another and the relationship of organisms to their environment. This subject incorporates principles from the scientific disciplines of biological sciences, physics, physiology, and chemistry.

The *ecosystem* is a rather loose concept that refers to a subdivision of the landscape or a geographic region that is relatively homogeneous. An ecosystem is made up of organisms, environmental factors, and physical or ecological processes. Hence, the concept of ecosystem comprises organisms, species and populations; soil and water; climate and other physical factors; and processes, such as nutrient cycles, energy flow, water flow, freezing, and thawing.

Although the two are related and are often confused, there is a clear distinction between environmental and ecological processes as well as between the environmental and ecological concerns: The ecological concerns always involve effects on ecosystems. For example, a hurricane will wash a great deal of soil into the sea and will change the coastline of an entire region. If we are only concerned with the physical process of soil erosion, the suspension sediment in the water, and its subsequent deposition on the bottom of the sea, three purely physical processes, then we have an *environmental concern.* If we are concerned about the effect of the erosion on the crops, the loss of habitat of subsurface organisms, or about the effect of the increased concentration of pesticides that accompanies soil erosion on the aquatic life, then we have an *ecological concern.*

Because ecosystems are closely connected to their environment, every environmental change has ecological consequences. The observed increase in carbon dioxide concentration and the expected global and regional climate changes are related environmental changes. Their consequences in the ecosystems include altered patterns of crop production as well as migration or disappearance of species from several regions. Similarly, the discharge of pollutants, such as dioxin or lead, on dump sites is an environmental event that has ecological consequences.[1] When one considers the effects of the pollutants on the subsurface organisms, the effects of the leaching of the pollutants in nearby aquifers, streams, or lakes and its ultimate effects on animals and humans that drink the water, then the concern is

---

[1] In most countries it is illegal to discharge pollutants in the environment.

ecological. Another example on the distinction of environmental and ecological effects and concerns may be made in relation to the Chernobyl nuclear power plant accident, which occurred in 1986. The steam explosions in the reactor released into the environment a great deal of radionuclides, which were accumulated in the region or were transported to other areas by atmospheric currents. As a result of run-off from the rainfall, a great deal of radioactive cesium and strontium is now physically buried in the bottom of rivers and lakes or in the subsurface of the land. These are environmental changes. The ecological effects that are consequences of these environmental changes include the mutations in the cells of living species that absorbed radionuclides via the food chain; the decimation of herds of reindeer in Lapland, which consumed grass contaminated with radionuclides; the forests with trees that have radioactive bark; and the significant increase of childhood leukemia and cancer incidents in the human populations, which were severely affected by the release of the radioactivity.

2.2 Global Climate Change

The most pressing environmental issue of the early twenty-first century is the accumulation of carbon dioxide (CO_2) and the expected global warming. Global warming has become an urgent political issue in many countries. The issue is often debated, frequently divides the experts and has affected several national elections. By the term *global warming* we define all the effects of the expected increase of the average temperature of the planet, which are due to the increase of the atmospheric concentration of CO_2 and other similar gases. The main cause of the CO_2 accumulation is the anthropogenic activities related to the fossil fuel combustion processes. All fossil fuels—coal, petroleum and natural gas—are composed of carbon and other atoms, typically hydrogen. The carbon atoms form CO_2 upon combustion, as for example in the following complete combustions of coal and methane:

$$C + O_2 \rightarrow CO_2 \quad ; \quad CH_4 + 2O_2 \rightarrow CO_2 + 2H_2O. \tag{2.1}$$

Since humans have increasingly used the fossil fuels for their energy needs, the amount of fossil fuels that are burned annually has increased exponentially and, as a result, the average concentration of CO_2 has reached very high, almost alarming levels and is expected to increase in the near future. Figure 2.1 shows the average concentration by volume of the CO_2 in the Earth's atmosphere since the beginning of the industrial revolution. While the concentration of this gas was almost constant for centuries before 1750, at approximately 280 ppm, the concentration started rising with the increased use of fossil fuels and reached the level 391 ppm in April 2010, a 40 % increase from its historical level. It is also apparent in this figure that the rate of increase of the CO_2 concentration has accelerated in the last sixty years. The increased CO_2 concentration and its rate of growth show a very

Fig. 2.1 Average atmospheric concentration of CO_2, since the beginning of the industrial revolution (data from Mauna Loa Observatory)

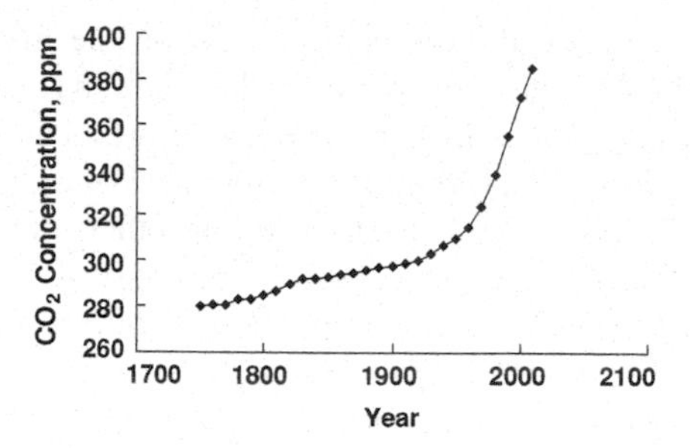

high correlation with the increased energy consumption by humans and, especially, with the significant increase of the combustion of fossil fuels since the 1950s, which is mainly due to the widespread adoption of personal transportation by automobile in the developed countries.

The 40 % increase of the CO_2 concentration represents a significant change in the composition of the planet's atmosphere, which is the outer "blanket" of the planet. The Earth is a complex, highly nonlinear, dynamic system, where small changes have the potential to cause significant local and global effects. Most climatologists and the vast majority of the scientific community expect that this significant change in the planet's outer "blanket" will also have a significant impact on the Earth's climate, globally and regionally as well as on human economic activities.

2.2.1 The Energy Balance of the Earth

We may consider that the atmosphere of the planet Earth is a thermodynamic system, the *system Earth,* which receives a rate of heat, $\dot{Q}_S$, primarily from the Sun, and simultaneously radiates heat, $\dot{Q}_E$ in all directions. In addition, because of the nuclear reactions that continuously occur inside the core of the planet, an additional quantity of heat power, $\dot{Q}_{\text{int}}$, is convected by magma to the surface of the planet. For this analysis, we may identify the atmospheric layer around the surface of the planet of total mass m_E, with average specific heat capacity, c_E, and average temperature T_E. For this thermodynamic system, which is schematically depicted in Fig. 2.2, one may write the energy balance equation as follows:

$$m_E c_E \frac{dT_E}{dt} = \sum_i \dot{Q}_i = \dot{Q}_S - \dot{Q}_E + \dot{Q}_{\text{int}}. \tag{2.2}$$

The Sun may be approximated as a black body, with absolute temperature, T_S, while the Earth is better approximated as a grey body with a radiation emissivity, ε_E, and absorbtivity, α_E. In a similar manner, one may approximate the conduction heat transfer from the interior in terms of an effective conductivity, k_E. Since the thickness of the atmosphere, H, is very small in comparison to the radius of the

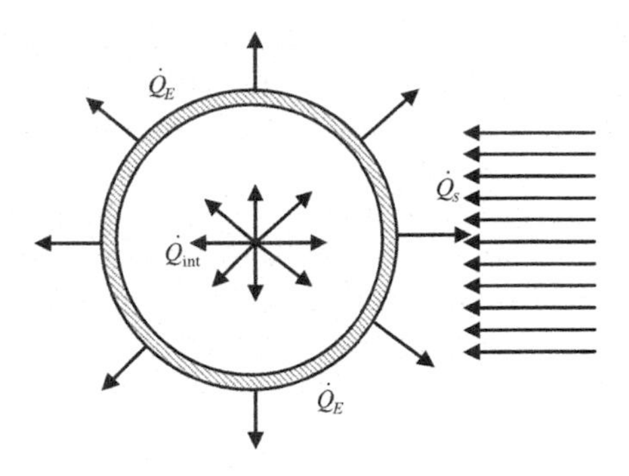

Fig. 2.2 The Earth's surface layer as a closed system and its heat balance

Earth R, that is $H<<R$, one may treat this layer as a thin layer on a sphere and then use the closure equations of radiation and conduction to write Eq. (2.2) in terms of the average Earth's surface temperature, T_E, as follows:

$$4\pi R^2 H\rho c_E \frac{dT_E}{dt} = \pi R^2 \sigma \alpha_E T_S^4 - 4\pi R^2 \sigma \varepsilon_E T_E^4 + 4\pi R^2 k_E (T_{\text{int}} - T_E)/H, \qquad (2.3)$$

where all temperatures are absolute temperatures, ρ is the average density of the atmosphere, σ is the Boltzmann constant $5.67*10^{-8}$ W/m^{2}K^4, and T_{int} is the interior temperature of the Earth. It must be pointed out that Eqs. (2.2) and (2.3) are approximations. Climatologists use less restrictive assumptions, better defined averages and more complex models that include geographic regions to derive quantitative predictions for the Earth's regional temperature and climate. The above approximation is sufficient to qualitatively demonstrate the global warming, to perform simple calculations for the average atmospheric temperature, T_E, and to elucidate the climate trends associated with global environmental change.

It is apparent from the last two equations that if the three rates of heat are constant over a long period of time, the surface temperature, T_E, will reach a constant value. Then the "Earth system" will be at steady state and the temperature T_E, will not vary. However, if any parameter of the system is perturbed, one or more of the three heat rates will change, the non-linear system will undergo a transient process and will reach a new equilibrium state with a different temperature T_E'. For example, if the radiation emissivity were to decrease, the rate of heat loss of the Earth, $\dot{Q}_E$, would also decrease. If the other parameters did not change at the same time, Eq. (2.3) shows that the surface layer's temperature T_E would gradually increase until it reached a new equilibrium temperature, $T_E' > T_E$, where the sum of the three rates of heat as shown in Eq. (2.2) is again zero. In a similar manner, an increase in the Earth's average radiative absorptivity, α_E, would cause an average temperature increase, while a decrease of α_E, would cause an average temperature decrease. Such variations of the average atmospheric temperature will inevitably result in regional and global climatic changes with significant and, probably, adverse effects on the environment, the ecosystems and the human population.

It must be emphasized that, because of the very high thermal inertia of the surface layer ($m_E c_E$) the characteristic time of these changes is on the order of centuries. Therefore, the permanent changes in the average temperature, T_E, will take decades until they are accurately measured. As a consequence any warming or cooling trends that may be predicted from Eq. (2.3) and more accurate climate models will take decades to be experimentally verified and confirmed. Similarly, and because of this high thermal inertia, the effects of any corrective action humans may wish to undertake, in order to influence the Earth's surface temperature and their climate, will also take several decades or centuries to be realized.

2.2.2 The Greenhouse Effect

The Sun's light is transmitted to the Earth in a wide spectrum as will be explained in more detail in Chap. 7. A very high percentage of the energy from the Sun lies in the high frequency and short wavelength part of the spectrum (Fig. 7.2) which has a maximum at 490 nm (0.49 μm). The Sun's radiation supplies the rate of heat $\dot{Q}_S$ which provides needed energy to the planet and enables the photosynthesis and other life-supporting processes on the surface of the planet. CO_2, and other similar gases allow this high frequency part of the spectrum to pass almost unimpeded. According to the laws of radiation, the Earth also radiates energy to the universe. A qualitative difference between the radiative spectra of the Sun and the Earth is in the wavelengths or frequencies of the bulk of the spectra. This difference is explained in terms of *Wien's Law*, which relates the wavelength at the maximum radiation density, λ_m, to the absolute temperature of the radiative black body, T:

$$\lambda_m T = 0.0029 \quad mK, \tag{2.4}$$

where the temperature T is in Kelvin (K). The Sun's surface temperature is approximately 5,900 K, which corresponds to a maximum wavelength of 490 nm. The Earth's average surface temperature is close to 300 K, which corresponds to a 9,700 nm wavelength. The consequence of this is that most of the Earth's radiation is in the infrared part of the spectrum and invisible to human eye.

There are several atmospheric gases, most notably H_2O vapor, CO_2, CH_4, N_2O, and O_3, whose molecules freely absorb infrared radiation. These gases are frequently called *greenhouse gases (GHG's)*. By absorbing the infrared radiation, the molecules of these gases reach higher, non-equilibrium energy states. Because individual atoms and molecules cannot exist for long at non-equilibrium states, the molecules of these gases reach equilibrium with their surroundings by imparting their excess energy to other atmospheric molecules. This is accomplished by molecular collisions or by radiating the excess energy to neighboring molecules. The net effect of this energy transfer is the warming of the other atmospheric gases and the reflection of part of the Earth's infrared radiation back to the surface of the Earth. Figure 2.3 shows in a schematic diagram this exchange of radiation between

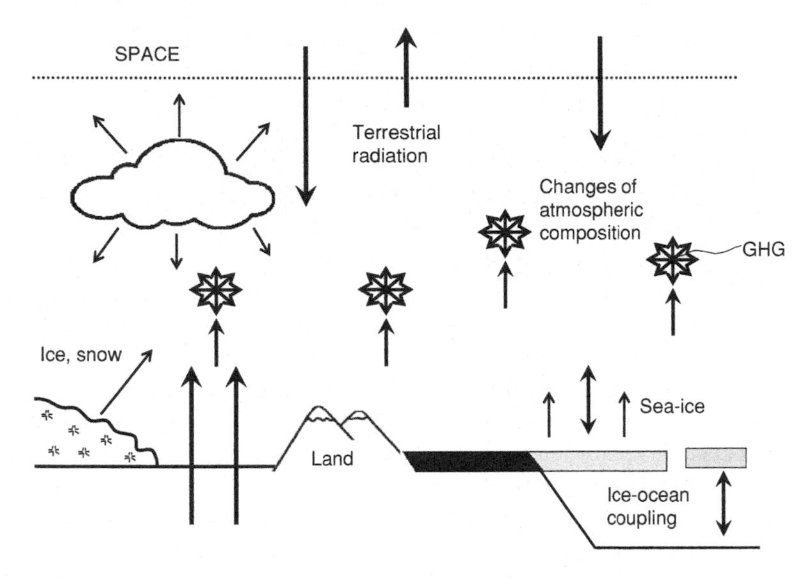

Fig. 2.3 Radiation exchange between the surface of the Earth, the atmosphere and the outer space

the greenhouse gases, the atmosphere and the surface of the Earth. The GHG's are depicted in the figure as absorbing preferentially the Earth's radiation and subsequently re-radiating it in all directions. This process results in the radiation of part of the energy back to the surface of the Earth and the atmosphere.

In terms of the simplified model described by Eqs. (2.2) and (2.3), the net effect of the presence of the GHG's is a decrease of the Earth's radiation emissivity, ε_E. This reflects a fraction of the heat rate $\dot{Q}_E$back to the surface layer of the Earth. The immediate consequence of this reflection is that the current temperature of the surface layer is higher than what would have been in the complete absence of the GHG's.

Actually, if the GHG's were entirely absent from the atmosphere, the rate of heat $\dot{Q}_E$ would have been significantly higher because the average emissivity of the Earth, ε_E, would have been much higher than its present value. Hence, Eq. (2.3) would predict a lower surface temperature T_E than the current average temperature. Accurate climatic models show that, in the absence of any GHG's, the average temperature of the atmosphere would have been approximately 33°C (59°F) lower than what it is at present. The beneficial effects of the current concentration level of the GHG's in the atmosphere are immediately apparent: without the GHG's (at their historical levels) most of the oceans and the surface waters would have frozen and the climate of the Earth would have been inhospitable to life in its current forms. Without the benign warming effect of the GHG's it is doubtful that human life would have evolved in this planet.

While the low concentration of the GHG's is necessary for the life on the planet Earth, significantly higher concentrations of these gases will have a detrimental effect to the ecology and the economic activities of the human society. The GHG's may be likened to a "blanket" around the Earth that keeps the planet warm. If this "blanket" becomes too thick, the inside temperature will increase and will cause several regional and global long-term effects that are unwanted and detrimental. The 40% increase of the CO_2 concentration in the last three centuries and, especially, the highly accelerated increase of this parameter in the last fifty years, as seen in Fig. 2.1, are causing the Earth's "blanket" to become significantly "thicker." Many climatologists and scientists as well as the Intergovernmental Panel on Climate Change (IPCC) of the United Nations have issued warnings about the uncontrolled increase of the CO_2 concentration and its consequences for the economic activity and life in our planet.

It must be noted that the term *Greenhouse Effect* is neither new nor a product of the twentieth century environmentalists. The effect was first predicted analytically by Jean-Batiste Jaques Fourier, the founder of the modern heat transfer theory, in 1824. The Greenhouse Effect was verified experimentally in the laboratory by the British physicist John Tyndall in the 1850s and was quantitatively validated for the atmospheric temperature in the 1890s by S. Arrhenius, a Nobel laureate and one of the founders of Physical Chemistry. Modern scientists and climatologists have developed sophisticated and accurate climate models to study quantitatively the Greenhouse Effect and its consequences on the planet Earth. While not all the climate models agree on the exact value of the average temperature rise, all the reliable models converge in predicting a significant average global temperature rise accompanied by significant regional changes of the temperature and severe weather changes that have the potential to disrupt the human economic activities.

2.2.3 Major Consequences of the Greenhouse Effect

The Greenhouse Effect threatens to change the entire global climate. Before we discuss the climate and its impending changes it is advisable to distinguish between weather and climate: The *weather* is the short-term product of all the complex interactions between the Sun, the atmosphere, the hydrosphere, the continents and their features, such as mountains, vegetation, and ice sheets. Weather is a short-term phenomenon that results from the temporary thermal interactions between the solar radiation, the atmosphere and the hydrosphere. The weather may be predictable over short times, e.g. a few days, but is unpredictable over long periods of time, e.g. months or years. The *climate* is the long-term result of the weather. It may be said that climate is the average weather, taken over a period of several years and decades.

The economic activities of humans have developed over the centuries based on the fact that the regional and global climate has been unaltered for millennia: For centuries one could rely on the fact that the January days in Hanover, Germany, will be exceptionally cold; the month of March in London will carry a significant amount of rain; and that July in Fort Worth, Texas, will be hot and dry. Weather may bring rain on a particular October day in the Sahara desert, but one can rely on the climatic fact that, during a given year or a given season, the rainfall in the Sahara will be significantly lower than the rainfall in the Amazon basin. Weather changes frequently, but the weather phenomena are brief and do not impact significantly the environment and the human activities. On the contrary, climate changes, whether regional or global, will affect significantly the environment, the ecosystems and the human economic activities. For example, a 2–3°C temperature increase in the American Midwest accompanied by drought will convert that area to a desert and will deprive the USA of its breadbasket. An increase of the average temperature of the Earth's surface is a *de facto* global climate change that will significantly impact all human activities and the global economy. This section enumerates some of the most important consequences of global climate change, several of which will be detrimental to the environment, the ecosystems and the lives of humans.

A. Melting of the polar ice caps: An increase of the average atmospheric temperature will result in an increase of the temperature of the Polar Regions. Actually, Global Circulation Models (GCM's) predict a higher temperature rise in the polar zones than in the equatorial and temporal zones. An immediate effect of the polar temperature rise will be the melting of part or the entire ice caps in the Polar Regions. Ice reflects 92% of the incident sunlight and the polar ice caps reflect a high percentage of the incident solar radiation. The disappearance or simply the size reduction of the polar ice caps will effectively increase the average global absorbtivity, α_E. A glance at Eq. (2.3) will prove that this will further increase T_E and will thus, accelerate the rate of global warming.

B. Sea level rise: The total or partial melting of the polar ice caps will free enormous masses of liquid water. Following the hydrological cycle, a very large fraction of the mass of this additional fresh water will eventually end in the oceans. Because water is incompressible, the volume conservation principle implies that the level of the oceans will rise to accommodate this additional mass of water. It is estimated that, if only the West Antarctic sheet ice melted, the average sea level would rise by 5 m (16.6 ft) and that if all the ice on the surface of the Earth melted, the sea-level rise would be approximately 60 m (200 ft). This will bring under water very large parts of the continents and will threaten other coastal parts with significant floods. Since, 76% of the planets population lives within 50 km from the coasts, even a moderate rise of the seal level will have catastrophic consequences on large parts of the human population and its economic activities: With a 5–10 m rise in the average sea level, not only cities such as Venice and New Orleans, which have been historically vulnerable to floods, will be underwater, but also modern and thriving economic hubs, such as New York, Shanghai, Los Angeles, Karachi and London, will be severely threatened. Large parts of the planet will become uninhabitable and close to one billion humans will need to be relocated with a 5–10 m rise of the average sea-level rise.

C. Regional climate change: The average temperature rise will be accompanied by regional temperature rises, some more significant than others. As a consequence, the climate of several regions will change. The GCM's are not sufficiently validated to make accurate predictions on regional climate changes and, for this reason, these predictions exhibit high variability. For example, model predictions, such as the development of a desert in the Midwestern and the Great Plains of the United States or of a significant rainfall in the Sahara, are not verified and may or may not materialize. However, scientific reasoning, common sense, and all the models agree that the global average temperature will increase and the regional climate will change. This will have unwelcome consequences in the agricultural and economic activities of the population, which depend on constant climate, predictable seasons and predictable rainfall.

Changes in the regional climate in combination with a sea level rise of any magnitude will disrupt the entire economic life of the planet, will necessitate the displacement of populations and will create severe socioeconomic problems. In 2005, following the hurricane Katrina, the entire world watched with horror the hardship of approximately 1 million persons who were displaced temporarily from the coastal area of the northern Gulf of Mexico. Global warming consequences may necessitate the permanent displacement of 1–2 billion inhabitants. Most of the contemporary nations and societies are not ready to respond to such dramatic consequences of global warming. There is a danger that several of today's societies and nations will crumble under the socioeconomic pressures that follow human displacements of this magnitude and that the human bonds that form the society and the nations will break and will be replaced by anarchy and destruction.

2.2.4 Remedial Actions for Global Warming

Even though all the scientific models predicted the temperature rise of the planet Earth since the time of Fourier in 1824, until the end of the twentieth century there have not been accurate measurements, independent and reliable confirmations of global warming. Reliable, scientific confirmations of the global average temperature rise came in the early twenty-first century. Most notable among them, the United Nations Intergovernmental Panel on Climate Change (IPCC) confirmed in 2007 that the average global temperature has increased during the Twentieth century by $0.74 \pm 0.18°C$ ($1.33 \pm 0.32°F$). This is a rate that is much higher than that of previous centuries. The pertinent IPCC[2] report also attributed most of the measured temperature rise to the observed increase of the GHG concentrations. These reliable and independent scientific confirmations of the global warming effect have alarmed the scientific community, which has called the global scientists and political leaders to action.

Since global warming is caused by the increased anthropogenic emissions of the GHG's, it is apparent that any mitigation of the problem is centered on the reduction of the rate of these emissions. Because CO_2 is the most abundant of the

---

[2] The members of the IPCC shared the 2008 Nobel Peace Prize.

greenhouse gases and because the atmospheric increase of CO_2 is the main reason for the acceleration of the global climate change in the last years, the reduction of the anthropogenic creation of this gas appears prominently in the set of remedial actions that humans will have to take. In any concerted action for the reduction of the anthropogenic emissions of CO_2, one must take into account that this is a global, not a national problem. A CO_2 molecule produced in Rome or Dallas has the same adverse effect as a molecule produced in Madras or in Beijing. For this reason the collaboration and the coordinated action of all nations is required to avert the potential adverse environmental effects of the global warming.

The *Kyoto protocol,* which was created within the *United Nations Framework Convention on Climate Change*, is an agreement reached between several nations, both developed and developing, for the reduction of the CO_2 global emissions. The protocol calls for the industrialized countries to reduce their collective greenhouse gas emissions by 5.2% from the level in 1990 and also has provisions for the transfer of energy conservation technology to the developing nations. The Kyoto protocol asked for a CO_2 reduction of 8% for the European Union countries, a reduction of 7% for the USA, 6% for Japan and 0% for Russia. The protocol has been signed and ratified by most countries, with two most notable exceptions: the U.S.A. and the People's Republic of China (PRC). While most of the signatories, and especially the European Union countries, have taken meaningful steps for the reduction of their GHG emissions, between 1997 and 2010, the USA has actually increased its emissions by 16% and the PRC by 130%. Because according to the provisions of the protocol, most developing countries did not have to reduce their own GHG emissions, this international agreement has only had a symbolic and not a real impact on the anthropogenic CO_2 and GHG global emissions. Simply, the Kyoto protocol was ineffective and the GHG concentration in the atmosphere has continued to increase at an alarming rate.

Since the beginning of the twenty-first century it has become apparent to the scientific community that a more concerted, stringent and inclusive global effort is necessary for the actual and meaningful reduction of GHG global emissions. The following list includes some of the actions individual nations and the global community may take to, first, reduce the growth of CO_2 emissions and, secondly, to reduce the actual concentration of the gas in the atmosphere.

1. **Reduction of energy consumption per capita**: this is the first of the actions the global community may take to at least reduce the growth of the CO_2 emission, especially in the wealthier, developed nations. This is the best, most inexpensive and most feasible alternative to counteracting global warming and may be simply accomplished with energy conservation and higher efficiency.
2. **Sequestration of CO_2 at the production sites, that is at the power plants, and subsequent storage**: while there is current technology and several proven and reliable methods for CO_2 sequestration, most sequestration methods involve the liquefaction of this gas, all the methods necessitate the use of large amounts of energy and, hence, are intrinsically very expensive. The cost of carbon sequestration on the price of the produced electric energy is significant:

estimates for coal power plants range from 130 to 230% higher electricity prices and those for natural gas power plants are in the range 40–90%. In addition, the safe, long-term storage of the produced CO_2, in liquid or superctitical form, may not be feasible with today's technology. Reliable, long-term CO_2 storage in the deep ocean, in depleted oil fields and in coal seams has not been demonstrated to be feasible during the time-frame required for the effective carbon sequestration, which is in the range of 1,000 to 5,000 years. It would be an environmental calamity if the CO_2 that is stored in 2015 starts leaking in 2020 through cracks in the geological formation that will be called "CO_2 Geysers."

3. **Substitution of coal with nuclear fuel for the production of electricity**[3]: While this is technologically feasible, nuclear energy has its own environmental problems, most notably the long-term storage of nuclear waste. At present, lack of nuclear reactor know-how and lack of safety standards in some developing countries make this option a very risky solution for the global environment. A concerted international program that would involve nuclear technology transfer to developing nations as well as international oversight of the nuclear reactors on a global scale would be a viable long-term solution for the reduction of the GHG emissions as well as for the reduction of other pollutants that emanate from the combustion of fossil fuels.
4. **Reforestation**: this process always removes some of the CO_2 from the atmosphere. While reforestation is always good for the regional and global environment, its impact on the atmospheric CO_2 concentration is very weak, because of the large magnitude of the daily CO_2 emissions. For example, it will take 8,900 fully grown pine trees to remove the CO_2 produced by a single 400 MW coal power plant during a single day. This plant requires 1,000 MW of heat input or $8.64{*}10^{10}$ kJ of heat per day. The latter is produced by the consumption of 2,637 tons of carbon and is accompanied by the production of 9,669 tons of CO_2. Similarly, it will take 8 fully grown eucalyptus trees to remove the CO_2 emissions caused by the engine of a single sport utility vehicle (SUV) which runs for 15,000 miles. Clearly, while reforestation is a desired activity and beneficial to the environment, it does not substitute for all the fossil fuels our society currently uses.
5. **Seeding large ocean regions with iron and nitrogen-rich fertilizer:** this will promote the rate of CO_2 absorption by biological organisms that form more complex organic compounds. While this may be an option on paper, the alteration of the ecological function of large tracts of the ocean will also produce many undesirable effects, such as eutrophication (abundance of organic food for all organisms), which leads to water hypoxia (low dissolved oxygen levels) that cause the death of fish and larger sea life.

[3] The USA would have been in compliance of the Kyoto protocol, in the first decade of the twenty-first century, if it simply diverted 15% of its electricity production from coal producing units to nuclear power plants. This would have been achieved with the construction of approximately 56 new nuclear power plants.

6. **Higher use of renewable energy sources:** such as solar, wind, hydraulic and geothermal energy not only for the production of electricity but also for other societal tasks, such as the heating of buildings, clothes drying, etc. Most of these energy sources are abundant in the developed as well as the developing countries and their increased use will alleviate the consumption of fossil fuels for energy production. Reliable and economical methods for energy storage will benefit significantly the increased use of these energy sources, which are to a great extend benign to the environment.

2.2.5 The Failure of the Copenhagen Summit

A very much advertised United Nations environmental summit took place in Copenhagen, Denmark during December 2009.[4] In the years leading to the summit, the environmental community developed the noble hope that the world leaders will finally adopt environmental principles and will take formal and legally binding measures to implement some of the actions, which were enumerated in the previous subsection and which are needed for the aversion of a major global environmental catastrophe. It is unfortunate that the entire world population became spectators of a well-orchestrated, politically-driven and acrimonious spectacle, where nothing of substance for the environment was achieved. Several reasons contributed to this, among which are the following:

1. The debate on global warming has been framed in many countries, including the EU and the USA, as a debate of "personal belief" rather than as an indisputable effect of well-researched scientific causes, supported by long-term scientific observations. There is still a great deal of dispute on the dependability of scientific predictions on global warming among many in the political circles and the news media. The immediate result is that several global powers do not espouse the idea that hard decisions and strict measures need to be taken for the "beliefs" of other people or nations.[5]
2. The 2008–09 severe global economic recession has diverted a great deal of the attention from remedies for the global warming to the economic realities of high unemployment and lower GDP in most nations. At the national levels, global environmental concerns were relegated to secondary issues in favor of

[4] A subsequent summit that took place in Cancun, Mexico in December 2010 was not attended by any world leaders. Its activities were largely symbolic and did not result in anything definitive, substantial or committing for the climate change.

[5] The scandal, which erupted from the unfortunate revelation of hundreds of "tongue in cheek" e-mail messages from environmental scientists at the East Anglia University, England three weeks before the summit, created confusion and doubts on the immediate need for action among many political leaders and citizens. It also fueled the objections of the opposition to remediation efforts.

national economic stimulus packages, which usually result in increased economic activity and higher levels of GHG emissions.
3. During the three months leading to the summit, it became apparent that the preliminary multilateral environmental negotiations, which were taking place for a couple of years, have failed to produce the political framework of a solution that would be acceptable to even a simple majority of the participating nations. The restrained tone of national announcements and press communiqués and the downplaying of the expectations set the eventual tone of the summit's failure.
4. On the OECD side, the leaders of the European Union and the North American countries came to the summit with other more important issues at home among which were the consequences of the global recession, high unemployment rates and overall citizenry dissatisfaction. Especially in the USA, during the summit, the attention of the government was at the economy and the impending passage of a health care legislation on which the national administration had spent a great deal of its influence and political capital.
5. On the side of the developing nations, there was an obvious linkage between economic aid, economic development and environmental action. With daily statements in the press and a short-duration walkout during the last days of the summit, the delegates of several African countries made it clear that, as a result of any agreement on the environment and global climate change, they expected increased economic aid from the developed nations, which was not apparently forthcoming.
6. China and India, two developing nations that have made great strides towards industrialization in the first decade of the twenty-first century and account for most of the growth rate in carbon emissions have refused to make any binding concessions for the long-term reduction of CO_2 emissions. In the absence of realistic concessions from these two countries with the highest growth rates of CO_2 emissions, the leaders of other nations were reluctant to make commitments that would appear to be unilateral.

The failure of the Copenhagen summit notwithstanding, the environmental threat from increased carbon dioxide concentration and the impending global climate changes are realistic and imminent. What is at stake here may be the continuation of human civilization as we know it. Action in national and international fora is needed and taking remedial measures is essential before changes become irreversible. The scientific and engineering community must play a leading role in the arena of global change by doing the following:

(a) Continuing to make accurate global measurements;
(b) Communicating these measurements and the pertinent conclusions to the public in an unbiased and honest (that is, scientific) manner;
(c) Developing reliable measures of accountability for GHG emissions;
(d) Developing methods and building meaningful engineering projects for the mitigation of the adverse effects of the global climate change;

(e) Improving the efficiency of power production plants and internal combustion engines; and
(f) Continuing the research and development efforts on alternative sources for energy that would reduce and would finally nullify the global GHG emissions.

2.3 Acid Rain

Acid rain or acid precipitation is the return to the terrestrial aquatic environment of the oxides of carbon, nitrogen and sulfur in an acidic form. Acid rain is closely related to the combustion of fossil fuels. Fossil fuels, and especially coal, contain large quantities of sulfur which forms SO_2 upon combustion. In addition CO_2 and a series of nitrogen oxides with the general formula NO_x (or, commonly, NOX) are formed during coal combustion. These oxides combine with water vapor in the atmosphere to form mild acids. For example the hypo-sulfuric and the carbonic acid, two weak acids, are formed in the atmosphere by the following reactions:

$$\begin{aligned} SO_2 + H_2O &\rightarrow H_2SO_3 \text{ and} \\ CO_2 + H_2O &\rightarrow H_2CO_3. \end{aligned} \tag{2.5}$$

Atmospheric SO_2 may also combine with ozone first and then with water vapor, to form the much stronger sulfuric acid:

$$SO_2 + O_3 \rightarrow SO_3 + O_2 \text{ and } SO_3 + H_2O \rightarrow H_2SO_4. \tag{2.6}$$

In addition, the several NO_x compounds that are formed during the combustion processes may also combine with water vapor in the atmosphere and finally form the weaker nitrous acid (HNO_2) or the stronger nitric acid (HNO_3) thus making NOX a contributor to acid precipitation.

Typically the acidic chemicals in the atmosphere are formed within small droplets or on the side of very fine particles, which are called aerosol particles. The sizes of these droplets and particles are in the submicron range. This implies that they settle extremely slowly and, may remain airborne in the atmosphere for weeks or months following the air currents and turbulence. During rain or snow precipitation, the aerosols combine with the larger rain drops or snow flakes, precipitate faster on the ground and, thus, are removed from the atmosphere. The rain or snow runoff, which eventually feeds rivers and lakes, contains higher concentration of the acids and for this reason it has been called *acid rain, acid snow* or in general, *acid precipitation.* In the aquatic environments, H^+ ions are released from these acids according to the following chemical reactions:

$$\begin{aligned} H_2CO_3 &\rightarrow H^+ + HCO_3^- \\ H_2SO_3 &\rightarrow H^+ + HSO_3^- \\ \text{and} \quad H_2SO_4 &\rightarrow 2H^+ + SO_4^{2-}. \end{aligned} \tag{2.7}$$

As a consequence of acid precipitation, the concentration of the H^+ increases significantly and the pH of these bodies of water drops from its natural range of 6.8–7.4 to significantly lower values. Some of the more dramatic acid precipitation observations are listed below [1]:

1. A storm in Scotland in 1974 dropped rain with pH 2.4.
2. The pH of rain in Kane, Pennsylvania on September 19 1978 was 2.32. This is lower than the pH of vinegar.
3. For the entire year of 1975, rains in Norway and Sweden recorded pH less than 4.6.
4. During the 1970s the pH of 80% of drizzles in Holland was less than 3.5, and sometimes as low as 2.5. This is the pH of common vinegar.

The drop of the pH has significant adverse effects on the ecosystems of the rivers and lakes, because many animal species cannot survive at these low (as well as very high) pH levels. As a result, several of the species may disappear, either because of the direct effect of a lower pH or because of lack of nutrients. The low pH resulting from acid deposition decimated the fish population in several lakes in the 1970s and 1980s. In addition, high acidity precipitation rendered the soil acidic with a significantly adverse effect on crops as well as on forests. Some of the environmental and ecological effects of acid precipitation are:

1. As the water of the streams becomes more acidic, a shift to acid-tolerant plants occurs, such as green algae.
2. Acid sensitive species, such as snails, clams and amphipods disappear.
3. Higher concentrations of Al^{3+} and other metal ions are observed. These ions damage the gills of fish and also enhance the precipitation of dissolved organic matter in the water, which is a source of food for fish. With decreased food supply, fish become emaciated or die.

These causes had a catastrophic effect on the aquatic populations of most rivers and lakes in northern Europe and North America. For example, salmon in several Norwegian rivers did not reproduce for years and became almost extinct. Also the fish disappeared from 190 lakes in the Adirondacks, Canada and more than 2,000 lakes in southern Norway.

What accentuated the environmental problems of acid deposition is that, in most cases, the production of SO_2 and the other oxides actually occurred in other, neighboring countries. The oxides or the acid laden aerosols are carried by the air currents over international boundaries and affect neighboring nations. Given that most of the economic activity of the world is produced in the northern latitudes between the 30th and the 60th parallels, where the predominant winds are south-easterlies—directed from southeast to northwest—acid rain produced in countries to the South was deposited in countries to the North. Thus, the acid oxides produced in Ohio and Michigan affected the lakes in the Ontario Province, Canada, while acid oxides produced in the industrial Ruhr of Germany affected the aquatic environment of Holland and Scandinavia. Pollution does not respect national boundaries and, when it occurs it becomes an international issue.

A concerted international effort to mitigate acid rain started in the 1970s and continued in the 1980s and 1990s with great success. Despite the protests of the coal industry and several electricity generating corporations, one after another, national governments enacted regulations to limit the emissions of SO_2. In the United States a goal was set to reduce the SO_2 emissions to less than 9 million tons per year by 2010. The Environmental Protection Agency (EPA) incorporated this program in an amendment to the *Clean Air Act*[6] and developed a market-based initiative to achieve the reduction of SO_2 emissions. This amendment sets annual upper limits (caps) for the emissions of SO_2 for all polluters and issues permits to these companies, which are called *annual allowances.* The *allowances* are consistent with the overall goals for the national reduction of the emissions. Corporations that exceed their targets may trade their *allowances* to others that do not meet their own goals. This is the so-called *cap and trade program.* It creates a market incentive for corporations to exceed their own goals and trade the differences of their annual allowances to others for a profit. The sulfur cap and trade programs have been immensely successful in Europe and North America, where the 2010 reduction goals were met before 2007. As a result, in the beginning of the twenty-first century, acid deposition has dropped by two-thirds from its peak and it is not any more the environmental and ecological threat that was in the 1980s. Because of this resolute international action, the ecosystems in most of the affected lakes, rivers and forests have recovered.

The strategies for compliance with the reduced SO_2 emission standards varied among countries and corporations. These strategies affected the choice of fuel for the production of electricity, implementation of new technologies for the removal of SO_2 and the location for the construction of new power plants. In the USA and the European Union the principal technical approach that was used to reduce the SO_2 emissions has been flue gas desulphurization (FGD) that removes SO_2 from the stack gases by scrubbers before they are discharged to the atmosphere. A FGD process is shown in Fig. 2.4. SO_2 laden gas enters the scrubber, where it is "showered" by a basic water solution, typically a limestone-water solution that contains the basis chemical of calcium hydroxide, $Ca(OH)_2$. The SO_2 is absorbed by the water to first form hypo-sulphuric acid as in Eq. (2.5) and then the weak acid reacts with the basis in the water solution to form calcium carbonate:

$$\begin{aligned} &SO_2 + H_2O \rightarrow H_2SO_3 \ and \\ &H_2SO_3 + Ca(OH)_2 \rightarrow 2H_2O + CaSO_3. \end{aligned} \tag{2.8}$$

$CaSO_3$ is a solid that precipitates. It is subsequently removed from the water and, since it is not a pollutant, it is buried or disposed of.

[6] The Clean Air Act of the U.S.A. was enacted in 1963 and significantly amended in 1970 and 1990. The NO_x emissions problem was tackled by the 1990 amendment.

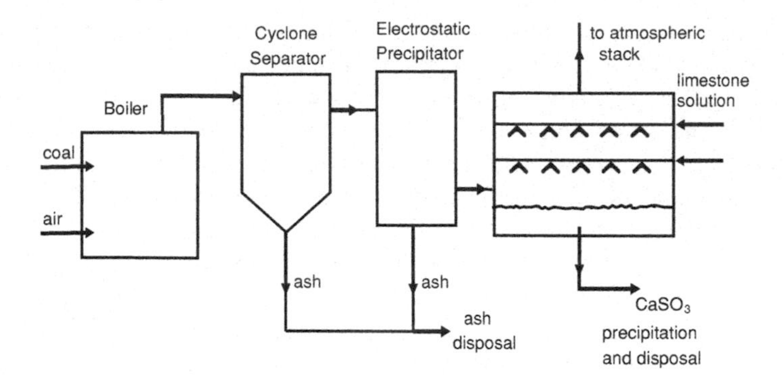

Fig. 2.4 The flue-gas desulfurization process

Some of the other methods that have resulted in the significant reduction of the SO_2 emissions are:

1. Using Fluidized Bed Reactors (FBR) in new plants, which employ limestone particles *in situ* to remove SO_2 during the combustion by converting it to solid $CaSO_3$.[7] The latter is removed with the solid materials of the ash.
2. Blending high-sulfur coal with low-sulfur coal.
3. Switching coal fuel to natural gas, or a mixture of coal and natural gas.
4. Retiring old electricity generation units and replacing them with FBR's or units with SO_2 scrubbers.
5. Purchasing or transferring emissions allowances from other units.
6. Increasing the demand-side management and conservation efforts to reduce the electric power consumption.
7. Power purchases from other utilities or non-utility generators that use low-sulfur coal or other fuels.

The acid-rain reduction programs that were implemented in Europe and North America have been an overwhelming environmental success. Figure 2.5 shows the dramatic drop of the emissions of SO_2 and NO_x in the U.S.A. as a result of the implementation of the Clean Air Act. It is apparent in this figure that the SO_2 emissions dropped to one-third of their values in the 1970. The significant reduction of the NO_x started after the 1990 amendment to the Clean Air Act and the emissions of this pollutant dropped to 60% of their values in 1990. More encouraging is the fact that the recent slope of the two curves is significantly negative, which implies that the emission reduction of the two pollutants will continue in the near future. Not only the long-term goals of the programs were achieved ahead of the deadlines, but also the costs of the programs' implementations to the businesses and the consumers were significantly lower than the

[7] In a FBR the SO_2 comes in direct contact with particles of $Ca(OH)_2$ and reacts as: $SO_2 + Ca(OH)_2 \rightarrow CaSO_3 + H_2O$. The solid $CaSO_3$ is removed with the ash.

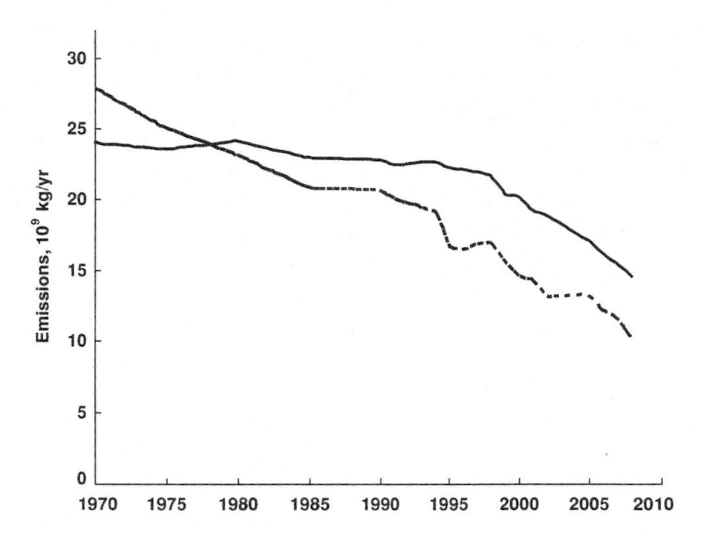

Fig. 2.5 Emission reductions of SO_2 (dashed line) and NO_x (solid line) in the USA

original predictions. It is estimated that, in the USA, the total cost of the SO_2 emissions reduction implementation strategies was in the range $1 billion to $2 billion. This is only one fourth of the original estimates by the coal industry and, most important, it did not cause any disruptions in the electric power production. By the concerted efforts of the international community, the detrimental environmental effects of acid rain have been mitigated and remediated in a short time at a very small cost to the electricity generation industry and the population.

2.4 Lead Abatement

Gasoline and diesel are mixtures of liquid hydrocarbons. While the diesel-air mixture in the diesel engines is designed to ignite by itself at the end of the compression stage, when high temperature is reached, the gasoline-air mixture is designed to ignite during the ignition stage by a spark. Because high temperatures are achieved during the compression stage, several of the hydrocarbons in the gasoline liquid mixture reach their own ignition point, auto-ignite and release heat prematurely. This has led to the "knocking" problem in gasoline engines where auto-ignition has caused premature engine detonation, severe vibrations, low cycle efficiency and subsequent engine damage.

Auto-ignition in gasoline engines may be prevented by chemical additives, the most common of which is tetra-ethyl lead, $Pb(C_2H_5)_4$. Tetra-ethyl lead, when added to the gasoline, prevents engine knocking and engine damage. The use of this chemical compound was widely adopted by the refining and automobile industries as an "anti-knock" additive to the gasoline in the early twentieth century. However, the $Pb(C_2H_5)_4$ burns with the fuel and its combustion releases lead oxides, primarily PbO and Pb_2O, as well as atomic Pb, to the environment. These chemicals were proven to be harmful to the health of the population. Lead compounds affect the synapses in brain cells, especially those of children. Prolonged exposure to lead has been proven to cause mental retardation and brain disorders.

The lead compounds are also incompatible with several types of catalytic converters and reduce significantly the useful life of these converters. For this reason, during the 1970s, regulations were enacted in all OECD countries to phase out the use of $Pb(C_2H_5)_4$ from gasoline additives. In addition other lead compounds were phased out from other commonly used materials, such as paints. In the USA and the countries of the European Union the sale of leaded fuel for automobiles has been completely banned since the 1990s, but it is still allowed for marine engines, racing cars and certain farm equipment. Most of the other countries have followed suit by 1990 and $Pb(C_2H_5)_4$ has now been replaced by other additives, typically made by aromatic hydrocarbons. Only in a few countries of South America, Asia, some of the countries of the former Soviet Union and the Middle East, leaded gasoline is still in use. Even in these countries emerging environmental regulations significantly restrict its use. It is anticipated that by 2020 the use of leaded gasoline will be banned globally. The restriction and ban on leaded gasoline, as well as leaded paint, resulted in millions of tons of lead not being released in the environment. The immediate effect was the lowering of lead in the human bloodstream, especially in children. This is expected to become the means to better public health, lesser neurological disorders and significant improvements of the quality of life.

The vast reduction of the acid rain environmental effects and the reduction of the lead concentration in the blood stream of humans are two significant environmental developments of the late twentieth century. Both took a great deal of effort by various citizen and scientific groups to bring to the attention of the public and, finally, to translate this attention to regulations and national legislation. Several years after their implementation, there is no doubt that their effects have been beneficial to the environment and to the public health and that there are very few individuals who would like to see these measures reversed. While at the start, the affected industries resisted the adoption and implementation of the measures to curb Pb and acid rain using the argument that the cost of energy and gasoline would rise significantly and that the general population would suffer economically, both measures were implemented at a fraction of the cost that was originally predicted by the industry. Furthermore, this cost was absorbed very well by the market and the consumers with no apparent regional and global economic effects. For example, SO_2 abatement methods only reduced the overall efficiency of modern coal-fired power plants by only 1–1.2%. This was counteracted by the use of new materials and processes that resulted in an actual increase of the overall efficiency of the modern electric power plants.

Similar arguments are currently used against measures to curb the global warming by reducing the GHG emissions. The success stories of lead and acid reduction convey optimism that, despite the resistance of the fossil fuel industry and some electric utilities, a significant reduction of GHG emissions will be achieved in the near future without an undue economic disruption.

2.5 Thermal Pollution and Fresh-Water Use

Thermal power plants reject a great deal of energy to the environment in the form of low-temperature heat. All processes in these power plants are subjected to the Second Law of Thermodynamics and, as a consequence, the power plants must reject a great deal of heat to their condensers and through their cooling system to their surroundings (Sect. 3.5). A typical 1,000 MW fossil fuel plant has an overall efficiency close to 40%. It receives 2,500 MW of heat power, of which 1,000 MW are converted to electric power and the remaining 1,500 MW are rejected to the environment. These numbers are slightly different for a typical nuclear power plant, which has an overall efficiency close to 33%. The reactor of this plant would produce approximately 3,000 MW of heat, of which 1,000 MW are typically converted to electricity and 2,000 MW are rejected as waste heat to the environment. These vast amounts of heat power are rejected at low temperatures, typically in the range 30–45°C and may not be used for any practical applications.

A common misconception, even among engineers, is that the waste heat from a power plant may be somehow used for the production of more power. This is impossible: The power plant is designed to produce the maximum possible amount of energy and any heat that needs to be rejected is a consequence of the Second Law of Thermodynamics. The waste heat is at such a low temperature that it is not possible to be of further use for the production of power. Small quantities of this waste heat may be used for heating buildings, for aquaculture or for agricultural purposes (heating of the soil to produce a higher yield). However, economic considerations limit significantly the amount of waste heat that is utilized. Several possible uses of the waste heat from power plants and the overall potential of waste heat utilization are discussed in Sect. 13.2.4.

The annual global electricity production from thermal power plants is approximately 17,500 TWh, which is equivalent to $63*10^{15}$ kJ. At an average efficiency of 35% these power plants reject $117*10^{15}$ kJ to the environment. The rejection of this quantity of heat to the environment is the total *thermal pollution* due to the power plants. An additional part of the thermal pollution is produced from the transportation industry as exhaust heat from automobiles, ships and airplanes. The thermal pollution energy is by all means a vast amount of energy that is released to the environment and causes concerns among some environmentalists. However, it must be noted that the waste heat is only a minor fraction of the total energy that enters the earth's atmosphere from the sun or the energy that is radiated from the earth itself. The annual energy received from the sun is equal to $5.46*10^{24}$ kJ and the annual amount of heat radiated by the earth is of a comparable magnitude. Both of these quantities by far surpass the total waste heat rejected annually by all the power plants on the planet. For this reason, the waste

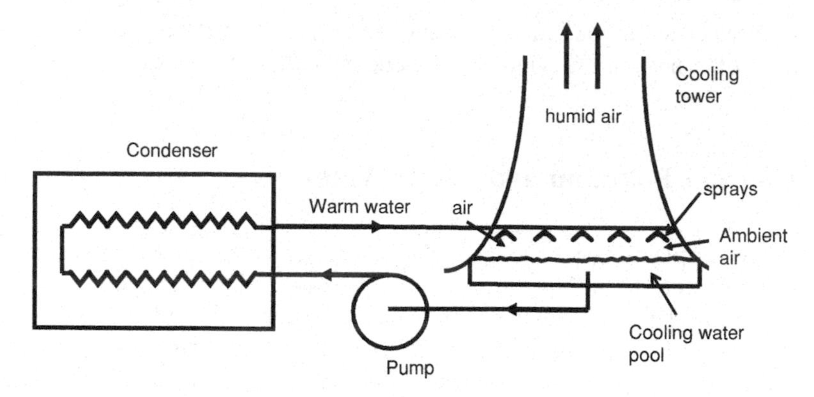

Fig. 2.6 The cooling system of a thermal power plant

heat rejection by power plants does not contribute in any sizable measure to the global warming and does not pose a threat to be such a contributor in the near future. Actually, the calculated rate of heat that is absorbed and diffused in the atmosphere by the GHG's is higher than the entire waste heat production caused by anthropogenic activities by several orders of magnitude.

Their insignificant contributions to global warming notwithstanding, thermal power plants and the environmental heat rejection processes make a significant claim on the fresh water resources of the planet. Heat is rejected primarily from the condensers of power plants, which are cooled by a closed or open circuit of *cooling water*. The process is shown schematically in Fig. 2.6. Relatively cold water enters the condenser, removes the waste heat and then enters a wet cooling tower, where part of it evaporates and cools the rest. The colder water from the cooling tower is directed back to the condenser (Chap. 3). The cooling process uses very high amounts of fresh water, because the temperature rise, ΔT, between the colder and the warmer is typically of the order of 5°C. A heat balance in the condenser yields the following equation for the rate of waste heat rejection$\dot{Q}_{wh}$:

$$\dot{Q}_{wh} = \dot{m}_{cw} c_p \Delta T, \tag{2.9}$$

where $\dot{m}_{cw}$is the mass flow rate of the cooling water that is needed and c_p is the specific heat capacity of water, 4.18 kJ/kgK. A quick computation proves that a typical nuclear power plant, which rejects 2,000 MW of heat through a cooling tower, would need close to 95,000 kg/s of water. If instead of a cooling tower the nuclear power plant rejected this rate of heat to a river or a lake, the maximum ΔT would be 2.7°C and the actual need for cooling water would be close to 176,000 kg/s.[8]

[8] This is the reason why large thermal power plants are usually built close to a natural source of fresh water, a river or a lake. In the USA, 42 of the 103 nuclear power plants in operation are located near the banks of the Mississippi river or one if its larger tributaries. 85% of the nuclear power plants in France are located at the banks of the Seine, the Loire and the Rhone rivers.

It must be emphasized that only a small fraction of the water used in the cooling system evaporates. The latent heat of evaporation, h_{fg}, of the water is approximately 2,400 kJ/kg. Assuming that all the waste heat, $\dot{Q}_{wh}$, is released to the environment solely by evaporation, the amount of water that evaporates, $\dot{m}_{ev}$, is given by the expression:

$$\dot{Q}_{wh} = \dot{m}_{ev} h_{fg}. \tag{2.10}$$

Therefore, a nuclear power plant that rejects 2,000 MW of heat will also cause the evaporation of approximately 833 kg/s of water in its cooling tower. This is equivalent to 72,000 tons per day, by all means a significant amount. Even the power plants that do not use cooling towers, but instead draw their cooling water directly from rivers and lakes in the "once-through" cooling systems, will cause significant water evaporation: The evaporation occurs because the water, which is returned to the river or the lake, is at an elevated temperature, typically 2–3°C higher. The partial pressure of warmer water is higher and, therefore, it evaporates faster. In the long run, the river or the lake will attain their equilibrium temperature by the evaporation of additional amounts of water. On average, approximately 1 kg/s of water is evaporated for every 1 MW of electric power produced by a thermal power plant. This is equivalent to 1 gallon of water per kWh produced.

Fresh water availability for the production of electric power is fast becoming an environmental issue in the twenty-first century. Even though 71% of the surface of the planet is covered by water, only 3% of the water on the planet is fresh water and 90% of it, or 2.7% of the total, is in the form of ice glaciers and underground water aquifers. The remaining 0.3% of the total water of the planet is fresh water in lakes (87%) swamps (11%) and rivers (2%). While the availability of fresh water was not a significant issue in the previous centuries, it is becoming a significant environmental and political issue in the twenty-first century. With the rise of the planet's population and the desired increase of the standard of living in all countries, there is a higher demand of fresh water for agricultural and domestic uses. Large water consumers, such as large thermal power plants, must compete for this resource, which is becoming scarce in several parts of the planet, as for example in the southwestern part of the USA and several regions of Asia and northern Africa. In the near future, the human society will have to make hard decisions on how to better allocate this precious resource among nations and among competing users. Alternative energy sources for the production of electricity are very promising in this regard, either because they do not need cooling (photovoltaics, wind, hydroelectric, tidal, etc.) or because they produce sufficient water for the needs of their own cooling systems (geothermal).

2.6 Nuclear Waste

The transportation and storage of the waste materials produced in the nuclear power plants around the world is a significant global environmental threat because the uncontrolled release of radioactive compounds is harmful to all living animals

Table 2.1 Nuclear waste isotopes and their characteristics

Isotope	Half-life (yrs)	Radioactivity (Bq)
Americium-231	433	$11.84*10^{10}$
Americium-234	7,900	$0.7*10^{10}$
Iodine-129	17,000,000	$5.9*10^{6}$
Plutonium-239	24,400	$0.23*10^{10}$
Plutonium-240	6,600	$0.81*10^{10}$
Technetium-99	210,000	$6.29*10^{8}$

on the planet. In 2009 there were 439 operating nuclear power plants in the world, 103 of which operated in the USA. Heat in a nuclear power plant is typically produced by the fission of the nuclear fuel, primarily uranium-235 and plutonium-239. The fission of the radioactive materials produces other isotopes, some of which are radioactive. Since the typical reactor is a closed system, the entire radioactivity remains inside the reactor until the next refueling period. During refueling, the spent fuel and fission products are removed from the reactor and stored temporarily, within the confines of the nuclear power plant. These materials constitute the *nuclear waste* from the reactor. Similar nuclear waste materials are produced in fuel reprocessing and fuel enrichment facilities. Table 2.1 shows some of the isotopes that are present in nuclear waste, their half-lives, in years, and the level of radioactivity in Becquerel, Bq, or disintegrations per gram per second. More details on the half-lives of isotopes and the theory of nuclear reactions and nuclear power plants are given in Chaps. 4 and 5.

It is apparent from Table 2.1 that nuclear waste will continue to be radioactive and will pose a health threat to the human population for millennia. For this reason permanent storage facilities must be constructed that will be capable to store the radioactive waste for thousands of years, until the eventual residue does not pose a public health threat. This presents a significant scientific and engineering problem, simply because of the timescale of the storage. There is not a proven and reliable method for the storage of such materials during the thousands or tens of thousands of years required for remediation of the nuclear waste materials. Any accidental or intentional (e.g. by an act of terrorism) release of radioactive materials from these sites may render whole regions uninhabitable.

The safe and permanent storage of nuclear waste is an environmental issue of paramount importance to the nuclear industry. However, the populations and governments of several countries, including the USA, have not come to grips with the magnitude of this problem and have not prepared permanent storage facilities for the nuclear waste that has been produced since the 1950s. At present, the nuclear waste is typically stored in temporary facilities in the vicinity of the power plant that produced the waste. A typical temporary storage facility is a water pool, where the nuclear waste is immersed. The heat produced by nuclear disintegrations is convected to the water of the pool, which is maintained at almost constant temperature by evaporation. Make-up water replenishes the evaporated water in the pool. These storage arrangements are temporary, mostly unsecured and pose a threat to the surrounding communities.[9] Permanent disposal of nuclear waste in safe and controlled sites will reduce significantly this environmental concern.

An environmental concern that needs to be addressed before the public accepts the reliability of sites for the long-term storage of radionuclides is the long-term structural integrity of the containers and compartments, where the nuclear waste is stored. Metal or composite tanks corrode, develop cracks from where leakage may occur and, in general deteriorate to the point that they are unsuitable to contain the nuclear waste. The release of heat from the decaying radionuclides only accelerates the deterioration of the containers. The further decay of the radionuclides in the nuclear waste and the reduction of radioactivity to natural levels occur after 10,000 to 1,000,000 years. At present, there is no known storage material that may be reasonably expected to preserve its structural integrity and keep the nuclear waste confined for such long periods of time. For this reason, the proposed nuclear waste management processes involves several stages and processes, of which the most important are summarized in the following sections.

2.6.1 Initial Treatment of the Waste

An initial treatment helps reduce the volume and radioactivity of the waste, while the waste is located in a controlled environment and the heat generated is removed in a controlled manner. Methods for the initial treatment are:

1. *Vitrification*, or *glassification* of the waste. The nuclear waste is mixed with sugar and heated until all the water and nitrates in it are evaporated. The mixture is then combined with glass and further heated until the glass melts. This melt is poured into stainless steel containers, where it solidifies and forms a glass-like substance, that is, "it vitrifies." The vitrified substance is then stored in a steel cylinder. Vitrified materials are very stable. They are hard, water resistant, have very low erosion or chipping and are believed that they arc capable to last unaltered for thousands of years.
2. *Concentration* of the waste. The volume of the nuclear waste may be reduced by concentrating it into a smaller volume, which may be disposed of or stored better and more economically. Flocculation (concentration of fine particles) with ferric hydroxide is often used to remove highly radioactive metals from aqueous solutions. After the removal of these isotopes, the resulting low-level radioactive materials are stabilized and immobilized by mixing with ash and cement to form concrete. The low radiation levels of this concrete do not pose any threats to the environment or the population and may be stored anywhere.

---

[9] The 2011 nuclear power plant accident in Fukushima Dai-ichi, where the stored nuclear waste was exposed and contributed significantly to the environmental pollution underscores this environmental problem.

3. *Synrock* is a complex chemical material of nuclear waste stabilization. Synrock consists of hollandite ($BaAl_2Ti_6O_{16}$), zirconolite ($CaZrTi_2O_7$) and perovskite ($CaTiO_3$). The zirconolite and perovskite become hosts and immobilize the actinide elements, that is elements with atomic number higher than 89, such as uranium and plutonium. The radioactive strontium and barium, which are produced in nuclear reactors, are also trapped and immobilized in the perovskite, while the hollandite immobilizes the caesium and similar lighter metals.

2.6.2 Long-Term Disposal

Immobilization stabilization or simply immobilization is the first stage in nuclear waste management. The long-term disposal of the nuclear waste includes the following suggestions:

1. *Geologic disposal,* either in deep and stable formations on the earth or in the deep sea. The proposed *Yucca Mountain* repository in the United States and the *Schacht Asse* repository in Germany, which operated briefly in the 1990s, are two examples of such ground disposal sites. These repositories are typically in stable, arid geological formations, where water leakage will not be a problem in the future. One of the impediments for permanent geologic disposal is the legal problem of *stewardship cessation* of the materials. This legal term implies the shifting of the burden for the safe maintenance and perpetual management of nuclear waste from the producer to the one who undertakes the storage. The latter is typically the government or a smaller receiver corporation, which does not have the financial resources to guarantee stewardship in the long term and to compensate for damages that may potentially be incurred. Several environmentalists do not believe this is prudent and recommend perpetual management and monitoring of the waste by the producer.
2. *Transmutation* implies the transformation of radionuclides to other materials that are not radioactive. Special nuclear reactors will be needed for the transmutation processes. In the United States research activity on the transmutation has ceased since the late 1970s because plutonium is a byproduct of the process. Since plutonium is used in atomic bombs, its production raises concerns of atomic weapon proliferation. Relevant research work has continued in the European Union, where the reactor *Myrrha* has been built and may be used for transmutation purposes along with other high technology applications.
3. *Waste re-use* usually accompanies the concentration process which was described in the previous section. The produced high-radioactivity materials may be re-used in a nuclear reactor for the production of additional power. Because a great deal of the current nuclear waste is the isotope uranium-238, it is envisioned that this isotope will be separated from the waste and will be used in the breeder reactors of the future.

4. *Space disposal* is a possible alternative that has been advocated by a few non-experts. Given that it costs more than $25,000 to lift a kg of mass to the space, this is extremely expensive and has not been proven to be a reliable way of nuclear waste storage. Considerations of the adverse effects of "space debris" to satellite communications make this a prohibited option.

It must be noted that, in many countries, the long-term or permanent storage of nuclear waste has become a political problem, because local and regional governments resist nuclear materials storage within the boundaries of their jurisdictions. A prime example in the United States is the permanent repository in the Yucca Mountain, Nevada, which was identified as a permanent storage site in the 1970s. A multitude of engineering and scientific studies showed that nuclear waste storage in the Yucca Mountain does not pose significant risks for the local population. Despite of this and the spending of billions of dollars for the studies, local population resistance and political and legal maneuvering delayed the construction of a permanent nuclear waste storage facility. After several years of studies and many legislation acts, in March 2009, the U.S. Department of Energy announced that the Yucca Mountain "…is not considered as an option for storing nuclear waste," without specifying an alternative location for nuclear waste storage. As a result, and for the foreseeable future, the tons of nuclear waste that have been generated since the 1950s and continue to be generated in the USA will be temporarily stored in makeshift pools next to the more than 100 power reactors that produced them. In the absence of a centralized facility to receive the produced nuclear waste, these poorly planned and "temporary" facilities have become a significant environmental threat to the surrounding area and its population.

2.7 Sustainable Development

Sustainable development or simply *sustainability,* is an all-encompassing concept, which basically advocates that the global economic development must be pursued without causing irreparable damage to the ecology and the environment. Sustainability includes all the global economic activities from the production of goods and services to the transportation and energy production. Measures of sustainability include the calculation of pollutant emissions per unit of the desired product or service. Most notable among these measures is the *carbon footprint,* which is defined as the amount of CO_2 produced for the completion of an economic activity. For example, the carbon footprint of driving 1,000 km in a small car with mileage 30 km/l (kilometers per liter) of gasoline (at the consumption of 23.3 kg of gasoline) is 9.0 kg of CO_2 while the carbon footprint of the same trip with a 5 km/l in a SUV is 54.0 kg of CO_2. Similar metrics have been adopted for other pollutants as well as for the use of scarce resources, such as water.

The concept of sustainability treats the pollutant emissions and reductions of emissions as a global economic phenomenon. Its advocates maintain that, for the long term environmental health, the economic activities must be re-engineered to ensure that their net effect on the environment is neutral. It must be pointed out that the sustainability concept does not necessarily advocate the banning of pollutants. Rather, it supports a pollution reduction counteraction for every pollution emission action. A simple example of the application of the sustainability concept is the removal of the 54.0 kg of CO_2 produced from the 1,000 km trip in the SUV of the previous paragraph: the owner of the SUV may plant a tree that will absorb this amount of CO_2 from the atmosphere. Actually, and if the tree grows to become a mature tree of approximately 500 kg, it would have removed enough CO_2 from the atmosphere to counteract the equivalent of 17 trips of 1,000 km. Therefore, one may drive for 17,000 km in an SUV and all the CO_2 that is emitted would be counteracted by the planting and growth of this tree. Similar counteraction measures may be taken in every field to remove the environmental effects of all economic activities. The subject of sustainability advocates that carbon sequestration and storage on a large, global scale is necessary to remove the adverse environmental effects of fossil fuel combustion.

The subject of sustainability is a product of the late twentieth century reaction to the environmental effects that are caused by anthropogenic activities and the realization that if these activities are continued unchecked and unmitigated, the planet may become uninhabitable to the vast majority of the population. Sustainability encompasses ideas and concepts from several disciplines including engineering, environmental science, ecology, economics, sociology, anthropology, political science, and public policy. Central to this subject is the realization that significant global threats, such as global warming and pollution prevention, may only be tackled by a combination of technological advances, social awareness, and public policy.

It has become apparent that the continuation of the current practices on energy consumption and the *laissez faire* or *market* energy policies in all countries are not sustainable for long and that, if continued unchecked, they will inevitably bring environmental disaster that may be followed by agricultural crop failures. This may lead to socio-economic, security, and cultural disruptions in the long run. The long-term sustainability of human economic activities is a subject that needs careful consideration. The subject of sustainability is still undergoing evolution, its definitions are mainly subjective, and its metrics are still debated among the scientists and policy makers. However, the adoption of at least some of the sustainability principles in the energy field appears to be a reasonable and realistic way for global economic expansion and equitability as well as for the long-term continuation of our current standards of living.

It must be noted that several sustainability promoters of the twenty-first century have advocated a "return to the fundamentals," where individuals become farmers and herders, withdraw in simple farms from urban centers, grow their own food and produce their own energy. These advocates often withdraw from urban centers and live in farms or rural communities. While a small fraction of the world population may be able to live such lives, the entire population reverting back to a farming society is not a solution to the global environmental problems. Simply, the finite surface of the Earth cannot support so many small, independent and inefficient farming enterprises. A small, family farm that would grow food as well as fuel for the needs of a family of four persons is approximately 150 acres (60 hectares). In most of the countries there is simply not enough arable land for each family to have such a small farm. For example, in the United States—one of the least densely populated countries of the world—such a societal system would only support 21% of its current population, even assuming that the entire land in the country is arable. In China it would support less than 5% of the current population and in Belgium—the most densely populated country in the world—a mere 1.9% of its current population. Clearly, the entire Earth is too small for all the humans living in the twenty-first century to return to a simple, agrarian and sustainable economy.

A more realistic alternative for sustainability, which may encompass the entire, current population of the planet, is the wider use of alternative energy sources combined with increased efficiency and energy conservation. The use of alternative energy sources, including the nuclear option, is fundamental to tackling several pollution problems, most important of which is the increased global CO_2 concentration. The substitution of a single 400 MW base-load coal-fired plant by twenty 20 MW geothermal plants will have the net effect of removing 3,530,000 tons of CO_2 annually from the atmosphere. Similarly, the substitution of a 60 MW gas turbine for peak power generation that operates for 20% of the year with solar power will have the effect of removing 20,400 tons of CO_2 annually. Permanent reforestation, not simply for biomass-based fuel production, removes tons of CO_2 and other pollutants from the atmosphere for several decades. Extensive use of the hydroelectric potential, tidal and wind power and a more widespread use of electric cars avert the further emissions of CO_2, SO_2, NO_x and other pollutants from an environment that 7 billion humans inhabit. Therefore, energy production from alternative sources, energy conservation and higher efficiency are the principal long-term solutions to achieving global and sustainable development.

Problems

1. A type of anthracite contains 90% carbon by weight. How much CO_2 is produced from the combustion of 1 metric ton of this anthracite? How much CO_2 is produced from the combustion of 1 Mcf (1,000,000 ft^3 at standard conditions) of propane?
2. A 400 MW electric power plant with an overall efficiency 38% uses bituminous coal, which contains 70% carbon, 2% sulfur with the rest being volatile matter and ash. The heating value of this coal is 26,500 kJ/kg. Determine: (a) how much heat the plant needs annually, if it operates continuously; (b) how much of this coal the power plant uses daily and annually; and (c) how much CO_2 and SO_2 the power plant produces annually.
3. Three coal power plants with a total power producing capacity of 1,000 MW and average thermal efficiency 36%, are substituted by one nuclear power plant with 33% thermal efficiency. What is the annual amount of CO_2 that is not emitted to the atmosphere? What is the increase of the waste heat produced? The heating value of the coal that was used is 28,000 kJ/kg and contains 80% carbon.

4. Three coal power plants with a total capacity of 1,000 MW and average thermal efficiency 36%, are substituted by ten smaller gas units that use methane. The new units have an average thermal efficiency 43%. What is the annual amount of CO_2 that is not emitted to the atmosphere because of this substitution? Assume that the heating value of coal is 30,000 kJ/kg and contains 90% carbon, and that of methane is 50,020 kJ/kg.
5. The coal power plant of problem 2 is fitted with a sulfur abatement system that has 99.6% efficiency. How much of the SO_2 mass is removed by the abatement system and how much is released in the atmosphere?
6. What effect the following parameters would have on the long-term temperature of the atmosphere? Write a short statement to explain your reasons.

   (a) A decrease of the earth's core temperature.
   (b) An increase of the average cloudiness.
   (c) A decrease of the earth's reflectance.
   (d) An increase of the amount of atmospheric methane.
   (e) An increase of the surface temperature of Venus.
   (f) Producing 10% of the total electric power from solar cells.

7. "The melting of the polar ice caps will be a major environmental calamity because it will increase the average temperature by 10.41°C." Comment by writing a short (250–300 word) essay.
8. A lake has a surface area of 18 square kilometers and an average depth of 6 m. The average pH of the lake is 6.9. How many tons of acid rain—in the form of H_2SO_3—would reduce the pH of the lake from 6.9 to 3.2?
9. Three Fluidized Bed Reactors (FBR's) consume bituminous coal with 65% carbon, 2.3% sulfur by weight and heating value $24*10^3$ kJ/kg. The FBR's supply with heat power a 600 MW coal power plant with 42% overall thermal efficiency. Determine: (a) How much heat the set of FBR's produce annually; (b) how much coal they consume; and (c) how much $Ca(OH)_2$ must be supplied to the FBR's in order to remove the SO_2 produced?
10. A certain type of leaded gasoline contains 1.2% of tetra-ethyl lead by weight. How much lead oxide, PbO, is released in the environment with every litter and with every gallon of this gasoline? Assume that all Pb is converted to PbO.
11. Calculate the amount of heat, in TJ (10^{12} J), rejected annually from the following types of electric power plants assuming that they operate continuously.

    (a) A 400 MW coal plant with thermal efficiency of 40%.
    (b) A 1,000 MW nuclear power plant with thermal efficiency of 33%.
    (c) A 35 MW geothermal power plant with 14% thermal efficiency.
    (d) A 10 MW thermal solar plant with 18% thermal efficiency.
    (e) An 80 MW natural gas power plant with 46% thermal efficiency.

12. The wet cooling towers of power plants use air and water for the cooling of the condensers. Air enters the cooling tower of such a plant at 22°C, 50% relative humidity and exits at 34°C, 90% relative humidity. Determine: (a) the amount of heat removed by 1 kg of (dry) air from this cooling system; and (b) the

amount of water (in kg) consumed for every kg of air that passes through the cooling tower.

13. The cooling tower of a 200 MW coal power plant with 40% thermal efficiency admits air at 15°C, 70% relative humidity and rejects it at 32°C, 95% relative humidity. Determine: (a) How much cooling air passes through the cooling system of this plant per minute? (b) How much water is used annually?
14. It is recommended that ten small geothermal power plants of 20 MW each (total 200 MW) substitute a coal power plant. If the average thermal efficiency of the geothermal power plants is 14%, how much water will be used annually?[10]
15. A 1,000 MW nuclear power plant with a 32% thermal efficiency discharges its waste heat in a lake with 22 km^2 surface and 3 m depth. If there is no other cooling effect for the lake, what would be the average increase of the water temperature annually? What other factors would nullify this temperature increase?
16. What is the Carbon footprint of the following activities? For all, you will need to find the wattage of the pertinent appliances in your residence. In the case of electric appliances, you may assume that 70% of the electricity comes from coal power plants with an overall thermal efficiency 38%.

    (a) Watching television for 1 hour.
    (b) Using a microwave oven for 10 minutes.
    (c) Forgetting to switch off a 100 W light bulb for 12 hours.
    (d) Driving for 2,000 miles in a SUV, which consumes 12 miles per gallon.
    (e) Driving for 2,000 miles in a compact car, which consumes 40 miles per gallon.

17. "The human society in its current form has failed to create a sustainable future for us and the next generations. Humans should abandon the cities and urban life and create a sustainable future for themselves and their children by producing their own food and fuel from products they have grown for millennia and can depend on in the future." Comment by writing a short (250–300 word) essay.

Reference

Gates DM (1985) Energy and Ecology. Sinauer, Sunderland Massachusetts

[10] This is not an environmental burden, because geothermal wells produce fresh water that may be used for the cooling of the geothermal power plant.

Chapter 3
Fundamentals of Energy Conversion

Abstract It is often taught in layman's terms that "energy may be neither created nor destroyed" and this brings into question the meaning of "energy conservation." Basic principles, such as energy conservation, stem from more general laws of Physics. The principles that govern the exchange and transformations of energy are succinctly examined in this chapter. The governing equations of energy conversion processes, or Laws of Thermodynamics, their corollaries and some of their applications to energy conversion processes are presented. A historical perspective is first given on the origins of modern energy conversion principles. The characteristics, significant variables/properties and types of Thermodynamic Systems are briefly explained. The two fundamental Laws of Thermodynamics (first and second) are postulated and the implications on the energy conversion processes are given succinctly. The operation of the simple gas and vapor power cycles is elucidated as well as several processes that are commonly used for the improvement of these cycles. Finally, the concept of exergy is introduced quantitatively and in detail. Based on exergy, several ways are presented on how this concept may be used to improve thermodynamic processes and cycles.

3.1 Origins of Thermodynamics and Historical Context

The development of the subject of Thermodynamics was the intellectual response to the technological advances that brought the Industrial Revolution in the mid eighteenth century. Following the invention of the steam engine that demonstrated the conversion of chemical energy through heat to mechanical power, the scientific and technical community of the nineteenth century developed the subject of Thermodynamics in order to understand and optimize the energy conversion processes and to better design heat engines.

E. E. (Stathis) Michaelides, *Alternative Energy Sources*,
Green Energy and Technology, DOI: 10.1007/978-3-642-20951-2_3,

The concepts of work, heat and energy were not understood by scientists of the eighteenth century in the same way as they are today. Work was essentially motion and heat and motion were considered entirely different and, certainly, not equivalent. The seventeenth and eighteenth centuries are sometimes called the *Phlogiston Era* because the scientific community held the opinion that heat is not an energy form, but is a "weightless fluid." When this weightless fluid passed from one body to another it cooled the first and warmed the second. Thus, the direction of heat transfer or *heat flow* was established. In today's scientific and technical literature there are still reminders of the *Phlogiston Era*: Heat transfer is often described as *heat flow* and the potential of a substance to store thermal energy (internal energy or enthalpy) is called *specific heat capacity* (either at constant pressure or at constant volume) in analogy to the capacity of a vessel to hold a liquid.

Because of the preconception that heat is a weightless fluid, the establishment of the heat-work equivalency and the development of the First Law of Thermodynamics took a very torturous path. Benjamin Thomson was probably the first scientist to demonstrate experimentally the conversion of what we now call "work" to "heat," and, thus, establish the roots of what we now call *the first law of Thermodynamics.* An American who supported the British forces during the Revolutionary War, Colonel Thomson was forced to leave the United States and was hired by the Duke (later King) of Bavaria. He managed to excel in the service of the King of Bavaria during the early Napoleonic Wars and was awarded the title of Count Rumford of Munich. He devised a public demonstration for the production of a significant amount of heat by friction and demonstrated to the public the conversion of "motion" to heat: In 1799 he used friction belts and converted the power of four horses careening in a circular path to raise the temperature of water in a metallic vessel. He continued the experiment until the water boiled. Count Rumford's experiment is the first recorded instance when work or motion was converted to heat in a controlled experiment. However, he was considered a charlatan by the scientific community, because he performed his experiments in public, did not take any measurements and did not try to combine his empirical demonstrations with even a rudimentary analysis or property measurements. For this reason, his demonstrations were largely neglected by the mainstream scientists of the early nineteenth century.

The conversion of the chemical energy, stored in the food to motion (or, equivalently, to work) was documented by an Alsatian physician, Dr. Robert Meyer. While serving as a ship doctor in the tropics, Dr. Meyer observed that the blood of the European sailors had a brighter red color than that of the natives. He attributed this difference to higher amounts of hemoglobin produced by the organs of these sailors. The hemoglobin is the "food" of the muscles since it transfers oxygen to the muscles that produce motion and, thus, work and mechanical power. Based on his observations, in 1822, he promulgated the theory that the chemical energy in the food is converted to chemical compounds that flow in the blood, and, finally, produce the muscle power. In effect, he established the equivalency of chemical energy and the ability to produce work, which is considered now one of the cornerstones of the

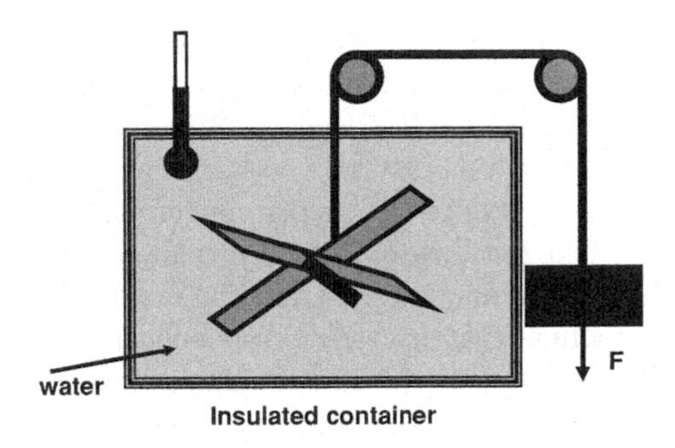

Fig. 3.1 Schematic diagram of Joules experiment

modern theory of Thermodynamics. It is rather unfortunate, that the established scientific community of the early nineteenth century was stuck with the Phlogiston theory and vehemently rejected Dr. Meyer's ideas. This drove him to commit suicide.

Thomas Prescott Joule was a professor at the University of Manchester and a well known academician when he performed the definitive experiment that demonstrated the equivalency of work and heat, in 1848. The experimental apparatus he used is shown schematically in Fig. 3.1. A closed and well insulated vessel contains a constant mass of water with a thermometer and a paddle wheel is inside the vessel. The paddle wheel is driven by a mechanism, which is powered by a falling weight. The energy of the falling weight is dissipated by friction in the water. Joule demonstrated that the mechanical power generated by the falling weight caused an increase in the temperature of the water and, thus, the mechanical energy was converted into heat, which resulted in the increased temperature. Joule's meticulous experiment and accurate measurements not only established the equivalency of work and heat, but also helped calculate with high accuracy the conversion factor of 1 calorie being equal to 4.184 units of energy that are now called "Joule." Joule's experiment was definitive and brought an end to the Phlogiston era. Even though several scientists of the middle and late nineteenth century stuck to the idea of heat as a weightless fluid (Lord Kelvin was one of the late converters) the experimental evidence in support of the heat-work-energy equivalency became overwhelming and the statement of this equivalency resulted in the formulation of the First Law of Thermodynamics.

Sadi Carnot[1] is considered the discoverer of the Second Law of Thermodynamics, which is a central principle for the conversion of heat into work. In his short memoir "On the motive power of heat..." published in 1824, Carnot stipulated that there are limits to the conversion of a quantity of heat into work.

[1] Sadi Carnot (1796–1832) was the fourth child of Lazare Carnot, a colleague and General of Napoleon, who was in charge of logistics during the early Napoleonic Wars and a member of the first Directorate. The father Carnot, a scientist in his own merit, contributed to the discipline of Mechanics and was often called "savant." Almost every city in the francophone world, including several in the United States, has a street or avenue named "Carnot," to honor Lazare Carnot (not Sadi). Other offspring of the same family became prominent statesmen in France.

His student, Clapeyron, wrote several commentaries on Sadi Carnot's work that emphasized the limitations of heat to work conversion. Later in the nineteenth century, Rudolf Clausius, William Thompson (Lord Kelvin) and Max Plank developed the several corollaries and implications of the 2nd Law on the property entropy and on the energy conversion processes. It must be noted that what is now called 2nd Law of Thermodynamics was formulated earlier and was more readily accepted than the 1st Law. The reasons for this acceptance are that the Law did not conflict with any pre-existing concepts and theories (e.g. heat is a weightless fluid) and that those who promulgated this principle (Carnot, Clapeyron and Clausius) were accomplished scientists, well known for their rigorous work, accurate measurements and came with excellent scientific credentials.

3.2 Fundamental Concepts of Thermodynamics

The concept of the *Thermodynamic system* (or simply *System*) is central to the understanding of the theory of Thermodynamics. The Thermodynamic System is enclosed by a *Boundary* and outside the boundary are the *Surroundings*. The latter is the part of the Universe that is affected by changes in the Thermodynamic System. Thermodynamic systems are *closed* or *open* Systems. The two types of systems are depicted in Fig. 3.2. Closed systems contain a fixed amount of molecules (and mass), while open systems have inlets and exits through which mass is allowed to flow. If the sums of the mass flow rates at the inlets and the exits are equal, the open system is at *steady state*. Otherwise it is *unsteady* or *transient*. Both closed and open systems receive or supply work and heat. In the case of closed systems we are usually interested in the total work and heat, $\dot{W}$ and $\dot{Q}$, while in the open systems we perform calculations based on the instantaneous rates of work and heat, $\dot{W}$ and $\dot{Q}$ as well as the mass flow rates that enter and leave the system.

The vast majority of energy conversion machinery is open systems: Pumps, boilers, turbines, compressors, nozzles and heat exchangers are all open systems. At typical operating conditions all these devices operate as open systems at steady state. Cylinders fitted with pistons are typical examples of closed systems.

The *properties* of the Thermodynamic system are all the measurable variables associated with that system. Temperature, pressure, volume, enthalpy, entropy, electrical conductivity, viscosity are typical examples of Thermodynamic properties. When the properties of a System do not change with time, the system is considered to be in *Thermodynamic Equilibrium*. Thermodynamic properties of systems at equilibrium may be measured, calculated via *Equations of State*, or computed from *Thermodynamic Tables*. For homogeneous substances (substances that have only one molecular composition) two independent properties are sufficient for the determination of all the other properties. The *Equations of State* are algebraic equations that give one property in terms of two other measurable properties, as for example with the ideal gas equation of state, $Pv = RT$. When a simple equation may not be adequate for the accurate determination of properties,

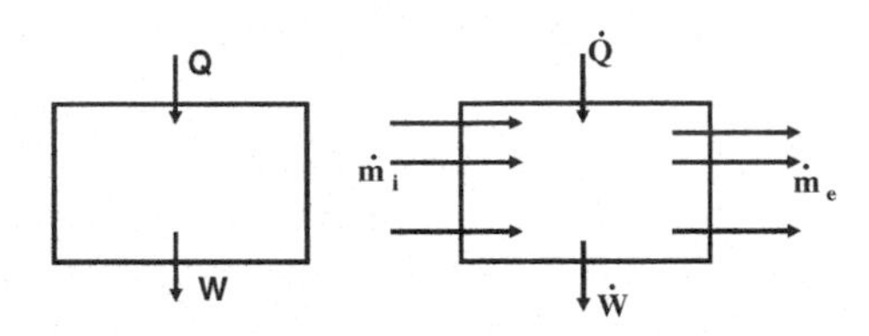

Fig. 3.2 Closed and open thermodynamic systems

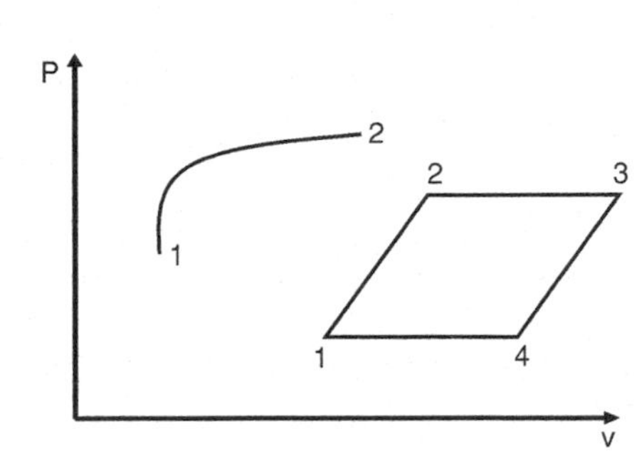

Fig. 3.3 A process, 1–2, and a cycle, 1–2–3–4, depicted on a P,v diagram

the latter are calculated by numerical computations and listed in Thermodynamic Tables. Such Tables for water/steam, refrigerants and several commonly used substances are routinely included in standard books of Thermodynamics.

The *State* is a fundamental concept of thermodynamics and is simply defined as the set of its properties. When the properties change, the state of the system changes and the system undergoes a *process*. The concept of the thermodynamic process is central to energy conversion. Processes take place within a finite amount of time. The system's properties change during a process as a response to the external changes imposed by the process. When the timescale of the change of the properties of the system is much lower than the characteristic time of the process, $\tau_{system} \ll \tau_{process}$, the system responds almost spontaneously to the imposed changes and the process is called *reversible*. Otherwise, the process is *irreversible*.

Processes are usually depicted in Thermodynamic diagrams, such as the pressure–volume (P,v) diagram, which is shown in Fig. 3.3. As a result of the details of a process the thermodynamic system may produce, absorb or reject work and heat. For example, a turbine produces work, a pump absorbs/consumes work, a boiler/burner produces heat by burning a fuel and a condenser rejects heat to the cooling system. Since the objective of the subject of Thermodynamics is the determination of the work and heat associated with processes that take place in a system, the understanding and good description of processes is of paramount importance. A combination of several processes, where the end point of the last process coincides with the starting point of the first, constitutes a *thermodynamic cycle* as depicted in Fig. 3.3. Thermodynamic cycles are routinely used for the production of electric power in thermal power plants as well as in engines that are used for propulsion and transportation (diesel, gasoline and jet engines). When the Thermodynamic cycles use an evaporating liquid/vapor fluid they are called *vapor cycles* and when they use a gas (ideal or real) they are called *gas cycles*.

3.3 Work, Heat and Energy

Work, heat and energy are considered three different forms of the layman's concept that is commonly referred to as *energy*. The three forms are measured by the same units, and each form may be transformed to one of the others. During these transformations the total *energy* is conserved.

3.3.1 Work

The concept of work is a primitive idea in the discipline of Mechanics and is associated with the concept of a force: A force performs work when its point of application moves. The amount of work done when the force moves by a distance described by the end points 1 and 2 is:

$$W_{12} = \int_1^2 \vec{F} \bullet d\vec{x}. \tag{3.1}$$

Work is a scalar and depends on the path followed between points 1 and 2. For a compressible substance enclosed in a cylinder that is fitted with a piston, this force is the external force acting on the piston. When the process 1-2 is reversible, the external force is equal to the product of the internal pressure and the area of the piston ($F_{ext} = PA$). Under these conditions, the last equation yields an expression of the work in terms of the two properties of the enclosed compressible substance, pressure, P, and volume, V:

$$W_{12} = \int_1^2 PdV. \tag{3.2}$$

A glance at Eq. (3.2) proves that the work depends on the process 1–2 and, therefore, it does not correspond to a potential function. The amount of work performed by a compressible substance, such as a gas or a vapor, during several commonly performed processes is given by one of the following expressions:

$$\begin{aligned}
&\textit{isobaric (constant pressure) process}: W_{12} = P(V_2 - V_1)\\
&\textit{isothermal (constant temperature) process, ideal gas}:\\
&W_{12} = mRTln(V_2/V_1) = mRTln(P_1/P_2)\\
&\textit{polytropic process } (PV^n = const): W_{12} = \frac{P_1V_1 - P_2V_2}{n-1} = m\frac{P_1v_1 - P_2v_2}{n-1}\\
&\textit{isochoric process (constant volume) process}: W_{12} = 0.
\end{aligned} \tag{3.3}$$

In the above expressions, m is the mass of the compressible substance that performs the work, R is the gas constant, $R = \tilde{R}/M$, and n is the *polytropic index*, a constant of the process, typically in the range 1–1.67.

3.3.2 Heat

Heat is transferred from an object at higher temperature to an object at a lower temperature. Thus, the prerequisite for the transfer of heat is a temperature difference, regardless of how low this difference may be. The transfer of heat takes place by one of the following transfer modes:

A. *Conduction*: Conduction occurs when a temperature gradient exists. The rate of heat conducted through an area A is given by the expression:

$$\dot{Q} = -kA\frac{dT}{dx}, \tag{3.4}$$

where k is the *thermal conductivity*, a property of the materials. The negative sign in the r.h.s of the last expression signifies that heat is transferred from the high-temperature part to the lower temperature part of the material.

B. *Convection*: Convection is caused by the motion of at least one fluid, which may pick or dissipate heat at a solid surface. *Forced convection* occurs when a fluid is pumped or blown by mechanical means with the purpose of causing heat to be transferred, as in the case of car radiators, where colder air is blown by the fan over a heat exchanger surface in order to cool the interior coolant. *Natural convection* occurs without mechanical forcing by the density difference, which is caused when a fluid's temperature changes. For example, an ice cube floating in a glass of warm water cools the water around it. Since the colder water around the ice cube has a higher density, it will sink into the glass. Thus, the colder ice cube creates a flow, which may be imperceptible to the naked eye, but nevertheless contributes significantly to the transfer of heat and to the melting of the ice cube. The rate of heat transferred by forced or natural convection from a solid object with an area A, which is at temperature T_H, to a fluid at temperature T_L is:

$$\dot{Q} = -hA(T_H - T_L) \tag{3.5}$$

where the coefficient h is the convective coefficient. This variable is a function of the flow conditions as well as of the properties of the fluid. The negative sign, again, signifies that the heat is transferred *from* the solid object at temperature T_H *to* the fluid at temperature T_L.

C. *Radiation*: Material contact is not necessary for the heat transfer by radiation. Radiation may pass through vacuum, fluids, and even some solids. The Sun heats the Earth by radiation to provide solar energy. Similarly, the Earth transfers heat to the rest of the Universe by radiation. Incoming radiation may be absorbed or reflected by an object, which emits radiation to all the other bodies around it by the same mechanism. The rate of total heat absorbed by radiation by an object at temperature T_L from another at temperature T_H is given by the expression:

$$\dot{Q} = \sigma A F_{HL} \varepsilon \left(T_H^4 - T_L^4\right), \tag{3.6}$$

where σ is the Stefan-Boltzmann constant, $5.67*10^{-8}$ W/m^2K^4; A is the surface area of the object at absolute temperature T_L; F_{HL} is a geometric factor that is related to the area the two bodies "see" through straight radiation rays; and ε is the emissivity of the surface of the same object, an empirical factor that characterizes surfaces.

3.3.3 Sign Convention

As it happens with the work performed during a process, the heat transfer also depends on the process and does not correspond to a potential function. The actual values of both the work and heat transfer are determined from the consideration and analysis of the underlying process. The scientific community has adopted a sign convention for the quantities of work and heat exchanged by a thermodynamic system for all processes, which is as follows:

- A quantity of work is positive when it is produced by a thermodynamic system and negative when it is absorbed by the system.
- A quantity of heat is positive when it enters the system and negative when it leaves the system.

Figure 3.4 illustrates schematically this scientific sign convention for the heat and work exchanged by a Thermodynamic system. The algebraic sign of the quantities of heat and work that enter or leave a Thermodynamic system are clearly shown in this figure.

3.4 The First Law of Thermodynamics: Energy Balance

The First Law of Thermodynamics is one of the fundamental and most general principles of science. It defines what is commonly referred to as the *energy conservation principle.* There are several formulations of the First Law, which are pertinent to the various types of systems and processes used. All formulations may be summarized by the general expression of energy conservation: *energy is neither created nor destroyed. It may only be transformed from one form to another.*

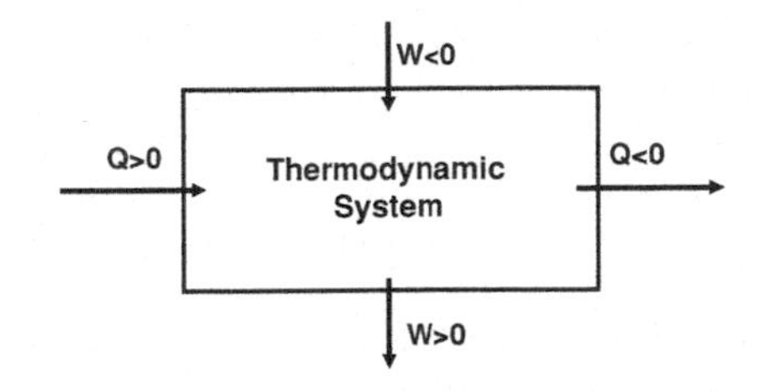

Fig. 3.4 Scientific convention for the algebraic signs of work and heat

3.4.1 Closed Systems

For a closed Thermodynamic system, the energy balance is best given in terms of a process leading from state 1 to state 2 and may be stated as follows: The heat entering a closed system minus the work produced by this system during a process 1–2 is equal to the difference of the total energy of the system between these two states. This energy conservation law is depicted schematically in Fig. 3.5. In symbolic form we may write:

$$Q_{12} - W_{12} = E_2 - E_1, \tag{3.7}$$

where the total energy of the system E is defined as the sum of the internal energy, U, the potential energy, mgz, the kinetic energy, 1/2 mV^2, and any other forms of energy the system may possess, and which may be described by potential functions as for example, electric charge energy, magnetic energy, surface tension energy, elastic energy, etc. Thus:

$$E_2 - E_1 = (U_2 - U_1) + \frac{1}{2}m\left(V_2^2 - V_1^2\right) + mg(z_2 - z_1) + \ldots \tag{3.8}$$

It must be noted that the symbol V (in italic) denotes the velocity, while the symbol V denotes the volume and v the specific volume. For most terrestrial thermal systems, the internal energy difference is typically very large and by far exceeds the changes in kinetic, potential and any other form of energy the system may possess. For example, when 10 kg of water boil to produce steam the internal energy change is 20,880,000 J. The same mass of water would gain approximately 1,000 J if it were raised to a height of 100 m; or 50,000 J if it were accelerated from rest to 100 m/s (or 360 km/hr or 225 mph). For this reason, one may approximate Eq. (3.8) in terms of the internal energy and the specific internal energy, u, as follows:

$$Q_{12} - W_{12} = U_2 - U_1 = m(u_2 - u_1), \tag{3.9}$$

The specific internal energy difference $u_2 - u_1$ is typically obtained from thermodynamic tables or by using a closure equation with the specific heat capacity at constant volume, e.g. $u_2 - u_1 = c_v(T_2 - T_1)$. Most of the engineering thermodynamics textbooks include thermodynamic tables as well as the numerical values for the specific heats for commonly used materials, such as steam, gases,

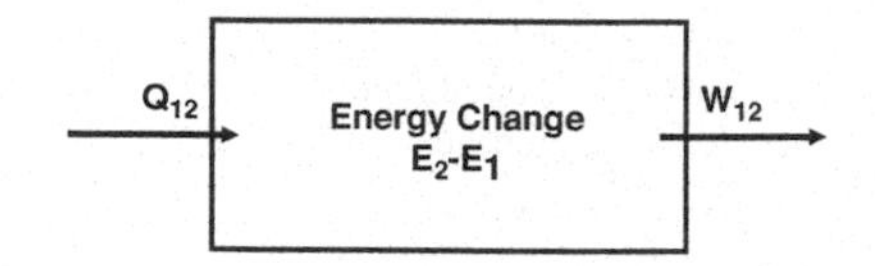

Fig. 3.5 The first law of thermodynamics as an energy balance

refrigerants and several hydrocarbons. Alternatively, one may obtain these properties from software.

The first part of Eq. (3.9) simply states that, for a closed system undergoing a process 1–2, the energy entering the system in the form of heat minus the energy exiting the system in the form of work is equal to the total energy change of the system.

3.4.2 Cyclic Systems

Thermodynamic systems undergoing a series of processes that constitute a cycle, as in Fig. 3.3 are commonly used for the production of power in thermal power plants, and in most of the engines used in the transportation industry. Let us visualize a thermodynamic system undergoing a series of n processes, 1–2, 2–3, 3–4, …, n − 1, where the last process, n − 1, ends at the initial state of the system. One may write the first law of thermodynamics (Eq. 3.1) for each one of the n processes and add the resulting n equations to obtain the following expression:

$$\begin{aligned} Q_{12} - W_{12} + Q_{23} - W_{23} + \cdots + Q_{n1} - W_{n1} = \\ U_2 - U_1 + U_3 - U_2 + \cdots + U_1 - U_n = 0. \end{aligned} \tag{3.10}$$

A glance at Eq. (3.10) proves that the left hand side represents the difference of the net heat entering the system and the net work produced by system during the cycle 1-2-3….n-1. The r.h.s of this equation is equal to zero. Therefore, Eq. (3.10) may be written more succinctly as follows:

$$Q_{net} = W_{net}, \tag{3.11}$$

or, the net amount of heat that enters a cyclic thermodynamic engine is equal to the net amount of work performed by this engine. This is another expression of the law of conservation of energy as applied to cyclic systems.

Example 3.1: 0.1 kg of air, enclosed in a cylinder is compressed by a piston during a polytropic process from 1 bar, 300 K to 7 bar. The polytropic index during this process is n = 1.3. Determine the work performed and the heat transfer during this process.

Solution: The air enclosed in the piston-cylinder assembly constitutes a closed system. Under these conditions, air may be approximated as an ideal gas, with a gas constant $R = 0.287$ kJ/kg K and $c_v = 0.719$ kJ/kg K.

Step 1—determine the properties of the air at states 1 *and* 2: From the ideal gas equation, $Pv = RT$, $v_1 = 0.861$ m^3/kg and from the polytropic relationship, which applies to the entire process including both states 1 and 2 one obtains: $P_1v_1^{1.3} = P_2v_2^{1.3}$. This yields $v_2 = 0.193$ m^3/kg and applying again the ideal gas equation $Pv = RT$ for state 2, $T_2 = 470.7$ K.
Step 2—determine the amount of work: Substitute in the third part of Eq. (3.3) for the polytropic process 1–2, the values for these properties to obtain: $W_{12} = -16.3$ kJ. The negative sign signifies that the system receives work (or the piston does work to the gas).
Step3—determine the amount of heat: From Eq. (3.9) $Q_{12} = W_{12} + m(u_2\text{-}u_1)$ and since air is an ideal gas $(u_2\text{-}u_1) = c_v(T_2\ \text{-}T_1)$. Therefore, $Q_{12} = W_{12} + m\ c_v\ (T_2 - T_1)$ and upon substitution, $Q_{12} = -4.05$ kJ. The negative sign signifies that the system (air) transfers heat to the surroundings.

3.4.3 Open Systems

Mass flow rates cross the boundaries of open systems, which typically have several inlets and exits. At the same time, a net rate of heat enters the system and a net rate of work, that is power, is produced by the system. The conservation equations for the open systems are described in terms of rates, which are typically denoted by a dot (.) above the symbol of the corresponding variable. The laws of mass and energy conservation apply to open systems. For the typical open system depicted in Fig. 3.2 the mass conservation law may be written as follows:

$$\frac{dm}{dt} = \sum_i \dot{m}_i - \sum_e \dot{m}_e, \tag{3.12}$$

where *i* denotes the inlets and *e* the exits of the system. The last equation may be interpreted as: the rate of increase of the mass in the open system (or the rate of mass accumulation) is equal to the difference of the mass flow rates that enter the system minus the mass flow rates that exit the system. When an open system operates at steady state, there is no accumulation of mass. Thus, for open systems at steady state the mass conservation equation stipulates that the sum of the mass flow rates entering the open system equals to the sum of the mass flow rates that exit the system. For simple, steady-state open systems with only one entrance and one exit, such as pumps, compressors, turbines, nozzles, etc., the mass conservation equation simplifies to the following expression:

$$\dot{m}_i = \dot{m}_e = \dot{m}. \tag{3.13}$$

For the development of the energy conservation equation, one should take into account that the effluent masses, which enter and exit an open system, perform a certain amount of work (called *the atmospheric* or *boundary work*) by flowing against the pertinent pressures at the inlets and outlets, which are denoted as P_i and

P_e. This work is incorporated as an addition to the property of internal energy u, which produces a similar property, the enthalpy: $h = u + Pv$. At steady state, the energy conservation law for an open system may be written in terms of the thermal and mechanical power exchanges by the system as follows:

$$\dot{Q} - \dot{W} = \sum_e \dot{m}_e h_e^o - \sum_i \dot{m}_i h_i^o, \tag{3.14}$$

where the property of specific total enthalpy, h^o (that is total enthalpy per unit mass) incorporates the specific enthalpy, h, as well as all the other potential forms of energy that are present in the specific total energy, e, or

$$\dot{Q} - \dot{W} = \sum_e \dot{m}_e \left(h_e + \frac{1}{2} V_e^2 + g z_e + \right) \ldots - \sum_i \dot{m}_i \left(h_i + \frac{1}{2} V_i^2 + g z_i + \ldots \right). \tag{3.15}$$

Since the *atmospheric* or *boundary work* has been taken into account in the property of enthalpy, the power, W, that appears in the above two expressions pertains only to the *useful power* that the system produces or absorbs. The useful power is power that may be used for the fulfillment of a task or a process. Of the systems that are routinely used in the power production industry, only pumps, compressors and turbines produce or consume useful power. For all the other open systems $\dot{W} = 0$.

For typical thermal systems, and as in the case of the closed systems that resulted in Eq. (3.9), the enthalpy changes are by far more significant than the changes in the other forms of energy. Hence, for typical open systems, such as pumps, turbines, compressors, heat exchangers, etc. the first law of thermodynamics at steady state may be simply written as:

$$\dot{Q} - \dot{W} = \sum_e \dot{m}_e h_e - \sum_i \dot{m}_i h_i. \tag{3.16}$$

For open thermodynamic systems at steady state with a single inlet and a single outlet, such as pumps, most turbines and compressors, Eq. (3.16) simplifies to the following:

$$\dot{Q} - \dot{W} = \dot{m}\left(h_e - h_i\right). \tag{3.17}$$

It must be noted that the first law of thermodynamics is applicable to open systems with chemical changes, such as burners and combustors. When chemical reactions are involved, a useful quantity is the heat of the reaction, or heat of combustion, which is expressed as enthalpy change per kg of the fuel, Δh, or as enthalpy change per kmol of the fuel, $\Delta \tilde{h}$. This enthalpy change is often called the *heating value* of the fuel. Burners and combustors do not produce or consume power. Hence, $\dot{W} = 0$ and the first law of thermodynamics may be used in the following simplified form to calculate the rate of heat released in the open system (combustor, boiler, burner, etc.) where a chemical reaction takes place:

$$\dot{Q} = \dot{m}\Delta h = \dot{n}\Delta\tilde{h}. \tag{3.18}$$

Nozzles and diffusers are the most commonly used thermodynamic systems in the power industry, where kinetic energy changes are significant enough to be accounted for in the energy balance equation. Both diffusers and nozzles do not consume or produce useful work (as mentioned above only pumps, turbines and compressors do so). Because fluids move very fast through these devises, the fluid properties are not influenced significantly by the, relatively slow, rate of heat transfer. Processes in nozzles and diffusers are usually approximated as adiabatic processes and, hence, it is assumed that the heat transfer during these processes is zero. Thus, for nozzles and diffusers, which always have one inlet and one outlet, one may simplify Eq. (3.15) to obtain:

$$V_e^2 - V_i^2 = \sqrt{2(h_i - h_e)} \tag{3.19}$$

In the numerical implementation of the last equation, care must be taken to use the appropriate units. For example, in the S.I. one must use J/kg for the specific enthalpy in order to calculate the velocity in m/s. Since, in most thermodynamic expressions, the specific enthalpy is given in units of kJ/kg, one must multiply the values obtained from the tables by the factor 1,000 J/kJ. Other conversion factors also apply in the English System of units.

Example 3.2: A small steam turbine admits 5 kg/s of steam at 10 bar, 360°C and exhausts saturated steam at 45°C. The turbine transfers 8.5 kW of heat to the surroundings. Determine the power produced by the turbine.

Solution: The turbine is an open thermodynamic system with a single inlet and a single outlet operating at steady state. The working fluid is steam.

Step 1—determine the properties at the entrance and exit: From the Steam Tables, $h_i = 3{,}178.9$ kJ/kg and $h_e = 2{,}583.2$ kJ/kg.

Step 2—determine the power: The algebraic value of the rate of heat is -8.5 kW, because heat leaves this open system. Using Eq. (3.17), $\dot{W} = \dot{Q} + \dot{m}(h_i - h_e)$, and substituting the enthalpy and mass flow rate values one obtains: $\dot{W} = 2{,}970\text{kW}$. The positive value signifies that power is produced by the turbine.

One observes in this example that the rate of heat loss is a very small fraction of the power produced by the turbine. This is typical of turbines and, for this reason, the typical turbine operation is often approximated as adiabatic $(\dot{Q} = 0)$.

Example 3.3: An adiabatic engine compresses 6 kg/s of helium from 1 bar, 300 K to 8 bar. During this process the temperature of helium rises to 753 K. Determine the power consumed by the compressor.

Solution: The compressor is an open system with a single inlet and outlet. Since it is also adiabatic, the rate of heat is 0. Helium is an ideal gas with $c_P = 2.5$ kJ/kg K.

Step 1—Determine the enthalpy difference: Because helium is an ideal gas, the change of the enthalpy of helium may be obtained from the temperature difference and the specific heat at constant pressure: $h_2 - h_1 = c_p(T_2 - T_1)$.

Step 2—determine the power: From Eq. (3.17) with $\dot{Q} = 0 : \dot{W} = \dot{m}c_P(T_i - T_e)$. Hence, $\dot{W} = -6,795\text{kW}$. The negative sign signifies that power is consumed by the compressor. Compressors usually get power (are driven) from electric motors.
Example 3.4: A nozzle admits 0.7 kg/s of steam at 5 bar, 400°C with a velocity 46 m/s and exhausts at 3 bar and 360°C. The rate of heat transfer from the nozzle is measured as 0.9 kW. Determine the exit velocity of the steam.
Solution: The nozzle is an open system with a single inlet and a single outlet. Nozzles do not produce or absorb power and, hence, $\dot{W} = 0$. The rate of heat output is negative, that is, $\dot{Q} = -0.9\text{kW}$.
Step 1—determine the properties at the entrance and exit: From Steam Tables we obtain: $h_i = 3{,}271.9$ kJ/kg and $h_e = 3{,}192.2$ kJ/kg.
Step 2—determine the exit velocity using the First Law: Using Eq. (3.15), which includes velocities, with one inlet and one outlet, one obtains the expression:

$V_e = \sqrt{2(h_i - h_e) + 2\dot{Q}/\dot{m} + V_i^2}$. Upon substitution of all the variables and conversion of the units from kJ/kg to J/kg—the square root of J/kg is equivalent to m/s—one obtains: $V_e = 399.6$ m/s.

It must be noted that the finite rate of heat transfer only makes a marginal difference in the final answer of this problem. If one neglects the heat transfer and assumes that the nozzle operates adiabatically, Eq. 3.19 yields: $V_e = 401.9$ m/s. The difference between the two values of the exit velocity is less than 0.5%.

3.5 The Second Law of Thermodynamics

It has always been observed that all natural processes have a directionality, that is they only proceed in one way: if left unsupported, an apple will always fall down and will not rise, a billiards ball will finally stop at a position on the table, fluids will flow from a higher pressure to a lower pressure and heat will be transferred from a hotter to a colder body. When these processes end, it is not possible to reverse them *spontaneously,* that is without spending work for the reversal. For example, if we allow two bodies one at high temperature, T_H, and the other at lower temperature, T_L, to come to thermal equilibrium, finally they will both assume a common temperature T_W, which is between the two original temperatures, $T_L < T_W < T_H$, as it is depicted in Fig. 3.6. During this process, from state 1 to state 2, the total internal energy that was originally contained in the two bodies, has been conserved, that is the total energy at state 1 is equal to the total energy at state 2. Now, if we wish to restore the two bodies to their original temperatures, T_H and T_L, we will soon find out that this cannot be done without the use of a refrigeration devise, which consumes work. Hence, despite of the fact that the energy of state 2 is equal to that of state 1, the process 2–1 is impossible without the addition of work. Similarly, if we wish to restore the apple that has fallen from a higher level 1 to a lower level 2, we must also perform work by lifting it back to its original level.

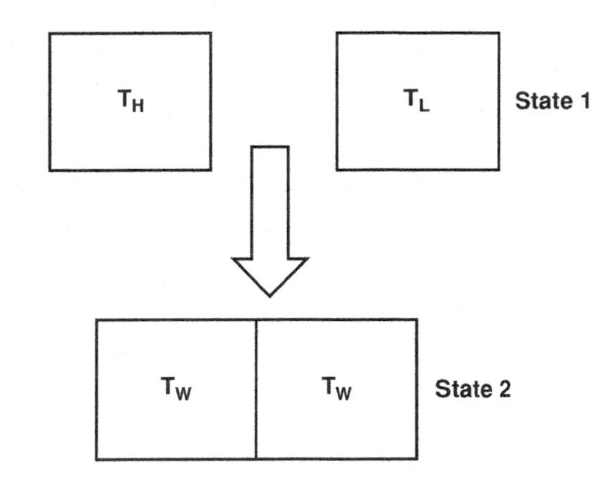

Fig. 3.6 The heat transfer process 1–2 that brought the two bodies to thermal equilibrium may not be reversed without the expense of work

The Second Law of Thermodynamics explains the directionality of all natural processes, by defining the property *entropy*, which increases in all natural processes undergone by all adiabatic systems. The Second Law may be expressed more generally by the following two simple statements:

1. There is a property of every system, entropy, which is defined as:

$$dS = \frac{dQ^O}{T} \Rightarrow S_2 - S_1 = \int_1^2 \frac{dQ^O}{T}. \tag{3.20}$$

2. For all natural process 1–2 occurring in adiabatic systems, that is systems with zero heat transfer to their surroundings:

$$S_2 - S_1 > 0. \tag{3.21}$$

The superscript "0" in the differential of heat denotes that this integral must be calculated during a *reversible* process. Although such processes are idealized processes, according to standard procedures of the theory of Thermodynamics, one may express the differential dQ^0 in terms of the system's properties (e.g. for a system containing a compressible substance $dQ^0 = dU + PdV$) and then carry the integration process using only the system's properties. The inequality (>) sign in Eq. (3.21) points out to the directionality of the natural processes: Any thermodynamic system is part of a (usually greater) adiabatic system. All thermodynamic systems that undergo processes would always proceed to a direction where the entropy of the adiabatic system increases. Since the Universe as a whole is considered to be adiabatic, the Second Law is sometimes expressed by the statement: *the entropy of the Universe tends to a maximum.*[2]

[2] This statement is often attributed to Rudolf Clausius who also coined the term "entropy".

Regarding the last inequality, it must be noted that there is not an, *a priory* established limit on how low or high the entropy change may be or should be. Processes where the entropy change is low and approaches zero are usually idealized as processes of constant entropy or *isentropic processes.* In all the other processes, the change in entropy is significant, e.g. in isothermal or isobaric processes.

3.5.1 Implications of the Second Law on Energy Conversion Systems and Processes

The most important implication of the Second Law on the subject of energy conversion is that work may not be produced spontaneously by a cyclic engine, when this engine receives heat from a single heat reservoir.[3] As a consequence of the Second Law, a cyclic engine—most power plants, gas turbines, jet and car engines are all cyclic engines—must be in contact with another heat reservoir where it rejects heat. Figure 3.7 shows a schematic diagram of the operation of the cyclic engines. During a cycle, the engine receives heat Q_H, rejects heat Q_L, and produces net work, $W = Q_H - Q_L$. Typically, the heat is rejected in the atmosphere (gas turbines, jet and car engine exhausts) or the hydrosphere (large steam power plants) and contributes to the *thermal pollution*.

A second consequence of the Second Law is that the thermal efficiency of such a cyclic engine may not exceed the *Carnot efficiency*, η_C:

$$\eta = \frac{W}{Q_H} \leq 1 - \frac{T_H}{T_L} = \eta_C. \qquad (3.22)$$

Typical thermal efficiencies of fossil fuel thermal power cycles are close to 40% and typical efficiencies of nuclear power cycles are close to 33%. This implies that a coal power plant, which produces 400 MW of electric power, must use approximately 1,000 MW of heat and rejects 600 MW of heat, usually to a river or a lake. For a typical nuclear power plant that produces close to 1,000 MW of electric power, the heat produced in the reactor is close to 3,000 MW and the amount of heat rejected is 2,000 MW. It is apparent, that the most important consequence of the Second Law of Thermodynamics is that, even though heat may be readily converted to work or power, it is only a fraction of the heat that is actually converted in today's thermal engines. The rest of this form of energy must be rejected to the environment as low temperature heat and becomes the *waste heat*.

In addition to the cycle thermal efficiencies, we also define *component efficiencies*, for turbines, compressors and pumps. The component efficiencies are

[3] This statement is attributed to Lord Kelvin and Max Plank.

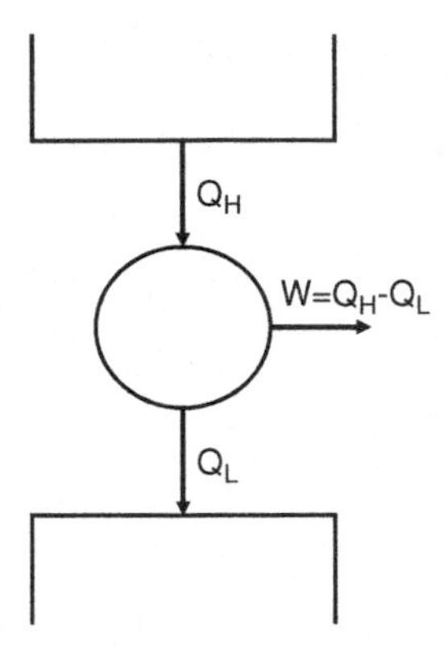

Fig. 3.7 The operation of a cyclic engine connected to two heat reservoirs

defined as ratios of the actual work, W_{act}, produced by the component and the isentropic work, W_s, which is achieved during the isentropic operation of the component. The component efficiencies are defined in a way that their numerical values are between 0 and 1 and are often given in percentages. For turbines, compressors and pumps these component efficiencies are respectively defined as follows:

$$\eta_T = \frac{W_{act}}{W_s}, \quad \eta_C = \frac{W_s}{W_{act}}, \quad \eta_P = \frac{W_s}{W_{act}}. \tag{3.23}$$

Typical turbine efficiencies are in the range 75–88%; compressor efficiencies are in the range 70–85%; and pump efficiencies in the range 70–80%. In practice, the component efficiency charts are supplied by the manufacturers of the equipment. In typical calculations, an engineer first calculates the isentropic work, W_s, from the theory of Thermodynamics using the First and Second Laws and then calculates the actual work, W_{act}, using the pertinent expression of Eq. (3.23).

3.6 Thermal Power Plants

The vast majority of the currently used thermal power plants utilize two main types of cycles: vapor and gas cycles. A succinct description of the essential components of these types of cycles will follow. Practical methods and processes for the improvement of the thermal efficiency of the cycles will also be discussed. For a more detailed description of the cycles one is encouraged to read a book on Engineering Thermodynamics, such as the one by Moran and Shapiro [1].

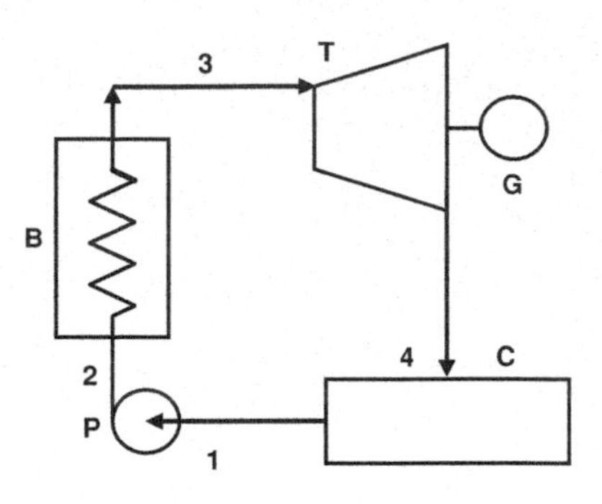

Fig. 3.8 Schematic diagram of the principal components used in a vapor power cycle

3.6.1 Vapor Power Cycles: The Rankine Cycle

The *Rankine cycle* is the most commonly used vapor power cycle. The working fluid of this cycle is usually water. The basic components of a typical Rankine cycle power plant are depicted schematically in Fig. 3.8 and the thermodynamic processes that comprise this cycle are depicted in Fig. 3.9a, b on the temperature-entropy and pressure–volume diagrams respectively:

1. Process 1–2 is the pressurization of the water effluent from the condenser. A pump (P) carries out this process, which is almost isentropic. Typical inlet conditions at state 1 for water are 0.06 to 0.10 kPa (6% to 10% of atmospheric pressure). The outlet conditions of the pump at state 2 vary from a few MPa to 30 MPa for supercritical cycles. A small amount of power, typically 1–2% of that produced by the turbine, is required for the operation of the pump. Since there is a small amount of power involved in process 1–2, the temperature of the pressurized water at state 2 is almost the same as that at state 1 ($T_2 \approx T_1$).
2. The boiler/superheater (B) receives the pressurized water at state 2 and, by imparting a significant amount of heat to the cycle, produces steam at high pressure and high temperature at state 3. The boiler is the high temperature reservoir for the cycle. The heating process 2–3 is considered to be isobaric (constant pressure).
3. The steam enters the turbine (T), or a system of turbines in large power plants and provides the motive power for its rotation and the simultaneous production of electricity in the generator (G). The pressure and temperature of the steam are reduced significantly in the turbine and the steam is exhausted in the condenser at the very low pressure range of 0.06–0.10 kPa and at low temperatures in the range of 40–50°C. The steam expansion process in the turbine, 3–4, is almost isentropic.
4. The function of the condenser (C) is to receive the steam from the turbine and condense it to liquid water. The condenser is a heat exchanger, where the steam from the cycle is condensed by warming up large quantities of *cooling water*. Heat is rejected at this part of the cycle to the cooling water and, finally, to the environment. The condensation process, 4–1, is considered to be isobaric.

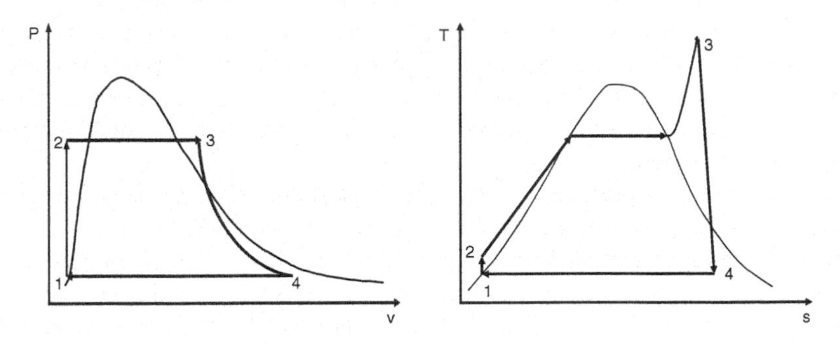

Fig. 3.9 a and **b.** A simple Rankine cycle depicted on the P–v and T-s diagrams

It must be pointed out that the condensation is not a necessary operation for the production of power and that the early power plants and locomotives operated without condensers by exhausting steam to the atmosphere. The condenser represents a significant improvement for the vapor cycle because, by reducing the steam pressure to sub-atmospheric values, it almost doubles the amount of power produced, thus, significantly improving the efficiency of the vapor cycle.

If the mass flow rate of water in the cycle is denoted by $\dot{m}$ then the amount of power consumed by the pump is: $\dot{m}(h_2 - h_1)$, the amount of heat power produced by the boiler/superheater is: $\dot{m}(h_3 - h_2)$, the amount of power produced by the turbine is $\dot{m}(h_3 - h_4)$, and the amount of heat rejected by the condenser via the cooling water to the environment is $\dot{m}(h_4 - h_1)$. The net electric power, which is produced by this plant in the generator, is: $\dot{m}(h_{3.}h_4) - \dot{m}(h_2 - h_1)$. Finally, the thermal efficiency of the power plant is defined as:

$$\eta_t = \frac{(h_3 - h_4) - (h_2 - h_1)}{h_3 - h_2}. \tag{3.24}$$

The enthalpy and other properties of the water and steam at the states 1–2–3–4 may be obtained from *steam tables* that are standard appendices of books on Engineering Thermodynamics. It must be also noted that the definition of the thermal efficiency of a cycle does not emanate from any of the Laws of Thermodynamics and that it is merely a benefit to cost ratio: the benefit is the power produced, which represents the revenue for the power plant operator. The cost is the heat addition to the cycle, which typically comes from a fuel that represents cost.

As with any benefit-to-cost ratio, engineers strive to improve the thermal efficiency of the power plants.[4] One may improve the efficiency of Rankine cycles by adopting several of the following processes:

---

[4] If the heat to the cycle is "free," as for example in the case of solar thermal power plants, one has to think carefully what the "thermal efficiency improvements" would mean for the operation of such a power plant.

1. Increase the upper cycle temperature, by superheating the steam in the boiler/ superheater to higher temperatures. There is a limit to this process, the *metallurgical limit*, where the temperatures are too high for the turbine materials to handle and fracture may occur. With the materials used at present, this limit is approximately 600°C.
2. Increase the upper pressure P_3. Again the materials used in the superheater and the turbines impose a limit, which is typically 35 MPa (350 atm).
3. Reduce the lower temperature and pressure of the cycle. This is limited by the temperature of the cooling environment, where finally all the heat must be rejected. When water is available, from a river or a lake, the temperature of the water is the limit temperature. Because water is colder than the ambient air, most of the modern power plants are close to lakes or rivers.
4. Use a reheat process: steam is partly expanded in the turbine to a lower pressure and is then directed back to the superheater, where it receives more heat and its temperature is increased to approximately T_3. This reheating process may be repeated 2–3 times and is facilitated considerably if the pressure P_3 is very high, as for example in supercritical cycles.
5. Use the *regeneration* or *bleeding* process: For this efficiency improvement process a fraction of the steam (10–20%) is extracted from the turbine at moderately high temperature. This part of the steam is used to heat up the water effluent of the pump at state 2 in a heat exchanger, which is called the *feedwater heater (FWH)*. This process may also be repeated 2–3 times by extracting steam at 2–3 points from the turbine. The effect of regeneration is to use heat at a lower temperature from the expanding steam in order to heat up the water at the lower temperatures of the cycle. In this case, high temperature heat is saved by not being used in the lower temperatures of the cycle.

Practical vapor cycles that utilize the above methods have upper cycle temperatures close to 560°C and may reach thermal efficiencies in the range 42–46%.

3.6.2 Gas Cycles: The Brayton Cycle

Gas cycles typically use air as the working fluid. However, advanced gas cycles that are used in some nuclear reactors have used carbon dioxide or helium. The arrangement of the components in a typical gas cycle is shown in Fig. 3.10. The processes that constitute this cycle may be enumerated as follows:

1. Air at ambient temperature and pressure, at state 1, enters a compressor where its pressure and temperature rise to state 2. Typical pressures at the exit of the compressor are 10–20 atm. Because a gas is compressed during this process, the compressor consumes a significant amount of power, typically one third of

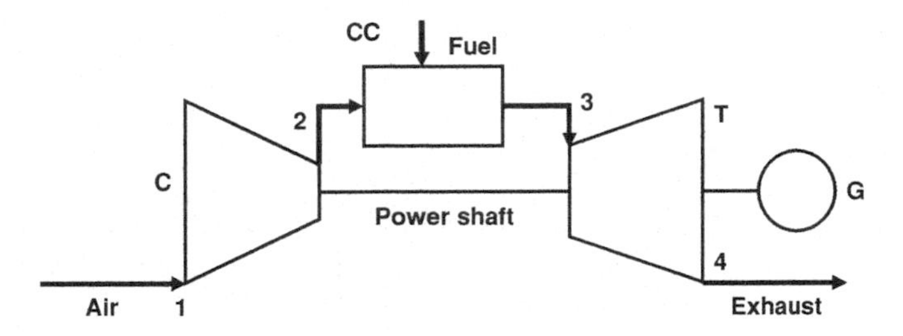

Fig. 3.10 Schematic diagram of the components used in a gas power cycle

the total power produced by the cycle. For this reason, the compressor is usually coupled mechanically and driven directly by the turbine.

2. The compressed air enters the burner or combustion chamber (CC). Fuel (typically natural gas or liquid hydrocarbons) is also introduced in this devise, where it burns by combining with the oxygen in the air. The combustion results in the significant elevation of the temperature of the remaining air and the combustion products. Temperatures at the exhaust of the combustion chamber (state 3) may reach as high as 2,000°C.
3. This high-temperature mixture is introduced to the gas turbine (T) where the air and the products of the combustion expand to atmospheric pressure and then are exhausted to the atmosphere at state 4. A generator (G), which is coupled to the turbine, produces electric power. The latter is equal to the difference of the power produced by the turbine minus the power consumed by the compressor.

Strictly speaking, the gas cycle is not a cycle *per se* because the turbine exhaust is at different state and composition from the compressor input. However, this series of processes is always treated as a cycle, because the input to the compressor is always air at ambient temperature and the rest of the atmosphere may be thought of as an enormous reservoir of mass and heat, which receives the hotter output of the turbine, cools and purifies it to the ambient temperature and composition and, finally, allows it to be fed to the compressor at the same, ambient pressure, temperature and composition. The temperature-entropy and pressure–volume diagram of this virtual cycle are depicted in Fig. 3.11a, b respectively. The thermal efficiency of the gas cycle is defined in the same way as in the vapor cycles and may be expressed as follows in terms of the states depicted in these two diagrams:

$$\eta_t = \frac{(h_3 - h_4) - (h_2 - h_1)}{h_3 - h_2}. \tag{3.25}$$

The enthalpy and other properties of the air at the states 1–2–3–4 may be found in *air tables* that are standard appendices of a book on Thermodynamics [1]. Alternatively, the enthalpy differences may be expressed in terms of a specific heat

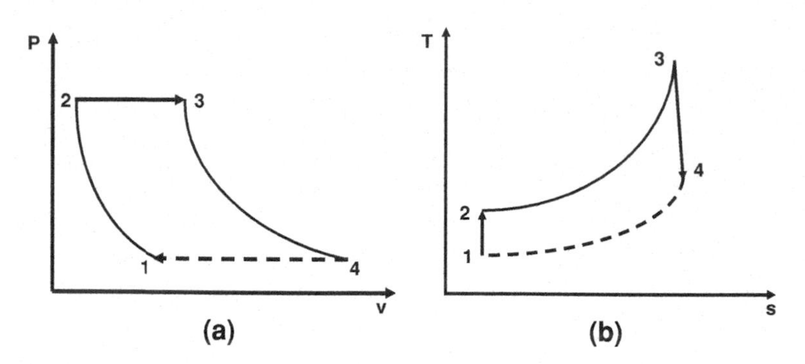

Fig. 3.11 **a** and **b.** A simple Brayton cycle depicted in a P–v and a T-s diagram

and the corresponding temperature differences. The efficiency of gas cycles may be improved by using the following processes:

1. Use a reheat process, similar to the vapor cycle: The gas expands in the first turbine to lower pressure and then is introduced to the combustion chamber, or a second combustion chamber, where more fuel burns. The temperature of the gas increases again to a value close to T_3. The gas is then directed to another turbine where it produces more power.
2. The temperature at the last turbine exhaust is rather high, typically at 400–500°C. This is oftentimes higher than the temperature, T_2, at the exit of the compressor. In such cases one may use a *regenerator*, which is a heat exchanger that heats up the compressor output by using the turbine exhaust mixture.
3. Use of intercoolers: the power requirement of the compression process is significantly lower if the compression is closer to isothermal than to isentropic. In this case, a single large compressor is replaced by a series of smaller compressors that operate in series. At the exit of each compressor, air is cooled to almost ambient temperature and is further pressurized in the next compressor. Intercooling in compression processes and its benefits in improving the efficiency of gas cycles are further explained in Sect. 13.2.2 and depicted in Fig. 13.1, which shows the use of intercoolers with a series of three compressors.
4. Use of a bottoming vapor cycle: The high-temperature gases that exhaust at the turbine may be used in a heat exchanger to generate steam (or another vapor) at high pressure, which in turn may produce power using another turbine-condenser-pump combination. The heat supplied to this cycle is the *waste heat* of the turbine and is virtually free of any cost because it comes from gases that would have been otherwise exhausted to the atmosphere. The additional power produced enhances significantly the thermal efficiency of the original gas cycle, which may reach values close to 50%.

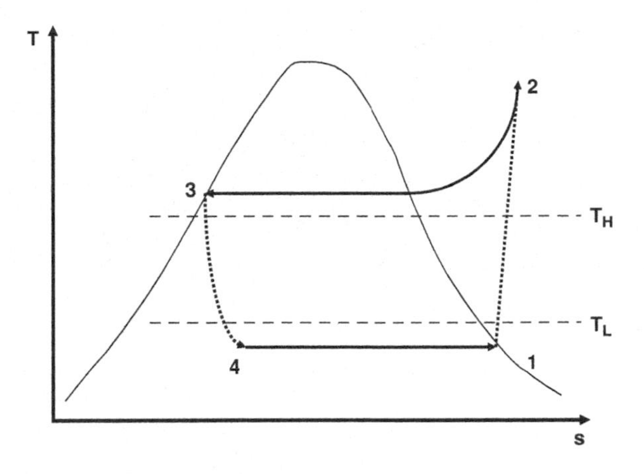

Fig. 3.12 Thermodynamic diagram of the operation of the heat pump/refrigeration/air-conditioning cycle

3.6.3 Refrigeration and Heat Pump Cycles

In principle, a reversed power generation cycle would absorb heat from a colder heat reservoir and will dissipate another amount of heat to a higher temperature reservoir. Depending on its use, such a reversed cycle is called a *refrigeration cycle* or a *heat pump cycle*. The former is used as an air-conditioning cycle too. The thermodynamic diagram in the T,s coordinates of a typical refrigeration cycle, which is a slightly modified, reversed Rankine cycle, is shown in Fig. 3.12. The four processes of the refrigeration cycle are as follows:

1. A compressor raises the pressure of the working fluid from state 1 to 2. The compressor consumes electric work, which is work given to the cycle and has a negative algebraic sign. Both the temperature and pressure of the fluid that exits the compressor are significantly high.
2. The refrigerant fluid is then passed through the condenser and undergoes the process 2–3. During this process the fluid dissipates heat to a heat sink at a higher temperature T_H. In a heat pump operation, the heat sink is the interior of a building. In a refrigeration operation the heat sink is the coils at the back of the refrigerator, which always feel warmer to the touch. In an air-conditioning cycle, the heat sink is the outside atmospheric air.
3. Process 3–4 is caused by an expansion valve and produces a mixture of liquid and vapor at significantly lower temperature ($T_1 = T_4$). The valve is insulated and the process is considered isenthalpic ($h_3 = h_4$).
4. After this stage, the refrigerant fluid passes through the evaporator and undergoes the process 4–1. Because the temperature of the fluid at this stage is very low, the fluid absorbs heat from the lower temperature source at T_L, and exits the evaporator at the original state 1. In an air-conditioning cycle, the lower temperature heat source, at temperature T_L, is the air in the interior of the building. In a refrigeration cycle this is the interior of the refrigerator.

In a heat pump the low temperature reservoir is the atmospheric air. The corresponding higher temperature reservoir, at temperature T_H, in all three cases is the outside air, the room where the refrigerator is located (e.g. the kitchen) and the interior of the building, respectively. It must be noted that the main difference between a reversed Rankine cycle and a refrigeration cycle is that process 3–4 of the refrigeration cycle does not produce work. This happens because the actual work that may be produced from the liquid refrigerant at state 3 is too low to justify the capital expense of an expander/turbine. The simple expansion valve that is used for the process 3–4 is by far cheaper and fulfills the task to lower the temperature of the refrigerant fluid at state 4.

The principal function of the heat pump cycle is to remove heat, Q_L, from a colder heat source and dissipate a significantly higher quantity of heat, Q_H, to a hotter environment. The addition of work, W, which emanates from electric power via a motor, is essential to the cycle and makes possible this heat transfer process. In the common household refrigerator, this cycle removes heat from the colder interior of the refrigerator and dissipates a higher quantity of heat to the coils in the back of the refrigerator. The energy conservation principle—the first law of Thermodynamics—applies to the transfers of heat and of work in the heat pump/ refrigeration cycle and may be written as follows:

$$Q_H = Q_L + W. \tag{3.26}$$

Depending on the function of the cycle, two benefit-to-cost ratios, the coefficients of performance (c.o.p.) have been defined for the heat pump and the refrigeration cycles. The c.o.p. is the ratio of the units of heat removed by the refrigerator (or added by the heat pump) to the electric work that must be spent for the operation of the cycle. Thus, the coefficients of performance of a heat pump cycle, β_{hp}, and a refrigerator or an air conditioning cycle, β_r, are defined as follows:

$$\beta_{hp} = \frac{Q_H}{W} \quad and \quad \beta_r = \frac{Q_L}{W}. \tag{3.27}$$

It is the absolute and not the algebraic values of the heat and work that are used in the calculation of the c.o.p.'s. The heat removed or added and the coefficients of performance of the cycles depend on the cycle temperatures and pressures. In general, the lower the difference between the condenser and the evaporator temperatures, T_1 and T_3 are (or correspondingly the pressures P_1 and P_3) the higher the coefficient of performance of the cycle will be. Thus, an air-conditioner would have a lower coefficient of performance during the hotter summer days, when heat is rejected to a higher temperature environment. Also, a heat pump has a lower coefficient of performance during the coldest days of the winter, when heat must be removed from a very cold environment.

3.7 Exergy: Availability

The concept of the *Carnot efficiency* is very useful in giving us a quantitative measure of the maximum efficiency when a given quantity of heat, Q, is to be converted to work. However, quantities of heat at high temperatures are not found in nature. Heat is normally released in burners and boilers from the consumption of fossil or nuclear fuels. Instead what is abundantly found are *natural energy resources*, such as fossil fuels, nuclear fuels, geothermal fluids, incident solar radiation, wind power, etc. Therefore, one may ask the question, "what is the maximum amount of work, one may obtain from a given quantity of an energy resource?" Clearly, the *Carnot efficiency* does not give an answer to this question, because the conversion of the energy resource to heat always involves thermodynamic irreversibilities and the subsequent conversion of this quantity of heat to work does not necessarily yield the maximum amount of work the original resource may provide. In such cases, one may use the concept of *exergy* or *availability,* which gives a quantitative measure of the maximum amount of work one may obtain from an energy resource.

One may make the following observations regarding the processes of the conversion of energy resources to work or power:

1. All energy conversion processes occur in the environment—the atmosphere and the hydrosphere—which act as reservoirs of mass, heat and work, that is, they may absorb or provide very large quantities of mass, heat and work without their state changing significantly.
2. Energy conversion processes are possible, because there are natural substances, the *energy resources,* which are not in thermodynamic equilibrium with the atmosphere. For example, coal is not in equilibrium with the atmosphere. Coal will come to equilibrium with the atmosphere when it combines with oxygen and forms carbon dioxide.
3. The maximum work is extracted from an *energy resource*, when, through several processes, the materials of the resource are brought in thermodynamic equilibrium with the environment. This is called the *dead state* of the resource.

Hence, one may conclude that the maximum work from a *energy resource* will depend on the following:

1. The type of the resource.
2. The thermodynamic characteristics of the environment, which include temperature, pressure and chemical composition.
3. The interaction allowed between the resource and the environment.
4. The rate of dissipation or irreversibilities in all the conversion processes.

Therefore, the maximum work that may be obtained from an energy resource may not be described by a single formula, such as the one of the *Carnot efficiency*.

One may obtain such formulae, which are applicable to specific resources by using the First and Second Laws of Thermodynamics in combination. This type of analysis leads to the thermodynamic concept of exergy. The following are some examples of the use of the exergy concept to three types of natural energy resources.

3.7.1 Geothermal Energy Resources

Geothermal reservoirs supply power plants with geothermal fluid, which is in the one of the following three conditions:

1. Superheated steam (e.g. the Geysers in California and Lardarelo in Italy).
2. Mixture of steam and water (e.g. Warakei in New Zealand).
3. Geopressured liquid water at high temperature (e.g. Port Aransas, Texas).

The geothermal power plant is an open thermodynamic system that receives a certain mass flow rate of the geothermal resource$\dot{m}$, at high temperature, T, pressure, P, specific enthalpy, h, and specific entropy, s. The plant operates in the atmosphere, which is at temperature T_0 and pressure P_0, and discharges the geothermal fluid as an effluent at temperature T_e, pressure P_e, specific enthalpy, h_e and specific entropy s_e. Therefore one may write the two laws of thermodynamics for this open system as follows:

$$\dot{Q} - \dot{W} = \dot{m}(h_e - h), \tag{3.28}$$

and

$$\dot{m}(s_e - s) - \frac{\dot{Q}}{T_0} = \dot{\Theta} > 0. \tag{3.29}$$

The 2nd Law inequality of Eq. (3.21) has been converted to an equation (Eq. 3.29) by introducing the rate of entropy generation, $\dot{\Theta}$, which is always a positive quantity ($\dot{\Theta} > 0$). Substituting the rate of heat from the second expression to the first, one may obtain the following expression for the power produced by the geothermal power plant:

$$\dot{W} = \dot{m}[(h - T_0 s) - (h_e - T_0 s_e)] - T_0 \dot{\Theta}. \tag{3.30}$$

It is now apparent that the power plant will deliver maximum work and power under the following two conditions:

1. The rate of entropy generation,$\dot{\Theta}$, is minimized to a value close to zero.
2. The state of the effluents, e, is the same as the state of the environment, at temperature T_0 and pressure P_0, as implied by observation 3, in the previous

Table 3.1 Specific work produced, in kJ/kg, by several types of geothermal resources

Steam at 280°C, 15 atm	955
Saturated steam at 220°C	929
50% water and 50% steam at 220°C	552
Saturated liquid water at 200°C	158
Saturated liquid water at 160°C	97
Geopressured liquid water at 140°C, 25 atm	73

section. At this state the specific enthalpy and entropy of the effluents will be denoted as h_0 and s_0 [2].

Hence, the maximum rate of work, or power, one may obtain from the geothermal resource is:

$$\dot{W}_{\max} = \dot{m}[(h - T_0 s) - (h_0 - T_0 s_0)] = \dot{m}(e - e_0). \tag{3.31}$$

In the last expression, the quantity $e = h\text{-}T_0 s$, is the *exergy* of the geothermal resource. It is apparent that the exergy of a resource utilized by an open system is a function of the thermodynamic state of the resource as well as of the state of the environment, where the energy conversion processes occur. In the case of geothermal resources, the maximum amount of specific work obtained by the geothermal fluid may be easily calculated from the properties of this fluid. Table 3.1 provides values of the specific work for several states of geothermal fluid, which is assumed to be water. For the calculations, the environmental temperature, T_0, is 298 K (25°C). It is observed in this table that, resources with significant amount of vapor (steam) would provide much higher amount of work than liquid water resources. Also, that the amount of specific work diminishes significantly when the temperature of the resource is relatively low.

It must be noted that, for vapor–liquid systems, such as steam-water, the numerical value obtained for the exergy at the dead state is very close to zero, that is $e_0 \approx 0$, which implies:

$$W_{\max} \approx \dot{m}e \tag{3.32}$$

3.7.2 Fossil-Fuel Resources

Fossil fuels may provide heat and work because they are not in chemical equilibrium with the atmosphere. An equilibrium state is reached when fossil fuels combine with the oxygen of the atmosphere and produce carbon dioxide and water. Figure 3.13 is a schematic diagram of the conversion of the hydrocarbon octane (C_8H_{18}) supplied to an engine at a mole flow rate $\dot{n}$ to produce power.

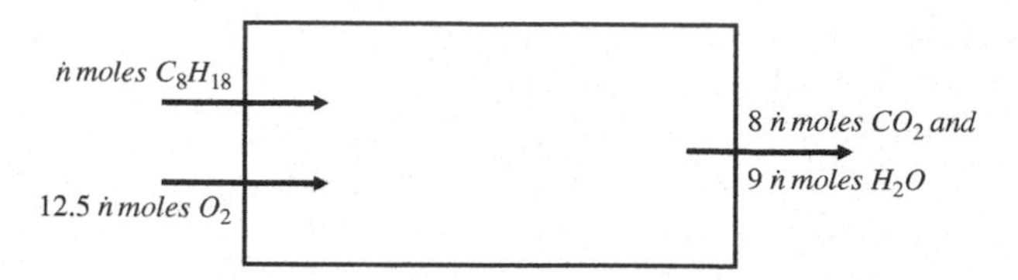

Fig. 3.13 Schematic diagram of the work production process from the chemical reaction of octane oxidation

Since fossil fuels undergo chemical reactions, the mole flow rates and properties of the reactants, denoted by the subscript *R*, and of the products, denoted by the subscript *P*, are used in the derivation of the maximum power that may be produced. The energy conversion process occurs in an open thermodynamic system. One may write the laws of Thermodynamics and the power delivered by the engine as follows:

$$\dot{Q} - \dot{W} = \sum_P \dot{n}_P h_P - \sum_R \dot{n}_R h_R, \tag{3.33}$$

and

$$\sum_P \dot{n}_P s_P - \sum_R \dot{n}_R s_R - \frac{\dot{Q}}{T_0} = \dot{\Theta} > 0. \tag{3.34}$$

After combining the last two expressions by eliminating the rate of heat, one obtains the following expression for the power produced by the engine that utilizes the fossil fuel:

$$\dot{W} = \left[\sum_R \dot{n}_R (h_R - T_0 s_R) - \sum_P \dot{n}_P (h_P - T_0 s_P)\right] - T_0 \dot{\Theta} \tag{3.35}$$

The maximum power will be produced by the engine when:

1. The internal irreversibilities are minimized to the point where the rate of entropy production is zero, and
2. When the products are exhausted and reach equilibrium with the atmosphere, that is, at temperature T_0 and pressure P_0.

Under these conditions of maximum power production, the terms in the square bracket may be written in terms of the *Gibbs free energy* of the reaction, ΔG, and the molar flow rate of the fuel, $\dot{n}_F$:

$$\dot{W}_{\max} = -\dot{n}_F \Delta G. \tag{3.36}$$

If the reaction takes place at atmospheric pressure, as most reactions do, the Gibbs free energy at atmospheric pressure, ΔG^0, should be used. From the last expression one may deduce that the maximum amount of specific work per kmol, which may be obtained from a fossil fuel at atmospheric pressure, is:

$$\tilde{w}_{\max} = -\Delta G^0 \text{in units of kJ/kmol or Btu/lbmol} \tag{3.37}$$

The *Gibbs free energy* of the reaction is a thermodynamic property for all reactions and may be found in standard thermodynamic or chemical tables. The superscript, *0*, denotes that this function is evaluated at the atmospheric pressure and temperature. One may easily observe that there are many similarities between Eqs. 3.35 and 3.31 and that the specific exergy of a fossil fuel may be defined as $e_F = h_F - T_0 s_F$.

In a strict sense, the theoretical maximum amount of work one may extract from a chemical substance is slightly higher than the amount indicated in the last two equations. For example for Eq. (3.37) the maximum theoretical work may be given by the expression:

$$\tilde{w}_{\max} = -\Delta G^0 - \tilde{R}T_0 \ln \frac{\prod_p x_p^{v_p}}{\prod_r x_r^{v_r}}, \tag{3.38}$$

where the operator Π denotes the algebraic product of the variables; x is the mole fraction of a chemical compound; v is the stoichiometric coefficient of the compound; and the subscripts p and r refer to the products and reactants of the pertinent chemical reaction. However, the last term may only be practically harnessed if semi-permeable membranes were available that would allow the reversible expansion of the reactants and products into the atmosphere. Because this technology is not available, it is best to use Eqs. (3.36) or (3.37) as expressions for the maximum work from fossil fuels and other chemical compounds.

3.7.3 Radiation: The Sun as Energy Resource

Contrary to popular thinking, not all of the solar radiation energy may be converted to work. The energy flux (in W/m^2) emitted by a source at temperature T_H is equal to: σT_H^4, where σ is the Stefan-Boltzmann constant. By combining the first and second Laws of Thermodynamics and substituting the appropriate expressions for the specific energy and entropy of radiation, one may obtain the following expression for the exergy flux of radiation energy emanating from a source at temperature T_H and utilized in an environment at temperature T_O.

$$E = \frac{\sigma}{3}\left(3T_H^4 - 4T_H^3 T_0 + T_0^4\right). \tag{3.39}$$

The units of this expression are power per unit area (e.g. W/m^2). In the case of solar energy, the temperature of the source of radiation, the Sun, is at approximately 5,900 K, while the temperature of the receiving object is that of the Earth, approximately 300 K. When these numbers are substituted in the last equation, it is concluded that approximately 93.5% of the amount of incident solar energy may be converted to work. It is of interest that, if one considers the Sun and the Earth as the two temperature reservoirs of a cycle connected by a cyclic thermodynamic engine, the maximum efficiency of this giant cyclic engine would be determined from Eq. (3.22) to be 95%.

3.7.4 Second Law Efficiency: Utilization Factor

The thermal efficiencies defined in Eqs. (3.22) and (3.23) are based on the thermal energy that could have been extracted from the heat source. Because heat is an integral part and is defined by the First Law of Thermodynamics, these efficiencies are often referred to as first law efficiencies. Essentially, the first law efficiency defines how much of the available heat in the energy resource is converted into work. It is apparent that the available heat from a resource is not the best indicator of the work that may be extracted from a particular resource and that the exergy of the resource is a better measure of the work that may be extracted from it. For this reason, for energy conversion processes and systems another figure of merit, which is based on the concept of exergy is defined as follows:

$$\eta_u = \frac{W}{m(e - e_0)} \quad or \quad \eta_u = \frac{\dot{W}}{\dot{m}(e - e_0)}. \tag{3.40}$$

This figure of merit is called *second law efficiency* or *utilization factor*. The second law efficiency is, essentially, the ratio of the actual work produced in an energy conversion process or energy cycle to the maximum possible work. Since it is based on the maximum work that *could* have been extracted from the energy resource, the second law efficiency gives a better indication as to the specific way the energy resource is utilized. A low numerical value for the second law efficiency indicates that the method of energy utilization is not optimum and that the energy conversion process or cycle would benefit from improvements. On the contrary, a high value of the second law efficiency would indicate that the process or cycle is near optimum and that any improvements would only result in marginal gains of the total work or power produced.

Problems

Thermodynamic tables are needed for problems marked with an asterisk (*).

1*. A cylinder fitted with a piston contains 2 kg of liquid water and 1 kg of steam at equilibrium at 0.2 MPa.

   a) What is the equilibrium temperature of the mixture?
   b) Heat is added at constant pressure until all the water has evaporated. Show the process carefully in a P – v diagram.

c) Determine the amount of heat added and the change in the volume of the system.

2. Ammonia at temperature $T_1 = 27.0°C$ and pressure 0.1 MPa (atmospheric pressure) flows into an apparatus where it is heated by an electric resistance of 100 ohm at a volumetric rate of 0.000041 m^3/s. When a current of 0.050 A (Ampere) flows through the resistance, the exit temperature of ammonia is 31.1°C.

   a) What are the specific heat capacities, c_p and c_v for ammonia?
   b) Note: Ammonia at ambient conditions may be considered to be an ideal gas and the power supplied by an electric heater is equal to I^2R.

3. Helium enters an isentropic nozzle at 3 MPa, 350°C with a negligible velocity and exits at 1.6 MPa. The mass flow rate is 0.5 kg/s. Calculate:

   a) The exit temperature.
   b) The velocity of the gas at the exit of the nozzle.
   c) The area at the exit of the nozzle.
   You may assume that helium is an ideal gas with constant specific heats, $M = 4$ kg/kmol, $k = 1.67$ and $c_p = 5.2$ kJ/kg.

4*. Steam at 2 MPa and 450° C enters an isentropic nozzle with a velocity of 10 m/s and exits at 1.5 MPa. The mass flow rate is 0.2 kg/s. Calculate:

   a) The exit temperature.
   b) The velocity of the steam at the exit of the nozzle.
   c) The area at the exit of the nozzle.

5*. A steam turbine admits 50 kg/s of steam at 10 MPa and 450°C. The turbine operates with an isentropic efficiency of 85% and exhausts at 0.08 MPa. Determine

a) The specific work of the turbine in kJ/kg.
b) The total power produced by the turbine in kW.

6*. For the operation of the turbine in problem 5 determine the amount of cooling water needed at the condenser, in kg/s, if the water is supplied from and returned to a river. The temperature difference of the supply and return water is 3°C and the specific heat of liquid water is constant and equal to 4.184 kJ/kg K. Note that the condenser receives the effluent of the turbine and exhausts it as saturated liquid water.

7. Steam Rankine cycle with superheat: The turbine entrance pressure is 15 bar and the temperature is 400°C. The condenser pressure is 0.08 bar. The mass flow rate is 850 kg/min. The turbine efficiency is 82% and the pump efficiency 75%. Calculate:

   a) The thermal efficiency of the cycle.
   b) The total power produced in MW.
   c) Suggest three ways to improve the efficiency of this cycle.

8*. Steam enters the turbine of a power plant operating with a Rankine cycle at 500°C and 100 bar. The turbine exhausts to a condenser at 0.08 bar. If the isentropic efficiency of the turbine is 84% and that of the pump is 75%, calculate:

   a) The thermal efficiency of the cycle.
   b) Suggest three ways to improve the efficiency of this cycle.

9. Air in a Brayton cycle enters the turbine at 1.8 MPa and 1200°C. The compressor entrance conditions are 0.1 MPa and 20°C. The isentropic efficiency of the turbine is 83% and that of the compressor 78%. Calculate:

   a) The thermal efficiency of the cycle.
   b) The back work ratio.
   c) Suggest three ways to improve the efficiency of this cycle.

10*. Propose a simple Rankine cycle that would utilize as a source of heat the exhaust gas from the turbine of problem 9. Determine the amount of additional work this bottoming cycle may produce and the efficiency of the combined cycle.

11. The exhaust gases of a gas turbine, are used in a well insulated regenerator (heat exchanger) to raise steam at 2 bar, which is subsequently fed to a steam turbine in order to produce additional power. The air enters the regenerator at 758 K and exits at 350 K. The mass flow rate of the air is 3.2 kg/s. Water at 45°C and 2 bar enters the regenerator and exits as saturated steam at the same pressure. Determine:

   a) The mass flow rate of the water.
   b) The rate of entropy production in the regenerator in kJ/s.

12. A small residential water heater uses methane gas (CH_4) as its fuel. The heater uses 92 m^3 of water annually and raises its temperature from an average 18–50°C. The density and specific heat capacity of water is 1,000 kg/m^3 and 4.18 kJ/kg K. Determine:

   a) The amount of energy consumed by this water heater.
   b) The amount of methane used annually and the amount of CO_2 released in the atmosphere from this heater if the heater burns methane. The heating value of methane is 50,020 kJ/kg.

13*. The cooling system of a power plant uses lake water. The mass flow rate of the plant cycle is 1,050 kg/min. Wet steam with quality 92% at 0.01 MPa (0.1 bar) enters the condenser of the power plant and saturated liquid water at the same pressure exits the condenser. The cooling water of the lake enters the condenser at 18°C and exits at 20.8°C. For this range of temperatures you may assume that the liquid water is incompressible and its specific heat is constant and equal to 4.18 kJ/kgK. Determine the mass flow rate of the water that is needed to cool this plant.

14*. Determine the exergy of saturated liquid water and saturated steam at the following temperatures: 100, 150, 200, 250, 300 and 350°C.

From the above values what can you conclude about the potential of steam and liquid water to produce work?

15. A 400 MW steam power plant uses coal as its fuel. The average efficiency of the plant is 38% and the plant operates for 84% of the time annually. Determine:

    a) The electricity in kWh and kJ produced by the plant annually.
    b) The amount of heat in kJ produced by the boiler of the plant.
    c) Assuming that the coal burned is pure carbon and that 1 kg of carbon when burned produces 32,770 kJ of heat and 44/12 kg of carbon dioxide, determine the mass of coal consumed by this plant and the mass of CO_2 released annually.

16. It has been suggested that, in order to reduce the amount of CO_2 that is released to the atmosphere, the boiler of the power plant in problem 14 is refitted to burn propane (C_3H_8), which produces 46,360 kJ/kg of heat.

    a) Determine the mass of propane that will be consumed by this plant and the mass of CO_2 released annually.
    b) What conclusions can you make about the substitution of coal with hydrocarbons for the reduction of CO_2 released? Are there other alternatives to produce electric power without CO_2 emissions?

17. A 600 MW coal power plant uses bituminous coal as a fuel, which has the following consistency by mass: carbon 62%, moisture 27%, ash 10% and sulfur 1%. You may assume that, only carbon provides heat with $\Delta h = 32{,}770$ kJ/kg when it is burned. The power plant operates with an average efficiency of 40% for 83% of the year. Determine:

    a) The annual amount of electricity in kWh the plant produces.
    b) The annual amount of heat produced by the boiler of the plant.
    c) The mass or carbon necessary to produce this amount of heat and the total mass of coal that produces it.
    d) The masses of carbon dioxide, ash and sulfur dioxide generated annually.

References

1. Moran MJ, Shapiro HN (2008) Fundamentals of engineering thermodynamics, 6th edn. Wiley, New York
2. Michaelides EE (1984) Exergy and the conversion of energy. Int J Mech Eng Educ 12:65–73

Chapter 4
Introduction to Nuclear Energy

Abstract One may ask the question: Why a book devoted to alternative energy should have three chapters on nuclear energy? After all nuclear power plants and nuclear energy have been the pariahs in every environmentalist's mind for decades. The accidents at the Three-Mile Island, Chernobyl and Fukushima Dai-ichi power plants have contributed to the horrific images of large-scale environmental disasters. The answer to this question is very simple: Global warming, caused by the anthropogenic emission of carbon dioxide is a very serious threat for our planet. Nuclear power plants have the capability to produce a significant part of our electric power at relatively low cost and without any carbon dioxide emissions. A case in point: If in 2008 the United States would have produced 50%, of its electric power from nuclear energy, instead of approximately 25% it actually produced, the country would have exceeded by 150% its quota from the *Kyoto Protocol* without any other changes in the rest of its energy mix. Other OECD countries, such as France and Japan produce more than 70% of their electricity from nuclear power plants. The fundamental concepts of atomic physics with emphasis on the nuclear fission reactions are given in this chapter. At first, the structure of the atom is explained, basic definitions of the atom and the subatomic particles are given succinctly and useful numbers pertaining to the atoms and the nuclear reactions are calculated. Secondly, the nuclear reactions are introduced and the physical principles governing these reactions are explained. Examples of nuclear reactions include radioactive decay and carbon dating. Thirdly, the several ways of interaction of neutrons with nuclei are explained and fission is introduced. The subjects of nuclear fission, chain reactions, nuclear fuels and thermal neutrons are explained in detail. The role of the cross-sections of the naturally occurring nuclear fuels is explained in the fission process as well as in the sustenance of the chain reaction in conventional reactors. The neutron cycle in a nuclear reactor and the striving for the production and conservation of the thermal neutrons are elucidated. Fourthly, the basic concepts of fuel conversion and breeding are given as

E. E. (Stathis) Michaelides, *Alternative Energy Sources*,
Green Energy and Technology, DOI: 10.1007/978-3-642-20951-2_4,

an introduction to breeder reactors. Finally, a few useful numbers are computed on the utilization of natural uranium as a fuel in the conventional nuclear reactors.

4.1 Elements of Atomic and Nuclear Physics

Engineers seeking an insight into the operation of nuclear reactors are interested to know only the results of nuclear reactions and do not need to be concerned with the details of the complex theory of subatomic physics. For this reason a simplified depiction of atoms and nuclei will be given in this chapter, which is sufficient for the understanding of the underlying principles that govern the release of nuclear energy. Central to the theory of energy obtained from nuclear reactions is the famous Einstein equation:[1]

$$E = mc^2, \tag{4.1}$$

where c is the speed of light in vacuum, which is approximately equal to $3*10^8$ m/s.

4.1.1 Atoms and Nuclei: Basic Definitions

Each atom is composed of a *nucleus* and of *electrons*, which are very light particles, have negative charge, $e = -1.602*10^{-19}$ Coulomb (Cb), and revolve around a nucleus in distinct orbits, which are very far from the nucleus. The nucleus is composed of *protons*, heavy particles with positive electric charge and *neutrons*, also heavy particles without any electric charge. Neutrons and protons together are called *nucleons*. The neutrons have almost the same mass as the protons and are electrically neutral. Electrons and protons have equal but opposite charges. The number of electrons and protons is the same in all atoms. Therefore, all atoms are electrically neutral. The number of protons in an atom is usually close to the number of neutrons, but the two numbers are not always equal. Figure 4.1 shows schematically the structure of the atom of carbon-12 ($_6C^{12}$). The nucleus of this atom is composed of six protons and six neutrons. Also, six electrons revolve around this nucleus at two distinct orbits.

The mass unit by which nuclei and subatomic particles are measured is the a*tomic mass unit, u,* which is equal to $1.6604*10^{-24}$ grams. The masses of the three elementary particles and their electric charges are as follows [1]:

- Mass of an electron: 0.000549 u charge: $-1.602*10^{-19}$ Cb
- Mass of a proton: 1.007277 u charge: $+1.602*10^{-19}$ Cb
- Mass of a neutron: 1.008665 u charge: 0

[1] This immensely popularized equation is actually a corollary of Einstein's *Special Theory of Relativity*.

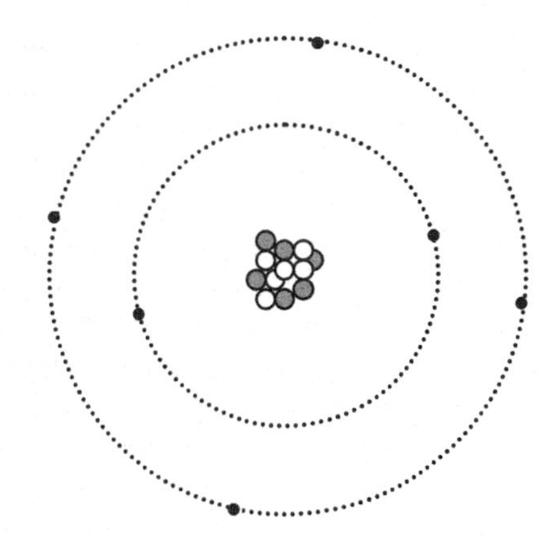

Fig. 4.1 Schematic representation of the atom of carbon-12. Protons (gray) and neutrons (white) are clustered inside the nucleus while the electrons (black) revolve around the nucleus

From Eq. (4.1), it follows that the equivalent energy of 1 u is $1.494*10^{-10}$ Joules (J). In nuclear reactions, the electron-volt (eV) is the commonly used unit of energy. This is equal to the charge of an electron moving through a voltage difference of 1 V, and, it is equal to $1.602*10^{-19}$ J. Therefore, the equivalent energy of 1 u is $9.31*10^{8}$ eV or 931 MeV. This implies that a mass of 1 u will produce 931 MeV if it were entirely converted to energy. For 1 kg, the equivalent energy is $5.62*10^{29}$ MeV or $9*10^{16}$ J, which is a very high amount of energy to be released. The latter is equivalent to the heat obtained from 3,100,000 metric tons of bituminous coal and is sufficient to produce 45,000,000 metric tons of steam.

Protons and neutrons are clustered in the nucleus of an atom. Almost all of the mass of the atom resides in the nucleus. In the case of the carbon-12 atom, which is depicted in Fig. 4.1, only 0.025% of the mass is outside the nucleus and rotates around it with the electrons. Between the nucleus and the electrons there is a great deal of space which is not occupied by any mass particles that is, vacuum. Another characteristic of the atom is that the radii of the orbits of the electrons are very long compared to the radius of the nucleus. If the radius of the nucleus in Fig. 4.1 were of the size of a tennis ball (approximately 7 cm in diameter) then the inner orbit of electrons would have a radius of approximately 600 m and the outer orbit's radius would have been at almost 2,000 m. These dimensions and mass proportions are typical of all atoms and lead us to the conclusion that all matter is concentrated in very small nuclei, with light electrons orbiting the nuclei at extremely long distances. Most of the space we see as a continuum is actually composed of vacuum.

Since electrons are very far from the nucleus, the electrostatic attraction at the outermost orbits is rather weak and an atom may lose one or more electrons, which may attach themselves to another atom or may be shared by several other atoms. In these cases, the atoms are electrically charged and are called *ions*. Ions with positive charges have a deficit of electrons and ions with negative charges have a surplus of electrons.

An atom is characterized by its *atomic number*, Z, which is equal to the number of protons in its nucleus. Oxygen has eight protons and its atomic number is

Z = 8. For carbon, Z = 6, for barium, Z = 56 and for uranium, Z = 92. Since the numbers of protons and electrons in an electrically neutral atom are equal, the atomic number of the atom is also equal to the number of the electrons. Another characteristic number of an atom is the *mass number*, A. The mass number is equal to the sum of the numbers of protons and neutrons in the nucleus. There are atoms that have the same number of protons and, hence, the same number of electrons, but differ in their numbers of neutrons. Such atoms are called *isotopes.* Isotopes have the same atomic number, Z, but different atomic mass, A.

In nuclear physics, an atom is denoted by its chemical symbol (O for oxygen, U for uranium, Pb for lead, etc.) which is accompanied by the atomic number as a left subscript and the mass number as a right superscript. The general notation for an atom, whose chemical symbol is X is: $_{Z}X^{A}$. For example, $_{92}U^{235}$ denotes the uranium isotope, which has an atomic number 92 and mass number 235. Since the difference A-Z is equal to the number of neutrons, this isotope has 92 protons and 143 neutrons. Similarly, the symbol $_{92}U^{238}$ denotes the uranium isotope with 146 neutrons and mass number 238, while $_{92}U^{233}$ denotes the uranium isotope with 141 neutrons and mass number 233. Other common isotopes are the following:

For Hydrogen: $_{1}H^{1}$ (common hydrogen); $_{1}H^{2}$ (deuterium); and $_{1}H^{3}$ (tritium), which have 1 proton and 0, 1 and 2 neutrons respectively.
For Oxygen: $_{8}O^{16}$, $_{8}O^{17}$ and $_{8}O^{18}$, all of which are met in the atmosphere.
For Plutonium: $_{94}Pu^{238}$, $_{94}Pu^{239}$, $_{94}Pu^{240}$ and $_{94}Pu^{241}$, all of which have been artificially formed by nuclear reactions.

In the macroscopic world, the mass of the elements is measured in grams (g), kilograms (kg) or gram-atoms. The latter is equal to the atomic mass, A, expressed in grams. Thus, a gram-atom of $_{8}O^{16}$ is equal to 16 g, while a gram-atom of $_{94}Pu^{239}$ is equal to 239 g. The number of atoms in a gram-atom is a constant for all the elements and is equal to $6.023*10^{23}$ atoms per gram-atom. This very large number is known as the *Avogadro number.*

4.1.2 Atomic Mass, Mass Defect and Binding Energy

The *atomic mass* of an isotope/atom is the mass of this atom, expressed in atomic mass units. Since most of the mass of the atom resides in the nucleus and the masses of protons and neutrons are approximately equal to 1 u, the atomic mass is approximately equal to the mass number, A. Table 4.1 from [1] gives the values of the atomic masses of a few common isotopes.

A glance at Table 4.1 proves that the actual mass of an atom is smaller than the sum of the masses of its constituent protons, neutrons and electrons. Thus, the boron-9 atomic mass is 9.01333 u, while the sum of the masses of its constituent five protons, five electrons and four neutrons is 9.07380 u. There is a small difference between the two numbers, 0.06047 u, which is called the *mass defect* of

Table 4.1 Atomic masses of common isotopes, in u

$_1H^1$	1.007825	$_6C^{12}$	12.00000
$_1H^2$	2.01410	$_6C^{13}$	13.00335
$_1H^3$	3.01605	$_7N^{14}$	14.00307
$_2He^3$	3.01603	$_8O^{16}$	15.99491
$_2He^4$	4.00260	$_8O^{17}$	16.99914
$_2He^5$	5.0123	$_8O^{18}$	17.99916
$_3Li^6$	6.01513	$_{92}U^{234}$	234.0409
$_3Li^7$	7.01601	$_{92}U^{235}$	235.0439
$_5B^9$	9.01333	$_{92}U^{238}$	238.0508

an atom. One way to understand the significance of the mass defect is to relate it to the equivalent energy, which is 56.3 MeV: If the boron atom were to be constructed by its constituents, there would be a mass defect of 0.06047 u. According to the conservation of mass-energy of Eq. (4.1), the mass defect must be released as energy during the formation of the atom. The energy released in such a process is often called the *binding energy.* Since a mass of 1 u is equivalent to 931 MeV, in the case of $_5B^9$, the binding energy is equal to 56.3 MeV, or 6.25 MeV per nucleon of the $_5B^9$ atom. Groups of nucleons that have lower energy are also more stable and do not disintegrate easily. The release of the binding energy holds the protons and neutrons together and makes the nuclei more stable than their atomic particle constituents. Actually, the higher the binding energy per nucleon, the more stable an atom is. For the atoms of $_8O^{16}$ and $_{92}U^{235}$ the binding energy per nucleon is, 7.97 MeV and 7.59 MeV, respectively. This partly explains why the oxygen-16 atom is more stable than the uranium-235 atom.

4.1.3 Nuclear Reactions and Energy Released

There is a plethora of nuclear reactions that take place naturally or artificially. Of relevance to the energy production processes are nuclear reactions that are caused by light subatomic particles, such as neutrons, protons and electrons. An example of an artificial nuclear reaction is the interaction of boron-10 with neutrons, which results in the formation of helium-4 and lithium-7:

$$_5B^{10} + {}_0n^1 \rightarrow {}_3Li^7 + {}_2He^4. \tag{4.2}$$

The nuclei of $_2He^4$ atoms are also called *alpha particles.* Electrons, which are denoted by the symbol $_{-1}e^0$, are called *beta particles* and strong electromagnetic waves are called *gamma particles.* When these three are considered as sources of radiation, one may call them *α-radiation*, *β-radiation* and *γ-radiation.*

A great deal of electromagnetic radiation, or gamma particles, is released in a typical nuclear reaction, because newly formed atoms are at an *excited state,* where the energy level is higher than normal. As these atoms decay to their lower energy level, strong electromagnetic radiation is released, which is rapidly converted to heat in the nuclear reactor.

All nuclear reactions are governed by four fundamental laws of physics:

1. *Nucleon conservation:* The total number of nucleons in the two sides of the reaction is the same. Protons may be converted to neutrons and vise versa by the absorption or emission of an electron. This implies that the sum of the right-hand superscripts in nuclear reactions is the same in both sides. In the example of Eq. (4.2): $10 + 1 = 7 + 4$.
2. *Charge conservation:* The sum of the charges of all nuclear particles in the two sides of the reaction is the same. This implies that the sums of the left-hand subscripts in the two sides of nuclear reactions are equal. In the example of Eq. (4.2): $5 + 0 = 3 + 2$.
3. *Momentum conservation:* The momentum of the particles before and after the reaction is the same because there are no external forces acting on the particles during the reaction.
4. *Mass-energy conservation:* Any mass defect during the reaction is accompanied by a release of energy. The sum of mass plus the mass equivalent of energy, according to Eq. (4.1), applies to the two sides of a nuclear reaction. In nuclear reactions the energy balance (or First Law of Thermodynamics) also includes the energy equivalent of mass.

In the case of the nuclear reaction shown in Eq. (4.2), the masses of the nuclei ${}_5B^{10}$, ${}_3Li^7$ and ${}_2He^4$ are 10.01294 u, 7.01601 u and 4.00260 u respectively, while the mass of the neutron, ${}_0n^1$, is 1.008665 u. Therefore, the reaction results in a mass defect of: $(10.01294 + 1.008665 - 7.01601 - 4.00260) = 0.002995$ u. This is equivalent to an energy release of 2.793 MeV. While this may appear to be an insignificant amount of energy, it must be recalled that it corresponds to only the conversion of a single atom and that the number of atoms corresponding to a macroscopically measured quantity of boron-10 is of the order of the Avogadro number, $6.023*10^{23}$ atoms per gram-atom. Therefore, when a mass of 50 g, or 5 gram-atoms, of boron-10 reacts according to Eq. (4.2) it would release an amount of energy equal to $8.411*10^{24}$ MeV or $1.347*10^{12}$ J. This is equivalent to the amount of energy released by 46.5 metric tons of bituminous coal.

The examples of the conversion of boron-10 and the fact that the conversion of an entire kg of any substance releases energy equivalent to the energy released by the combustion of 3,100,000 metric tons of bituminous coal, give an idea of the tremendous magnitudes of energy that may be released by nuclear reactions. Simply put, nuclear reactions release tremendous amounts of energy that are by far higher than the energy released by other primary energy sources. The reason for these high amounts of energy is the very large number of atoms in a gram-atom, $6.023*10^{23}$. When even small amounts of mass undergo nuclear reactions, the thermal energy released by far exceeds the chemical energy or any other form of energy stored in the mass. Nuclear reactions release very large amounts of energy, and for this reason, if properly and safely harnessed, they can play a very important role in meeting the energy demand of humankind.

4.1.4 Radioactivity

Every chemical element has a number of isotopes. Of these, some are stable and are naturally occurring, while other isotopes are unstable and undergo spontaneous nuclear transformations to become different elements. In general, stable isotopes have approximately the same number of protons and neutrons. When there is a significant imbalance in the numbers of protons and neutrons in the nucleus, the atom is unstable and may undergo a spontaneous transformation to become the nucleus of another element. This spontaneous transformation process is called *radioactivity.* In nuclear terminology, the original nucleus is often referred to as the *parent* nucleus and the product of the transformation as the *daughter* nucleus. The unit of radioactivity is one transformation per second or 1 Becquerel (1 Bq). Oftentimes the unit 1 Curie (1 Ci) is used, with 1 Ci $= 3.7 \times 10^{10}$ Bq. One Ci represents the radioactivity of one gram of radium-226.[2]

The types of transformations that occur with unstable nuclei may be categorized as follows:

1. *Alpha decay*: During this transformation, an alpha particle ($_2He^4$) is emitted by the nucleus. Alpha decay typically occurs with the heavier nuclei that have atomic numbers higher than 82. An example of alpha decay is the conversion of Plutonium-239 to Uranium-235:

$$_{94}Pu^{239} \rightarrow {}_{92}U^{235} + {}_{2}He^{4}. \tag{4.3}$$

   It is apparent that, the daughter nucleus, which is created in this transformation, has an atomic mass four units less than the parent nucleus and am atomic number two units less than the parent nucleus. Typically, the daughter nucleus will undergo a type of decay too and this will continue until the last daughter nucleus is a stable isotope. Equation (4.8) at the end of this section shows the long line of transformations, through which uranium-238 decays to the stable lead-206 isotope.
2. *Beta decay*: This is essentially the transformation of a neutron inside a nucleus to form a proton, with an electron being produced simultaneously. This type of radioactive decay occurs in nuclei where the number of neutrons is significantly high in comparison to the number of protons. A very small particle, which is

[2] Radium-226 and the phenomenon of radioactivity were discovered by Marie and Pierre Curie, who were students of Professor Henri Becquerel at the Grande Ecole de la Physique et Chemie Industriele, in Paris. The three shared the Nobel Prize in Physics in 1903.

called a neutrino and denoted by v, accompanies the emission of the electron as in the following example of the conversion of indium-115 to tin-115:

$$_{49}In^{115} \rightarrow {}_{50}Sn^{115} + {}_{-1}e^{0} + v. \quad (4.4)$$

The parent and daughter nuclei of the beta decay have the same mass number. The neutrino is lighter than the electron, interacts weakly with matter and may penetrate very thick layers of materials. Its presence in the nuclear reactions serves to satisfy the momentum conservation and mass-energy conservation principles.

3. *Positron emission decay*: A positron is a light particle of the same mass as the electron, but has positive charge and is denoted by $_{+1}e^{0}$. In general, positron decay occurs when the nucleus has too many protons in comparison to the number of neutrons. Again, the parent and daughter nuclei have the same mass number as in the following example of the conversion of iron-53 to manganese-53.

$$_{26}Fe^{53} \rightarrow {}_{25}Mn^{53} + {}_{+1}e^{0} + v. \quad (4.5)$$

The emitted positron is a short-lived particle among the plethora of electrons that surround the nucleus. Positrons are annihilated by combining with electrons and produce two gamma particles, which are very strong electromagnetic radiation:

$$_{+1}e^{0} + {}_{-1}e^{0} \rightarrow 2\gamma. \quad (4.6)$$

The combined energy of the two gamma particles is equivalent to the combined mass of the annihilated electron and positron. In this case, the equation $E = mc^2$ shows that 1.02 MeV of electromagnetic energy is produced from the reaction of Eq. (4.6).

4. *K-capture*: During this transformation, an electron from the K orbit is captured by the nucleus. The result is the transformation of a proton to a neutron. Simultaneously, a neutrino is produced as well as a significant amount of electromagnetic radiation in the form of gamma particles.

$$_{4}Be^{7} + {}_{-1}e^{0} \rightarrow {}_{3}Li^{7} + v + \gamma. \quad (4.7)$$

Oftentimes the daughter nucleus is unstable and transforms to another daughter nucleus. Thus, a parent nucleus may cause a series of transformations, which is called a *radioactive decay series.* It is typical of the trans-uranium radioactive elements, with $Z > 92$, to decay and produce a variety of daughter elements. The last element in this radioactive decay series is usually lead-206 or another stable isotope. Equation (4.8) shows the radioactive decay series of uranium-238 as a series of alpha and beta decay processes, which involve several isotopes of thorium (Th), protactinium (Pa), radium (Ra), radon (Rn), polonium (Po) and bismuth (Bi).

$$
\begin{aligned}
{}_{92}U^{238} &\rightarrow {}_{90}Th^{234} + {}_{2}He^{4} &,\quad & alpha\,decay\\
{}_{90}Th^{234} &\rightarrow {}_{91}Pa^{234} + {}_{-1}e^{0} &,\quad & beta\,decay\\
{}_{91}Pa^{234} &\rightarrow {}_{92}U^{234} + {}_{-1}e^{0} &,\quad & beta\,decay\\
{}_{92}U^{234} &\rightarrow {}_{90}Th^{230} + {}_{2}He^{4} &,\quad & alpha\,decay\\
{}_{90}Th^{230} &\rightarrow {}_{88}Ra^{226} + {}_{2}He^{4} &,\quad & alpha\,decay\\
{}_{88}Ra^{230} &\rightarrow {}_{86}Rn^{222} + {}_{2}He^{4} &,\quad & alpha\,decay\\
{}_{86}Rn^{222} &\rightarrow {}_{84}Po^{218} + {}_{2}He^{4} &,\quad & alpha\,decay\\
{}_{84}Po^{218} &\rightarrow {}_{82}Pb^{214} + {}_{2}He^{4} &,\quad & alpha\,decay\\
{}_{82}Pb^{214} &\rightarrow {}_{83}Bi^{214} + {}_{-1}e^{0} &,\quad & beta\,decay\\
{}_{83}Bi^{214} &\rightarrow {}_{84}Po^{214} + {}_{-1}e^{0} &,\quad & beta\,decay\\
{}_{84}Po^{214} &\rightarrow {}_{82}Pb^{210} + {}_{2}He^{4} &,\quad & alpha\,decay\\
{}_{82}Pb^{210} &\rightarrow {}_{83}Bi^{210} + {}_{-1}e^{0} &,\quad & beta\,decay\\
{}_{83}Bi^{210} &\rightarrow {}_{84}Po^{210} + {}_{-1}e^{0} &,\quad & beta\,decay\\
{}_{84}Po^{210} &\rightarrow {}_{82}Pb^{206} + {}_{2}He^{4} &,\quad & alpha\,decay.
\end{aligned}
\tag{4.8}
$$

It must be noted that radioactivity is a natural phenomenon and occurs naturally regardless of the man-made nuclear reactors and atomic weapons. There are several radioactive elements in the lithosphere, hydrosphere and atmosphere that contribute to the *natural* or *baseline radioactivity*. Cosmic rays also carry radioactive isotopes and radioactive particles in the earth's atmosphere. Of the naturally occurring radioactive isotopes several are absorbed by humans and animals and become part of their metabolism. For example, potassium-40 exists in the human body and contributes approximately 0.1 μCi of radioactivity or 3,700 transformations per second (3,700 Bq).

4.1.5 Rate of Radioactive Decay: Half Life

The decay of nuclei is a random process. A macroscopic sample of a radioactive isotope consists of a very large number of nuclei and, hence, the decay of the sample may be only described statistically. This implies that one does not know deterministically which atoms will decay within a given time interval or when an individual atom will decay. However, one is certain that, out of a large assembly of atoms, N, a number of them, δN, will decay within the time interval δt. For a large sample of atoms, the rate of decay, λ, is the proportionality constant, which defines the decay process:

$$dN = -\lambda N dt \Rightarrow \lambda = -\frac{1}{N}\frac{dN}{dt}, \tag{4.9}$$

where N is the number of nuclei in the sample, t is the time, and the negative sign denotes that the number of the original nuclei of the isotope decreases. The rate of decay, λ, characterizes an isotope and is a unique parameter for that isotope. Given an initial condition, the last equation may be easily integrated. For example if at $t = 0$ the number of nuclei in the sample is N_0, then at time t the number of remaining nuclei, N, is:

$$N = N_0 e^{-\lambda t}. \tag{4.10}$$

The *half life* of an isotope is also a commonly used parameter to characterize the decay process. The half life, $T_{1/2}$, is defined as the time for the number of nuclei to be reduced to half the original number, or:

$$\frac{N_0}{2} = N_0 e^{-\lambda T_{1/2}}. \tag{4.11}$$

A comparison of Eqs. (4.10) and (4.11) yields the following relationship between the half life and the rate of decay:

$$\lambda = \frac{\ln 2}{T_{1/2}} = \frac{0.693}{T_{1/2}}, \tag{4.12}$$

and for the number of the remaining nuclei after time t:

$$N = N_0 \exp\left(-\frac{0.693t}{T_{1/2}}\right). \tag{4.13}$$

Table 4.2 shows the half lives of several common isotopes. It is apparent, that the half lives span several orders of magnitude and range from millions of years to a few seconds. A glance at Eq. (4.13) proves that isotopes with small half lives are depleted rapidly, while isotopes with very long half lives are depleted very slowly. As a consequence, during a finite amount of time, which is much less than the half-life of an isotope, $t \ll T_{1/2}$, the concentration of this isotope may be considered to be constant.

Equations, such as (4.10) and (4.13) may only be interpreted statistically and only when the sample of the material contains a very large number of nuclei. They do not predict which nuclei will disintegrate, but how many nuclei out of a very large sample. These equations have been validated and are considered very accurate in predicting the total number of radioactive nuclei of a given isotope or the total mass of the isotope in a large sample.

It is apparent from Eq. (4.8) that many radioactive processes have several stages and one parent nucleus produces several daughter nuclei. Let us consider such a chain of decay of three radioactive elements: A → B → C with corresponding decay rates λ_A, λ_B *and* λ_C and initial number of nuclei N_{A0}, 0 and 0 respectively. Equation (4.10) yields the amount of isotope A present after a time t is:

Table 4.2 The half life and radioactivity emitted by some common isotopes

Isotope	Half life	Radioactivity
Uranium-235	7.1×10^8 yrs	α, γ
Uranium-238	4.51×10^9 yrs	α, γ
Plutonium-239	2.44×10^4 yrs	α, γ
Thorium-232	1.41×10^{10} yrs	α, γ
Krypron-87	76 min	β
Strontium-90	28.1 yrs	β
Barium-139	82.9 min	β, γ

$$N_A = N_{A0} e^{-\lambda_A t}. \tag{4.14}$$

For isotope B we may stipulate that the rate of change of its nuclei is equal to the rate of the decay of A (which results in the formation of B) minus the decay rate of its own. This yields the following differential equation:

$$\frac{dN_B}{dt} = \lambda_A N_A - \lambda_B N_B = \lambda_A N_{A0} e^{-\lambda_A t} - \lambda_B N_B. \tag{4.15}$$

Given the initial condition $N_{BO} = 0$, the solution of the last equation is:

$$N_B = N_{A0} \frac{\lambda_A}{\lambda_B - \lambda_A} \left(e^{-\lambda_A t} - e^{-\lambda_B t} \right). \tag{4.16}$$

Similarly, the rate of change of the nuclei of isotope C is:

$$\frac{dN_C}{dt} = \lambda_B N_B - \lambda_C N_C = \lambda_B N_{A0} \frac{\lambda_A}{\lambda_B - \lambda_A} \left(e^{-\lambda_A t} - e^{-\lambda_B t} \right) - \lambda_C N_C. \tag{4.17}$$

The solution of this differential equation, subject to the initial condition, $N_{CO} = 0$, would yield the number of nuclei of isotope C as a function of time. If the rate of decay of isotope C is very low in comparison to the rate of decay of B, that is if $\lambda_C << \lambda_B$, an approximate solution to this equation is:

$$\mathrm{N_C} \approx \mathrm{N_{A0}} \left(1 - \frac{\lambda_\mathrm{B}}{\lambda_\mathrm{B} - \lambda_\mathrm{A}} e^{-\lambda_\mathrm{A} \mathrm{t}} + \frac{\lambda_\mathrm{A}}{\lambda_\mathrm{B} - \lambda_\mathrm{A}} e^{-\lambda_\mathrm{B} \mathrm{t}} \right). \tag{4.18}$$

This equation becomes exact if the isotope C is stable.

Figure 4.2 shows the variation of N_A, N_B and N_C when the half lives of the isotopes A, B and C are 2 years, 0.5 years and 100 years respectively. It is apparent from this figure that the fraction of isotope B is significantly lower than the fractions of both A and C because the half life of this isotope is lower than that of the other two isotopes. At the end of the ten year period, isotopes A and B have almost completely decayed and converted to isotope C, which has significantly higher half life.

An interesting application of radioactive decay is *carbon dating*, or the determination of the age of a specimen that was produced by a plant or an animal. The atmospheric nitrogen interacts with cosmic rays and produces the isotope ${}_6C^{14}$,

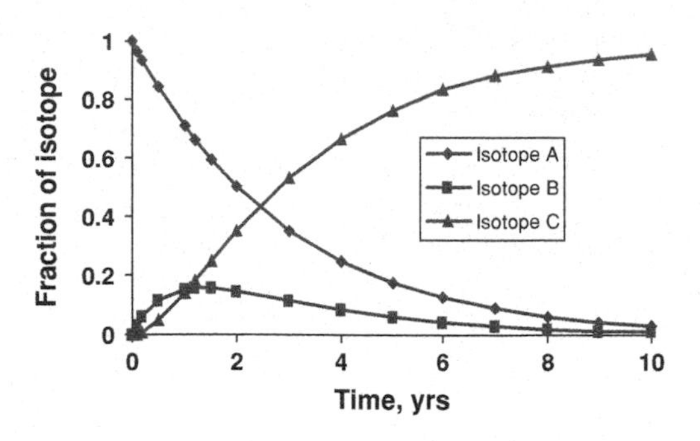

Fig. 4.2 Variation of the mass of isotopes A, B and C vs. time

which is radioactive with a half life of 5,730 yrs. This isotope is chemically converted to carbon dioxide and diffuses in the atmosphere. At any time, 0.1% of the carbon in the atmospheric CO_2 is composed of the isotope ${}_6C^{14}$. CO_2 is typically absorbed by plants and through the food chain it is also absorbed by animals. Therefore, at any time, 0.1% of the carbon atoms in living plants and animals is composed of ${}_6C^{14}$. When the absorption of ${}_6C^{14}$ ceases, due to the harvesting of the plant or the death of the animal, the fraction of this isotope starts decreasing by radioactive decay. The application of Eq. (4.14) yields the time that has elapsed from the harvesting of the plant, which produced the specimen. Following this technique, we may estimate the age of a manuscript written on an old papyrus, in which the current carbon-14 content was determined to be 0.072%. According to Eq. (4.13) the age of the manuscript, t, in years is given by the expression: 0.00072 = 0.001*exp(-0.693t/5730), or t = 2,716 years.

4.2 Nuclear Fission

The neutron was discovered in 1932 by Chadwick in England. Because it is an electrically neutral particle it is not repelled by the charge of the electrons or the nuclei and, thus, it may penetrate the atoms and interact directly with the nuclei. In 1938 Hahn and Strassmann in Germany observed that barium-139 was produced when uranium-235 interacted with a beam of neutrons. This was the first demonstration of a fission process. Fission occurs when neutrons are captured by heavy nuclei, such as ${}_{92}U^{235}$, ${}_{92}U^{238}$, or ${}_{92}Th^{232}$. The resulting compound nucleus is unstable and soon splits into two large fragments. A few—typically two or three—free neutrons are also released in the process. A typical fission process with uranium 235 as the fuel is shown schematically in Fig. 4.3. The four stages of this process show how the ${}_{92}U^{235}$, produces the unstable nucleus ${}_{92}U^{236}$, which almost immediately splits into ${}_{57}La^{147}$ and ${}_{35}Br^{87}$ plus two free neutrons according to the nuclear reaction:

$$ {}_{92}U^{235} + {}_0n^1 \rightarrow {}_{92}U^{236} \rightarrow {}_{57}La^{147} + {}_{35}Br^{87} + 2{}_0n^1. \quad (4.19a) $$

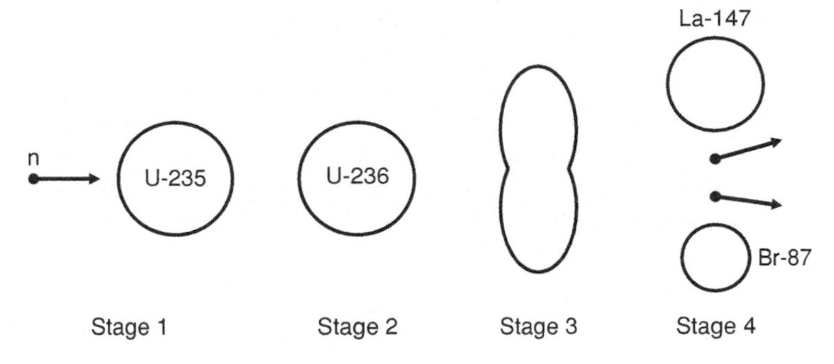

Fig. 4.3 Fission process of $_{92}U^{235}$ by neutrons

Another typical reaction of the $_{92}\mathrm{U}^{235}$ nucleus produces xenon-140 and strontium-94:

$$_{92}\mathrm{U}^{235} + {_0}\mathrm{n}^1 \rightarrow {_{92}}\mathrm{U}^{236} \rightarrow {_{54}}\mathrm{Xe}^{140} + {_{38}}\mathrm{Sr}^{94} + 2{_0}\mathrm{n}^1. \tag{4.19b}$$

Plutonium-239, which may be produced by the capture of a neutron by the $_{92}\mathrm{U}^{238}$ nucleus, may also undergo fission in several ways. One such fission reaction is:

$$_{94}\mathrm{Pu}^{239} + {_0}\mathrm{n}^1 \rightarrow {_{56}}\mathrm{Ba}^{137} + {_{38}}\mathrm{Sr}^{100} + 3{_0}\mathrm{n}^1. \tag{4.19c}$$

Conventional nuclear reactors use uranium, which contains both isotopes $_{92}\mathrm{U}^{235}$ and $_{92}\mathrm{U}^{239}$, as their fuel. Small amounts of plutonium-239 are always formed in conventional reactors and contribute to the thermal energy produced and the neutron flux. The so called *fast breeder reactors* produce larger amounts of $_{94}\mathrm{Pu}^{239}$, which later becomes the main fuel of the reactor, as will be explained in more detail in Sect. 4.3.

4.2.1 Interactions of Neutrons with Nuclei

When neutrons penetrate atoms and interact with the nuclei, three types of interactions are possible:

1. *Scattering collisions,* which may be elastic or inelastic. During inelastic scattering collisions, a neutron loses some of its energy to surrounding atoms and slows down. The slowing down of fast neutrons is necessary for the maintenance of the chain reaction in the commonly used reactors. Fast neutrons undergo inelastic collisions and progressively slow down. When the neutrons are in thermal equilibrium with the surrounding nuclei, they are called *thermal neutrons,* and their scattering collisions are considered elastic. It must be noted

that, after a scattering collision, elastic or inelastic, the neutron remains in the reactor and may interact again with nuclei.

2. *Capture*. During capture, a neutron enters and remains in the nucleus. In this case a relatively stable compound nucleus is formed, which has one extra neutron. The compound nucleus may be radioactive and may decay to form another nucleus, as in the following reaction of the production of ${}_{94}Pu^{239}$ from the neutron capture of ${}_{92}U^{238}$:

$$\begin{aligned} {}_{0}n^{1} + {}_{92}U^{238} &\rightarrow {}_{92}U^{239} \\ {}_{92}U^{239} &\rightarrow {}_{93}Np^{239} + {}_{-1}e^{0} \\ {}_{93}Np^{239} &\rightarrow {}_{94}Pu^{239} + {}_{-1}e^{0}. \end{aligned} \tag{4.20}$$

3. *Fission*. The fission of a nucleus results from the capture of a neutron and the formation of an unstable, bigger nucleus, which immediately splits into two larger fragments and a few neutrons, as shown in Eqs. (4.19a) to (4.19c) and depicted schematically in Fig. 4.3. The produced neutrons may cause further fissions.

The neutron-nuclei interactions of capture and fission are often called *absorption,* because the neutron is initially absorbed by the nucleus. During scattering the neutron does not enter the nucleus and is immediately rejected. After the scattering process the neutron is available to interact with other nuclei and to cause fission.

For the modeling of the interactions of neutrons with bigger nuclei, let us assume that a uniform beam of neutrons is incident upon a slab of a material with area A, where the density of the atoms is N. At a depth z inside the material the flux of neutrons is Φ and at a depth $z + dz$ the flux is $\Phi - d\Phi$, as shown in Fig. 4.4. The amount of nuclei in the volume Adz, with whom the neutrons may interact is $NAdz$. The number of interactions occurring in this volume, dI, must be proportional to the number of nuclei and the intensity of the beam:

$$dI = \sigma \Phi N A dz. \tag{4.21}$$

The constant of proportionality, σ, has the dimensions of area and is called the *microscopic cross-section* of the nucleus, or simply, the *cross section* of the nucleus. This rate of interaction is equal to the difference of the neutron flux from z to $z + dz$, and is equal to the product of $d\Phi$ and the area of the material, A:

$$\sigma \Phi N A dz = -A d\Phi. \tag{4.22}$$

The negative sign in Eq. (4.22) indicates the diminishing neutron flux as it penetrates the material of the slab. The solution of the resulting differential equation, with the boundary condition $\Phi(z = 0) = \Phi_0$, yields the following expression for the strength of the neutron flux at depth z:

$$\Phi = \Phi_0 e^{-\sigma N z}. \tag{4.23}$$

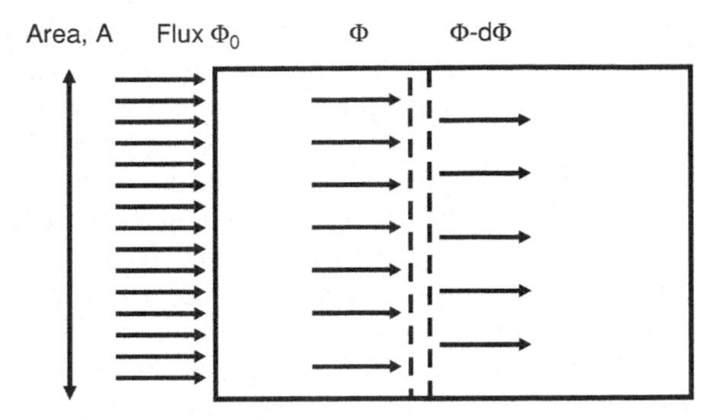

Fig. 4.4 Schematic diagram of the interaction of a flux of neutrons with a material

4.2.2 Cross Sections of Common Nuclei

The cross-section of an isotope is a unique parameter of that isotope. It has the units of area and its numerical value is of the order of magnitude of the actual cross section of a stationary nucleus. The radius of the nuclei is approximately equal to $r_n = 1.3*10^{-13}A^{1/3}$ cm, which implies that the actual cross sectional area of the nucleus of a heavy atom, such as $_{92}U^{238}$ or $_{92}U^{235}$, is approximately $2*10^{-24}$ cm^2. For this reason cross sections are expressed in units of 10^{-24} cm^2, which are called *barns*. One barn is equal to 10^{-24} cm^2 or 10^{-28} m^2.

The neutron-nucleus interaction is a complex phenomenon, where quantum mechanical interactions play an important role. The cross-section is an average phenomenological parameter that takes into account these interactions and expresses them in terms of parameters that are understood from concepts of classical mechanics. The cross-section is, essentially, the "target area" of the nucleus as seen by the neutron, which is the projectile. Therefore, neutrons have high probability of interacting with nuclei of large cross sections and lower probability to interact with nuclei of smaller cross sections. However, because of the quantum mechanical effects, the area of this "target" depends strongly on the characteristics of the nucleus, the type of interactions that occur, and also on the kinetic energy of the neutrons, which are the "projectiles." There are different cross sections for inelastic scattering, σ_i, capture, σ_c and fission, σ_f. These parameters are functions of the energy of the incident neutrons.

Figures 4.5a and b depict the fission and capture cross sections of $_{92}U^{238}$ and $_{92}U^{235}$ as functions of the neutron energy. It is observed in the first figure that for the nucleus of $_{92}U^{238}$ the fission cross section, σ_f, is approximately 0.6 when the neutron energy is higher than 2 MeV and drops to negligible values at neutron energies less than 1 MeV. The capture cross section of this isotope, σ_c, increases from very low values at 5 MeV, to approximately 1 barn at 1,000 eV. At lower neutron energies, there is a *resonance range* that extends from 1,000 eV to 5 eV, where the capture cross section may reach values as high as 1,000 barn. At electron energies below 2 eV, the capture cross section is approximately constant, with values in the range 10–13 barn. This has significant implications for the operation of a thermal nuclear reactor, where the neutrons must be slowed down to very low energy values (they become thermal neutrons) in order to cause fission. As the neutrons are progressively slowed to lower kinetic energies there is high probability that, in the *resonance range* of 1,000 eV > E > 5 eV, these neutrons will be captured by a $_{92}U^{238}$ nucleus. As a consequence, the captured neutrons will not become available to cause fissions and to produce energy.

Figure 4.5b, which depicts the cross sections for capture and fission of the isotope $_{92}U^{235}$, shows a different dependence of the cross sections on the energies of incident neutrons: At first, the fission cross section of this isotope is significantly higher than the capture cross section in the entire range of neutron kinetic energies. The cross section for fission of this isotope, σ_f, is less than 2 barn in the range of kinetic energy $E > 10^5$ eV and, thereafter increases at a faster rate to the value 35 barns at E = 80 eV. After a *resonance range* that extends from 80 to 4 eV, and where the resonance cross section values may reach up to 1,000 barn, the fission cross section of $_{92}U^{235}$ continues its monotonic increase to values close to 6,000 barn at 0.001 eV. Secondly, the capture cross section of $_{92}U^{235}$, σ_c, is negligibly small at energies E > 1 eV and increases almost monotonically at lower energies. However, in the whole range of neutron energy, the values of σ_c are at least one order of magnitude lower than σ_f. Therefore, in pure $_{92}U^{235}$, the fission has a significantly higher probability of occurrence than capture. It must be noted that the composition of natural uranium is far from this: natural uranium consists of 99.285% $_{92}U^{238}$ and 0.715% $_{92}U^{235}$. That is, in a given mass of natural uranium there is only 1 nucleus of $_{92}U^{235}$ for every 139 nuclei of $_{92}U^{238}$. Therefore, a neutron in natural uranium is more likely to be captured by a $_{92}U^{238}$ nucleus than to cause the fission of a $_{92}U^{235}$ nucleus. The data for the production of Fig. 4.5a and b were obtained from Bennet [1].

4.2.3 Neutron Energies: Thermal Neutrons

Neutrons are released from nuclear reactions at very high average velocities and, hence very high kinetic energies. For fission-released neutrons typical values of their kinetic energy are close to 2 MeV (two million eV). Because of the complex quantum mechanical interactions between neutrons and nuclei, the kinetic energy of neutrons plays a very important role in determining whether neutrons will cause fissions or what kind of interactions will occur between the neutrons and nuclei. For this reason, neutrons are classified according to their average energies as follows:

1. Fast neutrons with $E > 10^5$ eV. When neutrons are first released from fission reactions, they are fast neutrons.
2. Intermediate neutrons with $1 \text{ eV} < E < 10^5$ eV.
3. Slow neutrons with E < 1 eV.

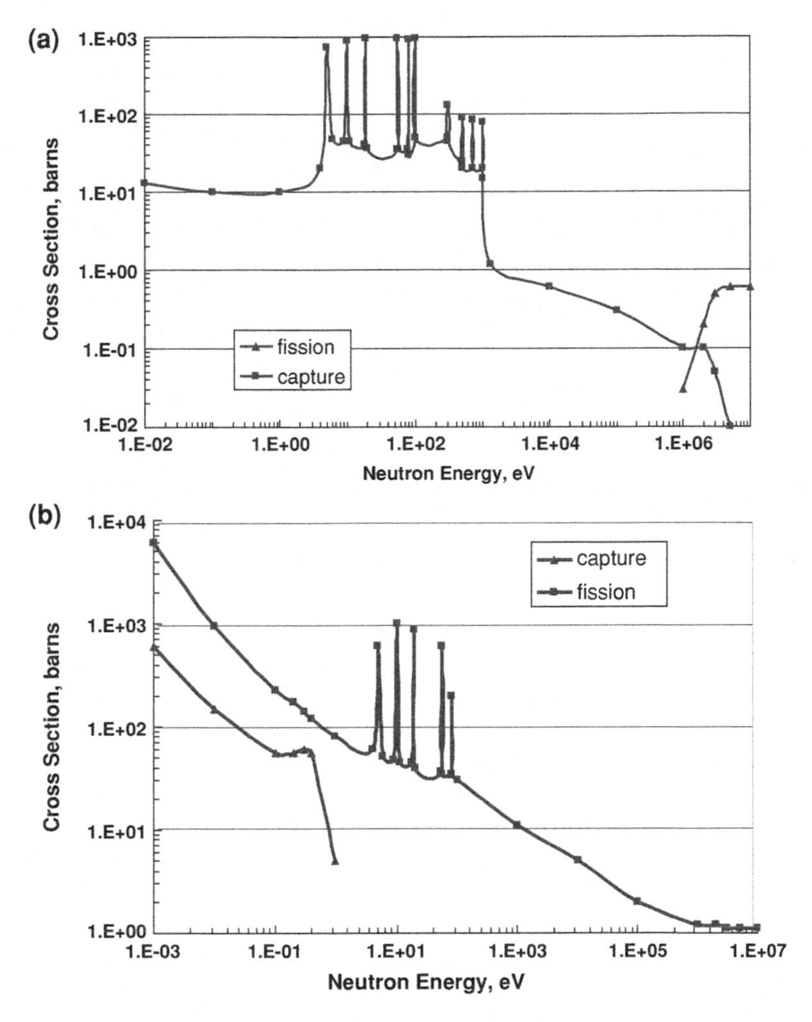

Fig. 4.5 **a** Fission and capture cross-section for uranium-238 **b** Fission and capture cross-section for uranium-235. Data from [1]

The *thermal neutrons* are a category of slow neutrons, which is important in nuclear reactors because they have the right amount of energy to cause a chain reaction of uranium-235 nuclei in nuclear reactors. These neutrons are in thermal equilibrium with the surrounding atoms in a reactor. The state of thermal neutrons is analogous to the state of gas molecules in a container: The molecules of the gas collide with the atoms at the walls of the container and the molecular velocities are significantly high. While individual gas molecules are slowed down or are accelerated by the collisions, there is a dynamic equilibrium of the molecular velocities, which exhibit a distinct distribution, known as the Maxwell-Boltzmann distribution. Similarly, there is a dynamic equilibrium of the velocities, and kinetic

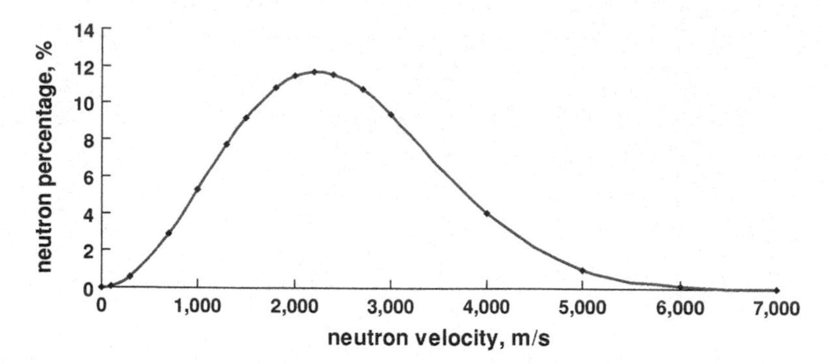

Fig. 4.6 The Maxwell-Boltzmann distribution for thermal neutrons

energies of thermal neutrons enclosed in any material. These neutrons also have a wide range of velocities, which is approximated by the classical Maxwell-Boltzmann distribution:

$$n(v)dv = 4\pi N\left(\frac{m}{2\pi kT}\right)^{3/2}\exp\left(\frac{-mv^2}{2kT}\right)v^2dv, \tag{4.24}$$

where $n(v)dv$ is the number of neutrons in the range of velocities v and $v + dv$; N is the total number of neutrons; m is the mass of a single neutron, $1.6748*10^{-27}$ kg; k is the Boltzmann constant, $1.38*10^{-23}$ J/K; and T the absolute temperature, in K. A diagram of the Maxwell-Boltzmann distribution, for neutrons at 300 K is shown in Fig. 4.6. It is observed that the distribution has a maximum, which denotes the most probable velocity of the neutrons. The latter may be calculated by differentiating the distribution function and equating the result to zero. Thus, the most probable velocity and the most probable energy of the thermal neutrons are:

$$v_{mp} = \sqrt{\frac{2kT}{m}}, \quad E_{mp} = kT. \tag{4.25}$$

Similarly, the average velocity of the neutrons, v_{av}, and the average energy, E_{av}, may be calculated from the Maxwell-Boltzmann distribution by integrating the function $vn(v)$ and dividing by the total number of neutrons, N, to yield:

$$v_{av} = \sqrt{\frac{8kT}{\pi m}} = 1.128v_{mp}, \quad E_{av} = \frac{4kT}{\pi}. \tag{4.26}$$

It must be noted that all general references to neutron velocities refer to the average velocity and all references to neutron energy refer to the kinetic energy that corresponds to the average velocity. The energy of thermal neutrons is 0.025 eV, their average velocity is 2,482 m/s and their most probable velocity is approximately 2,200 m/s.

Table 4.3 shows the values of the most significant cross sections of pure ${}_{92}U^{235}$, pure ${}_{92}U^{238}$, and natural uranium for fast, intermediate and thermal neutrons (data

Table 4.3 Neutron cross sections, in barns, for uranium isotopes

Neutron Energy		$_{92}U^{235}$	$_{92}U^{238}$	Natural U
Fast, 2 MeV	σ_c	0.09	0.14	0.14
	σ_f	1.20	0.018	0.026
	σ_i	2.87	2.3	2.84
Intermediate, 0.3 MeV	σ_c	0.15	0.2	0.2
	σ_f	1.3	0.01	0.02
	σ_i	0.5	0.7	0.7
Thermal, 0.025 eV	σ_c	101	2.75	3.47
	σ_f	577	0	4.16
	σ_i	10	8.3	8.4

from [1]). It is observed in this Table that, since natural uranium is composed mostly of the $_{92}U^{238}$ isotope, its pertinent cross-sections are very close to those of $_{92}U^{238}$.

4.2.4 The Chain Reaction: Probability of Fission

There is no naturally occurring flux of neutrons to cause fissions in a nuclear reactor. The few neutrons that may be produced during the natural radioactive decay of isotopes are not sufficient to maintain the desired reaction rate in a nuclear reactor. The neutron flux, which is crucial for fissions must be artificially created and maintained. Typical nuclear reactors are fueled with nuclear material once every 18 to 24 months and remain hermetically closed. These reactors produce approximately 3,000 MW. Given that a single fission reaction produces approximately 200 MeV of energy, the rate of fission reactions in the nuclear reactors is approximately $9.4*10^{19}$ fissions per second. As a consequence, these nuclear reactions use $9.4*10^{19}$ neutrons per second and these neutrons must be produced continuously inside the reactor.

At the beginning of the refueling process, a neutron source is introduced in the reactor, which produces the first flux of neutrons and acts as the spark for the "ignition" of the fission reactions. A common method for starting the neutron flux is the introduction of a small amount of radium-226, which disintegrates producing alpha particles, and beryllium-9. The alpha particles ($_2He^4$), which are produced during the radioactive decay of radium, are captured by the beryllium to produce cabon-12, a stable nucleus, and a free neutron according to the reaction:

$$_4Be^9 + {}_2He^4 \rightarrow {}_6C^{12} + {}_0n^1. \tag{4.27}$$

A mixture of 1 g of $_{88}Ra^{226}$ and 5 g of $_4Be^9$ constitutes a compact neutron source that provides approximately 10^7 neutrons per second. While this flux may be sufficient for the "ignition" of the reactor, it is simply not sufficient to maintain for long the operation of a typical reactor. For this reason, nuclear reactors are

Table 4.4 Average number of neutrons emitted per fission

Isotope	Neutron energy, eV	n_{av}
$_{92}U^{235}$	0.025	2.44
	1,000,000	2.50
$_{94}Pu^{239}$	0.025	2.87
	1,000,000	3.02
$_{92}U^{233}$	0.025	2.48
	1,000,000	2.55
$_{90}Th^{232}$	1,500,000	2.12
$_{92}U^{238}$	1,100,000	2.46

designed to produce neutrons and maintain a high flux of neutrons by a process called the *chain reaction.*

During the chain reaction the neutron source causes the first fission reactions in the nuclear fuel. As it is apparent from Eqs. (4.19a) typical fission reactions produce a few neutrons (two to three), which are called the first generation neutrons. Some of these neutrons may be used to cause further fissions, which would produce more neutrons, the second generation of neutrons. A fraction of the latter neutrons may cause a third wave of fissions and produce another generation of neutrons and so on. The necessary flux of neutrons is maintained in the nuclear reactor by the neutrons produced during the fissions that took place previously. Figure 4.7 shows schematically the chain reaction. A key parameter of the chain reaction is the *reproduction constant, k*, defined as the number of neutrons in a generation divided by the number of neutrons in the preceding generation. When $k = 1$, the chain reaction is maintained. When $k < 1$, the neutron flux and, hence, the fission reaction rate is decreasing and, if continued, it will lead to the extinction of the fission reactions. When $k > 1$, the neutron flux increases, the fission reaction rate increases and, if sustained, it may lead to an explosive situation. In the last two cases the reactor operators must take action to maintain the controlled chain reaction with $k = 1$. Another important parameter is the average number of neutrons produced by the fission of the isotopes that fuel the reactor. Table 4.4 shows the average number of neutrons emitted per fission reaction, n_{av}, for several isotopes that are commonly used as reactor fuels [2].

While it is rather easy to visualize the chain reaction, its practical realization in fuels other than pure $_{92}U^{235}$ poses several difficulties. The main problem for the establishment of a chain reaction is that when neutrons are produced from a fission reaction they are fast neutrons with energies approximately 2 MeV. A glance at the cross sections of fast neutrons, in Table 4.4, proves that the neutrons produced in each generation are more likely to undergo inelastic scattering or capture than to cause fissions. A concept that is helpful in the understanding of the likelihood of chain reaction realization is the *probability of fission,* which combines the concentration of the various nuclei in the fuel with the cross sections. For a fuel composed of the isotopes $_{92}U^{235}$ and $_{92}U^{238}$, such as natural or enriched uranium, the probability of fission, P_f, is defined by the expression:

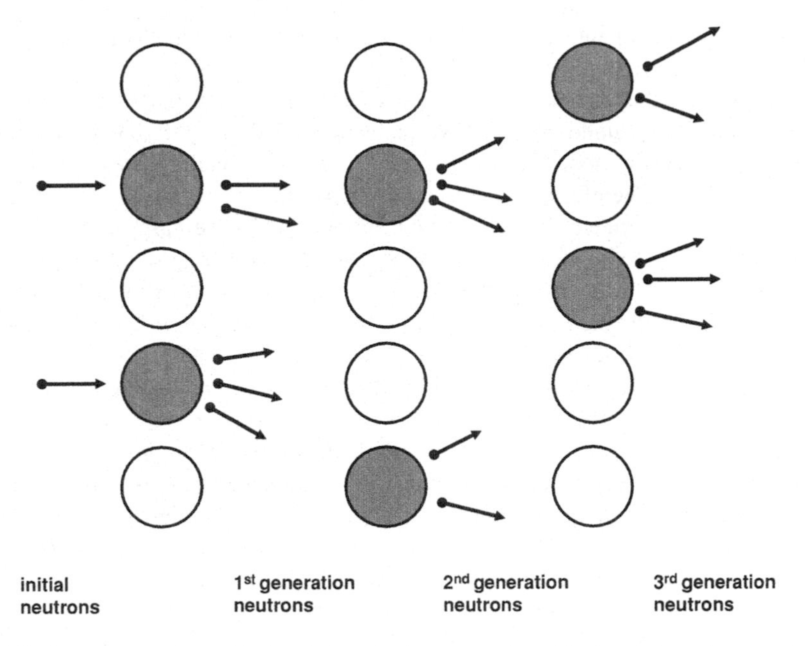

Fig. 4.7 A chain reaction with $k = 1$: Nuclei in grey undergo fission and produce neutrons. Neutrons that do not cause fissions are absorbed by other nuclei or leak outside the reactor

$$P_f = \frac{c_{238}\sigma_f^{238} + c_{235}\sigma_f^{235}}{c_{238}\left(\sigma_f^{238} + \sigma_c^{238} + \sigma_i^{238}\right) + c_{235}\left(\sigma_f^{235} + \sigma_c^{235} + \sigma_i^{235}\right)}, \tag{4.28}$$

where c denotes the concentration of the corresponding nuclei in the bulk of the reactor fuel and the superscripts in the several cross sections denote the pertinent isotopes of uranium. Several values of these cross sections are in Table 4.3 as well as in Figs. 4.5a and b. The probability of fission yields the probability that a given neutron, which is produced in generation j, will cause a fission reaction and, thus, contribute to the production of neutrons in the generation $j + 1$. Since there are n_{av} such neutrons produced by the nuclei in the fuel, and one neutron is needed to cause a fission, a chain reaction will be maintained only if the product $n_{av}{*}\ P_f$ is greater than one. In the opposite case, if $n_{av}{*}\ P_f < 1$, the number of neutrons in the reactor will diminish after each generation and the reactor would cease to operate.

Let us consider the possibility of establishing a chain reaction in a system that is composed entirely of natural uranium, where the concentration of isotope 238 is 0.99285 and the concentration of isotope 235 is 0.00715. When the neutrons are produced by fission reactions, they are fast neutrons with energies close to 2 MeV. For fast neutrons at this energy, and according to the cross section values listed in Table 4.3, the probability of fission by fast neutrons is 0.0076 and the value of the product ($n_{av}{*}\ P_f$) is 0.02, which is significantly lower than 1. This implies that it is not possible to sustain a chain reaction with fast neutrons in natural uranium.

At this energy most neutrons will likely be slowed down by inelastic collisions or will be captured by the nuclei of ${}_{92}U^{238}$.

The question then arises if it is possible to have a chain reaction with slower neutrons that have undergone a few inelastic collisions. At 0.3 MeV neutron energy the cross section values of Table 4.3 yield: $P_f = 0.021$ and $n_{av}{*}\ P_f = 0.052$. Therefore, a chain reaction would not be sustained in natural uranium at this intermediate level of neutron energy. In the case of thermal neutrons, however, the situation becomes different: Because the fission cross-section of the isotope ${}_{92}U^{235}$ is significantly higher than the other relevant cross sections for thermal neutrons, the probability of fission is calculated to be $P_f = 0.54$ and the product $n_{av}{*}\ P_f = 1.32$ is greater than 1. This implies that a sustainable chain reaction with natural uranium is possible with thermal neutrons. The challenge for the scientists and engineers is to build a large-scale system, the thermal nuclear reactor, that would realize and sustain the chain reaction for a long time.

Another way to sustain a chain reaction with faster neutrons is to use *enriched uranium* instead of natural uranium as the reactor fuel. Enriched uranium has a higher concentration of ${}_{92}U^{235}$ atoms than natural uranium. For example, if the reactor fuel had equal numbers of nuclei of the two uranium isotopes ($c_{238} = c_{235} = 0.5$) and neutrons of 0.3 MeV were used for fission, the probability of fission would have been $P_f = 0.46$ and $n_{av}{*}\ P_f = 1.15 > 1$. Hence, a sustained chain reaction would be possible with 50% enriched uranium and neutrons with 0.3 MeV energy.

It must be noted that the process of uranium enrichment requires a significant amount of power and is very expensive: The two isotopes of uranium have the same chemical properties and, thus, they may not be separated by chemical means. Since the isotope ${}_{92}U^{235}$ is slightly lighter, it diffuses slightly faster through a membrane than ${}_{92}U^{238}$, if the two isotopes were in gaseous form. For this operation, natural uranium is converted to the gaseous uranium hexafluoride (UF_6). In order to achieve a significant difference in the concentrations many diffusion stages are required and a significant amount of power to pump the hexafluoride through the system of membranes. Alternatively, one may use centrifuging for the separation of the two isotopes: The heavier atoms of ${}_{92}U^{238}$ in the gaseous hexafluoride will concentrate towards the outer walls of the centrifuge and the lighter ${}_{92}U^{235}$ atoms concentrate towards the inner part of the centrifuge. Because there is only a very small mass difference between the two isotopes, a large number of centrifuging stages is also required, which consumes large amounts of electric power. Because of this, enriched uranium is very expensive to produce compared to the natural uranium. This is a significant disadvantage for commercial nuclear reactors that require enriched uranium as their fuel.

The condition $n_{av}{*}\ P_f \geq 1$ is only a necessary and not a sufficient condition for the establishment of the chain reaction. A reactor contains several components other than the fuel itself (e.g. structural materials, fuel cladding, sensors, safety devices, etc.). The consideration of the fission and capture cross sections of these non fissile materials would lower the probability of fission of the fuel. The significance of these simple calculations is that a sustained chain reaction is possible to be established in natural uranium if the produced neutrons are slowed down and become thermal neutrons. A glance at Fig. 4.5b proves that this is not an easy task, because the neutrons, while slowing down, must pass through the capture resonance range of $_{92}U^{238}$. This presents another challenge for the design of the nuclear reactors: *to slow down the neutrons produced from fissions in order to yield as many thermal neutrons as possible, while avoiding the capture of neutrons by* $_{92}U^{238}$ *during the deceleration process.*

Contrary to popular thinking, a conventional nuclear reactor that operates with natural or slightly enriched uranium can not explode as an atomic bomb. The reactor simply does not produce the large number of thermal neutrons that are necessary to cause a nuclear explosion. Instead, the design of the reactor must be optimized to "thermalize" neutrons, while avoiding their capture by $_{92}U^{238}$ and to ensure that a high percentage of the produced thermal neutrons will cause the fission of the $_{92}U^{235}$ nuclei.

4.2.5 The Moderation Process and Common Moderators

The deceleration of fast neutrons in common reactors is accomplished by the *moderator.* Neutrons collide with the atoms of the moderator and impart a fraction of their kinetic energy to the atoms of the moderator. Hence, the deceleration of fast neutrons to thermal neutrons is accomplished by a series of collisions. From classical collision theory, we know that the reduction of the kinetic energy of a small particle is highest when the particle collides with another particle of the same mass. Collisions with much heavier particles are almost like collisions with hard "walls" and the kinetic energy of the small particle remains almost unchanged. Therefore, effective moderators are composed of very light nuclei. Typical moderator materials consist of light nuclei with very low capture cross section and significantly higher inelastic scattering cross section. Interactions of neutrons with the nuclei of these elements would predominantly be inelastic scattering collisions. These would slow down the produced neutrons, while keeping the neutron numbers almost constant because of the very low frequency of capture events. Such elements are hydrogen, both the $_1H^1$ and the $_1H^2$ isotopes, lithium, beryllium and carbon. Boron is not a suitable moderator because its capture cross section is significantly high. Actually, boron is used in the control system of the reactor as an element that captures neutrons.

The scattering and capture cross sections are important parameters of the moderators. Another important parameter that describes the effectiveness of a moderator is the logarithmic energy decrement, ξ, or the average decrease of the logarithm of the neutron kinetic energy per collision. A high value of ξ implies a low number of collisions/interactions needed to convert fast neutrons to thermal neutrons. A low number of collisions in the thermalization process is a desirable design attribute for moderators used in commercial reactors. Table 4.5 (data from

Table 4.5 Properties of common moderators

	H_2O	D_2O[a]	C (graphite)	Beryllium-9
σ_c, barns	0.66	0.001	0.0045	0.0095
σ_i, barns	50	10.6	4.7	6.1
ξ	0.92	0.509	0.158	0.209
Number of collisions to thermalize fission neutrons	20	36	115	87
$\xi\sigma_i$ (moderating power)	46	5.4	0.74	1.27
$\xi\sigma_i/\sigma_a$ (moderating ratio)	1390	5400	165	134

[a] D_2O is the commonly used symbol for heavy water. The molecule of heavy water is composed of one oxygen atom and at least one atom of $_1H^2$

[1]) lists some of the important parameters of several common materials used as moderators.

The diagram of Fig. 4.8 shows the events that may affect the number of one generation of neutrons in a reactor from the neutron production stage to the stage when the new generation of neutrons is produced. This may be called *the neutron cycle* in the reactor. The neutron cycle demonstrates that there are several complex processes that take place in the nuclear reactors, which result in the reduction of the number of neutrons that become available for fission. The choice of the fuel, the materials and the overall design of nuclear reactors strive to achieve the minimization of the number of neutrons that are leaked outside or are absorbed in the reactor.

4.2.6 Fission Products and Energy Released in Chain Reactions

The nuclear reactions in Eqs. (4.19a) and (4.19b) are typical of the fission of $_{92}U^{235}$, but they are not unique by any means. The fission products of $_{92}U^{235}$ cover a wide spectrum of isotopes with mass numbers from 70 to 165. Figure 4.9 depicts the percentage yield of the fission products, with the exception of neutrons and other very light particles, vs. the mass number of the isotopes produced. It is apparent in this figure that there are two almost symmetric peaks with mass numbers approximately 96 and 135, where the yield is maximum and approximately equal to 6.5%. The yield in this figure may also be interpreted as the probability that an isotope with a certain mass number will be formed from the fission of $_{92}U^{235}$. It is also observed that the yield in the middle of the spectrum, where the two nuclei have almost equal mass numbers, 117, is approximately 0.01%. This implies that a nuclear reaction that would produce two nuclei with equal masses has probability of occurrence equal to 0.0001 or 1 in 10,000. Given that approximately 10^{20} fission reactions occur every second in a typical nuclear reactor, Fig. 4.9 shows that there are very large numbers of nuclei of all mass numbers from 70 to 165 that are produced every second. The nuclei, which in their vast majority are radioactive, remain trapped in the reactor until the next refueling

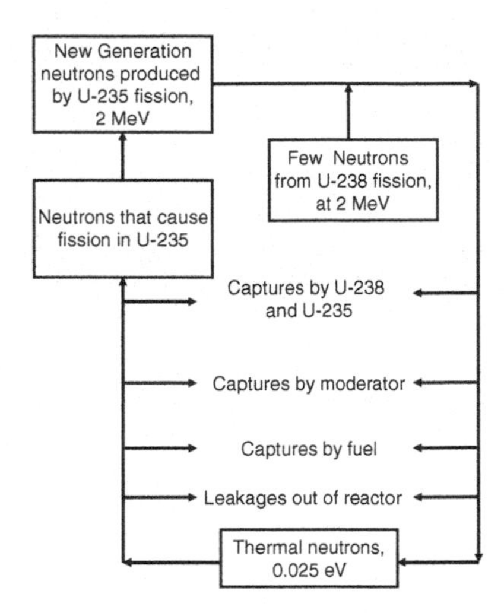

Fig. 4.8 The neutron cycle in a nuclear reactor

process, when the reaction products are removed for reprocessing. Other fissile materials, such as ${}_{94}Pu^{239}$ and ${}_{92}U^{233}$ have similar yield functions with two distinct peaks.

The amount of energy released in every fission reaction of ${}_{92}U^{235}$ depends on the products of the reaction and may be derived from the mass defect of the specific reaction. Given the yield of the product nuclei, as shown in Fig. 4.9, the average energy of the fission of ${}_{92}U^{235}$ may be calculated by averaging over all the probable combinations of the products. Using this method the average energy of the fission of ${}_{92}U^{235}$ is calculated to be approximately 200 MeV, a high percentage of which is manifested as kinetic energy of the fission fragments and is almost immediately dissipated by atomic/molecular collisions into heat. Additional energy is produced from the radioactive decay of the reaction products, also manifested as kinetic energy of the daughter isotopes. Table 4.6 [1] shows the average recoverable energy from the products of the ${}_{92}U^{235}$ fission.

Even though the recoverable energy of the nuclear products is in the form of the kinetic energy of the fragments, this kinetic energy at the molecular level is manifested as thermal energy (heat) at the macroscopic level. The conversion of this energy to electric power is accomplished in a thermodynamic cycle, usually a Rankine or a Brayton cycle and is subjected to the Carnot limitations of thermal energy conversion, which are described in Chap. 3.

4.3 Conversion and Breeding Reactions

The vast majority of nuclear reactors that are currently used are thermal reactors, which primarily use thermal neutrons and the isotope ${}_{92}U^{235}$ as their fuel. This isotope is supplied to the reactors in the form of natural or slightly enriched

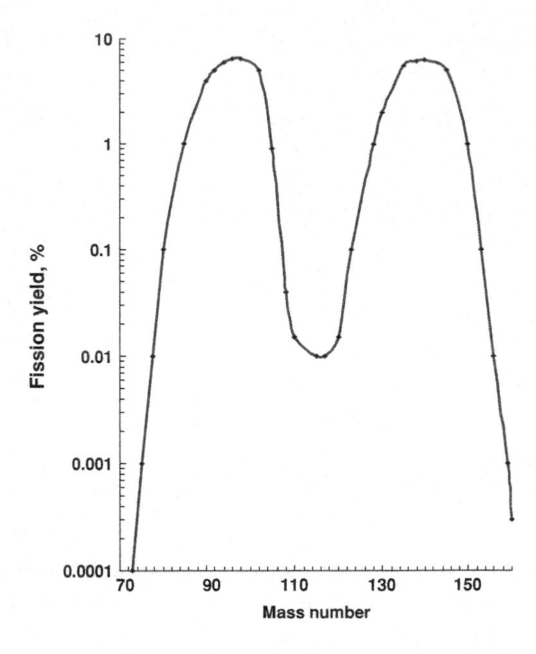

Fig. 4.9 The percentage yield of the fission of uranium-235

Table 4.6 The average recoverable energy of the ${}_{92}U^{235}$ fission

	Recoverable energy, MeV
Large fission nuclei	168
Fission neutrons	5
Gamma radiation	7
Fission product decay:	
Beta radiation	8
Gamma radiation	7
K-Capture radiation	5
Total energy released	200

uranium. It must be recalled that natural uranium only contains 0.715% of ${}_{92}U^{235}$. Only 1 out of 140 nuclei in natural uranium ore is a ${}_{92}U^{235}$ nucleus that may be undergo fission to produce thermal energy in a reactor. The vast majority of the remaining 139 nuclei of the isotope ${}_{92}U^{238}$ are not used in the fission process. The immutability of these nuclei does not only cause the underutilization of the nuclear energy resources, but also contributes significantly to the piling of the nuclear waste products, which are radioactive and must be stored for thousands of years. With the current mix of nuclear reactors, less than 1% of the nuclear resource is being used. The remaining, which is primarily composed of the isotope ${}_{92}U^{238}$ is stored as *depleted uranium* and contributes to the accumulation of the nuclear waste. Another naturally occurring isotope ${}_{90}Th^{232}$ is not utilized at all in thermal reactors for the same reasons. For the better utilization of the natural nuclear resources, scientists and engineers must devise ways to augment the percentage of natural uranium and thorium that are used in the reactors and to design future reactors that may use a larger proportion of the now wasted $_{92}U^{238}$ and $_{90}Th^{232}$ nuclei.

As it was seen in Sect. 4.2.4, the $_{92}U^{238}$ nuclei may undergo fission with fast neutrons. However, the fission cross section of this isotope is not significant enough to sustain the chain reaction in a reactor. A few $_{92}U^{238}$ nuclei actually undergo fission by the fast neutrons produced in the reactor, but inelastic scattering is the dominant process at these neutron energies. A large fraction of the fast neutrons produced would be slowed down before they cause the fission of $_{92}U^{238}$, and therefore, the fission of the $_{92}U^{238}$ nuclei is not a viable option in conventional reactors.

It is apparent from the study of the neutron cycle in a reactor that a significant fraction of the neutrons in a thermal reactor is captured by the nuclei of $_{92}U^{238}$ and form a short-lived isotope, $_{92}U^{239}$, which decays with a half-life of 23 minutes to neptunium-239 ($_{93}Np^{239}$) by electron emission. The latter is also radioactive and decays with a half-life of 55.2 hours to plutonium-239. Therefore, the capture of neutrons by the $_{92}U^{238}$ nuclei results in the production of $_{94}Pu^{239}$, within a relatively short time. Plutonium-239 is a fissile material, similar to $_{92}U^{235}$. The isotope $_{94}Pu^{239}$ has the additional advantage that its fission cross section is significant enough in the range of fast neutrons to allow for the establishment of a chain reaction at the fast neutron range. Similarly, thorium-232, which may be mined in several regions on the planet, when bombarded by neutrons may be finally converted to the fissile isotope $_{92}U^{233}$ according to the following reactions:

$$\begin{aligned} {}_{90}Th^{232} + {}_{0}n^{1} \rightarrow {}_{92}Th^{233} \rightarrow {}_{91}Pa^{233} + {}_{-1}e^{0} \quad (T_{1/2} = 22\ \text{min}) \\ {}_{91}Pa^{233} \rightarrow {}_{92}U^{233} + {}_{-1}e^{0} \quad (T_{1/2} = 27.4\, days). \end{aligned} \tag{4.29}$$

Both isotopes, $_{94}Pu^{239}$ and $_{92}U^{233}$, have significantly long half lives (25,000 years and 160,000 years) to be considered stable isotopes within the timescale of operation of a nuclear reactor. The two naturally occurring isotopes $_{92}U^{238}$ and $_{90}Th^{232}$ are not fissile, but may be converted in suitably designed reactors to produce the fissile isotopes $_{94}Pu^{239}$ and $_{92}U^{233}$, which in turn may be used as nuclear fuel. For this reason, they are called *fertile* isotopes. The process of the production of fissile materials from fertile materials is called *conversion* or *breeding* and the reactors, which enable this process, *breeder reactors.* Despite several technical difficulties with the current generation of the breeder reactors, many scientists have concluded that the wider use of the breeding reactors in the future will enable the utilization of the vast majority of the available nuclear resources for electricity production and will significantly reduce the nuclear waste produced. In addition, since breeder reactors will convert the $_{92}U^{238}$ isotope to the fissile $_{94}Pu^{239}$ nuclei, these breeder reactors will use the depleted and stored uranium and will contribute to the minimization of nuclear waste.

4.4 Useful Calculations and Numbers for Electric Power Generation

At the center of all the calculations on nuclear power is the fundamental equation of mass to energy equivalence $E = mc^2$. Based on this equation it was seen that the conversion of 1 u of mass to energy produces 931 MeV of thermal energy. Also, it was seen that the fission of fissile nuclei, such as those of ${}_{92}U^{235}$, results in the mass defect of the reaction products, which on average is equivalent to 200 MeV of thermal energy. Based on these numbers, we may calculate several useful parameters for the operation of the nuclear reactors.

At first, a typical large-scale nuclear reactor produces approximately 1,000 MW of electric power. Typical overall efficiencies of the nuclear power plants that utilize such reactors are 32–35%, which implies that the reactors produce approximately 3,000 MW of heat (thermal energy). In order to distinguish the thermal from the electric power output of the reactor the units of these two numbers are often denoted as MW-e and MW-t. Hence the power produced by the power plant would be written as 1,000 MW-e and the heat produced by the reactor as 3,000 MW-t. The latter is referred to as the *rating* of the reactor.

As it was observed in Sect. 4.2.6, at this thermal power the number of reactions per second is:

$3000*10^6/(200 * 1.6*10^{-13}) = 9.375*10^{19} \approx 10^{20}$.

While this appears to be a very large number of reactions, when compared to the Avocadro number ($6.023*10^{23}$ atoms/gram-atom) it represents a rather small number of atoms.

At this rate, the complete consumption of one gram-atom or 235 g of uranium-235, will take $6.023*10^{23}/9.375*10^{19} = 6{,}424$ seconds, or 1.78 hours. During this time the reactor produces 5,340 MWh of thermal energy and 1,780 MWh of electric energy. One single gram of pure ${}_{92}U^{235}$ produces 22.7 MWh of electric energy or approximately 1 MW for an entire day. This is in contrast to fossil fuel power plants, which require more than 3 metric tons of coal to produce this amount of electric energy.

Accounting for the fact that only 0.715% of natural uranium is the fissile isotope ${}_{92}U^{235}$, the reactor that produces 3,000 MW-t power would consume 238 grams of natural uranium every 45.9 seconds. During one day the reactor would consume the equivalent of:

$24*60*60*238/45.9 = 447.7$ kg/day of natural uranium.

The yearly consumption of nuclear fuel would be 447.7 * 365 kg or 163.4 metric tons per year. A typical nuclear reactor is fueled (charged) every 18-24 months and operates continuously under normal circumstances. Let us assume that the reactor is initially charged with 1,000 metric tons, or with 52.4 m^3, of natural uranium. At continuous operation, this reactor would consume 16.3% of its fuel in a whole year. The actual amount of mass deficit, Δm, which is converted to energy, is by far smaller than the amount of fuel consumed. During a whole year, the mass deficit for this reactor would be, from $E = mc^2$:

$3000*10^6*365*24*60*60/10^{16} = 9.46$ kg.

This amount of mass deficit is negligible compared to the original weight of 1,000 metric tons the reactor has been charged with. For this reason it is impossible to be measured by weighting the reactor at the beginning and at the end of the year with current measuring techniques.

A quantity of interest with the fuels is the *burnup* of the nuclear fuel, which is defined as the total energy released per metric ton of fuel. Since during one day the reactor under consideration would produce 3,000 * 24 MWh-t of thermal energy and would consume 0.4477 metric tons, the burnup rate of the reactor would be 160,821 MWh per ton of natural uranium. Enriched uranium has a higher percentage of fissile nuclei and its burnup rate is higher.

It must be noted that all the calculations in this section are based on the assumption that only ${}_{92}U^{235}$ undergoes fission in the reactor. In an actual reactor some of the ${}_{92}U^{238}$ undergoes fission by fast neutrons and, also some ${}_{94}Pu^{239}$ is formed from the ${}_{92}U^{238}$ conversion. The produced plutonium is fissile and produces an additional amount of energy. As a consequence, the amounts of the original fuel used during a day or a year would be slightly lower than that of the above computations and will depend on the small amounts of the ${}_{92}U^{238}$ that undergoes fission and the amount of ${}_{94}Pu^{239}$ formed.

Problems

1. Complete the following reactions (identify the elements X,Y,Z and W):

$$
\begin{aligned}
&{}_{92}U^{238} + {}_0n^1 = X +_{-1} e^0 = Y + 2_{-1}e^0 \\
&{}_{92}U^{235} + {}_0n^1 =_{38} Sr^{90} + Z + 3_on^1 \\
&{}_3Li^7 + {}_0n^1 = 2_2He^4 + W
\end{aligned}
$$

2. Use the periodic table of elements to identify the elements with the following atomic numbers: 8, 13, 21, 34, 55 and 89.
3. What is the energy equivalent of 1 lb (0.453 kg) of mass? What is the mass equivalent of the electric energy produced by a 1,000 MW nuclear power plant for an entire year?
4. If four nuclei of ${}_1H^1$ were to fuse to produces a nucleus of ${}_2He^4$, what is the mass defect and the corresponding energy released? What is the energy in kJ that would be released from the fusion of 1 kg of hydrogen?
5. What is the mass defect of the nuclei of ${}_3Li^6$ and ${}_{92}U^{238}$? What is the binding energy per nucleon of these two nuclei and which one of the two isotopes you expect to be more stable?
6. What is the amount of energy released during the reaction:

$$
{}_3Li^7 + {}_0n^1 \rightarrow 2_2He^4 + {}_{-1}e^0
$$

If one kmol of lithium reacted (6.023 * 10^{26} atoms), what is the total energy released in kJ? Given that the amount of heat released from the combustion of carbon is $\Delta h = 32{,}770$ kJ/kg, how many metric tons of carbon are needed to provide this energy?

7. It is suggested that nuclear waste, which includes $_{92}U^{238}$, be stored under controlled conditions in mountainous caverns. If one metric ton (1,000 kg) of $_{92}U^{238}$ is stored in the year 2020, how much of the material would remain in the year 3,000 and in the year 5,000? How long would it take for the amount of uranium stored to reach 10% of its initial amount?
8. The Chernobyl accident occurred on April 26, 1986 and all the strontium-90 from the reactor was released to the environment. What is the percentage of this isotope that remained on January 1, 2012 in the environment?
9. The following secondary reactions occur in a nuclear reactor:

$$_{56}Ba^{140} \rightarrow {}_{57}La^{140} + {}_{-1}e^{0}$$

$$_{57}La^{140} \rightarrow {}_{58}Ce^{140} + {}_{-1}e^{0}$$

The half-lives of the three radionuclides Ba, La and Ce are 12.9 days, 1.7 days and 34 days respectively. If at time t = 0, 100 kg of $_{56}Ba^{140}$ are released by accident, determine the approximate amounts of the three elements after 1 day, 5 days and 25 days.

10. Determine the fission cross-sections of uranium-235 at the following neutron energies: 1 MeV, 1 keV, 10 eV, 1 eV, 100 meV, 10 meV and 1 meV.
11. What are the most probable velocity and energy of neutrons at 600 K? At 2,000 K?
12. Is it possible to have a chain reaction with natural uranium and intermediate energy neutrons? Explain your answer thoroughly and state carefully under what conditions it is possible to have a chain reaction with intermediate energy neutrons.
13. What is the probability of fission of 5% enriched uranium (5% $_{92}U^{235}$ and 95% $_{92}U^{238}$) with neutrons at 0.1 eV?
14. Given that at 0.3 MeV the pertinent cross sections for $_{92}U^{235}$ are: $\sigma_f = 1.3$ barn and $\sigma_c = 0.7$ barn, while for for $_{92}U^{238}$ are: $\sigma_f = 0$ barn $\sigma_c = 0.5$ barn, estimate the minimum percentage of $_{92}U^{235}$ in the fuel for which a chain reaction with such neutrons would be possible.
15. Use Fig. 4.9 and the periodic table of elements to identify twelve of the most probable isotopes to be produced in a fission nuclear reactor.
16. A nuclear reactor provides 3,600 MW of thermal power. The reactor operates with 2% enriched uranium. Assuming that only $_{92}U^{235}$ reacts in the reactor, what are the daily consumption of this isotope in the reactor and the burnup rate of the reactor?

References

1. Bennet DJ (1972) The elements of nuclear power. Longmans, London
2. El-Wakil MM (1984) Power plant technology. McGraw Hill, New York

Chapter 5
Nuclear Power Plants

Abstract The building of more nuclear power plants is one of the proposed solutions to the increasing production of anthropogenic CO_2 and the mitigation of the global warming threat. Had the USA constructed an additional 56 nuclear power plants in the 1990s, the country would have been in compliance with the Kyoto protocol. Nuclear power plants are typically very large (1,000 MW) and very complex power producing units. The nuclear reactor itself contains large amounts of radioactive materials, which if released in the environment, may cause large-scale environmental accidents. For this reason, a nuclear reactor must have multiple levels of safety systems and its controls must be designed to shut down the reactor within a very short time. All nuclear reactors have six basic components, which are described in detail in this chapter. Several types of nuclear reactors and nuclear power plants have been developed by different countries. The special design characteristics and the operation of these plants are summarized. Because accidents in nuclear power plants are feared the most by the public, the three accidents that received the highest notoriety, at the Three-Mile Island in the U.S.A., at Chernobyl in the former Soviet Union and at the Fukushima Dai-ichi power plant in Japan are described in detail. The causes of the accidents are examined and early actions that could have been taken by the operators are presented. Finally, it is apparent that, if the world is to rely on nuclear energy in the long-term, the more abundant uranium-238 and other fertile nuclear materials must be utilized. This makes necessary the use of breeder reactors, which may become the next generation of nuclear reactors.

5.1 Basic Components of a Thermal Nuclear Power Plant

Nuclear power plants are similar to the conventional power plants that utilize the Rankine (steam) or the Brayton (gas) cycles. In nuclear power plants, the heat is supplied to the cycle from the nuclear reactor. Simply put, in a nuclear power plant

E. E. (Stathis) Michaelides, *Alternative Energy Sources*,
Green Energy and Technology, DOI: 10.1007/978-3-642-20951-2_5,

the reactor is a substitute for the boiler or the combustion chamber. The thermal energy generated by fission in the reactor is continuously removed by a fluid, the reactor coolant, and is used in the power plant for the production of electricity. A few other components are added to nuclear power plants that facilitate the heat transfer to the working fluid in the cycle or enhance the reactor's safety. There are two basic types of reactors, *thermal reactors* and *fast reactors*, named after the types of neutrons that are predominantly used for the fission of the nuclear fuel. Of these, thermal reactors are the ones that are predominantly used and, depending on the type of fuel that cools them, are divided in *water cooled* and *gas cooled* reactors. Even though the thermal nuclear reactors are very complex systems, the fundamental operation of the two types of reactors is very similar. Both types of nuclear reactors have the following common components:

1. Reactor fuel
2. Fuel moderator
3. Reactor coolant
4. Control systems
5. Safety devices and
6. Radiation shield.

The following sections are short expositions on the types and characteristics of these essential elements of a thermal reactor.

5.1.1 The Reactor Fuel

Uranium and thorium are the only naturally occurring minerals that may be used as fuel in nuclear reactors. Natural uranium is composed of ${}_{92}U^{235}$ and ${}_{92}U^{238}$ with traces of the radioactive isotope ${}_{92}U^{234}$. Thorium has one naturally occurring isotope, ${}_{90}Th^{232}$. Of these, ${}_{92}U^{235}$ is the only fissile material and the main fuel of the current generation of thermal nuclear reactors. The other isotopes, ${}_{92}U^{238}$ and ${}_{90}Th^{232}$, are fertile isotopes. With neutron bombardment these isotopes are converted to the fissile isotopes ${}_{94}Pu^{239}$ and ${}_{92}U^{233}$ respectively. Because of the high number of fission reactions and the significant amount of heat released in the reactor, the fuel in the reactor reaches very high temperatures. Therefore, desirable materials used for nuclear reactor fuel must have high melting points. The chemical oxides, UO_2, ThO_2 and PuO_2 have melting points above 2,000°C and are the most commonly used reactor fuel materials. Table 5.1 lists the density and several useful nuclear properties of these commonly used reactor fuels. The data were obtained from Bennett [1]. The cross sections are in barns and refer to thermal neutrons (0.025 eV energy). The last two columns contain the average number of neutrons released per fission, n_{av}, and the average number of fission neutrons produced per neutron absorbed in the fuel, n_{av}^{f}. It must be recalled that all neutrons absorbed in the fuel do not necessarily cause fission.

Table 5.1 Useful properties of commonly used nuclear fuels

	Density, kg/m^3	σ_c	σ_f	σ_i	n_{av}	n_{av}^f
Nat. Uranium	18,900	7.6	4.2	8.4	–	1.32
Uranium-235	18,700	101	577	10	2.44	2.08
Uranium-238	18,900	2.75	0	8.3	–	–
Plutonium-239	19,600	274	741	9.6	2.90	2.12

The reactor fuel is processed to form small cylindrical pellets with approximate diameter 1 cm and height 1.5 cm.[1] The fuel pellets are placed in the *fuel elements*, which are long and thin tubes with internal diameter slightly greater than 1 cm, as shown in Fig. 5.1. A small gap of air, which has very low heat conductivity, is formed between the sides of the fuel pellets and the cladding material. The length of the fuel rod is of the order of 5 m, which makes the fuel element a long cylinder with very high (500) aspect ratio. The material of this cylinder is called the *cladding*. Desirable properties of the cladding materials are high melting point; high strength and toughness; low capture cross-section for neutrons; immutability during severe temperature changes and cycles; and immutability to radiation. zirconium-steel alloy, or "zirkalloy," a high-strength material with melting point at 2,300°C, is often used for cladding. The choice of zirconium is primarily made because of its low cross-section for neutron capture.

Several fuel elements constitute a *fuel element bundle*, or *fuel assembly*, which is one of the elementary building blocks of the nuclear reactor. Typically, a fuel element assembly consists of a square arrangement of 12 × 12 to 16 × 16 fuel elements connected with metal spacers. The long fuel elements with an aspect ratio of 500 have limited bending strength. The square arrangements of the 144–256 tube bundles provide structural stability to the fuel assembly and reduce flow-induced vibrations during the convective heat transfer process. A cross-section of a 12 × 12 fuel bundle is shown in Fig. 5.2. The heat internally generated by the fuel increases significantly the temperature of the fuel and is radially conducted to the sides of the fuel element. The heat is conducted through the thin air gap and the highly conducting cladding material and is subsequently transferred to the coolant by convection. A typical temperature profile in a fuel rod, which drives this heat transfer process, is depicted in Fig. 5.3.

Fuel is not supplied to the reactor continuously. Reactors are re-charged with new fuel assemblies every 18–24 months. During the charging of the reactor, which lasts approximately two weeks, the entire power plant stops producing power, the reactor vessel is opened, and the depleted fuel assemblies are removed and replaced with new fuel assemblies. Partially depleted fuel assemblies are rearranged in the reactor. At the end of the recharging process, the reactor is shut

[1] This small pellet will produce heat that is equivalent to burning 564 L (149 U.S. gallons) of oil, or 809 kg (1,780 lb) of carbon, or 481 m^3 (17,840 ft^3) of natural gas.

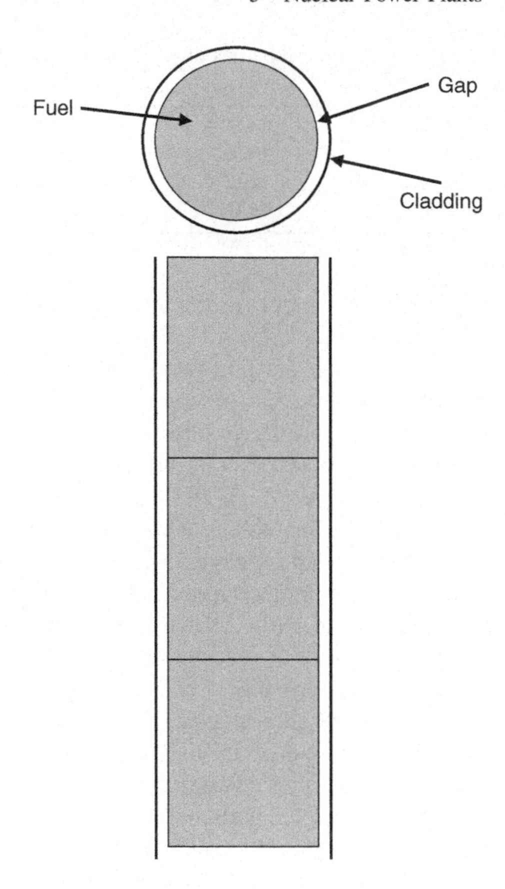

Fig. 5.1 Fuel pellet arrangement in a fuel rod

hermetically. All the fuel and reaction products remain in the reactor vessel until the next recharging period.

5.1.2 The Fuel Moderator

Slowing down the fast neutrons that are produced from fission is necessary for the maintenance of the chain reaction. The moderator accomplishes this and converts fast neutrons to thermal neutrons, which are used in common thermal reactors. Section 4.2.5 and Table 4.2 explain the properties of several of the materials that are used as moderators. Desirable characteristics of a moderator material are: (a) low cost; (b) extremely low neutron absorption cross-section; (c) high inelastic collision cross-section. Common water has excellent moderating properties and is very inexpensive. However, because the capture cross section of water is relatively high, water moderated reactors typically require the use of slightly enriched uranium as their fuel, which adds to the fuel cost. Heavy water (D_2O) is an excellent

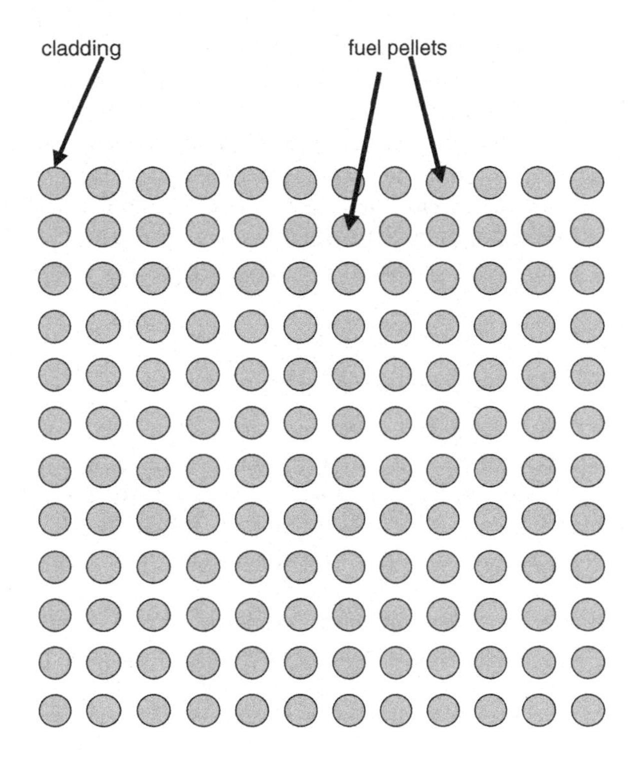

Fig. 5.2 Fuel pellet arrangement in a fuel rod

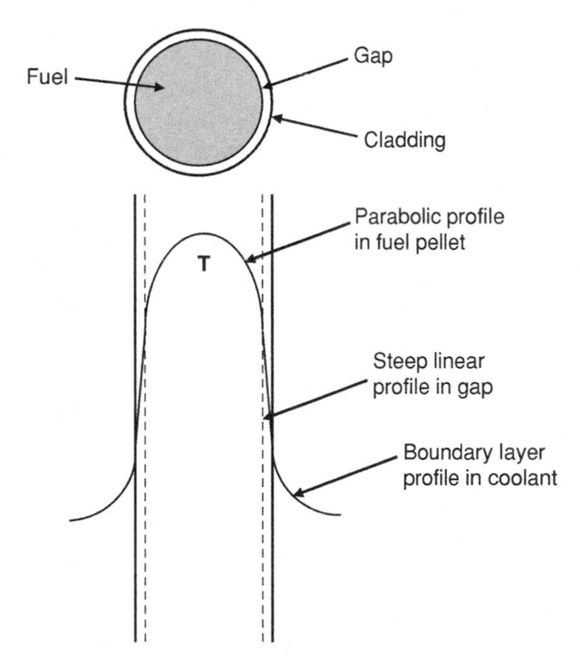

Fig. 5.3 Temperature profile in a fuel rod

moderator, but it is expensive to produce in large quantities. Reactors and power plants that operate with D_2O as moderator and coolant are specially designed to avoid the leakage of the expensive D_2O. Graphite does not have as good moderating properties as the other two, but it is inexpensive, it does not require the use of enriched uranium and may double as structural material that supports the fuel rods of the reactor. Several types of gas-cooled reactors use graphite as a moderator and as part of the reactor structure. Finally, beryllium is both expensive and toxic. For this reason, it is rarely used as moderator in reactors.

5.1.3 The Reactor Coolant

The reactor coolant removes the enormous quantities of thermal power generated inside the reactor. The function of the coolant is the most essential function in the reactor. It must be recalled that once a chain reaction commences, it continues for a long time. It is possible to slow down the chain reaction using the reactor controls. However, secondary reactions occurring from the radioactive decay of fission isotopes continue to produce thermal energy at a high rate. Thus, unlike conventional power plants, where the supply of the fuel to the boiler may be interrupted and the production of thermal energy stops instantaneously, the thermal energy production in the nuclear power plants may not be completely stopped. In order to avoid reactor melting, it is essential that the coolant continuously circulates in the reactor, even after the interruption of the chain reaction. Nuclear reactors need continuous cooling.

Desirable characteristics of reactor coolants are: a) high specific heat capacity; b) high thermal conductivity; c) low cross section for neutron absorption; d) low cost. Common water (H_2O) exhibits all of these characteristics and is very commonly used as a reactor coolant as well as the moderator. Heavy water (D_2O) has better moderating characteristics, but its production cost is high. Carbon dioxide (CO_2), helium (He) and argon (Ar) are the most commonly used gases in Gas Cooled Reactors. Mercury (Hg), potassium (K), sodium (Na) or a mixture of the last two are the most commonly used coolants in the Fast Breeder Reactors, where the rate of heat produced per unit volume of the reactor is very high.

5.1.4 The Control Systems

A control system is necessary in all nuclear reactors to perform the following essential functions:

1. Change the power level of the reactor in order for the plant to produce more or less electric power and to respond to the electricity demand.

2. Absorb or induce excess reactivity and compensate for short- and long-term reactivity changes due to: temperature effects; fission product absorption of neutrons, which is often called "reactor poisoning;" and fuel depletion.
3. Shut down the reactor in a case of emergency.

Desirable characteristics of control systems are: a) reliability; b) fast deployment in a case of emergency; c) avoidance of wasteful absorption of neutrons; d) low cost; and e) maintenance of a uniform flux in the reactor, to the extent possible.

Control of the thermal power in nuclear reactors is achieved by controlling the flux of the neutrons. An increase of the neutron flux also increases the reactivity and the thermal power produced, while a decrease of the neutron flux decreases its thermal power. It is apparent that this method of control would not stop completely all the reactions, because there would always be natural decay of the fission products in the reactor and, even, a small amount of fission reactions from decay-produced neutrons. The elements boron, indium, cadmium and hafnium have the ability to absorb neutrons and may be used in control systems. Of these, the most commonly used is boron, which absorbs neutrons to produce lithium according to the reaction:

$$ {}_5B^{10} + {}_0n^1 \rightarrow {}_3Li^7 + {}_2He^4. \tag{5.1}$$

A frequently used method of reactor control is to insert into the reactor long rods, the so called *control rods*, made of boron carbide, BC. The control rods may be completely or partly inserted in the reactor. The level of the partially lifted control rods determines the reactivity in the reactor and controls the thermal power produced. Lifting the rods decreases neutron absorption, increases the neutron flux and also increases the thermal power of the reactor. Lowering the rods, thus decreasing their volume in the reactor, has the opposite effect to reduce the thermal power. The system of control rods is suitable for emergency shutdown of the reactor since all the rods may be quickly lowered into the reactor to drastically reduce reactivity. In several reactor types the control rods are held in their positions by electromagnetic coils. During an emergency, the current in the electromagnets is interrupted and the controls drop instantaneously in the reactor. This is frequently called the "SCRAM" system.

One of the drawbacks with the use of control rods is that they intercept neutrons mostly in the upper part of the reactor. Hence, the neutron flux is not uniform in the reactor. This causes an uneven consumption of the nuclear fuel and uneven fuel rod depletion. One way to avoid this with water cooled reactors is to use boric acid (H_3BO_3), which forms $(H_4BO_4)^-$ ions and dissolves in the coolant water. This is called *chemical shim control*. The solution of boron in the coolant water ensures that the concentration of the neutron absorbing material is uniform in the reactor. When a decrease of the reactivity is desired, the boron concentration in the reactor is increased by feeding a concentrated solution of boric acid. Conversely, the concentration of boron is decreased and the reactivity is increased by feeding pure water into the coolant and by removing some of the original boric acid solution. The acid is regenerated by water evaporation. This method of the boric acid concentration adjustment as the nuclear reactor control system is called the *bleed and feed* method.

The chemical shim method of reactivity control is a slow process and is used to slowly vary the reactivity effects, such as effects of increasing the reactor temperature; the buildup of xenon-135 and samarium-149, which are short-lived isotopes that absorb neutrons; and for the reactivity decrease resulting from the depletion of the fuel. Control rods are used to handle rapid changes in reactivity and to shut down the reactor during a perceived emergency. Given that the routine operation of reactors involves slow transient processes, several large, water-cooled types of reactors have all the control rods completely withdrawn during their normal operations and use only chemical shim for reactivity control.

5.1.5 The Shield

Throughout its operation, the nuclear reactor contains a medley of radioactive isotopes. Besides the radioactive fuel and neutrons, there is a continuous buildup of fission reaction products, neutron capture products, products of radioactive decay, etc. All these materials constitute potent sources of α-, β-, and γ-radiation, as well as of neutron radiation, which cause severe damage to all living organisms. The principal function of the reactor shield, which is sometimes referred to as the *biological shield*, is to prevent the escape of significant dosages of radiation from the reactor to the environment. A secondary function of the shield is to thermally insulate the reactor, so that the thermal energy generated does not escape.

The containment of α- and β- particles is relatively easy. These are charged particles, they are stopped by a thin layer of matter and do not penetrate thick barriers. On the other hand, γ-radiation, which is purely electromagnetic radiation and the neutrons are not easily stopped and require thicker barriers. Neutrons need light nuclei to be slowed down and be stopped, while γ-radiation is stopped by materials of high density, such as steel and lead. Concrete is relatively inexpensive, contains hydrogen atoms, which slow down the neutrons, and its density is high enough to be a barrier to γ-rays. Concrete satisfies the biological shield requirements and thick layers of concrete are typically used as the biological shields. One particular type, the *Barytes concrete* contains barium, a heavy element, which is particularly effective in stopping γ-radiation. Thick walls of Barytes concrete form the shield of most land-based reactors. The biological shields of marine reactors, which are used for the propulsion of nuclear ships and submarines, consist of alternate layers of water for slowing the neutrons and steel for the attenuation of γ-radiation.

In order to prevent damage and extend the life of the biological shield from the flux of energy, which accompanies all forms of radiation, a thin thermal shield is used. The thermal shield is usually a relatively thin sheet of steel, which has high thermal conductivity and the ability to withstand the induced thermal stresses. The thermal shield bears the entire flux of the radiation, stops the vast majority of α- and β-particles, and attenuates significantly the γ-radiation and the neutron flux. A high fraction of the radiation energy from the reactor is dissipated in the thermal shield, which is continuously cooled by the reactor coolant.

5.2 Nuclear Reactor Types and Power Plants

The reactor is the heat source of the nuclear power plant: Heat generated from fissions and secondary reactions is removed by the coolant, which is either used by itself in a turbine or transfers the heat to a secondary fluid that powers the turbine. Usually the secondary fluid is water/steam, which operates in a Rankine cycle, but a few nuclear power plants use a gas (helium or carbon dioxide) in a Brayton cycle. As the nuclear programs evolved in different countries, several reactor types have been developed. The power plants that use these reactors are named after the types of reactors that are being used. It is typical in nuclear industry to use several acronyms for the reactor types, a practice that is confusing to the newcomer. The most common types and acronyms of nuclear reactors are listed here:

- PWR: Pressurized Water Reactor
- BWR: Boiling Water Reactor
- GCR: Gas Cooled Reactor
- AGR: Advanced Gas Reactor
- HTGR: High Temperature Gas Reactor
- CANDU: Canadian Heavy Water Reactor
- HWR: Heavy Water Reactor
- LWR: Light Water Reactor
- RBMK (acronym in Russian): Channel, Graphite-moderated Boiling-water Reactor
- IGG (acronym in Russian): Pressurized Water Reactor (former USSR)
- FBR: Fast Breeder Reactor
- LMFBR: Liquid Metal Fast Breeder Reactor

Typical dimensions and other characteristics of several types of reactors are presented in Table 5.2 [2]. It is apparent that, among the thermal reactors, the PWR's have the highest power density among the conventional nuclear power plants and provide the most thermal power per unit mass of fuel in the reactor. Also, that the FBR's are the most compact, with power density and fuel ratings exceeding by far those of the thermal reactors.

Table 5.2 Typical dimensions and characteristics of the most common types of thermal reactors

Type	Reactor	Thermal Power (MW)	Core Diam. (m)	Core Height (m)	Core Volume (m^3)	Power Density (MW/m^3)	Fuel Rating (kW/ton)
LWR	PWR	3800	3.6	3.8	40	95	39
	BWR	3800	5	3.8	75	51	25
AGR	Hincley	1500	9.1	8.3	540	2.8	11
HWR	CANDU	3425	7.7	5.9	280	12	26
RBMK	Chernobyl	3140	11.8	7	765	4.1	15.4
FBR	Phenix	563	1.4	0.85	1.4	406	149
	Prototype	612	1.5	0.9	1.6	380	153

5.2.1 The Pressurized Water Reactor (PWR)

This is the most common type of nuclear reactor in the world. PWR's make up for 75% of all the reactors in the United States and the majority of the nuclear reactors in France and Japan. Figure 5.4 is a schematic diagram of the layout of a typical PWR horizontal cross section with the fuel assemblies/bundles. The fuel is usually slightly enriched uranium, which typically contains 2.0–3.5% of the isotope ${}_{92}U^{235}$. The fuel bundles are arranged in three regions within the reactor as shown in Fig. 5.4, with the bundles that contain a lesser amount of the ${}_{92}U^{235}$ isotope placed on the inside of the reactor, where the neutron losses are minimum and the neutron flux highest. Control rod clusters, shown by the dotted circles in the Figure, are controlled by electromagnets and are inserted in several of the fuel assemblies. In an emergency, current is cut to the electromagnets and the control rods fall by gravity to disrupt the neutron flux and shut down the reactor. Liquid water, which is the moderator as well as the coolant, enters the reactor vessel through a number (two to four) of nozzles and exits through an equal number of nozzles. A small part of the water circulates between the *reactor barrel* and the *baffle* to cool the barrel and the attached *thermal shield* of the reactor. The whole assembly is hermetically enclosed in a stainless steel vessel, the *reactor vessel*, which is approximately 6-inch (15 cm) thick.

A simplified schematic diagram of a typical PWR nuclear power plant depicting the essential equipment is shown in Fig. 5.5. Water circulates in the reactor vessel at high pressure, typically 155 bar, which is regulated by the *pressurizer*. Its temperature is low enough (280–320°C) to be significantly below the saturation temperature at 155 bar (345°C). Therefore, water does not boil in a PWR. A system of *coolant pumps* (at least four and as many as eight) circulates the cooling water through the reactor in what is called *the primary loop* or the *Nuclear Steam Supply System (NSSS)*. The heat extracted from the reactor is transferred to the *secondary loop* via several heat exchangers, which are often called *steam generators (ST)*. The latter produce the steam needed for the operation of the turbine. The other parts of the PWR power plant, which are shown in Fig. 5.5

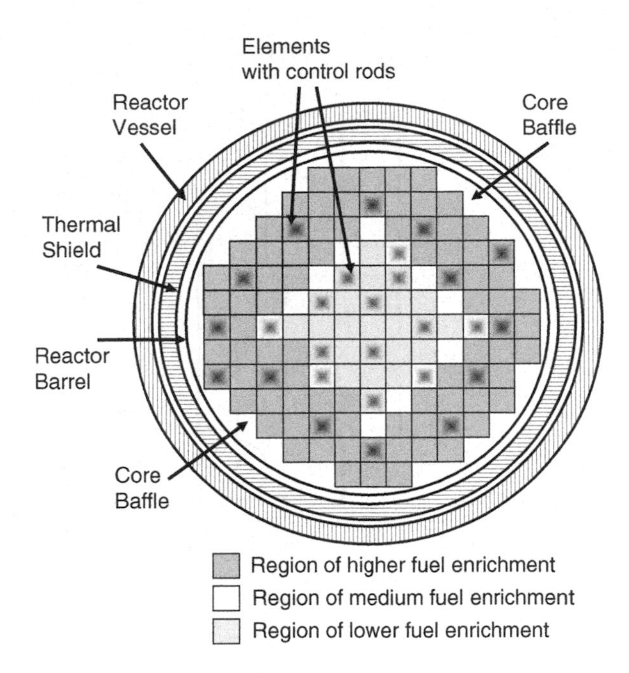

Fig. 5.4 Schematic layout of the components of a Pressurized Water Reactor (not in scale)

(condenser, C; feedwater heater, FWH; electric generator, G; turbine, T; and associated pumps, P) are the same as the similar equipment used in typical Rankine steam cycles.

Because the reactor coolant is in the liquid state and practically incompressible, small changes in its volume may cause significant changes in the pressure of the reactor. If the liquid water expands, e.g. by excess heat power production and subsequent temperature rise, the systems in the reactor and the surrounding reactor vessel would be subjected to enormous internal pressure changes that may undermine the structural integrity of the reactor and cause fracture. Also, if the pressure decreases significantly, boiling may occur in the bulk of the reactor. Boiling would cause pump cavitation and reactor malfunction.

Avoidance of large pressure variations and effective control of the reactor pressure is achieved by the *pressurizer*, an essential element of the PWR power plants. The function of the pressurizer is to accommodate volume changes and the accompanying pressure fluctuations. The pressurizer is a closed vessel, which contains approximately equal volumes of water and steam. Since water is in equilibrium with steam, the temperature in the pressurizer vessel is equal to the saturation temperature that corresponds to the nuclear reactor pressure. Several electric heaters at the bottom of the pressurizer vessel supply heat to increase the amount of steam when needed. Spray nozzles at the top of the pressurizer supply colder water from the primary cooling loop, to condense part of the steam. The lower part of the pressurizer is connected to the hot part (leg) of the primary cooling loop.

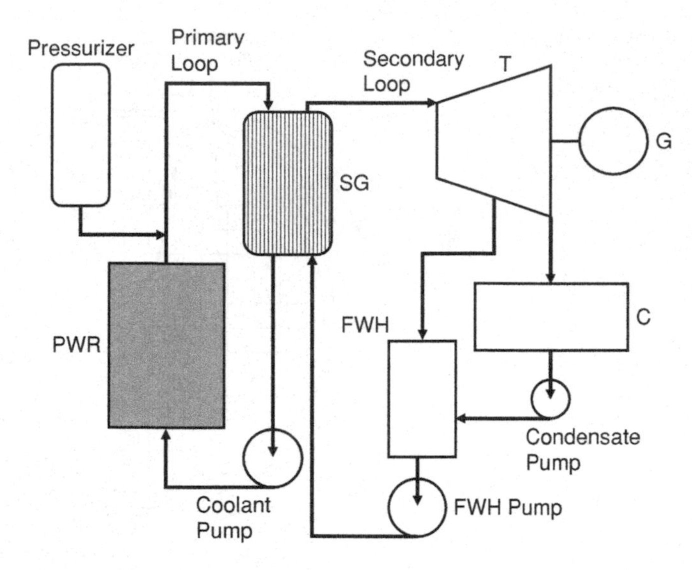

Fig. 5.5 Simplified schematic diagram of a PWR power plant

When there is an **increase of the pressure** in the reactor, the cooling water enters the pressurizer, the steam at the top of the pressurizer is compressed and, in order to maintain thermodynamic equilibrium, some of the steam condenses. In addition, the spray valves at the top of the pressurizer are activated and a colder spray causes more steam to condense. This action "frees" volume in the pressurizer and allows some of the coolant to enter and remain in the pressurizer. The removal of coolant contributes to the fast lowering of pressure in the entire coolant circuit and the reactor. If the pressure surge exceeds the normal capacity of the pressurizer, a relief valve will also open to discharge some of the steam to the *pressurizer relief tank*. The pressure in the latter is maintained by compressed nitrogen.

When there is a **decrease of the pressure** in the reactor, liquid water from the bottom of the pressurizer enters the primary coolant loop and helps increase the pressure of this loop. The water discharge causes a temporary, small reduction in the pressure of the pressurizer and this causes the water to flash and produce more steam at the top of the pressurizer. Simultaneously, the electric heating elements are activated to produce more steam. The production of the steam raises the pressure in the main part of the pressurizer and the reactor within acceptable pressure limits. The volumes of both the pressurizer vessel and of the relief tank are approximately 20–25% of the total volume of the nuclear reactor. Because of this, the pressurizer in its normal operation may accommodate a significant expansion or contraction in the nuclear reactor fluid.

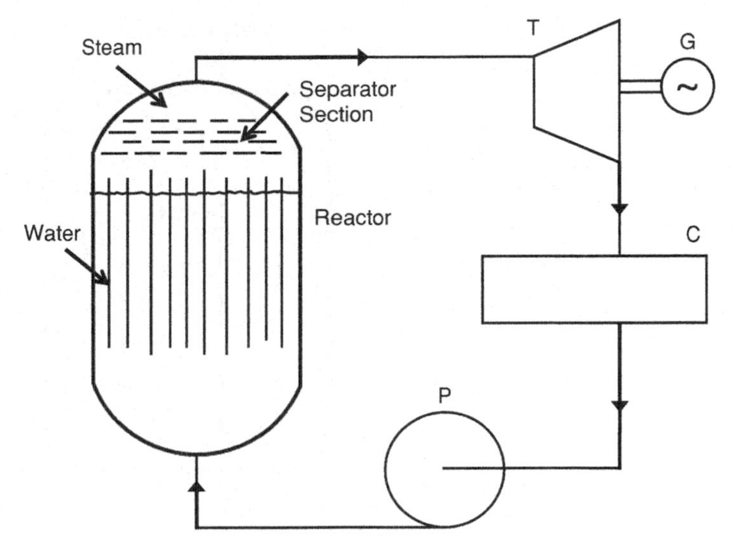

Fig. 5.6 Schematic diagram of a BWR power plant

5.2.2 Boiling Water Reactor (BWR)

Boiling of the water inside the reactor considerably simplifies the operation and lowers the cost of the nuclear power plant, because the need for steam generators is eliminated. Steam is generated in the BWR and may be directly fed to the turbine. Effectively a BWR power plant substitutes the boiler with a nuclear reactor. Because the steam generated in the reactor carries a percentage of moisture (droplets) with it, BWR plants use a *steam separation* section, to separate liquid water from the steam, which is subsequently fed to the turbine. Figure 5.6 shows the schematic diagram of a BWR power plant.

One of the shortcomings of the PWR and BWR power plant cycles is that the temperature of the steam supplied to the turbine is rather low, close to 300°C and there is typically no superheating. Comparable temperatures in fossil fuel plants, which use superheated steam, are close to 550°C. It follows from 2nd Law of Thermodynamics considerations that the thermal efficiency of these nuclear power plants is significantly lower than that of fossil fuel plants. Actually, PWR and BWR cycle efficiencies are in the range 30–34%, while those of modern fossil fuel cycles are in the range 40–45%. It must be pointed out that, in principle, superheating is not impossible in water-cooled nuclear power plants. The steam may be superheated in a conventional fossil fuel superheater (the Indian Point PWR plant in USA uses such a method) or a part of the reactor may be designed for superheating. Figure 5.7 depicts the schematic diagram of a BWR, where the interior part of the reactor becomes the superheater. The saturated steam that rises to the top of the reactor reverses direction to the superheater section at the center of the reactor and is fed to the turbine from the bottom part of the reactor.

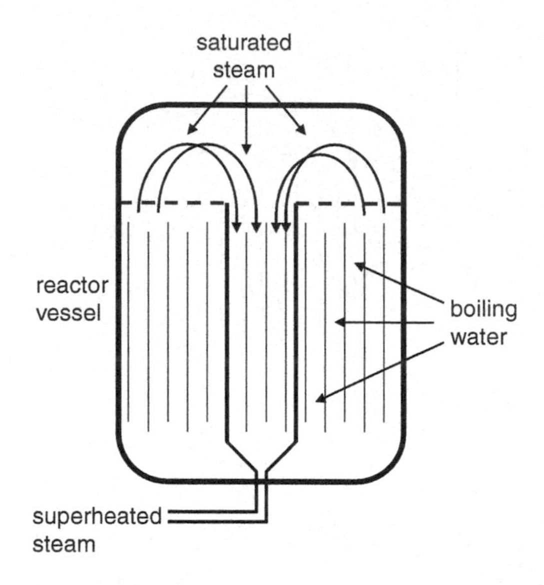

Fig. 5.7 Proposed layout of a BWR with a superheat section

5.2.3 The CANDU Reactor

Light water (H_2O) is very inexpensive and well-suited to be both a reactor moderator as well as coolant. However, its capture cross-section, 0.66 barns, is high enough for light water to absorb a significant number of neutrons. This makes it necessary for both types of LWR's (PWR's and BWR's) to use slightly enriched uranium (2–4%). Enriched uranium is produced by the gasification of natural uranium and a series of centrifugal processes, which add significantly to the cost of the nuclear fuel. On the contrary, heavy water (D_2O) has an extremely low capture cross-section, 0.001 barns, and absorbs a very small number of neutrons. Reactors that are moderated and cooled with D_2O operate with natural uranium, which is significantly cheaper than enriched uranium. However, D_2O is expensive to produce, because the isotope must be separated from common water, where it exists at a molecular ratio of 1:6,000.

The *CANDU* is a HWR that was developed by the Canadian nuclear program. It is essentially a PWR operating with D_2O as coolant and moderator. The Canadian nuclear program determined that it is less expensive overall to produce electricity from HWR reactors, because, unlike the fuel the coolant/moderator D_2O is only supplied once to the reactor and, if there are no leaks, it remains there for the life of the reactor. Thus, the CANDU reactor has higher capital cost associated with the production of D_2O, but lower fuel cost, because it uses natural uranium. As with the PWR's, the coolant is directed to a steam generator, from where the generated dry saturated steam is fed to the turbine at typical conditions of 41 bar and 250°C. Because of these low design temperatures, the CANDU power plants have lower efficiencies, which are typically close to 30%.

It must be pointed out, however, that while the lower cycle efficiencies may be a thermodynamic drawback for the LWR and HWR nuclear power plants, it is not a practical disadvantage, because the electricity production industry strives for lower costs per kWh produced rather than higher thermal efficiency: The price of nuclear fuels (per Btu's or MJ) has been historically significantly lower than that of fossil fuels. Water cooled nuclear power plants have higher capital cost than fossil power plants and, thus, require significantly higher initial investment. The lower cost of the nuclear fuel makes the nuclear plants very much competitive. Several older nuclear power plants, which have been in operation for more than 30 years, and their capital cost has been amortized, produced in the year 2010 significantly cheaper electric energy, close to $0.01/kWh, than fossil fuel plants.

5.2.4 The Gas Cooled Reactors (GCR)

Gas-cooled reactors have been built mainly in Britain, where the *Magnox* (from the Magnesium alloy fuel cladding) and the *Advanced Gas Reactor (AGR)* types were developed. Both types of these reactors are cooled by a gas (CO_2 and He are the most commonly used) and use graphite as their moderator. A typical arrangement for a GCR is to have several blocks of graphite with long bores drilled. The fuel is natural uranium for the Magnox type and slightly enriched uranium for the AGR type. The fuel, clad in canisters, and the control rods, typically made of boron carbide, are inserted in the graphite bores. A high fraction of the channels in each graphite block are reserved for the circulation of the gas coolant. The coolant is directed to a steam generator and produces steam that is fed to the turbines.

The main advantage of the GCR's is that the coolant is not limited by a saturation pressure. The temperatures that may be achieved are significantly higher than those of the water-cooled reactors. Typical coolant temperatures leaving the reactors are 650°C and this allows superheated steam to be produced at approximately 400–500°C in the AGR's. Thus, the temperature of the generated steam is significantly higher than that of the PWR, BWR and HWR plants. Hence, the thermodynamic efficiency of the GCR power plant is higher, typically in the range 33–36%. A second advantage of the GCR power plants is that they may operate with natural or slightly enriched uranium, while their graphite moderator is not as expensive as the D_2O of the CANDU reactors. Because of the lower heat conductivity of the coolant gases, higher areas and, hence, higher reactor volumes are necessary for the transfer of the heat power. Thus, the dimensions and the volume of the GCR's are significantly higher than those of comparable water-cooled reactors. Another disadvantage of the GCR's is that the circulation of the compressible coolant gas requires higher electric power for the gas blowers and compressors than that, which drives the water pumps in the LWR's and the HWR's.

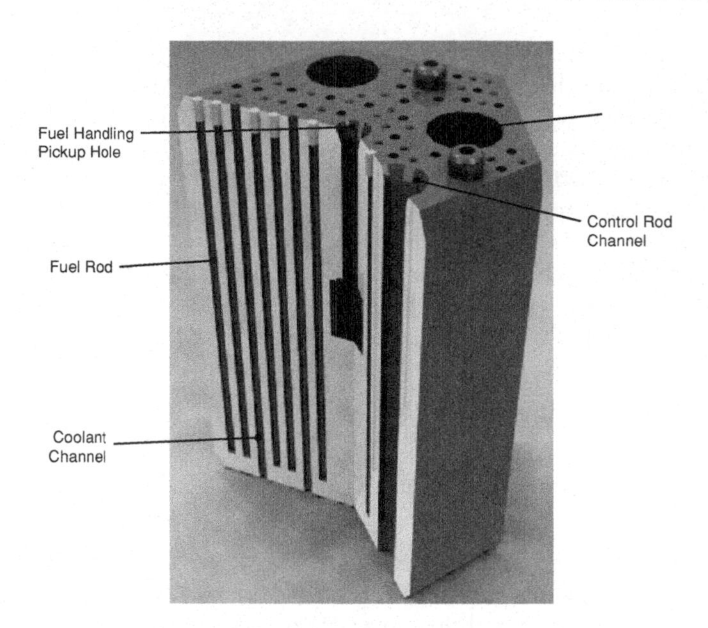

Fig. 5.8 HTGR hexagonal block

Another version of GCR, the *High Temperature Gas Cooled Reactor (HTGR)* design has been proposed for development in the United States. The proposed fuel for the HTGR is fine (0.5 mm) uranium oxide or uranium carbide spheres coated with an impervious thin layer (0.1–0.15 mm) of pyrolytic carbon, which may contain the fission products. These small particles are bonded in a graphite matrix to form a homogeneous mixture of fuel. Figure 5.8 shows the hexagonal arrangement of the graphite, fuel bores and coolant bores in a HTGR. The hexagonal geometric shape of the blocks spans the space and, hence, a large number of these blocks may be arranged geometrically without leaving any gaps between them. Several hundreds of these hexagonal blocks make up the reactor. This reactor is modular and, depending on the number of graphite block/units, the reactor will have high or lower rated thermal power.

Because of the high melting point of graphite, very high temperatures may be achieved in a HTGR. Helium, which does not absorb neutrons and has very good thermodynamic characteristics, is used as the reactor coolant and is subsequently used in a heat exchanger to produce superheated steam. Exit temperatures of helium up to 1,000°C are possible in a HTGR. This implies that the steam temperature would be equal to those of the most modern fossil fuel power plants. Therefore the thermal efficiency of a HTGR would approach that of the fossil fuel plants. Another advantage of the HTGR is that, because of the absence of metallic cladding materials that absorb neutrons, the available neutrons interact with fertile nuclei to breed nuclear fuel. With good neutron conservation and with the fertile

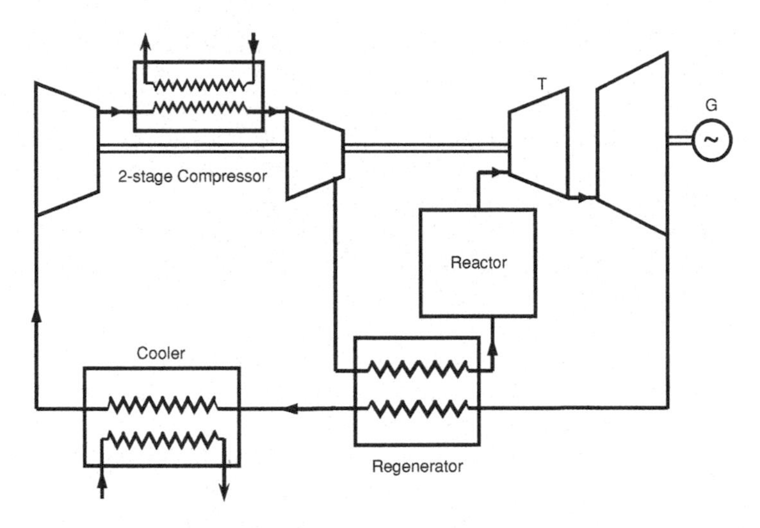

Fig. 5.9 Arrangement of the major components of a typical HTGR/HTGCR power plant

isotopes $_{92}U^{238}$ and $_{90}Th^{232}$ mixed with the fuel, the HTGR's will partly function as thermal breeder reactors.

Another variation of the design of the GCR is the High Temperature Advanced Gas Reactor (HTGCR), proposed to be built in Great Britain. These reactors will operate at significantly higher temperatures than the GCR temperatures and will have a number of regenerators and intercoolers. As with the HTGR, the absence of cladding materials makes this reactor a low-level breeder reactor. The arrangement of the essential components of a typical HTGR/HTAGR power plant is shown schematically in Fig. 5.9.

5.2.5 Other Reactors

Other, less common types of nuclear reactors have been designed. Several of these types of reactors are in operation, either as experimental reactors or as pilot plants. Among them are the following:

The *Pebble Bed Reactor (PBR)* has been designed in Germany and is essentially a GCR. The fuel in the PBR is a mixture of uranium and thorium oxides (UO_2 and ThO_2) and the moderator is graphite. Thorium is gradually converted to the fissile $_{92}U^{233}$ isotope. Fuel and moderator are fussed in 6 cm spheres and these spheres are placed at random in the reactor. Helium gas at 40 bar enters from the top of the reactor and leaves at the bottom at approximately 750°C. The helium gas raises superheated steam at 550°C in steam generators. An innovative feature of the PBR is that the spent fuel elements, which are of the same size as tennis balls, are easily removed from the bottom of the reactor, reprocessed and reinserted at the top. Thus, the reactor is continuously recharged and does not need to contain a large amount of fuel. One of the main advantages of the PBR is that small and economical size reactors may be produced. Also, because of the small size of the fuel elements and the high melting point of graphite, a loss-of-coolant accident in the PBR will not cause a reactor meltdown. Therefore, the PBR's do not pose the danger of a large-magnitude environmental disaster.

The *RBMK* (Channel, Graphite-moderated Boiling-water Reactor) was designed and operated in the former Soviet Union. This type of reactor became notorious following the Chernobyl accident, which is described in Sect. 5.3.2. Similarly, the *IGG* reactor is essentially a PWR reactor that was designed and operated in the Soviet Union. Several countries that were part of the former Soviet Union and their allies (e.g. Lithuania, Russia, Kazakstan, Poland, Bulgaria, etc.) still have RBMK and IGG reactors in operation.

Several types of the *Fast Breeder Reactor (FBR)* have been developed in several countries and one of these types, the *Phenix* in France produces commercially electric power. The breeding process and the breeder reactors are examined in more detail in Sect. 5.5.

5.3 Cooling of Nuclear Reactors

Regardless of the type of the nuclear reactor, the adequate and continuous cooling of the reactor core is the most essential safety feature of any nuclear power plant: Unlike other power plants, where the fuel supply may be interrupted at an instant, all the fuel in nuclear reactors is contained inside the core of the reactor and may not be readily removed. Even if the reactor is "shut down" by the complete insertion of the control rods, there is still a fraction of residual neutron flux to cause a number of fissions and produce heat. In addition, the radioactive decay of the nuclear reaction products continues to produce a significant amount of heat power several hours after the shut-down of the reactor. For example, a typical nuclear reactor that normally produces 3,400 MW-t thermal power would continue to produce heat at the following rates:

- 170 MW-t at 10 s after the shout-down,
- 68 MW-t 1000 s after the shut-down,
- 26 MW-t 10 h after the shut-down, and
- 4 MW-t 40 days after the shut-down.

It is apparent from the above data that, if the thermal power produced is not promptly removed from the reactor by the continuous circulation of the coolant, the reactor temperature will increase fast to the point, when melting of the reactor systems may occur. Continuous production of heat may also cause melting of the reactor vessel and the concrete foundation of the reactor building. Such a *reactor meltdown* will bring the radioactive material of the reactor into the environment with uncontrolled and catastrophic consequences for the human and animal populations. Because of this, nuclear power plants have a primary cooling circuit and one or two emergency cooling circuits that are designed to operate continuously, even when the plant does not produce any electric power. The readiness and reliability of the primary and emergency cooling systems of a reactor are paramount for the safe operation of any nuclear power plant. In all nuclear power plants a *Loss Of Coolant Accident (LOCA)*, may have catastrophic consequences for the population of a vast geographic region and represents an unacceptable risk.

5.3.1 Accidents in Nuclear Power Plants: Three-Mile Island, Chernobyl and Fukushima Dai-ichi

Since the very beginnings of nuclear energy utilization, several safety incidents and accidents occurred in the nuclear power plants that operate worldwide. The incidents at Windscale (now Selafield) Brown-Ferry, Hunterson, St. Lauent and Ginna nuclear power plants, were very serious but did not sway the public opinion against nuclear energy. On the contrary, the accidents at Three-Mile Island, Chernobyl and the most recent Fukushima Dai-ichi plant attracted worldwide attention and relentless press and public reaction. The first two accidents happened within seven years, were due to operator (human) error and involved loss of coolant. The enormous attention the accidents received exposed the general public to the dangers of nuclear power, whether real or perceived. This swayed the public opinion in Europe and North America away from nuclear power and caused an effective moratorium in the building of reactors in several industrialized countries. The following subsections describe these three accidents. Even though the description is as simple as possible, the reader may discern the complexity of the nuclear power plant design and the difficulties associated with controlling such a complex system.

5.3.2 The Accident at the Three-Mile Island

Background: Unit 2 of the Three-Mile Island power plant (TMI-2) received construction permit in 1969, was constructed in the early 1970s and was licensed to operate in February 1978. The power plant, which had three units, is located in a small island of the Susquehanna River, approximately 10 miles from Harrisburg, the capital of the State of Pennsylvania. A PWR reactor, built by Babcock and Wilcox, supplied approximately 2,700 MW-t of heat and the unit produced approximately 900 MW-e of electricity. Figure 5.10 shows a schematic diagram of the essential components of the TMI unit as they are related to the accident.

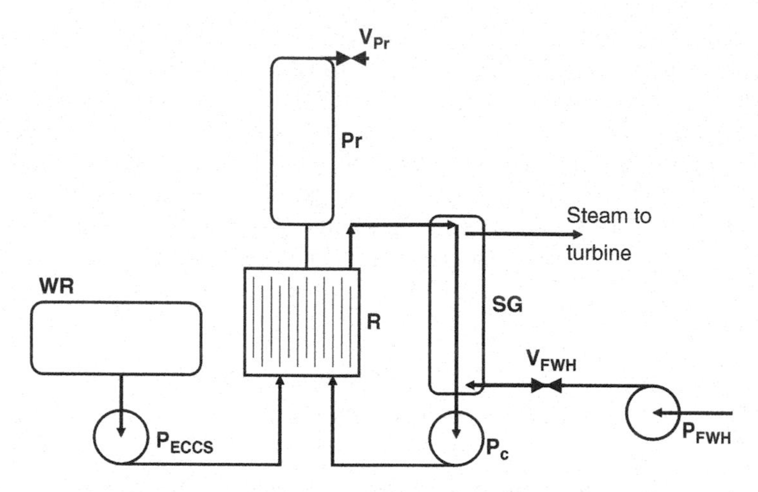

Fig. 5.10 Schematic diagram of the TMI-2 power plant. *Pr* pressurizer; *R* reactor; *SG* steam generator; *WR* water reservoir; P_C coolant pump; P_{ECCS} ECCS pump; P_{FWH} feedwater heater pump; V_{FWH} feedwater heater valve; V_{P_r} pressurizer valve

The accident: A couple of days before the accident safety tests of the emergency feedwater system were carried out. During these tests, the feedwater valves ($\mathrm{V_{FWH}}$) were closed and remained inadvertently closed after the test, thus making the emergency feedwater system inoperative. At 4:00 am on March 30, 1979 a condensate pump failed. According to the design of the unit, this led to the automatic tripping of the main steam generator feedwater pumps and the tripping (stopping) of the turbine. The reactor coolant pumps continued to operate and maintained the flow of coolant in the reactor core. Because the heat generated by the reactor was not removed at sufficiently high rate, this caused a small rise in the reactor pressure, which in turn triggered the opening of a pressure relief valve ($\mathrm{V_{Pr}}$) in the reactor pressurizer (Pr). However, this remedy was not sufficient and the reactor pressure reached 2,390 psi within 8 s from the beginning of the incident, a significant deviation from the 2,200 psi of normal reactor operation. At this point a SCRAM operation was successfully carried out: the control rods were driven into the reactor and the chain reaction came to a halt. All these procedures were automatically initiated and carried by the control system of the reactor, without the interference of operators. Despite the stop of the fission process, and as described in the beginning of Sect. 5.3, the decay of the secondary products in the reactor continued to produce a smaller but significant amount of heat power in the reactor core. This heat power was removed by the cooling system, which operated as it was designed to.

The stop of fission and the drastic reduction of heat power that followed reduced the reactor pressure to the point when the relief valve in the reactor pressurizer should have closed. However, the pressurizer valve ($\mathrm{V_{Pr}}$) failed to close and, unknown to the operators, water continued to flow from the reactor to the pressurizer and, through this valve, into the containment building. This caused the depressurization of the reactor, which prompted the automatic start of the emergency core cooling system (ECCS) pumps (P_{ECCS}) at 2 min and 4 s into the accident. Because of concerns that the vapor level at the top of the pressurizer may become too low for the effective control of the unit, one of the ECCS pumps was shut off by the operators at 4 min 38 s. The high pressure pump operation caused the complete filling of the pressurizer system. At the same time, the emergency feedwater pumps in the secondary system started to operate too, but no water reached the steam generators because the feedwater valves (V_{FWH}) were left inadvertently closed after the test. At this stage, the steam generators did not receive any water to vaporize and, hence, did not help remove any heat from the reactor. This was discovered by the operators 8 min after the start of the accident, the valves were duly opened and flow to the steam generators resumed.

During this stage of the accident, more water flowed out through the open valve at the pressurizer than was pumped in by the ECCS pumps. As a result, the water pressure in the reactor fell to approximately 1,030 psi, where it was stabilized by the operators and remained at that level between 20 min and 1 hr 14 min into the accident. At that point, excessive vibrations in the pumps and low coolant flow in one of the two cooling loops prompted the operators to shut off one of the reactor coolant pumps. The second coolant pump was shut off at 1 hr 40 min into the accident for the same reasons. Shutting of the two pumps was a poor choice by the operators: Steam and water separated in the two cooling loops and further coolant circulation by the pumps became impossible. At that time, the operators were still unaware of the massive loss of coolant water from the open pressurizer valve and, apparently, believed that the reactor core was completely filled with liquid water and that natural circulation would be sufficient to remove the heat generated by the fission decay products. However, it was later observed that the level of the water in the reactor was only 30 cm above the top of the core. The water soon evaporated from the decay heat and the top of the reactor core was exposed to steam, which has very low conductivity. As a result, the exposed, upper part of the reactor overheated significantly.

Only at 2 h and 18 min into the accident did the operators realize that the pressurizer valve (V_{Pr}) was open and they immediately closed it. The reactor pressure began to rise again but, by this time the upper third of the reactor core was exposed to the low conductivity steam and overheated to 2,100°C. At this temperature the fuel cladding (zirconium alloy) partly melted and also reacted chemically with the steam to produce hydrogen gas according to the chemical reaction:

$$Zr + H_2O \rightarrow ZrO + H_2. \tag{5.2}$$

It was calculated that approximately 1,000 kg of hydrogen gas was produced to form the "hydrogen bubble" at the upper part of the reactor core. It was feared that any sudden release of this gas into the reactor containment building would form an explosive mixture. The hydrogen gas was bled off by the circulating water to a regenerator where it combined with oxygen in a controlled reaction and a large

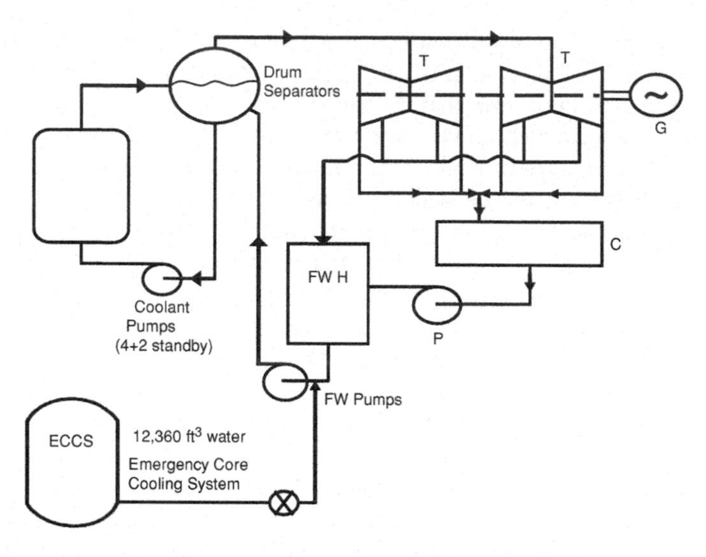

Fig. 5.11 Schematic diagram of the RBMK power plant at Chernobyl

explosion was avoided. It was estimated that the partial melting of the zirconium alloy cladding exposed at least 70% of the reactor fuel and that 30–40% of the fuel melted because of the high temperatures. However, the 6″ (15 cm) thick steel reactor core vessel contained the molten material in their entirety and prevented a reactor meltdown. In the case of TMI all the damage was contained in the now defunct reactor and the only environmental consequence was the release of small amounts of radioactive steam.

5.3.3 The Accident at Chernobyl

Background: The Chernobyl power plant is located in the Ukraine, approximately 80 km to the north of the capital Kiev and very close to the border of what is now the country of Belarus. At the time of the accident, both Ukraine and Belarus were parts of the Union of Soviet Socialist Republics (USSR). The power plant had four units, each producing approximately 1,000 MW-e and the accident occurred in unit IV. The reactor of unit IV was a RBMK reactor, which was designed in the 1960 s and constructed in the early 1980s. Figure 5.11 is a schematic diagram of this power plant. The unit operated with two turbine-generator systems, each producing approximately 500 MW-e. The reactor was graphite moderated and composed of 1,661 channels, which were actually 9-inch stainless steel tubes with the fuel assemblies inside. Each channel was inserted in the graphite pile. Water, which was forced inside the tubes by four primary pumps, removed the heat produced by the reactor. Two more pumps were on stand-by as emergency pumps.

The reactor was controlled by 211 boron carbide (BC) rods that were inserted at different depths of the reactor. The design of the reactor dictated that a minimum of 28 control rods had to be fully immersed in the reactor at all times.

In the case of power failure, the stand-by coolant pumps of the RBMK reactors were designed to operate with diesel motors. However, as everyone who has operated a car with a diesel engine knows, during cold weather diesel motors take some time to produce full power. In the cold weather of the USSR it was determined that the diesel motors for the emergency pumping system would require up to 180 s (3 min) from start-up until they are capable to provide full power to the pumps. When the power plant needs to shut down, these 3 min may be critical in the safety of the reactor. Therefore, a *bridging power* is needed to provide the power to run the pumps during the warm-up of the diesel engines. The Soviet engineers determined that this amount of power may be extracted from the kinetic/rotational energy of the turbine, which turns idle when the steam supply is interrupted. Effectively, the turbine/generator system would have been used as a flywheel that stores energy. They developed a voltage regulator, which when attached to the slowing turbine, would produce enough power to run the emergency pump system, until the diesel motors would provide full power. Even though the voltage regulator system could have been tested in a conventional (e.g. fossil fuel) power plant, it was decided to test the system at the Chernobyl nuclear power plant on Friday April 25, 1986. The testing was to be accomplished by a team sent from Moscow (the capital of the USSR) which was to fly back to Moscow on the same day. The choice of the date was probably the first of many mistakes that led to the accident: since the major holidays in the USSR were May 1 (workers day) and May 8 (2nd World War victory day) the 1986 celebrations were to start in the weekend of April 26/27 and to continue for two weeks. Most Soviet citizens desired to be near their home, friends and relatives for these celebrations.

The accident[2]: The voltage regulator test was to be performed in the afternoon of Friday, April 25. At 1:06 pm the reactivity and steam supply were reduced to 1,600 MW-t and the first turbine-generator system of the plant was shut down. In preparation for shutting down the reactor, the Emergency Core Cooling System (ECCS) of the reactor, which contained 12,360 ft^3 of cold water, was manually deactivated. The reason for the deactivation was that the anticipated low power in the reactor might trigger its automatic activation and the injection of the cold water may damage the reactor. However, at 2:00 pm the electricity dispatcher from Kiev asked for the continuing supply of power and, at 2:06 pm, it was decided to postpone the test until the electric power in Kiev was not needed. The reactor operated until 11:00 pm at half power (1,600 MW-t) with the ECCS disconnected.

At 11:10 pm the test was resumed and the team of reactor operators set the power controls to produce approximately 800 MW-t, at which point the supply to the remaining turbine-generator systems was to be cut-off for the voltage regulator

[2] This account of the Chernobyl accident is a summary of the judicial proceedings and investigation that followed the accident as reported in Medvedev [3].

test to commence. At midnight, the shift of the team of operators was completed and a new team of operators took over the control of the reactor. However, the reactor control rods were set erroneously and the power of the reactor dipped to the low level of 30 MW-t, from which it had to be increased. At this low thermal power level, there is very low neutron flux in the reactor and there is a danger of "iodine-xenon poisoning:" Iodine is an element that is formed from the fragments of the ${}_{92}U^{235}$ fission (e.g. Eq. 4.19a or b) and preferentially absorbs neutrons to form xenon according to the reactions:

$$\begin{aligned} {}_{53}I^{135} &\rightarrow {}_{54}Xe^{135} + {}_{-1}e^{0} + \nu \\ {}_{54}Xe^{135} + {}_{0}n^{1} &\rightarrow {}_{54}Xe^{136}. \end{aligned} \tag{5.3}$$

The iodine isotope ${}_{53}I^{135}$ is short lived with a half-life of 6.7 hr and the half-life of ${}_{54}Xe^{135}$ is 9.2 hr. Because the latter has a very high absorption neutron cross section, it absorbs preferentially thermal neutrons and the control rods of the reactor have little effect in controlling the power produced. In the absence of available neutrons, ${}_{54}Xe^{135}$ decays naturally to ${}_{55}Cs^{135}$, which does not absorb the neutrons.

At this point, common sense would have dictated that the reactor be shut for approximately 24 hr with the emergency cooling system on until the levels of iodine-135 and xenon-135 decayed to a level, where the control rods would have been effective. Instead, there was a heated argument in the reactor control room between the team of operators and the visitors' team that designed and performed the test. Tempers flared and the reactor operators were browbeaten to completely raise all but 28 of the control rods in order to increase the thermal power of the reactor. This was the absolute minimum number of rods required to be immersed by the design of the reactor.

Despite of this allowance, the thermal power of the reactor only increased to 200 MW-t at 1:00 am of Saturday April 26. The amount of steam produced fell to a low level and pressure in the steam separator (the drum separator) dropped significantly. The reactor operators tried to adjust the pressure in the separators manually, while at the same time lifting ten more of the control rods. This led to a warning by one of the operators (Mr. Toptunov) at 1:22 am that the reactor was operating at a negative margin of safety, because only 18 and not the statutory 28 rods were inserted, many of them only partially. The warning was disregarded by the chief of the visiting team. Instead, at 1:23:04 am, the experiment was initiated by shutting down the steam supply to the turbine. At the same time they over-rode the reactor automatic shut-down system, in order to keep the reactor operating and repeat the experiment if necessary (1:23:10 am).

This was a grave mistake by the operators: Shutting the steam supply eliminated the last system that removed heat from the nuclear reactor, at a stage when the statutory number of control rods was not even fully inserted. This action caused a slight increase in the pressure of the reactor core, corresponding to an increase of the temperature and a temporary reduction of the steam production. However, the decrease of the coolant flow initiated a significant steam generation in the reactor and an increase of the reactor power, which was recorded at 1:23:31 am. At 1:23:40 vigorous boiling occurred in the reactor and a swarm of bubbles appeared. In the RBMK reactor, which relies mostly on graphite for neutron moderation, the water is primarily a coolant but also acts as a weak thermal neutron absorber (σ_c of water is 0.66 barns). The appearance of steam reduced the density of this neutron absorber and, as a result, the neutron flux increased. In the nuclear engineering terminology steam is referred as "void" and the effect of increasing neutron flux with the appearance of steam is called "positive void coefficient" for the reactor.

What followed the appearance of the "void" in the reactor is a testament to the complexity of nuclear reactors and the explosive non-linearity of the phenomena that take place in them: While the power produced until 1:23:40 am was close to 200 MW-t, at 1:23:43 am the power suddenly increased to 530 MW-t and at 1:23:48 to an estimated 300,000 MW-t, approximately 100 times more than the reactor's rated power. At this rate of power, the attempted insertion of the control rods was ineffective, and the control rods were jammed in the graphite matrix, which exhibits anisotropic expansion (the so-called "Wagner" effect). The heat and steam produced was so much that the temperature and pressure in the reactor channels rose rapidly. A first steam explosion occurred at 1:23:48 am and destroyed a large part of the reactor. As in the Three-Mile Island accident, the high temperature also caused steam to react chemically with the zirconium alloy in the cladding and to produce large quantities of hydrogen. The hydrogen formed an explosive mixture with the oxygen in the reactor building and a hydrogen explosion, at 1:23:58 am completed the destruction by lifting and permanently tilting the biological shield of the reactor, thus exposing the interior of the reactor core to the atmosphere and releasing a great deal of the radioactive materials.

The accident at Chernobyl was probably the most severe industrial accident of the modern era: The exposure of the reactor core released the total inventory of noble gases (Xe, Kr, Rn) representing $0.7*10^{18}$ Bq; of the liquids and solids in the reactor, approximately 20% of iodine, 12% of cesium (caesium), 8% of uraniuim and 3% of the rare earth metals were released, representing an additional $2*10^{18}$ Bq. Of all the materials released, 0.3–0.5% was deposited on site, 1.5–2% was deposited within 20 km and the remainder was dispersed beyond 20 km and was carried by the weather patterns. Several parts of Europe and Asia were contaminated. In the early days of the accident, increased background radioactivity was observed as far as the eastern shore of the U.S.A., the islands of Japan and the Persian Gulf.

The effects of the disaster touched many: There were 31 immediate fatalities and 24 others permanently disabled. An area of 25 km around the plant was evacuated, 131,000 citizens of the former USSR were displaced officially (and many others unofficially). The released radiation, which has long-term effects, is expected to cause between 100,000 and 1,000,000 additional cancers in Europe. And, 600,000 citizens of the former USSR were placed in a special register to be monitored throughout their lives for radiation-related illnesses.

Table 5.3 Radioactivity released at TMI-2 and Chernobyl

	TMI-2 to containment bld.	TMI-2 to environment	Chernobyl to environment
Noble gases	48%	1%	100%
Iodine	25%	0.00003%	20%
Cesium	53%	0	12%
Uranium	0.5%	0	8%
Rare earth metals	0	0	3%

The major safety drawbacks of the design of the RBMK reactors are the absence of a reactor core vessel and the absence of a strongly built containment structure, which would contain the released radioactivity. The effect of the containment building in containing the released radioactive materials becomes apparent from a cursory comparison of the environmental effects following the Chernobyl and the TMI-2 accidents: While both reactor accidents were severe and both reactors were permanently damaged, the environmental and human impact of the TMI-2 accident was minimal because most of the released radioactive elements were trapped in the reactor containment building. Table 5.3 shows the fraction of radioactivity released by the two accidents to the environment and to the containment building.

Even though the environmental releases from the TMI-2 unit were minimal, the possibility of a nuclear meltdown, which would have contaminated the water supply and would have rendered most of central Pennsylvania uninhabitable, is truly alarming. By examining with hindsight the accidents at Chernobyl and the Three-Mile Island, one may conclude that both accidents were preventable. Both happened either during a test or a short time after a test. In both cases (and especially in the Chernobyl accident) there were a number of warnings signs for operator actions that should not have occurred. There were several actions of the operators, which were in direct violation of written procedures, design standards, engineering judgment and even of common sense. Should the operators have not over-ridden the emergency systems and allowed them to automatically shut down the reactor, the accidents would not have occurred.

It is strongly believed that the main reason the accident happened and so many rules were violated, was because of the unwavering conviction by the operators and the visiting team that the reactor is safe and an accident would never be possible. Several years' successful operation had lulled those in the operating room into overconfidence and a false sense of security. Overconfidence with such complex systems and disregard of operating procedures can only be destructive. A thorough study of the devastation that followed the Chernobyl accident and the scenario of what could have followed a meltdown at TMI-2, makes it apparent that, in a world of ever-increasing technical complexity, with tremendous economic pressures for higher efficiency, simplicity and profitability, our society cannot afford to relax its vigilance on the operation of complex engineering systems, such as nuclear reactors.

5.3.4 The Accident at Fukushima Dai-ichi

The Fukushima Dai-ichi (Dai-ichi means one in Japanese) was composed of six units, each with its own nuclear reactor. The units were constructed between 1967 and 1979 and were operated by the Tokyo Power and Electric Company (TEPCO). Two more units were in the planning stages at the site. All power units were of the BWR type. On March 11, 2011 an earthquake, measuring 9.0 in the Richter scale struck the coastal area of Japan. At the time of the earthquake the first three units (I, II and III) were in operation and the last three units (IV, V and VI) were shut down and undergoing seasonal maintenance. Immediately after the earthquake the three operating units were shut down, according to established operating procedure. Emergency generators were switched on to remove the decay heat from the reactors. The latter were not damaged by the earthquake.

However, the earthquake was followed by a 14 m tsunami, while the power plants were designed to withstand only a 6.5 m tsunami. The tsunami caused a widespread flooding in the entire area of Fukushima and also disabled the generators that drove the emergency cooling systems of the nuclear power plants. The only emergency cooling was supplied by batteries, which run out of power after a few hours. Without adequate cooling, the water in the reactors formed more steam, which caused higher pressure in the reactors. This triggered the relief valves to open and vent the radioactive steam. At the same time the liquid water level in the three reactors dropped significantly and the temperature rose locally to high levels. This caused partial meltdown in all three reactors that were initially in operation. Hydrogen explosions were also observed in units I and III, and very likely in unit II. The partial meltdown and the explosions damaged the three units significantly enough, so as not to be expected to be commissioned again.

Partial cooling water flow in the reactors was restored several hours after the accident and this stopped a wider meltdown of the nuclear units. Unit I remained without cooling water supply for 27 h and unit II for 7 h. A further complication for the accident arose from the fact that the reactors buildings also contained the spent fuel, or waste fuel, of the reactors, in local water pools. The latter also require emergency cooling for the decay heat to be removed. Almost the entire nuclear inventory of the reactor of unit IV was stored in the spent fuel pool for the maintenance operations. The absence of adequate cooling in the spent fuel pool caused a significant increase of the temperature and a fire in the fourth unit of the power plant.

Most of the released radiation was contained in the four containment buildings. However, the venting of the steam and the fires that followed the accident increased significantly the radiation levels close to the reactor and in the entire Fukushima region. The government of Japan declared first a 10 km and subsequently a 20 km evacuation zone. Isolated pockets of high radioactivity outside the 20 km radius were also included in the evacuation.

Unlike the accidents at TMI and Chernobyl, the Fukushima accident was not caused by operator error. At Fukushima, the engineering systems performed according to their design specifications. Simply, the natural disaster—the earthquake and the following tsunami—that caused the accident was too powerful for the plant to endure. At the time of writing of this book, the full environmental and ecological impacts of the Fukushima Dai-ichi accident have not been determined. However, the environmental effects of this accident do not appear to be as severe as those of the Chernobyl accident, primarily because of the presence of the containment buildings that trapped a great deal of the released radioactivity. Serious concerns have arisen about the food grown in the area, water supplies, and the medical treatment of the nuclear workers and other citizens that were affected.

5.4 Environmental, Safety and Societal Issues for Thermal Nuclear Reactors

More than any other form of energy, safety and environmental issues are of paramount importance to nuclear industry. The general public got its first glimpse of nuclear energy in the form of atomic bombs that dropped out of airplanes to devastate the cities of Hiroshima and Nagasaki. The average citizen of the globe still associates nuclear energy with nuclear weapons. From the outset, it must be emphasized that, given the design of all thermal nuclear reactors, a nuclear explosion in a thermal reactor is not possible. The number of neutrons created in the thermal nuclear reactor is simply not sufficient for a critical mass to be created and an explosion to occur.

The main concern on the operation of nuclear reactors is a serious accident that may contaminate a whole region with radioactive materials, as it happened for example after the Chernobyl and the Dai-ichi accidents or what *might* have happened at TMI-2 had a core meltdown actually occurred. Contamination of a whole state and evacuation of large regions are risks that many modern societies are not willing to take. For this reason, there are several groups of citizens that actively oppose the construction a new nuclear power plants throughout the world. Following pressures from such citizens groups in the wake of the Fukushima Dai-ichi accident, the German government announced plans to entirely abandon nuclear energy by 2022. The German plan includes the immediate closure of eight nuclear power plants that had been temporarily shut down after the Fukushima Dai-ichi accident. The remaining nine nuclear power plants in Germany will be shut down before 2022.

In 2009 there were 439 nuclear power plants in operation throughout the world and most of them have operated without a serious incident. During the routine operation of a nuclear power plant very small doses of radioactivity, which are considered to be harmless to humans, may be discharged to the environment from the cooling system of the condenser. Other than these small doses of radioactivity, the routine operation of nuclear units does not have any adverse environmental effects. There are other issues, related to the fuel cycle and the overall effects of the nuclear industry, which may have significant environmental impacts:

First, is the disposal of the *nuclear waste materials (NWM)*, which are mainly produced from three sources: the spent fuel, the contaminated internal reactor structure, and the mechanical equipment. A thermal reactor is refueled every 18–24 months and produces a significant amount of NWM, which are radioactive and will remain radioactive for hundreds of thousands of years. For example, the non-fissile isotope ${}_{92}U^{238}$ has a half life of 4.5 billion years. Other reactor products have half lives in the thousands and hundreds of thousands of years. Equally important is the disposal of contaminated equipment after the decommissioning of the nuclear reactor. The material of the entire reactor vessel, including cladding, control rods, pumps etc. is highly contaminated. Similarly, most of the equipment that process the reactor coolant, such as coolant pumps, emergency pumps, steam generator piping, etc. become contaminated after a few years of operation and must be safely stored after the decommissioning of the reactor.

The spent nuclear fuel may be reprocessed and several radionuclides may be removed, but the non-fissile ${}_{92}U^{238}$ remains and becomes nuclear waste. The safe reprocessing and storage of NWM is part of the national nuclear programs of most countries, but has been dealt with differently by the governments of countries that operate nuclear reactors: In the United States most of the NWM are stored at the site of the nuclear power plant, usually in water pools that help remove the heat produced from radionuclide decay. After a detailed feasibility study that cost several billion dollars, a federal NWM repository has been proposed, at the Yucca Mountain, Nevada, where nuclear waste would be stored in artificial tunnels and monitored. However, there are many administrative and legal obstacles before the Yucca Mountain nuclear waste repository starts operating.[3] Reprocessing is used in the United Kingdom with the disposal of low-level radioactive materials at Sellafield. A permanent disposal site will be studied and decided upon in the future. In Russia there are several temporary storage facilities. A recent study advocated the permanent storage of NWM in the remote arctic peninsula of Kola. Similarly in France, there are reprocessing and temporary storage facilities with a permanent deep geological disposal site proposed in Boure. In Belgium, NWM are reprocessed in the city of Mol and are stored in a central facility at Dessel. Germany already operates two small underground repositories at the Ahaus and Gorleben salt domes.

Even after reprocessing, the radioactive products must be safely stored for many generations. Safe storage should include significant cooling for the removal of any decay heat and continuous monitoring for thousands of years. While several national teams of meritorious scientists and engineers are diligently working to safely store nuclear waste, the sheer magnitude of the problem of storing tons of

[3] In March 2009 the U.S. Department of Energy announced that the Yucca Mountain "…is not considered as an option for storing nuclear waste." This was a political, not scientific decision. Given the longtime political maneuvering around NWM in USA, very few think that this is the final word on this project.

radioactive isotopes for thousands of years is a major societal concern and a significant impediment to the further development of nuclear energy.

A second issue is the safe transportation of the NWM materials from the power plants, where they are produced, to the reprocessing facilities and, finally, to the permanent repository. A transportation accident may have an enormous environmental impact over very large areas. For example, most of the spent fuel from the Japanese reactors is shipped to be reprocessed in Europe. Waste in the form of glass (vitrified waste) and the recovered uranium and plutonium are returned to Japan to be recycled in the reactors. A shipping accident in this case would cause a major environmental disaster.

A third issue related to the production, use and transportation of nuclear materials is the safety issue associated with an act of war or an act of terrorism. While the containment buildings of typical nuclear power installations in the United States have been designed to withstand the impact of a large jet airliner, their ability to actually withstand such an impact has never been tested. In addition, most of the other countries do not have such stringent requirements as it was amply demonstrated in the Chernobyl accident. Following the terrorist acts in the United States and Europe, there have been elevated levels of security in all the nuclear power plants, worldwide. Given the magnitude of the disaster that may follow, a terrorist attack in a nuclear power plant has been at the forefront of public awareness and concern, at least in the OECD countries.

A related issue is a possible act of war between two nations that have nuclear power plants, and, possibly, nuclear weapons. In 2010 there were 31 countries with operational commercial nuclear reactors and another 25 with experimental reactors. Since 1980, several of these countries have had minor wars with their neighbors or participated in acts of war. A nuclear reactor is a primary target in the event of war that may have devastating environmental consequences for the host country and the whole geographical region. As the number of nuclear reactors increases worldwide, the risk of bombardment or sabotage of nuclear reactors becomes significantly higher. An international treaty against such acts is often advocated. However, given the experiences in the use of chemical and biological weapons (which have been banned by international treaties since the 1930s) by rogue countries during recent wars, such a treaty may not be effective at all.

A fourth issue is the proliferation of nuclear weapons: Several types of nuclear reactors produce plutonium, which may be used for the production of nuclear bombs. In particular, the FBR's and to a lesser extend the AGR's convert a significant fraction of ${}_{92}U^{238}$ to ${}_{94}Pu^{239}$, which is the main charge of a nuclear weapon. Therefore, the proliferation of nuclear reactors may also lead to the proliferation of nuclear weapons, thus posing a significant risk worldwide. The society of nations, through the *United Nations* and the *International Atomic Energy Agency* has tried to restrain the proliferation of nuclear weapons by enticing countries with bilateral and multilateral treaties, international aid and trade agreements (as in the case of North Korea and Iran in 2007–2010). However, the results of these actions have not been evaluated yet and, it is too early to determine whether or not such strategies will be persuasive and effective in the future.

On the positive side, nuclear power plants do not emit any GHG and do not contribute at all to global warming. Nuclear reactors do not produce other pollution such as nitrogen and sulfur oxides, ash and other particulate matter. In 2007, nuclear power plants produced $2{,}608*10^{12}$ kWh of electricity without any significant emission of greenhouse gases or other chemical pollutants. Had this entire amount of electric power been produced by coal, there would have been an additional $265*10^9$ tons of CO_2 released to the atmosphere in addition to millions of tons of SO_2, and NO_x. In a world where global warming has become a significant environmental issue and a primary societal concern, the safe and reliable production of nuclear energy is an excellent alternative to the combustion of fossil fuels.

5.5 Breeder Reactors

Natural uranium consists of 0.715% ${}_{92}U^{235}$ and 99.285% ${}_{92}U^{238}$. The conventional nuclear reactors, which only utilize the isotope ${}_{92}U^{235}$ with a minute fraction of ${}_{92}U^{238}$, waste a very large portion of the natural uranium fuel. With the finite amount of nuclear fuel reserves in the world, the utilization of natural uranium by conventional reactors is a significant waste of a scarce resource. The current, high grade natural uranium reserves in North America are estimated at 8,000 Quads ($8{,}000*10^{15}$ Btu). These are sufficient to supply all the energy needs for the continent for approximately 100 years, if only ${}_{92}U^{235}$ is utilized. However, the same high grade ore would be sufficient to supply the North American continent with energy for 5,500 years if the ${}_{92}U^{238}$ isotope, which is a fertile material, were to be converted to ${}_{94}Pu^{239}$ and used as fissile material. In addition, if all the known reserves (high and low grade) of uranium were used in breeder reactors, this amount would be sufficient to supply the energy needs of North America for more than 30,000 years [2]. To this amount, one may add the contribution of another naturally occurring fertile material, ${}_{90}Th^{232}$, which may be converted to the fissile isotope ${}_{92}U^{233}$. It follows that the widespread use of breeder reactors may solve the energy needs of the entire world. The two fertile isotopes, ${}_{92}U^{238}$ and ${}_{90}Th^{232}$ are converted to the fissile isotopes ${}_{94}Pu^{239}$ and ${}_{92}U^{233}$ according to the following nuclear reactions:

$$\begin{aligned} {}_{92}U^{238} + {}_{0}n^{1} \rightarrow {}_{92}U^{239} \rightarrow {}_{93}Np^{239} + {}_{-1}e^{0} \quad (T_{1/2} = 23\text{ min}) \\ {}_{93}Np^{239} \rightarrow {}_{94}Pu^{239} + {}_{-1}e^{0} \quad (T_{1/2} = 55.2\text{ h}) \\ {}_{90}Th^{232} + {}_{0}n^{1} \rightarrow {}_{92}Th^{233} \rightarrow {}_{91}Pa^{233} + {}_{-1}e^{0} \quad (T_{1/2} = 22\text{ min}) \\ {}_{91}Pa^{233} \rightarrow {}_{92}U^{233} + {}_{-1}e^{0} \quad (T_{1/2} = 27.4\text{ }). \end{aligned} \tag{5.4}$$

Two nuclear processes take place in a breeder reactor, conversion and fission, both of which consume neutrons. As the last equations show, the first neutron is necessary to convert a fertile nucleus to a fissile nucleus and a second neutron is necessary to cause fission. Therefore, in a breeder reactor the average number of neutrons produced by the fission reactions must be significantly higher than 2 to allow for neutron losses, such as capture by the moderator and coolant nuclei in the reactor, as well as other parasitic losses. The *conversion ratio or breeding ratio*, *C*, is an important number in the breeder reactors and is defined as the number of fissile atoms produced in a reactor per atom of the existing fissile fuel. If $C = 1$, then the amount of fissile fuel in the reactor remains constant during the operation of the reactor, while if $C > 1$, the amount of fissile fuel increases with time and the reactor produces more fuel than the amount it was charged with originally.

The following example shows the importance of the breeding ratio: Consider a reactor initially charged with a number *N* of fissile ${}_{92}U^{235}$ nuclei, and a much higher number of fertile ${}_{92}U^{238}$ nuclei. When all the fissile nuclei have been consumed, the reactor would have also consumed a number, *CN* nuclei, of the fertile ${}_{92}U^{238}$. The latter would have produced *CN* nuclei of the fissile ${}_{94}Pu^{239}$. When these fissile nuclei are consumed too, the reactor would contain NC^2 nuclei of ${}_{94}Pu^{239}$ and when these are consumed, the reactor would contain NC^3 fissile nuclei. If the process continues, the amount of fertile ${}_{92}U^{238}$ that has been used is: $N(1 + C + C^2 + C^3 + \ldots)$. When $C > 1$ the series has no limit and, hence, at the last stage in the fission process all the fertile ${}_{92}U^{238}$ nuclei would have been consumed and the entire nuclear resource would have been utilized. However, if $C < 1$ the sum is equal to $N/(1-C)$, and the part of the fertile material that is utilized depends on the exact value of *C*. For example, if the conversion ratio is equal to 0.75, the series converges to $1/(1-C) = 4$ and the amount of ${}_{92}U^{238}$ that is used is only four times the original amount of the fissile ${}_{92}U^{235}$.

For a reactor that was originally charged with natural uranium, only 2.86% of the fertile material ${}_{92}U^{238}$ would have been utilized. While this is an improvement over the practice of using solely the ${}_{92}U^{235}$ it is also a very small number and would not extend significantly the life of the planet's nuclear reserves and resources. Therefore, for the breeder reactors to be effective in providing a permanent solution to the energy problem, the breeding ratio, *C*, must be very close to 1 or even to exceed 1. In the latter case there is a positive *gain* in the number of fissile nuclei produced per fissile nucleus consumed. The gain, *G*, of a breeder reactor is defined as follows in terms of the average number of neutrons produced per fission from the aggregate amount of fuel, η, the breeder ratio, *C*, and the other parasitic losses in the reactor, *L*:

$$G = C - 1 = \eta - 2 - L, \tag{5.5}$$

The maximum gain of a breeder reactor is achieved when the parasitic losses vanish and: $G_{max} = \eta - 2$. Table 5.4 depicts this and other important parameters of breeder reactors. It is apparent in this Table that, allowing for reasonable parasitic capture and leakage, only ${}_{92}U^{233}$ fueled thermal reactors may realistically

Table 5.4 Significant parameters for breeder reactors. Cross sections are in barns and the other parameters are dimensionless [2]

	Thermal neutrons			Fast neutrons				
	U-233	U-235	Pu-239	U-233	U-235	Pu-239	Th-232	U-238
σ_f	527	577	790	2.2	1.4	1.8	0.025	0.112
σ_a	580	675	1185	2.35	1.61	2.05	0	0.273
ν	2.51	2.4	2.9	2.59	2.5	3	2.04	2.6
η	2.28	2.08	2.12	2.35	2.09	2.53	2	2.27
G_{max}	0.28	0.08	0.12	0.35	0.09	0.53	0	0.17

achieve a breeding ratio greater than 1. There is a significant difference with fast neutrons: both ${}_{92}U^{233}$ and ${}_{94}Pu^{239}$ fueled reactors may realistically achieve $C > 1$ and, for this reason almost all of the breeder reactors are designed to use fast neutrons. Actually the value of C in fast breeder reactors is slightly higher because of the contribution of the fertile ${}_{92}U^{238}$ and ${}_{90}Th^{232}$ nuclei in the fission process and the production of neutrons, since both nuclei have finite σ_f at the fast neutron range. Thus, it is possible to achieve $C > 1$ in a fast neutron reactor fuelled by ${}_{92}U^{235}$ and ${}_{92}U^{238}$.

When the gain of a breeder reactor is positive, the reactor produces more fuel than it consumes. The number density of the new fissionable nuclei produced in the breeder reactor during a time interval, δt, is:

$$\delta N_b = F_c N_{0ff} \sigma_a \varphi(\eta - 2 - L)\delta t, \tag{5.6}$$

where N_{0ff} is the number density (nuclei/cm^3) of the fissile nuclei present in the reactor at the beginning of the time interval; F_c is a dimensionless fraction for the latter nuclei that are present in the core of the reactor; φ is the average neutron flux in neutrons/(s*cm^2) during the time interval, δt; and σ_a the absorption cross-section of the fissile material produced. The *doubling time*, t_d, of such a reactor is the time it takes for the fissile nuclei to double within the reactor. Doubling of the fissile nuclei will occur when $\delta N_b = N_{0ff}$. Under this condition, Eq. (5.6) yields the following expression for the doubling time of the fissile material:

$$t_d = \frac{1}{F_c \sigma_a \varphi(\eta - 2 - L)}. \tag{5.7}$$

Typical doubling times for the current generation of the breeder reactors range between 10 and 20 years. The shorter the doubling time, the more effective the reactor is in converting and utilizing the fertile material. It is desirable to have short doubling times, so that fissile Pu^{239} may be provided for new breeder reactors. It is expected that new breeder reactors in the future, will be started and originally fuelled with plutonium harvested from older breeder reactors and reprocessed.

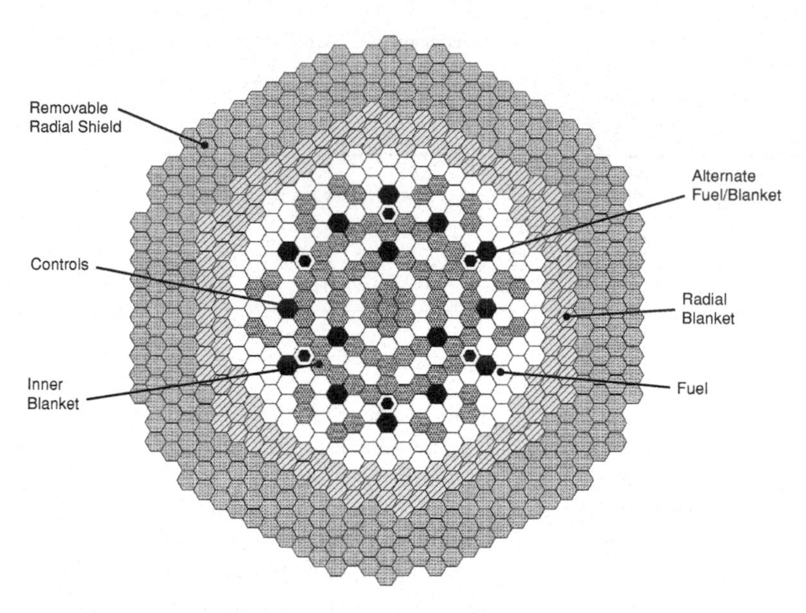

Fig. 5.12 Horizontal cross-section showing the essential elements in the core of a breeder reactor

5.5.1 Fast Breeder Power Plants

A fast breeder reactor has the dual purpose of heat/steam production and fissile material production. Typical reactor core arrangements use hexagonal patterns with the fissile material (fuel) in the inner part surrounded by the fertile material, which uses the neutrons produced by the fission reactions. Control and safety elements also follow this hexagonal pattern and are arranged strategically across the matrix. Figure 5.12 is a horizontal section that depicts the core arrangement of a typical breeder reactor.

Because of the need to conserve fission neutrons, the core of a fast breeder reactor is significantly smaller and more compact than the core of a thermal reactor. This results in a very high rate of volumetric heat production, of the order of 120 MW/m^3. At such high heat power rates, water or gases are not effective coolants because of their lower thermal conductivity. Liquid metals, such as molten sodium (Na) or potassium (K) or a mixture of the two, are used for the removal of the heat from fast breeder reactors. The melting point of both metals is below 100°C and the breeder reactor is designed to always maintain temperatures sufficient to keep the metals in the liquid state.

The irradiation of sodium with neutrons produces the highly radioactive isotope $_{11}Na^{24}$, which decays to $_{12}Mg^{24}$ by the emission of an electron. Because of the high radioactivity in the core of the reactor, two cooling loops are necessarily used for steam generation in a Liquid Metal Fast Breeder Reactor (LMFBR) power

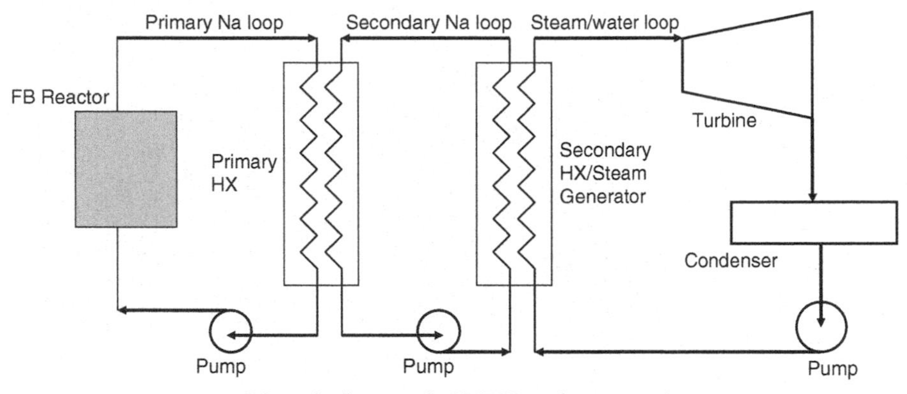

Fig. 5.13 Schematic diagram of a LMFBR nuclear power plant

plant as shown in Fig. 5.13. The primary loop extracts heat from the reactor core and, via a closed circuit primary heat exchanger, transfers it to a secondary loop, which also circulates a liquid metal. The secondary loop liquid metal emits significantly lesser amounts of radioactivity because it does not enter the reactor and, does not come in contact with the high neutron flux and radioactivity of the reactor. The fluid of the secondary loop transfers the heat to produce steam in a second, closed circuit heat exchanger. This is the steam generator for the power plant, which typically operates with a Rankine cycle. The steam produced in the second heat exchanger is directed to the turbines and, after it condenses, it is recirculated to the steam generator. One or more feedwater heaters may be also used in this part of the power plant, to improve the efficiency of the Rankine cycle.

It is apparent that the main difference of the LMFBR power plant from a thermal reactor power plant is the secondary loop of the liquid metal, whose function is to keep most of the radioactivity confined to the primary loop. The rest of the LMFBR power plant is similar to the conventional thermal power plants and is composed of condenser(s), feedwater heater(s), pumps, safety equipment, etc. For simplicity, not all of the generic equipment (e.g. feedwater heaters and pumps) are included in Fig. 5.13, which only depicts the equipment that are necessary in the basic Rankine cycle.

5.6 The Future of Nuclear Energy: To Breed or Not to Breed?

When, in 1955, the then President of the United States Dwight Eisenhower inaugurated the first commercial nuclear power plant in Glassborough NJ, he also declared that nuclear energy would be the way of the future and that nuclear power

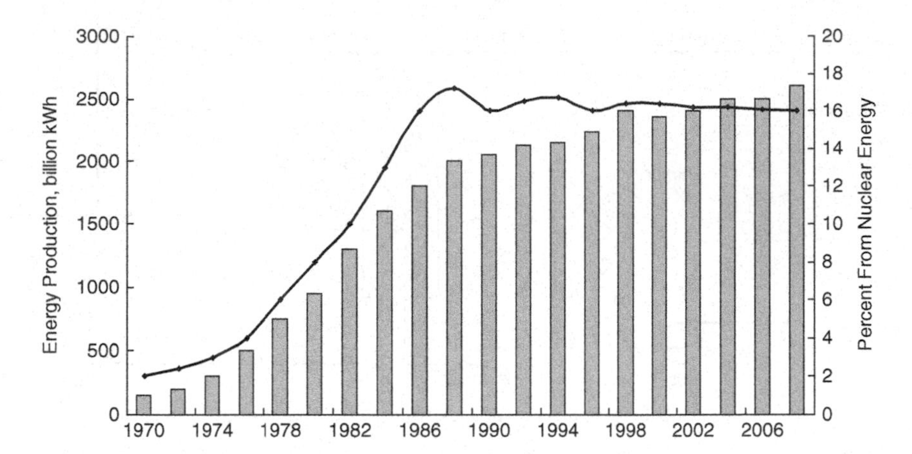

Fig. 5.14 Total global electric energy production from nuclear power plants (*bars*) and fraction of total electricity produced (*continuous curve*). Data from the website of the International Atomic Energy Agency (IAEA) 2011

will produce "…electricity too cheap to measure." However, his prediction never materialized. The construction of nuclear power plants has proven to be very much capital intensive, thus making the electricity produced anything but cheap; nuclear fuels, although reasonably priced have never been very cheap; the problem of the disposal of nuclear waste materials has not been solved in the USA and several other countries; and, most important, well-founded concerns of the citizens on the safety of these power plants have cast a dark shadow over nuclear energy.

In the late 1980s, following the accident at Chernobyl, one after the other the citizens of several OECD countries, such as Germany, Austria, Holland and Belgium, voted for moratoria on further construction of nuclear power plants. In the USA, licensing and construction of nuclear power plants has effectively stopped between 1985 and 2010. If in the 1950s the citizens welcomed nuclear power plants as a source of needed electricity, in the twenty-first century they are skeptical and their collective attitude may be summarized as: "not in my back yard" (NIMBY).

In 2010 there have been 439 operating nuclear power plants in 31 countries, with installed capacity approximately 375,000 MW-e. These plants produced $2{,}600 \times 10^9$ kWh or approximately 16% of the total electricity produced in the entire world. The fraction of electricity produced ranges from close to 80% in Lithuania and France to approximately 22% in the USA, to less than 5% in emerging nuclear nations, such as China, Brazil and Pakistan. Figure 5.14 shows the total electricity production in the world since the 1970s and the fraction of the world electricity produced by nuclear power. It is remarkable that the fraction of global electric power produced has peaked in the late 1980s, at approximately 18%, and that the fraction of the electric energy production from nuclear power plants has slowed down since 1989. This decline is due primarily to the increased electricity production in the world and the slow-down of new nuclear power plants, especially in the OECD countries. Since the early 1980s, the construction of nuclear power plants have all but stopped in most OECD countries (except France and Japan) and any increase of electricity production from nuclear power is almost entirely due to new plants in developing countries. In 2008, India had 6 nuclear power plants under construction and another 10 on order or planned; the Peoples Republic of China had 9 nuclear power plants under construction and another 24 on order or planned; and Russia had 8 nuclear power plants under construction and another 11 on order or planned. The corresponding numbers were 0 and 2 for the USA; 0 and 0 for the UK and Germany; and 1 and 0 for France (IAEA, 2011). Given the numbers of new nuclear power plants, it is apparent that, in the near future, the fraction of electric power produced from nuclear power plants will decrease significantly in the OECD countries and will most likely increase in the developing countries.

The nuclear debate in the OECD countries hinges on two issues:

(a) How safe are the current nuclear power plants and
(b) What can or what will be done with the accumulated nuclear waste materials that are radioactive and must be stored for several hundreds of thousands of years.

The expected decommissioning of the older nuclear power plants will significantly increase the amount of nuclear waste, because several equipment, including reactor vessels, heat exchangers, tubing, pumps and several turbines have become contaminated with radioactive materials.

The scientific community has been divided on the first issue: while many scientists and engineers advocate that the current PWR, AGR and BWR nuclear power plants are safe and that the accumulated experience of their operation warrants the safe operation of these plants in the future, there are other scientists who point out similar statements that were made months before the TMI-2, Chernobyl and Dai-ichi accidents. The latter advocate that the risk associated with a severe accident is too high for a society to bear: A nuclear reactor meltdown would contaminate the aquifers of a very large area and has the potential to make thousands of square miles of land uninhabitable. From the statistics of the currently operating and planned nuclear power plants, it appears that the societies in most of the OECD countries are not willing to accept such an enormous risk. The societies in the developing nations are more willing to accept the nuclear energy risk as they plan to construct and operate more nuclear power plants.

The energy shortage experienced by the world since 2006 and the accompanying enormous spike in fossil fuel prices have mitigated the unwillingness of several OECD nations to accept the nuclear risk. This is manifested by the number of nuclear reactors that have been proposed (these numbers are separate from those under construction or planned): 20 in the United States, 6 in the United Kingdom, 6 in Canada, 10 in Italy. However, these new power plants are in the early planning stages and far from construction or licensing. The citizenry in the OECD countries is apprehensive and suspicious of the nuclear industry, which is in a precarious position. The actual construction and completion of these plants in the OECD countries, as well as of many of the power plants that are to be built in the developing countries may be summarily cancelled, not only if there is a new serious accident, such as the TMI-2 or the Chernobyl accident, but also if there is any incident that has *the appearance, perception or portrayal by the international information media of a serious nuclear accident.* The 2011 accident in the Fukushima Dai-ichi power plant is such a serious accident. In the wake of this accident, several OECD governments have put on hold the further expansions of their nuclear industry. Germany even announced that it will abandon nuclear energy by 2022. While the full effects of the Fukoshima Dai-ichi accident have not been apparent at the time of writing of this book, it appears that the nuclear industry is facing a major challenge in the second decade of the twenty-first century.

Regarding the second issue of nuclear waste disposal, the scientific community has provided a viable solution: nuclear waste may be safely processed and stored in caverns or subterranean tunnels. The processing may involve densification and vitrification of the waste. Several European countries, including France, Belgium and Germany, have central waste reprocessing and storage facilities, such as those in Dessel in Belgium and the Ahaus salt mine in Germany, but no permanent disposal facilities in operation. Only the UK has a low level storage facility, in Sellafield, the site of the 1957 AGR Windscale accident. Low-level radioactive waste is first vitrified and then stored in this facility.

In the USA, where there are 104 nuclear power plants in operation, most of the nuclear waste materials are stored in temporary facilities (water pools) close to the power plant they were produced. It has been proposed, and it was approved by an act of Congress, in 2002, that a Federal permanent repository facility be constructed in the Yucca Mountain, Nevada. A great deal of research and several feasibilities studies have shown that the repository at Yucca Mountain would be safe to operate and would not pose a major environmental threat to the region. However, local citizens groups opposed the construction of such a repository on the grounds that the preliminary studies are not complete, while other environmental groups have raised objections to the transportation of thousands of tons of radioactive nuclear waste from their current locations to the remote location of Yucca Mountain. The storage issue in USA has become more of a political problem than a scientific or engineering problem. For example, after several years of studies, the spending of at least $4 billion, a great deal of judicial and political activity and several legislative acts, in March 2009, the U.S. Department of Energy announced that the Yucca Mountain "…is not considered as an option for storing nuclear waste." Apparently, this was a hastily made political decision that did not consider the national need to permanently store the nuclear waste. Given the lack of alternative solutions or even proposals, very few consider this statement to be the final decision on this project.

Most of the radioactive waste, which is currently in temporary storage, consists of the ${}_{92}U^{238}$ isotope, which is not used in conventional thermal reactors. It has been suggested that Fast Breeder Reactors (FBR's) would solve the problem of

${}_{92}U^{238}$ waste by converting it to the fissile ${}_{94}Pu^{239}$ and will consume the latter as a fuel. FBR's, would not only use all of the material in natural uranium, but also would convert some of today's nuclear waste to tomorrow's nuclear fuel. While this conversion is possible and its feasibility has been demonstrated in numerous experimental and practical engineering systems, the FBR's that have been constructed have had several engineering problems. Their widespread and safe use for the production of electric power and conversion of fertile materials is still in the arena of future engineering developments because of the following issues:

At first, FBR's produce ${}_{94}Pu^{239}$, which is the fissile material used in nuclear bombs and one of the risks associated with their operation is a small nuclear explosion. While the typical thermal reactor, such as the PWR, BWR and AGR, does not have the nuclear fuel in concentrations that may become "critical" for a nuclear explosion to occur, the isotope ${}_{94}Pu^{239}$ is formed almost indiscriminately in a FBR at an uncontrolled rate. Because of this, it is possible to form a sufficient amount of the ${}_{94}Pu^{239}$ fuel at a location within the reactor to attain "critical mass," which would lead to a small nuclear explosion. While this explosion may not have the power of an atomic bomb, it would be sufficient to destroy the reactor and disperse a great deal of the radioactivity in the environment.

Secondly, and related to the above, the production of ${}_{94}Pu^{239}$ contributes to global political and military instabilities and poses a threat by itself. A FBR produces sufficient mass of ${}_{94}Pu^{239}$ annually for the construction of several nuclear weapons. The danger of nuclear weapon proliferation alone, would dictate the severe restriction of the use of the FBR's and the export of FBR technology to several nations, as well as the strict accounting and regulation of the FBR fuels and products. This would be very difficult to accomplish on a global scale under the auspices and the infrastructure of the existing international organizations.

Thirdly, the very high power density of FBR's and the necessity to use Na or K as coolants has been the cause of minor accidents. When designing large heat power removal systems, such as the ones used in nuclear reactors, minor leakages of the coolant fluid are almost inevitable, they are often expected and are always anticipated in the engineering design of a reactor and its safety systems. Leakage of water or gases, which are most prevalently used as coolants in conventional reactors, has occurred in several occasions in conventional power plants, but was always contained and did not pose a major threat. However, Na and K react violently with air and water and any leakage of them would cause local fires in the nuclear reactors. The reaction

$$2Na + 2H_2O \rightarrow 2NaOH + H_2, \tag{5.8}$$

and its equivalent for potassium, are highly exothermic, may cause local fires and releases enough heat to melt steel piping. This would release more of the light metal coolant, feed the fire, raise the local temperature to a higher level and thus, start a catastrophic effect that may lead to a severe accident in the Fast Breeder nuclear power plant.

Fourthly, even though FBR's have been constructed since the 1950s, at least for experiments, our experience with them is not sufficient to warrant their safe operation. Of the more than 300 FBR's that have been built for commercial or experimental purposes, only a handful are in operation today, most notably the *Phenix* FBR in France and the *Jojo* FBR in Japan, which operate since 1973 and 1978 respectively. There have been several unexpected accidents and notable "incidents" with the other FBR's: The experimental EBR1, EBR2 and Fermi-1 reactors in the USA had several problems recorded and were shut down. In 1973 France's 250 MW prototype FBR at Marcoule went critical, followed in 1974 by the British prototype FBR at Dounreay. There is surveillance satellite evidence that a severe accident occurred in the Shevchenko plant in the former Soviet Union, although the Soviet authorities at the time gave no details or explanation for it. The operation of the 375 MW-e Clinch River Breeder Reactor in the USA, was marred with several mishaps until, in 1980, President Carter issued an executive order that the plant be shut down and the Breeder Reactor program in the USA be discontinued. The more recent accident, caused by coolant sodium leakage and followed by fire, at *Monju*, Japan, in 1995, rendered the reactor inoperable for more than a decade and required a decision by Japan's Supreme Court to clear the way to its restarting. All these incidents do not inspire confidence neither in the engineering community, which advocates the construction of more prototypes and the performance of more tests and definitive research on FBR's, nor in the general population, which feels that FBR's pose a significant health and environmental risk.

In conclusion, the Fast Breeder Reactors may hold the answer to the solution of the energy challenge of the human society. Not only are they capable to use completely the available nuclear resources in their entirety, since they consume both ${}_{92}U^{238}$ and ${}_{90}Th^{232}$, but also may contribute to the elimination of the current nuclear waste materials. However, these reactors have exhibited a variety of severe operational problems and may contribute to nuclear weapons proliferation. The risk of their operation is significantly higher than that of the conventional thermal reactors. In the near future, nations and the entire human community may have to seriously debate the nuclear solution to the energy problem, weight the advantages and disadvantages of the Fast Breeder Reactors and answer the hard question: *to breed or not to breed*?

Problems

1. Nuclear energy is often touted as a solution to the problem of CO_2 emissions and global warming. A 1,000 MW coal power plant with 41% efficiency is to be replaced by a 1,000 MW nuclear plant with 36% efficiency. Assuming that the coal plant burns only carbon (C) determine:

   a) The amount of carbon (in metric tons) used by the coal plant during a year.

b) The amount of CO_2 (in tons) produced during a year.
c) The amount of natural uranium to be used by the nuclear plant in a year.

2. A reactor fuel element is filled with natural uranium pellets. The inner diameter of the fuel element is 1 cm and its height 5 m. The pellets fill the inner volume of the element. What is the total weight of the uranium in this fuel element?
3. In a short essay (250–300 words) explain why the moderator is an essential part of the nuclear reactor.
4. A small, experimental fission reactor produces 10 MW of thermal power. The reactor is placed in a pool of 68 tons of water, which is cooled by a cooling tower. The pool of water is normally at 54°C and at pressure equal to 1 bar. If the cooling system of the pool looses power, how long will it take for the pool of water to start boiling? In how much longer will the pool be completely dry? For water, $c_p = 4.2$ kJ/kgK and $h_{fg} = 2{,}340$ kJ/kg.
5. What are the advantages and disadvantages of the PWR compared to the BWR?
6. What are the advantages and disadvantages of the GCR compared to the LWR?
7. The pressurizer of a large PWR power plant is maintained at 20 MPa. The vessel has 120 m^3 capacity and is half-full with steam. How much is the mass of the steam and the mass of the water in the pressurizer?
8. The pressurizer of a PWR power plant is maintained at 18 MPa, has 150 m^3 capacity and, originally, is half-full with steam. During a pressure transient in the reactor, the pressure rises to 18.5 MPa and, because of the pressurization and the spray action, the volume of the steam is reduced to 35% of the total volume. Calculate:

   a) The mass of the steam that condensed.
   b) The mass of the spray used in the condensation process, if the spray water enters at 300°C and comes into an equilibrium state in the pressurizer.

9. Explain what is the iodine-xenon poisoning of a nuclear reactor. What is the best course of action when such a condition appears in a reactor?
10. Is it possible for an accident similar to the Chernobyl accident to occur in a typical PWR or a BWR? Explain carefully your reasons.
11. What are the issues associated with nuclear waste disposal and why there has been no solution for the waste in the USA?
12. In your opinion, what is the best way to dispose of the nuclear waste and why.
13. "Fission reactors are only a short-term solution to the energy problem and it is not worth debating and pursuing this option. We should move full-steam ahead with the breeder reaction option." Comment on this statement.
14. Use your knowledge of heat transfer and thermodynamics as well as realistic values of thermal properties to justify why liquid metals are better cooling materials for breeder reactors than water or helium.

15. A fast breeder reactor uses a mixture of plutonium and uranium oxides as fuel and fertile materials. The reactor has a conversion ratio equal to 1.5 and a doubling time equal to 12 years. The fuel in the core represents 80% of the core fuel charge. Calculate the average neutron flux in the reactor.
16. A fast breeder reactor uses a mixture of plutonium and uranium oxides as fissile and fertile materials. The average neutron flux in the reactor is 10^{16} neutrons/(s*cm^2) and the number of neutrons lost by parasitic losses is 0.15 per neutron absorbed. Calculate the reactor doubling time.

References

1. Bennet DJ (1972) The elements of nuclear power. Longmans, London
2. El-Wakil MM (1984) Power plant technology. McGraw Hill, New York
3. Medvedev Z (1990) The legacy of chernobyl. Norton, New York

Chapter 6
Fusion Energy

Abstract Fusion is the energy that powers the stars, including our own Sun. Deep in the mass of the stars common hydrogen is converted first to deuterium. Two nuclei of deuterium combine to form helium, while releasing a large amount of energy. Since the fusion reactions were first discovered, it was realized by the scientific community that the controlled harnessing of fusion may provide enough energy to the human society for millions of years. Fusion energy may solve the "energy problem" for humanity. Fusion energy is a panacea whose realization may not be too far in the future. Multinational research and development efforts have produced plasma confining methods, which may lead to continuous and controlled fusion during the twenty-first century. While several technological breakthroughs are still needed, enough progress has been made to elucidate the basic mechanisms of the fusion reactions and to point fusion research to the right direction. Because of fusion and its great promise, the time when the energy of the stars will be harnessed on Earth and when the produced power will be abundant and inexpensive may not be too far in the future.

6.1 The Energy of the Stars

Fusion is the form of nuclear energy that is released when light nuclei combine to form heavier nuclei. The Sun and other stars are powered by fusion reactions and radiate most of the energy they produce. As with the fission reactions, the fundamental equation for the energy released is the mass-energy equivalency principle:

$$E = mc^2. \tag{6.1}$$

Scientific research on fusion started in the late 1910s. Francis Aston, a Cambridge chemist and Nobel laureate of 1922, developed the mass

E. E. (Stathis) Michaelides, *Alternative Energy Sources*,
Green Energy and Technology, DOI: 10.1007/978-3-642-20951-2_6,

Table 6.1 Atomic masses of commonly met isotopes in fusion reactions

Isotope	Mass, u	Isotope	Mass, u
$_1H^1$	1.007825	$_2He^4$	4.00260
$_1H^2$ (D)	2.01410	$_2He^5$	5.0123
$_1H^3$ (T)	3.01605	$_3Li^6$	6.01513
$_2He^3$	3.01603	$_3Li^7$	7.01601

spectrograph that enabled precise measurements on the mass of the various elements and the discovery of isotopes. His meticulous and accurate measurements on the mass of light isotopes showed that the mass of four hydrogen nuclei is greater than the mass of one nucleus of helium. As a consequence, if the four hydrogen nuclei were to combine and form helium, there would be a "mass defect" and, hence, a large amount of energy would be released during the reaction for the formation of helium. This was caught by a British Astronomer, Arthur Eddington, who in 1920, shortly after the publication of Aston's results announced that "...we need not look further for the source of a star's energy."

Table 6.1 (data from [1]) shows accurate measurements of the mass, in atomic mass units, u, of the most important isotopes in fusion reactions ($1 \text{ u} = 1.6604*10^{-27}$ kg).

It is apparent from this table that the mass defect of the helium formation reaction,

$$4\,_1H^1 \rightarrow\,_2He^4, \tag{6.2}$$

is 0.0287 u and, since 1 u produces 931 MeV of energy, the production of a single nucleus of helium-4 from light hydrogen releases 26.72 MeV of thermal energy. When 1 kmol of helium-4 is produced (4.0026 kg) the energy released would be $257.5*10^{13}$ J. This is a very high amount of energy. The energy produced by the formation of 41,000 kmols of helium-4 would be sufficient to satisfy the entire energy demand in the USA for an entire year. A few tons of helium-4 produced per second provides sufficient energy to power a star.

The technique of mass spectrometry was further developed in the first part of the twentieth century: more accurate instruments were developed and the mass of light isotopes was measured with a higher degree of accuracy. It became apparent that many isotopes were significantly lighter than the simple sum of their proton and neutron constituents. As a consequence, when protons and neutrons combine to form heavier nuclei, they become lighter than if they existed as independent particles. This gave rise to the concept of "mass defect per nucleon" for the formation of heavier nuclei, or equivalently to the "binding energy per nucleon." In the example of the formation of helium-4, the mass defect per nucleon is 0.007175 u (= 0.0287/4) and the binding energy per nucleon is the equivalent energy of 6.68 MeV. One may plot the binding energy per nucleon vs. the atomic mass of the heavier nuclei to obtain the graph depicted in Fig. 6.1. This figure shows qualitatively the binding energy of several light nuclei such as deuterium

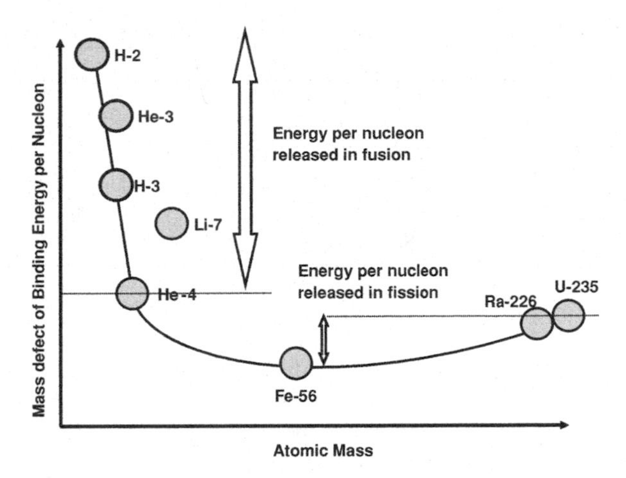

Fig. 6.1 Binding energy per nucleon versus atomic mass

and tritium ($_1H^2$ and $_1H^3$ respectively) lithium-7, helium-3 and helium-4 as well as heavy nuclei, such as uranium-235 and radium-226. The curve has an apparent minimum, which corresponds to the mid-level atomic masses in the range 50–60 u. Common metal isotopes such as iron-56 and nickel-56 correspond to these mass numbers. When reactions proceed from the left to the right of the binding energy curve and atoms combine to form heavier nuclei, fusion occurs. When reactions occur from the right to the left of the curve and atoms split, then we have fission. It is apparent from Fig. 6.1 that the amount of energy released per nucleon in the fusion reactions is by far higher than the amount released per nucleon in the fission reactions. Therefore, fusion reactors have the capability to produce significantly more energy than fission reactors per unit mass of their fuel.

Fusion reactions and Fig. 6.1 may also be used to explain the formation of the Universe and the life of the stars: Shortly after the Big Bang, the first protons and neutrons were formed as well as helium atoms from the fusion of the smaller nucleons. A few minutes after the Big Bang, the primordial matter was formed, composed of 75% hydrogen and 25% helium. Stars were formed from the gravitational attraction of these elements and their radiative energy and high temperatures are a consequence of fusion reactions, such as the one shown in Eq. (6.2). Astrophysicists have proven that stars like our own Sun will cease to radiate when their hydrogen is depleted. This type of stars will contract and become *white dwarves.* More massive stars, where the gravitational attraction is significantly higher than that of our Sun, will continue to shine with reactions that combine helium atoms to form more massive isotopes, such as those of carbon and oxygen. When helium is depleted, the light isotopes will fuse to form heavier elements. The chain of fusion reactions will continue until isotopes are formed at the region of minimum energy of the curve of Fig. 6.1. Further fusion reactions would not release energy to keep the star's material apart and the gravitational attraction will cause the star to first contract to much higher density and then to explode with a tremendous energy release. The star will then be known as a *supernova.*

Table 6.2 Fusion reactions, temperature and density in the six stages of the life of a massive star

Stage	Principal reactions	Temperature°C	Density kg/m^3
I	$_1H^1$ or $_1H^2 \rightarrow {}_2He^4$	1*10^7	10
II	$_2He^4 \rightarrow {}_6C^{12}$ or $_8O^{16}$	2*10^8	1*10^6
III	$_6C^{12} \rightarrow {}_{10}Ne^{20}$ or $_{12}Mg^{24}$	5*10^8	6*10^6
IV	$_{10}Ne^{20} \rightarrow {}_8O^{16}$ and $_{12}O^{24}$	8*10^8	3*10^7
V	$_8O^{16} \rightarrow {}_{14}Si^{28}$	3*10^9	2*10^9
VI	$_{14}Si^{28} \rightarrow {}_{26}Fe^{56}$ or $_{28}Ni^{56}$	8*10^9	4*10^{12}

Astrophysicists distinguish six distinct stages of a massive star before it explodes. The stages in the life of a star are chosen according to the type of fusion reactions that occur and the nuclei present. Table 6.2 shows these six stages as well as the most common fusion reactions taking place during each stage. The reactions feed the star's radiation activity. The temperature and density conditions that sustain the fusion reactions in the stars are also shown in Table 6.2 [2]. One may immediately observe that the fusion reactions take place at extremely high temperatures, where matter is in the state of plasma.[1] Also, that the reactions of stages II through VI take place at tremendously high densities that have never been naturally observed on Earth.

6.2 Man-Made Fusion

The very first, large scale demonstration of artificial fusion was the development and detonation of the hydrogen bomb (H-bomb). Fusion in the H-bomb occurs by the combination of two atoms of deuterium ($_1H^2$ or simply D) to form a single atom of $_2He^4$. The fusion reaction is crudely accomplished by the detonation of a conventional nuclear weapon (the U-bomb). The blast and shock waves produced by the detonation of the U-bomb create the extremely high temperatures and pressures that are necessary for the fusion to occur. The destruction caused by the H-bomb is due to both the fusion reactions that take place as well as the action of the conventional nuclear weapon.

The uncontrolled and destructive explosion of the H-bomb may hardly be used for the production of heat and electricity in a safe and orderly manner. Even though humans were able to reproduce the reactions and unleash the energy of the stars, they fell short of using this energy for their own needs. Sustained and controlled fusion is needed for all peaceful uses of nuclear fusion and this has been the objective of research in several laboratories worldwide since the 1920s.

[1] Plasma is considered the fourth state of matter, in addition to solid, liquid and gas. In the state of plasma the electrons acquire high energies and break free from the nuclei. Thus, the plasma state resembles a gas composed of nuclei and free electrons.

6.2.1 The Paths to Form Helium-4

The reaction depicted in Eq. (6.2) does not occur instantaneously in the stars, because the simultaneous collision of four hydrogen nuclei has extremely low probability to occur. Rather, the reaction occurs in two stages, with deuterium being the intermediate isotope:

a) The formation of deuterium by the combination of two nuclei of hydrogen.
b) The combination of two nuclei of deuterium to form helium-4 or

$$\begin{aligned} 2{}_1H^1 &\rightarrow {}_1H^2 +_{+1} e^0 \quad \textit{or simply} \quad 2H \rightarrow D \\ 2{}_1H^2 &\rightarrow {}_2He^4 \quad \textit{or simply} \quad 2D \rightarrow He^4. \end{aligned} \tag{6.3}$$

This is neither the only, nor the most probable path to the formation of helium-4. The isotopes of tritium (${}_1H^3$ or simply T) and helium-3 may also lead to the formation of helium-4 by reacting with a nucleus of deuterium while emitting a neutron and a proton respectively. These reactions may be written as follows:

$$\begin{aligned} {}_1H^3 +_1 H^2 &\rightarrow {}_2He^4 +_0 n^1 \quad \textit{or simply} \quad T + D \rightarrow He^4 + n \\ {}_2He^3 +_1 H^2 &\rightarrow {}_2He^4 +_1 H^1 \quad \textit{or simply} \quad He^3 + D \rightarrow He^4 + p. \end{aligned} \tag{6.4}$$

Tritium is a radioactive, artificial isotope of hydrogen with a half-life of 12.3 years. Helium-3 is a stable but rare isotope of helium. The two isotopes may be formed by the fusion of two deuterium nuclei:

$$\begin{aligned} 2{}_1H^2 &\rightarrow {}_2He^3 +_0 n^1 \quad \textit{or simply} \quad 2D \rightarrow He^3 + n \\ 2{}_1H^2 &\rightarrow {}_1H^3 +_1 H^1 \quad \textit{or simply} \quad 2D \rightarrow T + p. \end{aligned} \tag{6.5}$$

Another source of tritium is the light element lithium. Two naturally occurring isotopes of this element, with mass numbers 6 and 7, when bombarded with neutrons produce tritium and helium-4 as in the following reactions:

$$\begin{aligned} &{}_3Li^7 +_0 n^1 \rightarrow {}_2He^4 +_1 H^3 +_0 n^1 \quad \textit{and} \\ &{}_3Li^6 +_0 n^1 \rightarrow {}_2He^4 +_1 H^3. \end{aligned} \tag{6.6}$$

The first of the reactions in the last equation requires energy of 2.5 MeV and occurs with fast neutrons that have this level of energy, while the second reaction, releases 4.8 MeV and occurs with thermal (slow) neutrons. From the energetic point of view, the second reaction is preferable for the formation of the fuel tritium, because additional energy is produced, which may sustain the reactions.

In summary, there are three paths for the fusion to occur and for the formation of helium-4 from the reactions of the following isotopes:

1. Two deuterium nuclei (DD).
2. One deuterium nucleus and one tritium nucleus (DT).
3. One deuterium nucleus and one helium-3 nucleus (DHe^3).

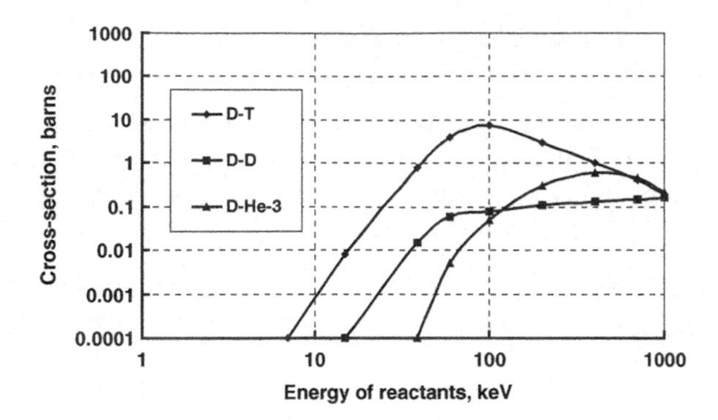

Fig. 6.2 Cross-sections of three fusion reactions: 1 barn = 10^{-28} m^2

As with the fission reactions, the probability that a fusion reaction will occur is determined by the cross-section of the reacting isotopes. Figure 6.2 depicts the pertinent cross-sections for the three reactions as a function of the energy of the reactants. The data in the figure were reproduced from McCracken and Stott [2]. It is apparent in this figure that for the reactions to occur, the reactant isotopes must have very high energies, which correspond to extremely high temperatures (10 keV of energy corresponds to approximately 100 million degrees K). Also, that the DT reaction has an advantage over the two other reactions because its cross-section is significantly higher than the other cross-sections at the lower range of fuel energies. Because of these practical considerations, the DT reaction has been pursued by more laboratories internationally for the achievement of controlled fusion than the alternative reactions.

6.2.2 The Deuterium–Tritium (DT) Fusion Reaction

The development of fusion reactors and the harnessing of controlled fusion power is an awe-inspiring and very expensive task, which is still in the research area with many nations participating, either separately or jointly, for its accomplishment. The principal reason for its pursuit is that fusion may provide mankind with an inexhaustible energy source. Unlike other energy sources that may be exhausted in the near or far future, if appropriately harnessed, fusion will provide mankind with energy for millions of years for the following reasons:

At first, deuterium is a naturally occurring isotope of hydrogen and comprises 0.015% of hydrogen atoms. With all the water on the earth's surface the total amount of deuterium is estimated to be $1.3*10^{18}$ kg. If this quantity were used in fusion reactions, the entire population of the Earth will have enough electricity for 1.3 trillion years. It is the sheer magnitude of such numbers that makes fusion energy research a worthwhile endeavor for the human society. Secondly,

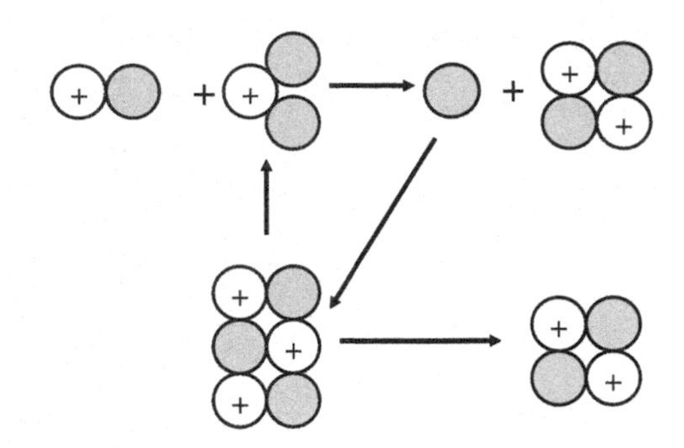

Fig. 6.3 Nuclear fusion reactions that lead to DT fusion. Nucleons with the symbol + denote protons and shaded nucleons denote neutrons

deuterium may always be produced from common hydrogen, which is one of the most common isotopes on earth. Thirdly, even if deuterium is not used for the production of tritium, there is sufficient metal lithium on the surface of the earth to supply with enough Li-6 or Li-7 for the production of tritium. Natural lithium is abundant on the crust of the earth and contains 92.6% of Li-7 and 6.4% of Li-6. The abundance of both deuterium and lithium on the earth make the proposition of producing power via the DT reaction very attractive. If the controlled DT reaction becomes a reality, mankind's energy problem will be solved.

A schematic diagram of the basic reactions that would lead to the DT fusion is shown in Fig. 6.3. Lithium-6 and deuterium are used as the fuel. Tritium is produced from the neutrons released during the DT reaction and two nuclei of helium-4 are produced as the final product. If lithium-7 were used as the fuel, an additional neutron would have been produced in the final product.

A glance at Fig. 6.2 and the values of the cross-sections of the DT reaction proves that the reaction has an optimum and significant probability to occur at the extremely high temperatures of a few hundred million degrees Kelvin, where plasma is the only state of matter. This brings the first and rather formidable challenge facing the scientists and engineers who endeavor to construct a fusion reactor: To maintain the reactor fuel at these extremely high temperatures for the reaction to occur, when it is known that any common material container would evaporate at temperatures below 3,000 K. Obviously, the fuel for the fusion reactors cannot be conventionally contained in a material container, because any contact of the fuel would vaporize the container. Scientists and engineers have agreed that the confinement of plasma in a vacuum by a strong magnetic field and the confinement of a rapidly compressed gas by its own inertia are the best methods to contain the plasma for a sufficiently long time for the fusion to occur. These two methods will be briefly described in the next section.

The second challenge scientists and engineers are faced with is that the fuel must be kept in the state of plasma for a short but significant amount of time, so that the fusion reaction will have the time to take place. This is called the *ignition condition.* A simple formula describes the ignition condition in the range of temperatures $1.0{*}10^8$ K < T < $2.0{*}10^8$ K, which is the optimum range for the DT fusion to occur [2]:

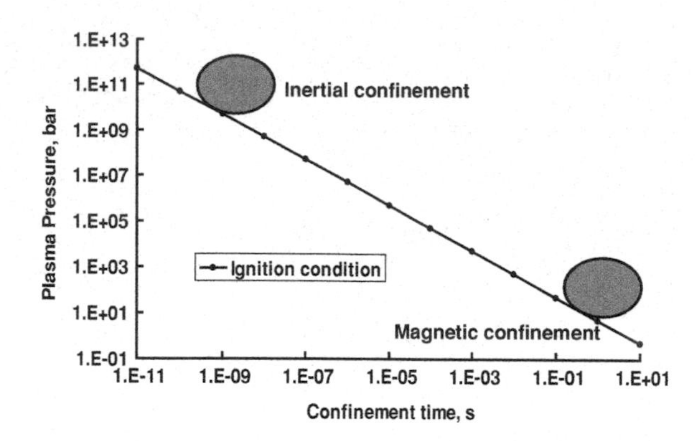

Fig. 6.4 The ignition condition in the confinement time-pressure diagram

$$nT\tau_c > 3 * 10^{21}\, keV * s/m^3, \tag{6.7}$$

where n is the density of all the particles in the plasma, and is measured in particles per cubic meter; T is the temperature of the plasma, measured in energy units of keV (1 keV $\approx$ 10^7 K); and τ_c is the confinement time that is necessary for the reaction to proceed. When one treats the plasma as a gas composed of nuclei and electrons, the product nT is proportional to the pressure of the plasma, P. Hence, Eq. (6.7) may be re-written in terms of the plasma pressure as follows:

$$P\tau_c > 5\, bar * s. \tag{6.8}$$

The last equation is valid under the condition that the temperature is in the range $1.0*10^8$ K $< T <$ $2.0*10^8$ K. Figure 6.4 shows this trade-off between pressure and confinement time for the ignition condition. The conditions achieved by the two confinement methods, magnetic and inertia confinement are at the two opposite ends of the line that defines the ignition condition. The two regions for magnetic and inertia confinement ignition are shown by the shaded ovals. It is apparent from this figure that, under the conditions prevailing during the magnetic confinement, the plasma must be kept at the confinement conditions for approximately 1 s. During inertial confinement, when pressures on the order of 10^{10} bars must be produced, the plasma needs to be confined for only 10^{-10} s. It must be noted that, in both types of confinement the plasma temperature must be between 100 and 200 million Kelvin. The confinement of any material at this temperature presents a formidable technical problem.

The third challenge fusion scientists and engineers must overcome stems from the relatively low cross-section of the DT reaction, which implies low probability of fusion. A comparison of the fusion cross-sections in Fig. 6.2 and those of the fission reactions in Fig. 4.5a, b shows that the cross-sections of the DT reaction are several orders of magnitude lower. Despite the fact that the cross-sections of the DT reaction are the most favorable of the three possible fusion reactions, the significantly lower cross-sections imply that the probability of fusion is much lower than a comparable fission reaction, even when the DT mixture is at the most favorable conditions of plasma. At the ignition condition, the probability of fusion is only on the order of 10^{-8}.

It takes a great deal of energy to produce the high temperature and pressure conditions of plasma. For sustainable fusion reactions, this energy must be recouped from the energy released during the DT reactions. If fusion is to be used routinely as an energy source, a given mass of fuel must provide several times the energy used to bring it to the state of plasma. During sustainable fusion reactions, the plasma particles should not be allowed to "cool off" significantly between fusions and must be maintained at their high energies.

6.2.3 Magnetic and Inertial Confinement of Plasma

The enormous mass of a star exerts a sufficient gravitational force to keep the hydrogen and helium atoms confined to a region around the star's center, to reach the state of plasma, and to achieve ignition conditions. In the case of man-made fusion reactions, plasma must be created from the common states of the known gases and must be kept confined for sufficient time for the fusion reactions to occur. The two most commonly used methods for the confinement of plasma are:

A. Magnetic Confinement

The magnetic confinement of plasma is based on the *pinch effect* of electric conductivity: basically, a high electric current density squeezes the material that conducts it. Since plasma is composed of a vast number of free electrons and positively charged nuclei, it conducts electricity very well.[2] Actually, the conductivity of deuterium plasma is approximately 10 times more than that of copper at ambient conditions, which makes plasma an excellent conductor. In addition to the *pinch effect*, the electric current creates a magnetic field around the conductor. These magnetic forces are tangential to the surface of the conductor. Therefore, electrically charged plasma creates a tangential magnetic field as well as a centripetal force that keeps it confined. However, if the plasma flows in a straight line, it would escape at the exit of the linear conductor and a constant supply of plasma would be needed with all the thermal energy expense to create the fresh plasma supply. A simple solution to this problem is to create a closed circuit for the flow of plasma by bending the conductor to form a closed, circular path and, thus, to induce the plasma to flow in a toroidal, or donut-shaped, path. The toroidal path does not have an exit. Highly energetic plasma circulates continuously in the torus and is not lost. Figure 6.5 shows schematically the circular motion of the plasma, the magnetic field induced and the centripetal confining force, which is created and keeps the plasma confined in the torus. It is apparent that plasma may circulate *ad*

[2] The conductivity of plasma is oftentimes referred to as the *Spitzer conductivity*, in honor of the late Professor Lyman Spitzer of Princeton.

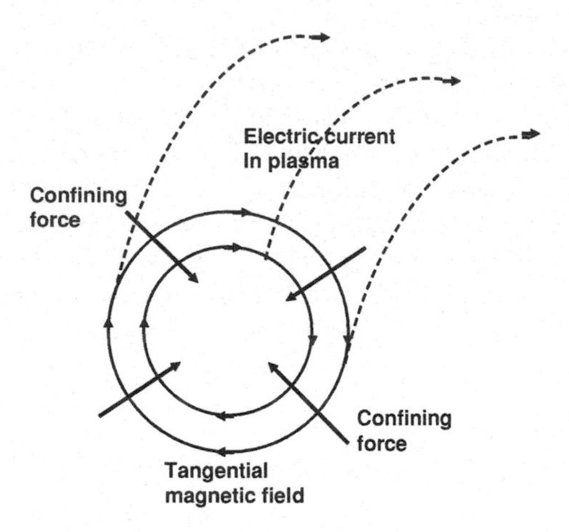

Fig. 6.5 Induced magnetic field and confinement "pinch" force in the toroidal motion of plasma

infinitum in this toroidal path as long as the driving electric and magnetic fields are maintained.

A complication in this picture of smooth toroidal plasma flow is that the charged plasma particles in the presence of the induced magnetic field do not simply move in a closed circular path. The induced magnetic force makes the charged particles to follow solenoidal paths. The radii of the paths are called the *Larmor radii* and are given by the expression [2]:

$$r_L = 0.00457\frac{\sqrt{AT}}{BZ}, \tag{6.9}$$

where A is the mass number of the particle, Z the charge number of the particle, T is the temperature in keV and B the strength of the magnetic field in Tesla. In the case of an electron, the Larmor radius is:

$$r_{Le} = 0.000107\frac{\sqrt{T}}{B}. \tag{6.10}$$

The mass of a deuterium nucleus is approximately 2 u and the mass of an electron is approximately 0.00055 u. Hence, deuterium moves in a solenoidal path whose radius is 60 times greater than that of the electron.

A more serious problem with the confined toroidal motion of plasma is the flow instabilities that are associated with the motion of the plasma. These instabilities are of two kinds:

(a) "Kink" instabilities, which cause the buckling of the torus, and
(b) "Sausage" instabilities that constrict parts of the torus and swell other parts.

Both types of instabilities are undesirable, because they bring the plasma in close proximity or in contact with the confining walls and thus, may destroy the walls that are designed to contain the plasma. The two types of instabilities are suppressed by inducing another current to flow on the toroidal walls that contain the plasma. This solenoidal current induces a stabilizing magnetic field, which suppresses both instabilities and allows the plasma to flow in a smooth torus. The toroidal *Tokamak*, reactor, invented in the former USSR in the 1960s and by now replicated in several different versions by other countries and international laboratories, is the best device that suppresses the kink and sausage instabilities and keeps the plasma well confined in a symmetrical torus. Experimental Tokamak reactors have been built in several countries for the study of nuclear fusion, including one in Princeton, NJ, which operated successfully between 1982 and 1997. A schematic diagram of the arrangement of the essential equipment and the formation of the toroidal plasma field in a Tokamak reactor is shown in Fig. 6.6a (image courtesy of the Lawrence Livermore National Laboratory). An actual photograph of the United Kingdom Tokamak reactor, which currently operates near Oxford is shown in Fig. 6.6b (image courtesy of the UK Atomic Energy Authority). The last photograph also shows the scale and complexity expected in a fusion reactor, even of experimental size.

B. Inertial Confinement

Inertial confinement fusion (ICF) of plasma follows the method used in the hydrogen-bomb: when the deuterium–tritium mixture is compressed and heated quickly, it becomes plasma and reaches fusion conditions before the rising temperature allows it to expand and escape. The inertia of the compression process keeps the plasma confined. In this case, safety dictates that the mass of the compressed fusion fuel is by far smaller than the mass used in a hydrogen bomb. The mass of fuel is also constrained by the amount of energy needed to heat the fuel in a sufficiently rapid way, in order to reach fusion conditions. Because of the tremendous amount of energy released by fusion, burning only five milligrams of a deuterium–tritium mixture would produce energy of the order of 10^8 J. This is equivalent to the energy released from three gallons of gasoline. According to the ignition condition of Fig. 6.4, the DT fusion would occur when the DT mixture reaches a temperature on the order of 200 million K and pressure in excess of 10^{10} bar during only a few billionths of a second.

While the magnetic confinement aims at establishing a steady-state for the plasma in its toroidal motion, when fusion will occur, inertial confinement is a periodic, pulsed process. During ICF the fusion occurs when the fuel reaches favorable conditions and starts to "burn." At this point the temperature is further increased and the fuel starts to expand. Normally, following the laws of gases, the fuel's temperature and pressure would drop significantly during the expansion process. However, the continuous burning of the fuel releases a significant amount of energy and may sustain the high temperature and pressure during the expansion, thus sustaining the fusion process, which spreads to the entire mass of the fuel. It is this released energy that prevents the mass of the stars from collapsing because of gravitational attraction.

A commonly used method for heating and compressing the DT fuel is by *laser ablation*, which is shown in Fig. 6.7. The DT fuel is supplied in solid form as a spherical capsule, with a layer of an ablating material on its surface (a). An intense

(a)

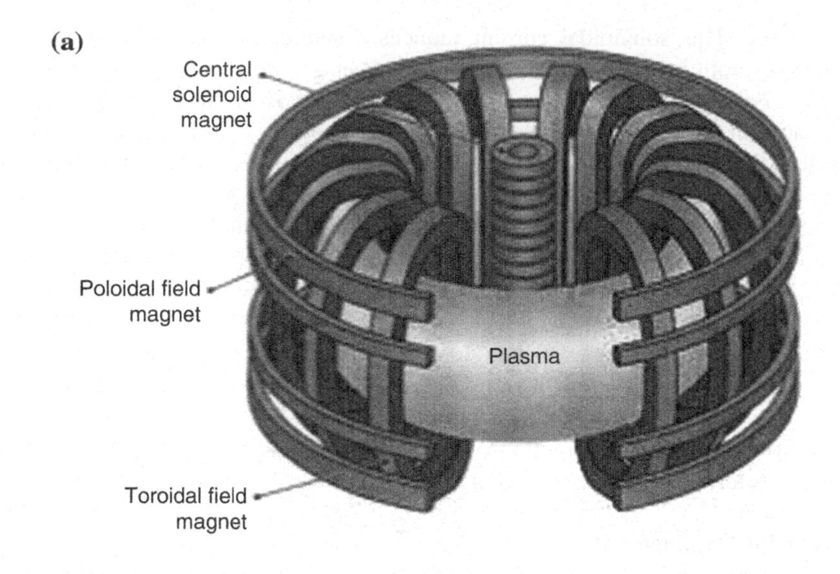

(b)

Fig. 6.6 **a** A schematic diagram of the Tokamak fusion reactor (courtesy of LLLNL) **b** A photograph of the MAST Tokamak reactor at Oxford, UK (courtesy of UKAEA)

laser beam evaporates rapidly (ablates) the material. As the rapidly produced vapor leaves the surface of the fuel in a radial direction, momentum conservation causes a rapid rise of the pressure on the surface of the capsule, which starts to compress (b). If the surface of the fuel capsule is heated uniformly from all sides, the uniform and rapid increase of the pressure causes a fast increase of the density and temperature of

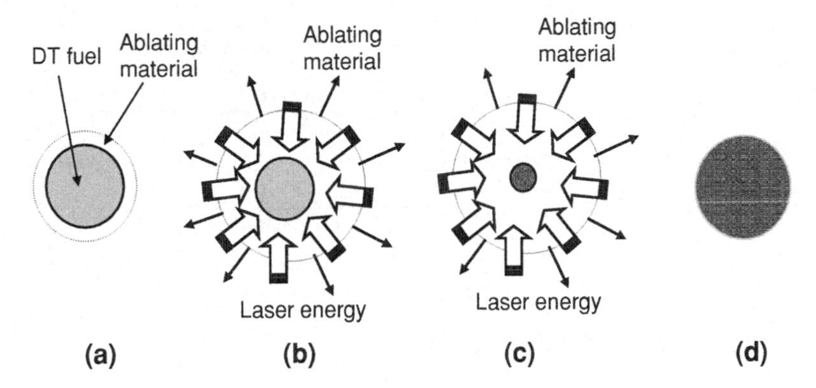

Fig. 6.7 The four stages of the laser ablation process. **a** Fuel capsule and ablating material, **b** Laser heating and compression, **c** Onset of fusion, **d** Fusion sustained expansion

the fuel material. The fuel attains the state of plasma, reaches the ignition condition and the fusion reaction commences (c). The fusion process generates a great deal more energy, which sustains the conditions for the fusion reaction. When the ablation stops, the inertial effects cease and the capsule starts expanding (d).

Neodymium lasers that produce intense bursts of light are used for the laser ablation process. The capsules are typically made of glass or plastic material, which encloses a few milligrams of the DT fuel. Recently, multiple layers of encapsulating materials have been proposed to make the ablation/compression process more powerful. The laser ablation process may create pressures on the order of 10^8 bar. If applied to a capsule of 1 mm diameter of DT solid fuel with an initial density 300 kg/m^3, the rise in pressure will cause a density increase by a factor of 1,000. Under these conditions the fuel will reach a density of 300,000 kg/m^3, which is sufficient to start the fusion process in the fuel mixture [2].

A key detail in the laser ablation process is that the heating of the ablating material must be uniform. Non-uniform heating and ablation create an imbalance of the forces acting on the capsule. With non-uniform heating the ablation forces will become propulsion forces, the capsule will simply move away from the laser's focal point and the fusion will not occur. Even a 0.1% non-uniform heating will cause the capsule to fly away. The *hohlraum* (cavity in German) was invented to prevent heating instabilities and loss of the fuel. The hohlraum is a small cylindrical cavity made of gold, with the spherical fuel capsule supported at its geometric center as shown in Fig. 6.8 [2]. It is arranged for the laser light to enter symmetrically through the two sides of the cylinder. Part of the laser light is reflected on the gold surface and reaches the capsule from all directions. Another part of the laser light evaporates the golden surface and excites the atoms, which produce x-rays that also bombard the capsule from all sides. The combination of many side reflections and x-ray bombardment from all the sides causes uniform heating of the ablation surface, the capsule remains inside the cavity and fusion takes place.

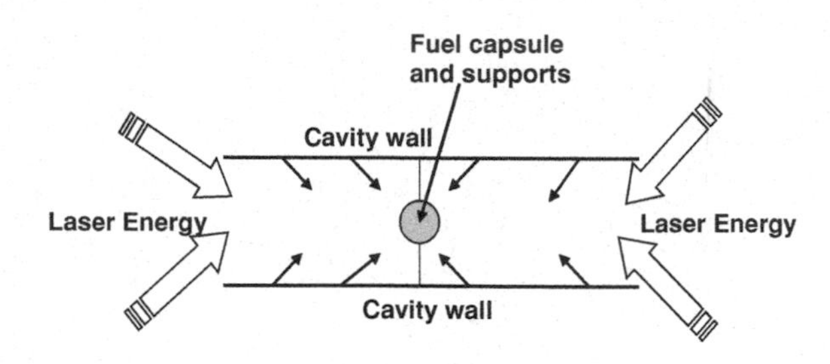

Fig. 6.8 The *hohlraum* or "cavity" causes uniform heating of the ablation material

6.3 A Fusion Electric Power Plant

Controlled nuclear fusion is still in the embryonic stages of research and development. Although significant progress has been made in the understanding and the realization of the fusion process, a power plant that would solely use fusion energy for the production of electricity is still in the realm of futuristic development. Two of the main problems for the realization of a fusion power plant are the sustenance of plasma conditions at the right state for the required durations, and the harnessing of the tremendous amounts of energy released even from a very small amount of fuel. It is expected that, with the concerted and synergistic effort of several nations, technological advances will be made in the next 50–60 years, which will enable humans to successfully harness the energy of the stars. At that point, several fusion power plants will be built that will be capable to convert the thermal energy release by fusion to electricity. Although a fusion power plant is still a futuristic concept, it is not difficult to discern the main parts of this type of thermal power plant, which are depicted in Fig. 6.9:

1. The central part of the plant is the plasma reactor, which is fed with a mixture of deuterium and tritium. The reactor may be of the toroidal type (Tokomak) or may use the pulsed laser ablation method. The central part of the reactor, where fusion occurs is surrounded by a lithium-6 blanket, which absorbs the neutrons and creates tritium, with helium as a by-product. The tritium and helium products are removed and separated. Helium is stored to be used for other commercial processes. Tritium is also stored to be later combined with the deuterium and be used as the DT fuel of the reactor.
2. A radiation shield, made of lead or a similar radiation-absorbing material, surrounds the central part of the reactor.
3. A thermal blanket, which surrounds the reactor and removes the heat generated in the center of the fusion reactor. The thermal blanket is cooled by the circulating coolant, which is water, helium or a molten metal (liquid Na or K).
4. A number of steam generators (SG), essentially large heat exchangers, where the reactor coolant transfers its energy to raise steam.
5. The turbine or series of turbines (T), generators (G), condenser (C), recirculation pump (P) and cooling system (not shown in the figure), are similar to the equipment used in the conventional thermal power plants. In this part of the

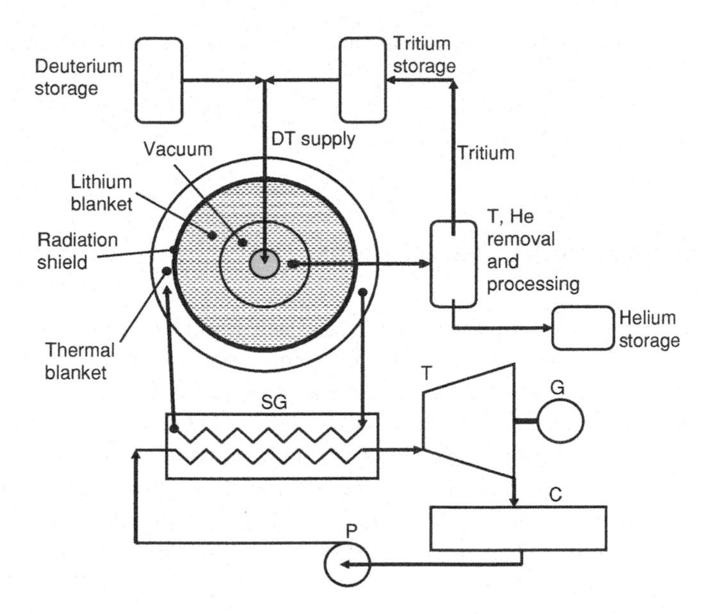

Fig. 6.9 A schematic diagram of a conceptual design of a fusion power plant

plant, steam expands, and electricity is generated. Subsequently, steam is condensed and follows the path of a closed cycle that feeds it back to the steam generator as liquid water.

Perhaps the most important difference of the conceptual design of the fusion power plant to the design of the other thermal power plants is that the fusion plant must produce tritium, which is one of the essential elements of its fuel. Because of this, the fusion reactor is also a breeder reactor. Unlike the fission breeder reactors where the produced plutonium stays and is used in the encapsulated reactor, in the fusion power plants tritium is produced with another gaseous product, helium. Because the two form a homogeneous gaseous mixture, it will be necessary to continuously treat the fuel and remove the helium product. The only practical and available technique at present to accomplish this separation process involves centrifuging the gaseous mixture, a process that consumes high amounts of energy.

The exact design of the fusion power plant will greatly depend on the method used to achieve fusion. If the magnetic confinement method is chosen, the plasma torus that will be used will have to combine most of the parts that are included inside the radiation shield of Fig. 6.9, with an inlet for the fuel and outlets for the helium product. If the inertial confinement method is chosen, the lasers that would provide the ablation power for the fuel pellets will have to be incorporated in the design of the power plant. However, fusion research still needs to overcome several technological barriers before a fusion power plant is to be developed for the immediate future and, for this reason, at present there are no immediate plans to build a fusion power plant. It is almost certain that a future fusion power plant will have the basic components 1–5 as listed above. However, as it often happens with futuristic concepts, when fusion plants become a reality, their actual design and processes will somehow diverge from the conceptual designs that have been promulgated in the scientific literature.

6.4 Environmental Considerations

Fusion energy produces significantly less and at a lower level nuclear waste than fission energy: The main product of the fission reactions is ${}_2He^4$, which is not radioactive. Also, the fuels for the fission reactions, deuterium and lithium are not radioactive either, and are abundant and well-distributed on the surface of the earth. Unlike the fission reactions, there are no radioactive daughter isotopes that are produced in large quantities. The only clearly radioactive isotope, tritium, is subsequently used as a fuel and, in addition, has a relatively short half-life of 12.3 years. This implies that there is virtually a very small amount of waste that will come from the fuel cycle and the fusion products and, this waste will decay in a relatively short amount of time.

Another significant difference between fission and fusion power plants is that the fuel will be continuously supplied and in very small quantities to the reactor. Fission reactors are charged once every 1½–2 years and all the fuel to be used during this time is enclosed in the fission reactor. In contrast, fusion reactors may be continuously supplied with small amounts of fuel. Safety systems may be easily implemented in the fusion reactors to interrupt the supply of fuel and to immediately stop the fusion reaction, if there is any perceived threat to the environment. It is apparent that a fusion power plant and the surrounding region will not be subjected to the dangers of a loss of coolant accident or a meltdown.

The fusion reactor itself and the material of the reactor will become contaminated and will contribute to nuclear waste. As shown in Fig. 6.9 a fusion power plant will comprise several systems that may become contaminated. At first, the radiation shield will be subjected to a high neutron flux that will contribute to its contamination. Secondly, some of the neutrons will inevitably pass to the thermal blanket and the circulating coolant. This will lead to the contamination of the heat exchanger material and, possibly, to the contamination of the steam, which will carry the contaminants to the turbine and the condenser. Contamination of the turbine-condenser system may be avoided to a large extent by including an intermediate coolant circuit as in the case of the Breeder Reactor power plants that are shown in Fig. 5.13. Even though the nuclear waste from a fusion power plant is significantly less than the waste from a fission plant, fusion energy is not environmentally benign and, before we embark on large-scale applications of this type of energy we must, first, solve the problem of plant decommissioning and disposal of the produced nuclear waste.

6.5 "Cold Fusion," Other Myths and Scientific Ethics

As mentioned in several parts of this chapter, if successfully harnessed, fusion energy has the potential to solve the energy needs of humanity for the entire foreseeable life of our planet. Unlike other alternative energy sources that may supply only a small fraction of the energy needed, fusion, in combination with a transition to a hydrogen economy for transportation fuels, has the potential to supply the entire energy demand of the planet and, therefore, is ***the answer to the energy problem of humanity***. For this reason, there are many scientists, who find irresistible the role of "inventor of fusion" and the accompanying title of "savior of humanity." As a result of this ambition, since fusion was discovered, there have been a number of false scientific claims, which have attracted international attention, including the following:

6.5.1 Muon Atomic Fusion

While studying the properties of muons that are produced by cosmic rays in the upper parts of the atmosphere, Professor Frank of Bristol University, in 1947 promulgated the opinion that if the electron in the deuterium atom is substituted by a negative muon,[3] the orbit radius of the muon would be roughly 200 times shorter than that of the radius of the electron and, hence, the new *muon deuterium* would be 200 times smaller than the normal deuterium atom. With such diminutive atoms, the repulsive forces developed between atoms would be significantly weaker and fusion would be much easier to achieve.

Upon closer examination of this fusion concept, two facts became quickly known: first, the muon is unstable with half-life on the order of nanoseconds and, second that the energy produced in a DT reaction is significantly less than the energy required for the production of a free muon. Even if it were possible to use muon deuterium for fusion, the net energy that would be generated would be negative. However, the latter problem could be resolved if the produced muons would be re-used in several reactions. Considerable analytical and experimental research was carried on this subject, which involved scientists from Britain, the Soviet Union and the United States. The conclusion of the combined research efforts is that, although the concept of fusion with *muonic atoms* is tantalizing, its application for the production of large amounts of energy is not a possibility.

6.5.2 Sonoluminescence

Sonoluminescence is the light/illumination produced from sound waves. It is well known that most liquids contain extremely fine bubbles of less than one

---

[3] A muon is an unstable subatomic particle, about 200 times heavier than the electron.

micrometer size. These tiny bubbles respond to sonic or supersonic waves, which are essentially pressure fluctuations. When the gas in a bubble is compressed, its temperature rises significantly. The reason for this rise is that, because of the very small area of the bubble and the short duration of the compression, heat does not escape from the tiny bubble. The compression process takes place almost isentropically according to the well-known expression for isentropic processes:

$$\frac{T_2}{T_1} = \left(\frac{P_2}{P_1}\right)^{\frac{\gamma-1}{\gamma}}, \tag{6.11}$$

where γ is the ratio of the specific heats of the gas in the bubble. Since the pressure ratio in sonic and supersonic waves is very high, the temperatures produced in the tiny bubbles may be very high too. For several years it was speculated that the high temperatures produced in a dissolved bubble may produce plasma. This hypothesis was put forward in many scientific events, including the 3rd International Conference on Multiphase Flow (ICMF-98) in Lyons, France. The discussions in all these scientific events were inconclusive, mainly because there were not any direct observations of neutrons or any other radioactive materials, which are the signatures of all fusion reactions. However, speculation on the attainment of fusion reactions by sonoluminescence was rampant in the 1990s and several large-scale experiments were undertaken to demonstrate the claims.

Reporting on one of these experiments in 2002, Rusi Taleyarkhan of Oak Ridge National Laboratory (ORNL) and several of his colleagues claimed to have observed both the neutron production of the right energy (2.45 MeV) as well as the tritium in their sonoluminescence experiments. They used acetone made of deuterium instead of normal hydrogen atoms (C_3D_6O). This team claimed that neutrons were not observed when they substituted normal acetone, C_3H_6O, in their experimental facility. Because of the enormous implications of the claim, the director of ONRL took the correct scientific stand and asked two independent teams to verify the claim. Both of these teams did not find any evidence of neutrons or tritium in the reproduced experiments with C_3D_6O and concluded that the originally observed neutrons from C_3D_6O were due to an experimental error. Later, a team at the University of Illinois showed analytically that several endothermic chemical reactions, which take place during the sonic compression, consume a significant amount of thermal energy and make it impossible for plasma to be formed and fusion to occur with C_3D_6O.

6.5.3 Cold Fusion in a Test-Tube

Perhaps the most notorious scientific scandal was perpetrated with the claim of "cold fusion in a test-tube" by two researchers of the University of Utah: Stanley Pons and Martin Fleishman held a press conference during March of 1989 and announced to the world they have achieved fusion in an electrolysis cell, kept at room temperature and that the University was filing for patents for this process. During electrolysis an electric current separates water into its constituents, oxygen and hydrogen, which are collected at opposite metal electrodes. Pons and Fleishman used heavy water, D_2O, and palladium electrodes. It is well-known that certain metals, including steel and palladium, absorb hydrogen in their atomic structure. Essentially, Pons and Fleishman claimed that the absorption forces in the palladium metal were strong enough to compress the produced deuterium atoms to a state of plasma and to cause a fusion process. In the beginning, their claim was solely based on the temperature rise they observed in their electrolysis cell. Pons and Fleishman claimed that more energy was released than was put in the cell and attributed the difference to fusion that occurred in the electrolytic cell.

This observation was very surprising to the scientific community, because hydrogen absorption in metals is a very well known process, which has been studied by many others, without any indication of a fusion reaction. An explanation was immediately offered by Pons and Fleishman that the onset of tiny cracks in the metal would release sufficient energy for the fusion to occur.

A major complication in the whole process was that the research results were presented in a press conference and not in a scientific forum, where the normal peer review process would take place. No scientific questions were asked in this conference. The journalists, who are not scientists trained to ask the right questions, accepted the claim unquestionably and lauded the two professors as the greatest innovators of the twentieth century. The story was rapidly disseminated by the popular media, by e-mails and the internet, which at the time was in its embryonic stage. A few weeks later Pons and Fleishman submitted a paper to the leading scientific journal *Nature*, but when the scientific editor asked for details of their experiments they claimed that they did not have the time to provide the information and withdrew the paper. Instead, they gave several press conferences to the impressionable journalists that flocked to Utah and made their paper available through the internet, by e-mail in text file. At the same time, the President of the University of Utah announced in another press conference the initiation of a "Center for Cold Fusion" with funds provided by an "anonymous donor." It was later revealed that there was not such a donor and that the funds came from within the endowment of the University, without the knowledge of the University Trustees.

Many accomplished scientists in several countries around the world had several reasons to doubt this claim: At first the experiments were conducted without any reference to control experiments with normal water, H_2O. Secondly, the peak of the gamma-ray radiation emitted by the neutrons during the experiment was shown to be 2.5 MeV, when the correct energy would have been 2.2 MeV. Pons and Fleishman never explained this discrepancy, but in a later press release the peak energy was "corrected" to 2.2 MeV. Thirdly, when measurements of the neutron flux were communicated, its values were determined to be too small for fusion to have occurred.

With such an important scientific discovery at stake, the scientific community was mobilized and many laboratories around the world attempted to reproduce the

Pons and Fleishman experiments. In the beginning, a few laboratories showed positive or inconclusive results, but soon afterwards all laboratories agreed that the experiment could not be duplicated as the authors claimed. Pons and Fleishman themselves, when challenged to reproduce their experiment, admitted that they were unable to reproduce their results and substantiate their claims. At the same time, the U.S. Energy Research Advisory Board investigated the matter, but they could not find anything that would substantiate the claim of "cold fusion." After a great deal of investigation, a skeptical scientific community looked carefully at the multitude of new scientific data, came to the conclusion that the claim of "cold fusion," as presented by Pons and Fleishman, was false and absurd and disseminated these conclusions to the world in no uncertain terms. The aftermath of this scientific ordeal, which lasted for almost four months, was rather sad: after several inquiries, the two scientists were asked to leave their University positions. The then President of the University of Utah, who blindly backed their claim with the investment on a "center," was forced to resign and was soon replaced.

6.5.4 Ethical Lessons from the "Cold Fusion" Debacle

Securing abundant food and energy are the most important conditions for the assurance of the humanity's future. When it comes to energy, there have been several recorded claims that have proven to be mere exaggerations or even patently false. A search in the popular press or the internet will prove that, even today, there are several devices, engines and processes whose claims violate the laws of thermodynamics. Behind each one of these engines or processes is usually a greedy "scientist," a gullible "company executive" or a misinformed and poorly educated "engineer." The popular press adores and strives on sensational news, such as "cold fusion," "cars running on water," or "an engine that defies gravity." However, the popular press and the internet are not subjected to any rigorous peer review processes. They are not scientific fora, and not everything that is reported in them is necessarily correct or true.

There is a true scientific forum that is open to all scientists, the scientific journals and conferences, which depend on the peer review process. This forum has served science and scientists for more than three centuries and reported all the significant scientific advances, inventions and failures. This scientific forum, and not the popular press, has promoted and created the Nobel laureates and it is to this forum of scientists/peers that every scientist is responsible and held accountable.

Unlike other more esoteric scientific subjects, the production of energy and its environmental effects are very important subjects to the human society and have generated a great debate. Parts of this debate have been carried by the popular press and disseminated fast by an uncontrollable internet. Oftentimes, opinions in this debate have been framed by the popular press as "scientific beliefs" of certain individuals. One must resist this framing of the scientific debate and must recall that modern science has progressed since the days of Galileo, not by a set of "beliefs," not by blindly sticking to a religious, political or philosophical framework of ideas, but by accurate measurements of the natural phenomena and the rational interpretation of the measured data. Good scientists take accurate and verifiable measurements, draw rational conclusions, and report them first to the scientific community through peer-reviewed publications or scientific conferences. Any deviation from this time-honored scientific method of inquiry and reporting should be met with doubt and suspicion of motive.

Problems

1. How many hydrogen atoms are there in 1 m^3 of water? Assuming that these atoms combine to form helium-4, how much thermal energy may be produced?
2. A secondary source of energy for the stars is the combination of three helium-4 nuclei to form a C^{12} nucleus. The atomic mass unit (u) is defined as 1/12th of the mass of C^{12}. How much energy is released from the combination of 100 kmols of He^4 to form C^{12}?
3. The decomposition of water to hydrogen and oxygen molecules consumes 286,000 kJ/kmol and the decomposition of molecular hydrogen to atomic hydrogen an additional 216,000 kJ/kmol. Assuming that the conversion of hydrogen atoms to deuterium and, subsequently, to helium-4 were possible, how much net energy may be produced by the hydrogen atoms in 1 kg of liquid water? What conclusions do you draw from this number?
4. Assume that the conversion of hydrogen atoms to deuterium and, subsequently, to helium-4 is made a reality in thermal power plants with 35% overall efficiency and that all the hydrogen is derived from water. How much hydrogen would be needed to supply the entire annual electric demand of the United States? How many m^3 of decomposed water does this hydrogen quantity correspond to?
5. The DT reaction appears to be the best candidate for sustained fusion. If the tritium is obtained from naturally occurring lithium and the deuterium is obtained from water, how many tons of Li and H_2O will be needed annually to produce the entire electric demand for France, which in 2008 was 490 TWh? Assume that the fusion thermal power plants' overall efficiency is 32%.
6. Equation (6.6) for the production of tritium requires that lithium be irradiated with neutrons. Suggest a method for the production of neutrons and describe an engineering system to be used for the collection of the tritium produced.
7. What are the temperatures of deuterium, tritium and helium-3 at energies, 10, 30, 100, 300 and 1,000 keV?
8. Explain by using a great deal of quantitative information why the deuterium–tritium reaction is considered the best reaction for sustained fusion.
9. Plasma pressures of 10^4 bar may be routinely created during an isentropic compression. What is the minimum confinement time (in ms) that would be necessary for fusion reactions to occur during isentropic compression?
10. Calculate the Larmor radii of the deuterium and tritium ions at 10^6 K and $B = 100$ Tesla.

11. Explain in detail what is a “hohlraum” and how fusion is facilitated by it.
12. Explain in an essay of 450–500 words the advantages and disadvantages of magnetic and inertial confinement processes for the attainment of sustained fusion. What engineering problems do you foresee?
13. What are the ethical lessons for an engineer from the several false claims on the achievement of “cold fusion?”
14. “Fusion energy may supply the energy needs of the entire human population for millennia. This fact alone justifies the huge investment required for the realization of sustained fusion.” Comment in a 250–300 word essay.

References

1. Bennet DJ (1972) The elements of nuclear power. Longmans, London
2. McCracken G, Stott P (2005) Fusion—the Energy of the Universe. Elsevier, Amsterdam

Chapter 7
Solar Energy

Abstract At an average rotational radius around the Sun of $1.49*10^{11}$ m and with an average radius of 6,378 km, the planet Earth receives solar radiation power of approximately $1.73*10^{14}$ kW, a quantity that surpasses by far all the power requirements of the Earth's inhabitants. This continuously received power, which is called *incident solar radiation* and shortened to *insolation*, integrates to a total energy of $5.46*10^{21}$ MJ per year, or more than 100 million times the total energy used by earthlings in a year. This tremendous amount of energy is abundant, free of charge, almost uniformly distributed and available to all nations and inhabitants of the planet. However, only a very small fraction of the incident solar radiation is used by the Earth's population. Passive solar heating systems provide with space heating and hot water a low fraction of the buildings, primarily in OECD countries, while thermal solar power plants and photovoltaic cells provide a small fraction of the electricity consumed. Despite its low utilization at present, and because of the enormous amounts of power reaching the Earth, solar energy is a prime alternative energy source and has the potential to supply a very high fraction, if not all, of the power used by the Earth's population. This chapter starts with a short exposition on the amount of solar energy available and continues with the exposition of the two main families of systems that are currently used for power production from solar energy: solar thermal systems and photovoltaic systems. Photovoltaic solar cells and solar thermal systems utilize the solar energy in entirely different ways and are examined separately. Emphasis is given on the electric power production by solar energy as well as the current and proposed systems used for electric power production. A brief mention of the passive heating systems and the environmental effects of solar energy utilization are also included in the chapter.

E. E. (Stathis) Michaelides, *Alternative Energy Sources*,
Green Energy and Technology, DOI: 10.1007/978-3-642-20951-2_7,

7.1 Earth-Sun Mechanics and Solar Radiation

With a period of one year, which is approximately defined as 365.25 days, the Earth rotates around the Sun in a slightly elliptical orbit. This orbit has major and minor axes equal to $1.54*10^{11}$ and $1.45*10^{11}$ m. The Earth also rotates daily around its polar axis, which is directed from the North Pole to the South Pole. The polar axis is at a constant inclination of 23.45° to the plane of the elliptical orbit. The Earth's inclination causes the differential energy absorption by the two hemispheres of the Earth and causes the seasonal variations of solar radiation, the local temperature variations, the local wind patterns and the local seasonal weather. Figure 7.1a shows the position of the Earth on four days of the year, which characterize the beginning of the four seasons in the northern hemisphere: winter, spring, summer, and autumn (fall) respectively: a) the winter solstice (it usually occurs on December 21); b) the vernal (spring) equinox[1] (usually on March 20); the summer solstice (usually on June 21) and the autumnal equinox (usually on September 23). Figure 7.1b shows the polar axis (PA) and the axis of the elliptical orbit (OA) during these 4 days of the year. The slightly different duration of the four seasons is due to the elliptical orbit of the Earth, which makes the position of the Earth to be closer to the Sun during the winter season for the northern hemisphere. According to Keppler's laws the translational speed of the Earth is also highest during the winter. As a consequence, the winter season in the northern hemisphere is slightly shorter and the summer season is longer than the other seasons. The opposite holds for the southern hemisphere.

It is apparent from Fig. 7.1a that, with the exception of the two equinox days, the two hemispheres of the Earth receive different amounts of solar radiation. This causes the development of the seasons and the seasonal variations of climate and weather. Given the combination of the two motions, the rotation of the Earth around its polar axis and the motion of the Earth around the Sun, the intensity of the solar energy received by any point on Earth varies significantly, but varies in a predictable way. A stationary receiver of solar energy would receive power in a periodic manner, based on two timescales: diurnal and annual. This receiver will only receive energy during the daytime hours. If the receiver is in the northern hemisphere, it will receive more energy during a summer day than during a winter day, simply because the summer daytime is longer. This type of reasoning has led many to characterize solar energy as *intermittent.* A moment's reflection, however, would prove that the variability of insolation at a given terrestrial location is periodic and is based on the distance of the Earth from the Sun and its orientation.

---

[1] *Solstice* may be roughly translated as the day when the sun stands still in the sky and *equinox* (equal night) implies that daytime and nighttime are equal. During the two solstices the sun is at the northernmost and the southernmost points. At these points the sun appears to stop moving and reverses its direction, thus receding towards the equator. Despite the meaning of the term, during the two equinoxes, day and night are not exactly equal because of the refraction and diffusion of the sunlight, which is caused by the atmosphere of the Earth.

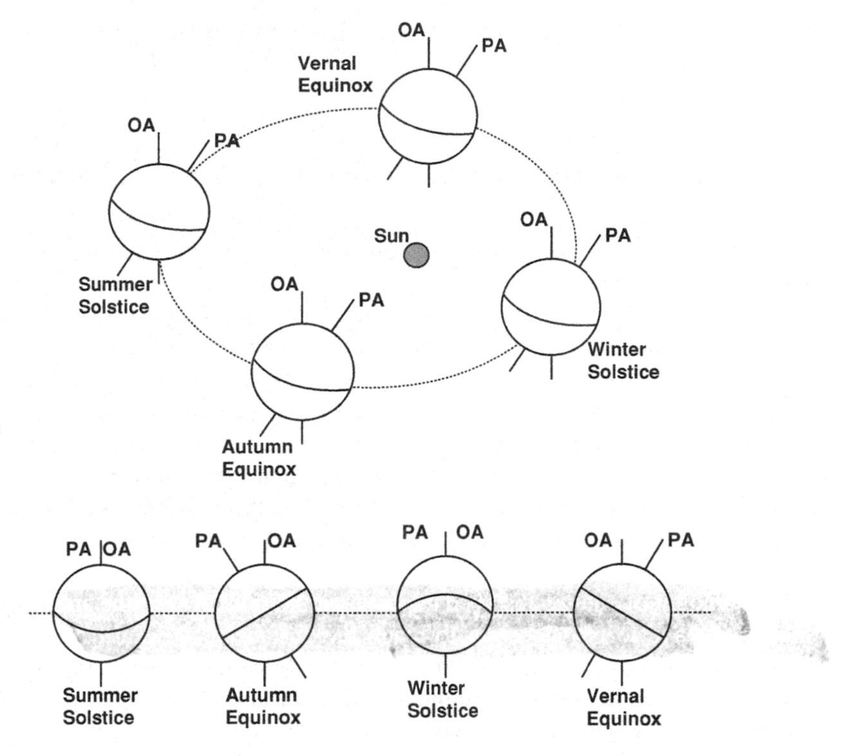

Fig. 7.1 **a** The elliptical orbit of Earth around the Sun and **b** The Earth's orientation relative to its orbit around the Sun. PA is the polar axis and OA is the axis of the plane of the elliptic orbit

The latter are periodic functions of the annual and diurnal rotations of the Earth. Cloudiness, humidity, pollution and other temporal variables that affect the insolation at a specific location provide a small amount of almost random perturbation to the local insolation. As a consequence, we know with a high degree of certainty that during the week of July 10–17, solar collectors in Arizona will collect a great deal of incident radiation, while during the week of December 7–14 the solar energy that will be collected by the same collectors is significantly lower. Therefore, a better way to characterize insolation and solar energy is *predictably variable* or *periodic.*

The power emitted by the Sun is homogeneous in all directions. A fundamental quantity that characterizes the incident solar radiation or insolation is the *solar constant*, which is the power received by a unit area, perpendicular to the solar rays and is located outside the Earth's atmosphere. The solar constant, *S*, which is approximately equal to 1.353 kW/m^2 or 4.871 MJ/(hr*m^2) or 438 Btu/(hr*ft^2) or 1.940 Langley/min, is the yearly average power a collector would receive outside the Earth's atmosphere, when it is faced perpendicularly to the Sun's rays. S is not exactly a constant, because the distance of the Earth from the Sun is variable, with a period of one year. However, the amplitude of the variation of the Earth's radius, or distance from the Sun, is small enough (3%) for the insolation to be considered almost constant.

7.1.1 Solar Spectrum and Insolation on a Terrestrial Surface

It is well known that the Sun's radiation is not monochromatic. The Sun's radiation has the characteristics of a black-body radiation at approximately 5,762 K. The full spectrum of power that is emitted by the Sun may be intercepted at the outer atmosphere. When the Sun's radiation enters the terrestrial atmosphere, the various elements that constitute the atmosphere absorb preferentially parts of this spectrum. Among these, oxygen, water vapor and carbon dioxide absorb a great deal of the solar radiation at discrete wavelengths in the visible light and the infrared part of the solar spectrum. Therefore, when the solar radiation is received at the surface of the Earth, part of the energy of the radiation has already been absorbed and several parts of the solar spectrum are missing because of this preferential absorption.

The amount of energy absorbed depends on the thickness of the atmosphere the solar rays have to pass in order to reach the surface of the Earth. For example, during the equinox, where the Sun is directly above the equator, the Sun's radiation has to pass through the minimum amount of air mass to reach the surface of the equator. This is usually denoted as m = 1, with the parameter m indicating the "mass" of air the Sun rays must travel to reach a specific location. During the winter days, when the Sun is in the southern hemisphere, the Sun rays have to travel through a thicker air path to reach locations in the northern hemisphere, such as New York or Germany. During winters the parameter m for New York City is close to 3 and for some locations in Germany it may become as high as 4.2, which indicates that a great deal of the insolation has been absorbed by the atmospheric constituents. According to this notation, the solar radiation outside the atmosphere and the solar constant $S = 1.353\ \mathrm{kW/m^2}$ would correspond to $m = 0$. Clearly, the parameter m is closely related to the geographic latitude of the location.

The Sun may be accurately approximated as a *black-body*, which has several convenient radiation properties. Among these are: a) the power emitted or absorbed is homogeneous and constant in all directions and b) the energy density is a function of the temperature of the black body and given by the expression:

$$I_\lambda = \frac{2\pi hc^2}{\lambda^5\left(\exp\left({}^{hc}/_{\lambda kT}\right) - 1\right)}, \tag{7.1}$$

where λ is the radiation wavelength, T is the absolute temperature, c the speed of light ($3*10^8$ m/s), k is the Boltzmann constant ($1.380*10^{-23}$ J/K), and h is Plank's constant ($6.62*10^{-34}$ Js). Since the speed of light is equal to the product of the frequency of radiation v and the wavelength ($c = \lambda v$) the energy density may also be written in terms of the frequency of radiation, Iv as follows:

$$I_v = \frac{2\pi h v^5}{c^3\left(\exp\left({}^{hv}/_{kT}\right) - 1\right)}. \tag{7.2}$$

It is apparent from the last two equations that the energy density of the solar spectrum is approximately zero at the two extremities of the spectrum—where v and λ are zero and very large respectively—and that the function exhibits a maximum. The maximum of the energy density may be obtained by differentiating either Eq. (7.1) or Eq. (7.2) with respect to λ or v. Thus, the wavelength λ_{max} where the maximum energy density of the spectrum occurs is given by the expression:

$$\lambda_{max} T = 0.0029\ mK. \tag{7.3}$$

This expression is known as *Wien's Law*. The total radiation power, $\dot{W}_{rad}$, emitted by a black body is calculated by integrating the appropriate expression for the energy density, Eq. (7.1) or (7.2) over all the wavelengths or the frequencies and around a sphere that surrounds the black body. The final expression of this integration is:

$$\dot{W}_{rad} = A\int_0^\infty I_\lambda d\lambda = A\int_0^\infty I_v dv = A\sigma T^4. \tag{7.4}$$

where σ is the Stephan-Boltzmann constant: $5.67*10^{-8}$ W/(m^2K^4) and A is the area of the receiver. If the radiating body may not be considered a "black body" then it is a *gray body* and the power emitted by it is multiplied by an empirical correction factor, the emissivity, ε. Equation (3.6) for radiation heat transfer which uses the concept of emissivity and the geometric factor of interaction of two radiating objects has been derived from Eq. (7.4).

Figure 7.2 shows this transformation of the solar spectrum as it passes through the terrestrial atmosphere by depicting the energy density, I, of radiation for each wavelength in W/(m^2 nm). The figure shows three graphs:

a) The solar spectrum outside the atmosphere;
b) The approximate solar spectrum received at the surface of the Earth for $m = 1$ after all the atmospheric effects on the incident radiation; and,
c) The radiation spectrum of a black body at 5,762 K for comparison.

The total insolation in a given location on the Earth's surface may be determined by integrating the local solar spectrum along all the wavelengths or frequencies. This may be accomplished, simply, by a summation process. Following Eq. (7.4) the insolation or total power per unit area, $\dot{W}/A$, received is equal to the integral under the energy intensity curve, that is:

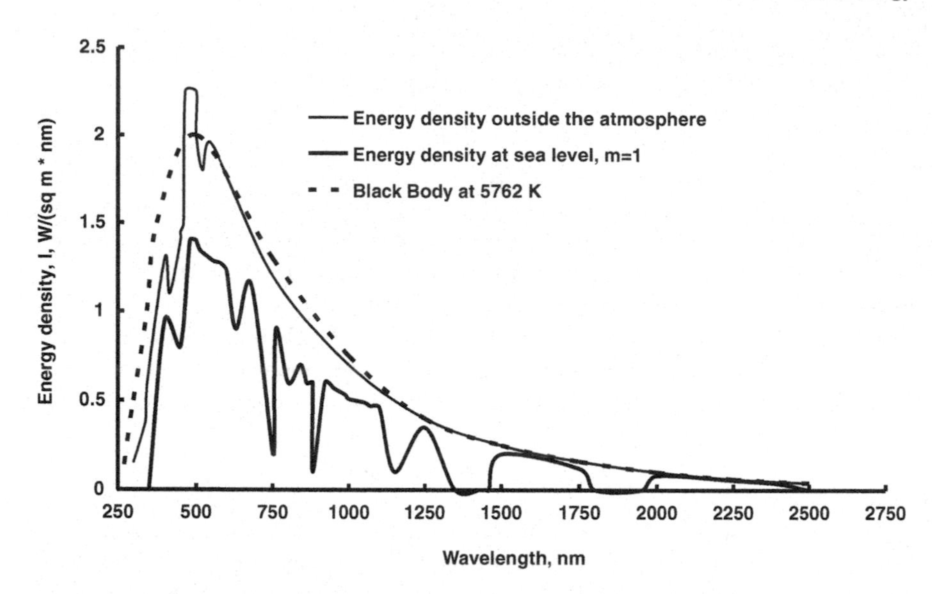

Fig. 7.2 The solar spectrum in the outer atmosphere and on the terrestrial surface for $m = 1$. The black-body radiation for $T = 5{,}762$ K is given for comparison

$$\frac{\dot{W}_{\mathrm{rad}}}{A} = \int_{0}^{\infty} I_{\lambda} d\lambda. \tag{7.5}$$

It is apparent that $\dot{W}_{\mathrm{rad}}/\mathrm{A}$ is a variable with spatial and temporal dependence.

It is observed in Fig. 7.2 that the solar radiation, which reaches the outer atmosphere, is very close to and may be well approximated by a black body at approximately 5,800 K. It is also seen that the spectrum at the surface of the Earth is significantly different and that the total power that reaches the surface of the Earth—the integral under the graph for $m = 1$—is significantly lower than the total power received by the outer atmosphere. Because of this, the power that reaches the surface of the Earth is significantly lower than the solar constant S. On a clear day, the power that reaches the Earth varies from approximately 1 kW/m^2 at noon on the equator to <0.2 kW/m^2 in Canada and the UK.

Another factor that affects the local insolation is the cloudiness or clarity of the atmosphere. Clouds reflect a great deal of the radiation they receive, while both clouds and fog disperse some of this radiation. Hence, locations with high average cloudiness during the year receive lower average power from the Sun. The diffusion of the solar radiation from clouds, dust and fog in the atmosphere as well as the radiation, which is absorbed by the atmospheric gases, is emitted back in all directions homogeneously. This secondary emission is known as the *diffuse radiation* and contributes to the total radiation received by a terrestrial collector. While absorption and diffusion reduce the power that reaches a location, the secondary emission of the diffuse radiation contributes, sometimes

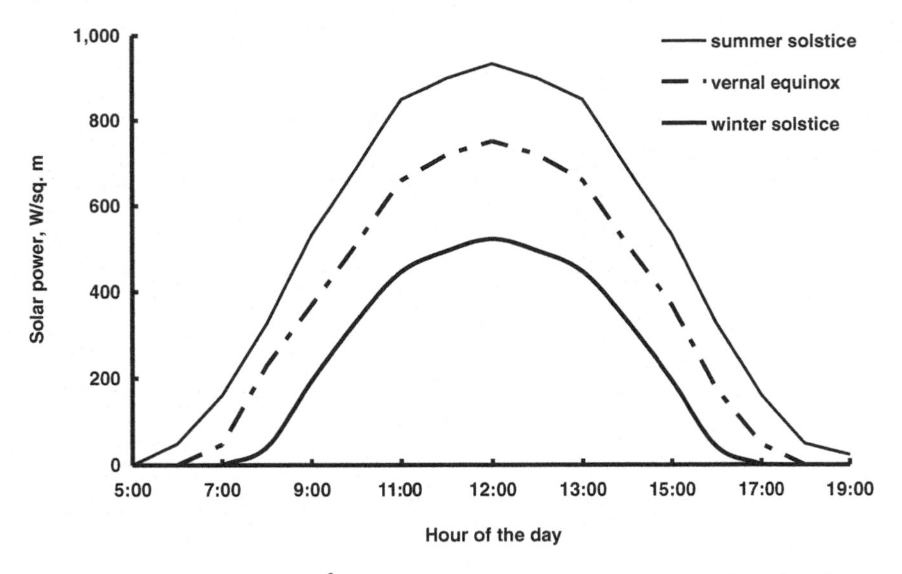

Fig. 7.3 Solar insolation, in W/m^2, received in San Antonio, Texas by a horizontal surface

significantly, to the total power received in several locations, especially on foggy days. For example, in Oxford England, a location known for its high cloudiness, on a bright summer day the direct radiation contributes 7.0 kWh/m^2 and the diffused radiation approximately 1.4 kWh/m^2 for a total of 8.4 kWh/m^2 (the instantaneous insolation was integrated for the entire day in this example). On a dull winter day in the same city the corresponding amounts of direct and diffuse radiation are 0.5 and 0.3 kWh/m^2 for a total of 0.8 kWh/m^2 during the entire day. Typically, the diffuse radiation amounts from 10 to 40% of the direct radiation on a given location, but, as this example shows, this fraction may exceed 60%.

A third factor that affects the radiation received on a terrestrial location is the time of the day. There is significantly less power during the morning and the afternoon and there is zero solar power during the night. The *solar noon* for a location is the time of the day when the Sun is at its highest point on the horizon and the solar power that reaches the location is at a maximum. Figure 7.3 depicts the solar power per unit area, in W/m^2, received in San Antonio, Texas, on 3 days of the year, the two solstices and an equinox. It is evident that the power per unit area received on all days of the year will fall between the curves that correspond to the two solstices. The total energy received during a day is equal to the area under the daily curves:

$$E = \int_{-12}^{12} \dot{W}_{\mathrm{rad}} dt, \tag{7.6}$$

where $t = 0$ represents the solar noon and the time units are hours.

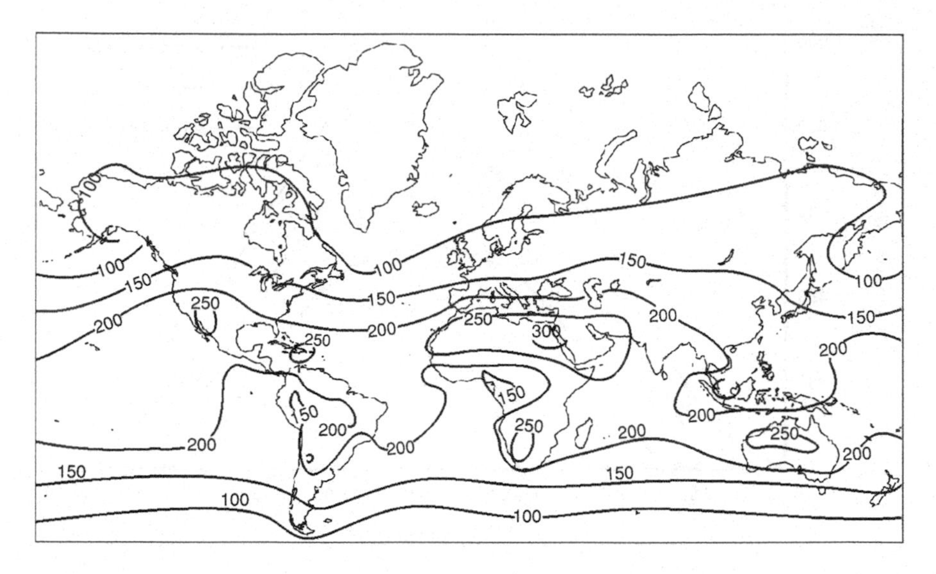

Fig. 7.4 Contours for the annual average solar radiation on a horizontal surface in W/m^2

7.1.2 Average Annual Solar Power: Solar Energy Potential

It is apparent that the total amount of solar energy received by a collector on the surface of the Earth during a day or during a year depends on several factors, most important of which are:

1. The geographic latitude of the location of the collector.
2. The average cloudiness or coverage of the location.
3. The day of the year.
4. The angle of the collector with the horizontal.

Several average quantities are being used to determine the potential of a location to convert solar power to useful energy. Among these are:

1. The maximum daily average power on a horizontal surface (W/m^2).
2. The maximum daily energy received by a horizontal surface (J/m^2).
3. The annual average incident power on a horizontal surface (W/m^2).
4. The annual energy received by a horizontal surface (W/m^2).

A solar installation that would convert the incident solar power to thermal or electrical energy operates throughout the year and, therefore, its merits must be evaluated based on the total energy it produces for an entire year, not just a few bright days. For this reason, the yearly averages of energy and power are the best indicators of the solar energy potential in a given location. A global map of the annually averaged incident solar power on a horizontal surface is shown in Fig. 7.4, where the contours depict the average solar power levels.

The contours on this map represent the solar power potential of the Earth's locations, averaged annually, and take into account all the factors that affect the incident solar radiation. A few conclusions may be drawn from this map:

1. Despite the fact that the solar constant is equal to $S = 1.353$ kW/m^2 and the maximum incident radiation at an equatorial location is close to 1.0 kW/m^2, when the solar radiation is averaged over the entire day and night and over the whole year, the best locations on Earth receive an average of only 250–300 W/m^2. These locations are in the Sahara and the Arabian Peninsula, which are considered prime regions for the utilization of solar energy.
2. While the best locations for solar energy conversion are in the equatorial zone, there are several temperate locations where the average annual solar power is close to that observed in the equatorial zones. Most notable among these locations are Mexico and the Southwest of the United States, all the Mediterranean countries, South Africa, Australia, Argentina and Chile. Lower cloudiness and less rainfall in these regions are the reasons for an annually averaged power close to 200 kW/m^2.
3. Almost all the OECD countries, where the capital to develop and build solar energy systems is readily available and most of the solar facilities exist today, are located in areas with annually averaged power between 100–200 kW/m^2.
4. The geographical area from Pakistan in the East to Korea and Indonesia to the west, where 60% of the Earth's inhabitants live and which is experiencing a high economic growth and very high energy demand growth is within the contours of 150–200 kW/m^2. This region has an excellent potential for solar energy utilization. Higher solar energy utilization in these developing economies is not only feasible and advisable but will ease significantly the energy shortage in the world as well as the environmental effects of fossil fuel combustion.

The engineering systems that have been developed for the harnessing of solar energy are divided into two categories:

A. Solar-Thermal systems, which include systems for the storage of heat as well as the production of electricity.
B. Direct Solar-Electric or Photovoltaic systems.

7.2 Solar-Thermal Systems

The common characteristic of all Solar-Thermal systems for energy conversion is that they capture the solar radiation as heat and subsequently use this heat primarily for two purposes:

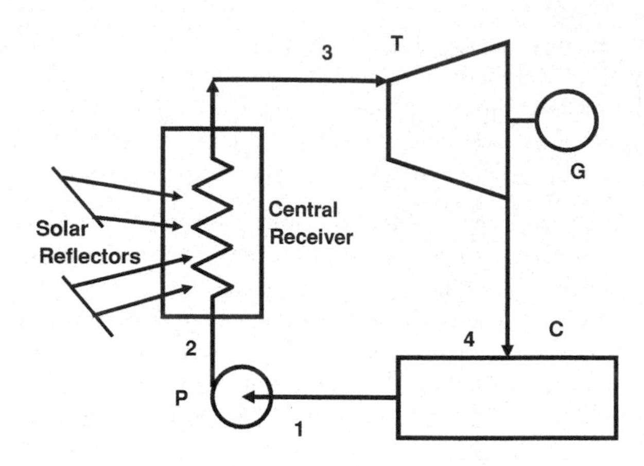

Fig. 7.5 Essential components of a solar thermal power plant

a) The production of electricity via a conventional thermal cycle, and
b) Residential space or water heating and the supply of heat for commercial and industrial processes.

Solar thermal collectors are required for all these processes. Solar thermal collectors intercept and absorb the Sun's rays and transfer this heat to the pertinent energy conversion system.

7.2.1 Power Cycles

A Rankine cycle (simple or with superheat) is typically used for the electricity production from solar energy. The boiler in this cycle is replaced by a system of solar collectors or reflectors that impart the solar energy to water or another working fluid, which undergoes the Rankine cycle and produces electric power in a conventional turbine. Figure 7.5 depicts the essential components of such a solar-thermal power plant, which are:

a) The pump that circulates the working fluid;
b) The system of *solar reflectors* and a *central receiver* that receives the solar energy and raise vapor;
c) The vapor turbine, usually working with steam and
d) The condenser.

It is observed that the components of Fig. 7.5 are identical to those of the simple Rankine cycle of Fig. 3.7 with the exception that the boiler has been substituted by the reflector-receiver combination. It follows that the thermodynamic diagram of this solar powered cycle is identical with the T-s diagram depicted in Fig. 3.8a and the P–v diagram depicted in Fig. 3.8b. Similarly, all the Rankine cycle improvements that have been enumerated in Sect. 3.6.1 may be applied to the solar powered cycle. Typically, the receivers of a solar power plant may produce steam at 550–600°C. Irreversibilities of the practical Rankine cycle cause the cycle efficiencies to be in the range 38–42%. The overall thermal efficiency of the solar plant is significantly lower (20–28%) because of other thermal losses associated with the collection and transmission of solar power. These are enumerated in Sect. 7.2.2.

A variation of the Rankine cycle working with steam is to use another fluid, such as a refrigerant or a hydrocarbon. The choice the Organic Rankine Cycle (ORC) allows the use of lower temperatures, which are usually accompanied by lower efficiencies. It also adds to the cost of construction of the solar power plant.

The operation of the solar thermal power plant cycle is very similar to that of any other small, Rankine cycle, power plant. Steam is raised at high pressure and is directed to the turbine at pressures in the range 10–50 bar temperatures in the range 550–600°C. The turbine and the condenser are located at the foot of the receiver tower and operate in a way that is similar to turbines and condensers of fossil fuel plants. However, cooling water is not always available in locations with high potential for solar thermal power plants. For example, the Southwestern part of the United States and the Saharan countries are excellent regions for solar power production. Both regions are also arid, with very few rivers or natural lakes that could provide enough cooling water to a power plant. For this reason, solar thermal plants are usually designed with dry cooling towers, that is, they are air cooled. As a consequence, their condensers operate at temperatures in the range 50–60°C. This is detrimental to the overall thermal efficiency of the power plants.

Striving for higher cycle efficiency and more power out of a given amount of insolation, engineers have developed several *Stirling Cycle engines* in the first part of the 21st Century. The T,s diagram of the Stirling cycle, which works with air as the working fluid, is shown in Fig. 7.6. Process 1-2 is an isothermal compression of the gas accomplished by a compressor with an embedded heat exchanger. Process 2-3 is an isochoric (constant volume) compression, where the addition of heat increases the pressure of the gas. Process 3-4 is an isothermal expansion in a gas turbine, also embedded with a heat exchanger. Finally, process 4-1 is an isochoric expansion, where the removal of heat causes the depressurization of the gas. A great deal of the heat rejected during the process 4-1 may be transferred to the heat addition process, 2-3 by a heat exchanger, which is called a *regenerator*, thus lowering significantly the overall heat requirements of the cycle.

The four processes of the Stirling cycle may be accomplished in a cylinder-piston engine. A thermodynamic analysis of this cycle shows that its theoretical efficiency is equal to the Carnot efficiency, which is the maximum possible efficiency a thermal engine may achieve:

$$\eta_{St} = \eta_C = \frac{T_3 - T_2}{T_3}. \tag{7.7}$$

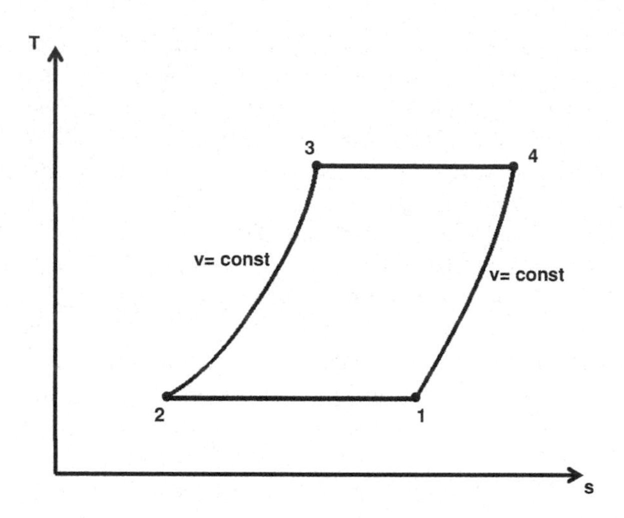

Fig. 7.6 The T-s diagram of the Stirling cycle

However, practical Stirling engines exhibit efficiencies significantly lower than this upper limit, because of internal cycle irreversibilities. One of the reasons for the lower efficiencies is that the two isothermal processes, 2-3 and 4-1, are difficult to achieve in finite time, because heat diffuses slowly in and out of the gas. In addition, the fouling of the heat transfer surfaces contributes to the quick deterioration of the Stirling engine because it slows more the heat transfer processes. Despite of these limitations, the Stirling cycle has been extensively studied and refined since the 1990s and several engines that operate accordingly have been constructed. As a result of this research, several Stirling engines are now in operation with practical efficiencies in the range 25–35%. Technological advances in this area, which are focused in the areas of almost isothermal compression and expansion, may deliver engines that are more suitable for solar energy utilization.

Figure 7.7 shows an aerial photograph of *Solar Two* a small, 10 MW pilot thermal power plant that was built near Barstow, California. Solar Two resulted from the conversion of a smaller thermal power plant, Solar One, by the addition of several rows of heliostats and was decommissioned in 1999. The receiving tower, the turbine housing building, the solar reflectors and the large area covered by the reflectors of this rather small power plant are visible in the photograph. One of the attractive features of the Solar Two plant was the use of a pool of molten salts (60% $NaNO_3$ and 40% KNO_3) as heat storage medium. This allowed the power production to continue normally during brief reductions of insolation due to cloud coverage. It also allowed the plant to produce power up to 3 h after sunset. While the Solar Two power plant was decommissioned, this type of project of solar thermal energy conversion is considered successful: *Solar Tres*, a new, scaled up 15 MW pilot project with energy storage capacity is under construction in Andalucía, Spain.

Fig. 7.7 An aerial photograph of Solar-Two, a 10 MW solar thermal plant near Barstow California (courtesy of Sandia National Laboratories)

7.2.2 Solar Reflectors and Heliostats

One of the salient characteristics of solar is that it is a diffuse form of energy. This means that the energy density incident on a given area is relatively low, typically less than 1 kW/m^2 and, oftentimes, much less. Let us consider a small solar-thermal power plant that produces maximum power[2] of 15 MW with an overall thermal efficiency 25%. We would need approximately 60 MW of heat power for the operation of this plant. A simple calculation shows that even at the maximum rate of incident radiation of 1 kW/m^2, we would need at least 60,000 m^2 of reflecting area. This area corresponds to a large number of solar reflectors, which must be placed in a predefined pattern with all of them reflecting the Sun's energy to the central receiver.

Ideally, the solar reflectors would be slightly parabolic—in order to direct the solar energy to a point in the solar receiver—but, oftentimes flat reflectors are used because they are less expensive and easier to manufacture. The dimensions of the

[2] Solar thermal power plants are rated based on the maximum power they may deliver. The average power produced by a solar power plant is always significantly less than the rated power.

solar reflectors are squares or rectangles with typically 4–5 m^2 reflecting area. Hence, the hypothetical 15 MW electric power plant would utilize a minimum of 12,000 reflectors. Taking into account all the losses and other practical factors, the actual number of reflectors would be closer to 20,000. This implies that a solar thermal power plant would utilize a great deal of land area and also, since the power received depends on the solar insolation of a region, this land area would depend strongly on the region the power plant is built.

The land area required by solar thermal power plants depends on the number of solar collectors that must be placed. Typical solar thermal power plants in the southwest part of the United States utilize 5–6 acres of land per peak MW (2–2.5 ha/MW). Power plants in areas with lower average insolation require significantly more land area.

One of the important considerations for the solar thermal power plants is that the solar energy reflectors may not be left stationary on the ground. As the Sun apparently moves in the sky, each reflector must receive the solar energy and re-direct it precisely at the central receiver. Therefore, the reflecting angle of each receiver must be constantly adjusted to compensate for the apparent motion of the Sun. The process of adjustment of the reflecting angle is called *tracking the Sun* and is accomplished continuously by two methods:

a) Computer control on the reflector side, which orients the reflector surface according to the position of the Sun in the sky
b) Active control with sensors that track the reflected beams so that they are continuously received at particular areas on the central receiver.

Both tracking systems require two degrees of freedom for the effective tracking of the Sun in the azimuthal and elevation directions. In order to minimize the costs of the systems, several reflectors (from 4 to 16) are mounted on the same tracking system. When the reflectors are fitted with tracking systems, their aim is almost constant and directed consistently at the stationary central receiver. For this reason, the reflectors are called *heliostats*[3] and the field where they are positioned, the *heliostat field*. It becomes apparent that the directional control of the heliostats must be precise, or the energy received will not reach the collector. This is a factor that limits the size of the heliostat field and, consequently, of the maximum power of the power plant. Very large solar power plants, e.g. on the order of 400 MW, would require correspondingly very large heliostat fields, where thousands of heliostats must be placed very far from the receiver. Small errors in the controls of the tracking system would result in larger errors in tracking and the loss of a high fraction of the insolation received. Because of this inherent limitation, the size of the solar thermal power plants is limited to a few tens of MW.

---

[3] "Heliostat" literally means "one that keeps the sun still." Because the heliostats rotate with the sun, from the point of view of the central receiver the sun remains still at the same point in the horizon.

The cost of the heliostats is a significant part of the cost of the solar power plant and ranges between 30 and 50% of the capital plant's investment. Because of this, it is of interest to design inexpensive heliostats by large-scale manufacturing processes. Other factors that affect the economic considerations of the heliostats are solar reflectivity, durability and environmental impact. Given the enormous size of typical heliostats, the use of lighter but high-strength and highly-reflecting materials is one of the prime factors that come into the design of these devices. The high strength of the materials is necessary for the many-reflector heliostats to withstand the wind drag.[4] The originally used glass materials are gradually substituted with reflecting polymeric films that are coated with very thin silver or copper films, which reflect the incident solar energy well. Light composite materials for the support of the reflectors and thin silver or aluminum reflective films have been tried recently as reflector surfaces. Problems with these newer designs include delamination of the coatings and a gradual deterioration of the composite that results in reflectivity reduction over long periods of Sun exposure.

On the receiver side, two main types of central receivers are used, *cavity and external.* Both types are essentially heat exchangers that receive heat and impart it as enthalpy to the working fluid of the cycle, typically water. The cavity receivers have a few cavities that face outside, and where the power of the heliostats is concentrated. Coolant tubes line the interior of the cavity to absorb the power. The rest of the cavity receiver is enclosed to minimize radiation and reflection losses. In the external receivers, the coolant tubes are on the outside and receive directly the incident power. Both types of receivers are placed on relatively high towers, on the order of 100 m, which enable all or most of the reflected solar energy to reach the receiver. Figure 7.8 depicts a typical arrangement of the tower and heliostat field in the Northern hemisphere, where the Sun provides power from the Southern direction. The dimensions in this figure are based on the height of the receiver tower, H. It is observed that most of the heliostat field lies to the North of the receiver, and that the closest heliostats are placed at distance $0.75H$ from the foot of the receiver.

7.2.3 Energy Losses and Thermal Power Plant Operation

The physics of solar radiation; the design of the heliostats; the desired compactness of the heliostat field; and the design of the central receiver introduce several energy losses, which are added to the loss of energy from the outer atmosphere to the solar power plant. These additional losses are:

1. *Cosine losses*: because the function of the reflectors is to direct the solar energy to the central receiver, their reflective surfaces are not perpendicular to the Sun rays. Therefore, the incident radiation on the heliostats is reduced by the cosine

[4] Wind is generally stronger in areas that have high insolation, because of the strong ground heating and the natural convection that follows.

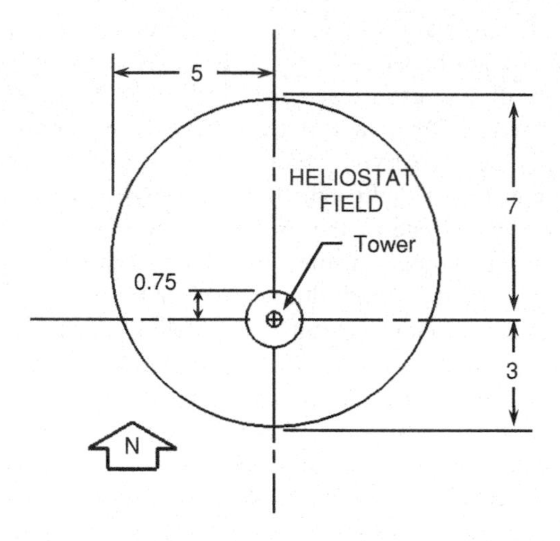

Fig. 7.8 Typical arrangement of a solar heliostat field and receiver for a power plant in the North hemisphere. All dimensions are based on the height of the central receiver

of the angle that is formed between the Sun's rays and the perpendicular direction of the reflecting surface.

2. *Shadowing*: shadows are cast from the central receiver as well as from one heliostat to another. While the receiver has a relatively small area and its shadowing is not significant, the shadowing from heliostats may become a very important loss of power, especially when the Sun is close to the horizon.
3. *Blocking*: is similar to shadowing and refers to reflected sunlight blocked by other receivers or heliostats.
4. *Reflective losses*: no reflective surface has 100% reflectivity. The surfaces of currently used mirrors reflect between 85 and 95% of the incident solar energy.
5. *Attenuation*: these are the atmospheric losses between the heliostats and the central receiver and are caused by water vapor, particulates and smog in the air between the receiver and the heliostats.

On the receiver end, the most important losses of the reflected solar power are:

1. *Reflection:* the material of the central receiver has a finite reflectivity and, as a consequence, some of the incident radiation is reflected. Painting the surface of the receivers with absorptive paint may minimize the reflection losses to 4–6% of the total power. Cavity receivers have lower reflection losses than external receivers because they "trap" the Sun's rays.
2. *Heat losses at the receiver*, by radiation conduction and convection. The central receivers are essentially heat exchangers that operate at very high surface temperatures to produce high temperature steam. The outside surface temperature of the central receivers is in the range 650–800°C and, because the function of the receiver is to absorb as much of the transferred power as possible, there is no thermal insulation to minimize any heat transfer losses. With external receivers, the heat transfer losses from the receiver may be as

high as 20% of the transmitted power. Cavity receivers have typically lower, but not insignificant, heat transfer losses.

3. *Spillage:* the energy reflected by the heliostats that does not reach the central receiver is the spillage. Spillage is caused by control errors in the tracking system or high winds and may be minimized by the better design of the heliostat controls. A good design of the heliostats would minimize the spillage losses to less than 5% of the total power. The spillage in external receivers is lower than in cavity receivers, which are more compact.

As long as the heliostats aim at the central receiver of the plant, the latter must be supplied with cooling water that raises steam and removes the heat transmitted. Interruptions of the supply of cooling water, or of a similar cooling fluid, would cause the overheating of the central receiver and damages associated with high temperatures. The interruption of cooling water for a few minutes may cause the partial melting of the materials of the receiver and its partial or total destruction. This type of scenario is similar to the Loss of Coolant Accident (LOCA) in nuclear reactors. In order to avoid LOCA accidents an emergency supply of water by fast-activated emergency pumps is a necessary safety system in solar thermal power plants. However, while a LOCA accident in a nuclear power plant may have very adverse environmental consequences, a similar accident in a solar power plant has limited destructive potential and, at worse, is limited to the replacement of the central receiver.

The energy losses at the heliostat and receiver ends are lumped together and appear as the *collection efficiency,* η_c, of the plant. The collection efficiency is the ratio of the energy absorbed by the working fluid of the thermal solar plant cycle to the insolation energy at the location of the plant. If the collection efficiency of a thermal power plant is 60% and the insolation in the location is 800 W, the working fluid in the cycle receives heat power of 480 W.

Another efficiency of interest is the *overall thermal efficiency* of the plant, η_a, which is defined as the net electric power produced by the thermal power plant divided by the insolation at the location of the plant. The latter is equal to the product of the collection efficiency and the thermal efficiency of the cycle:

$$\eta_a = \frac{\dot{W}_{\text{net}}}{\dot{W}_{\text{rad}}} = \eta_c \eta_t. \tag{7.8}$$

Typical thermal solar power plants have peak collection efficiencies in the range 50–70% and their Rankine cycles have thermal efficiencies close to 40%. Therefore, the overall thermal efficiencies calculated for peak power of well-designed thermal solar power plants are in the range 20–28%. Given that the power plant does not operate continuously from sunrise to sunset, the time-average overall efficiency of a solar power plant is even lower, in the range 14–22%.

One of the significant limitations of a thermal solar power plant is that it does not operate continuously. Its actual operation is limited by the available solar power, which is periodic and predictable. However, the solar thermal plants do not commence their operation at sunrise and they do not stop operating at sunset when the solar incident radiation approaches zero. Because of the placement pattern of the heliostats and the long shadows, the plant typically operates when the Sun is at least 15° from the horizon. This limits the hours of operation of the plant to significantly fewer hours than the sunlight hours and, as a consequence, limits the total energy produced by solar power plants. The hours of the daily operation of a thermal solar power plant range from a maximum of 11 h during the summer solstice, to a minimum of 5 or 6 during the winter solstice. Of course these values depend to a great extend on the latitude, where the solar power plant is located.

This is shown in Fig. 7.9, which has been produced with data corresponding to San Antonio, Texas, during the vernal equinox. The data pertain to collection efficiency 60% and overall thermal cycle efficiency 40%. The figure shows the effects of the heliostat and the receiver losses as well as the effect of the finite operation of the power plant. It is apparent in the figure that only a small part of the total incident solar energy is transformed to electric output. In this case the total, daily incident energy on 1 m^2 is approximately 18.1 MJ, the energy that is transmitted and becomes available at the collector is 10.9 MJ and the energy produced by the plant during its hours of operation is 3.7 MJ. Thus, only 20.5% of the incident solar energy is converted to electricity by the power plant.

The following conclusions may be drawn from Fig. 7.9 and the discussion of the collection and transformation of solar energy:

1. Because of the inherent losses in the heliostats and central receiver, only 50–60% of the incident solar power is transferred to the working fluid of the thermal cycle.
2. Because the plant does not operate continuously, the total daily thermal energy transmitted to the working fluid of the cycle is 35–50% of the daily solar energy that is incident on a flat surface. The total energy is calculated from the area/integral under the corresponding power curves. This fact, in combination with the thermal efficiency of the solar thermal power plant, which is approximately 35–40%, limits the average overall efficiency of utilization of solar energy to 14–22%, even for the well-designed units. Herein lays the importance of a more efficient thermal cycle for solar plants, such as a well designed and well-functioning Sterling cycle.

This average efficiency, η_{av}, which is based on the average energy (not power) a solar plant may produce during a period of time, T, typically one day or one year is defined by the following expression:

$$\eta_{av} = \frac{\frac{1}{T}\int_0^T \dot{W}_{net}dt}{\frac{1}{T}\int_0^T \dot{W}_{rad}dt} = \frac{1}{T}\int_0^T \eta_c \eta_t dt. \tag{7.9}$$

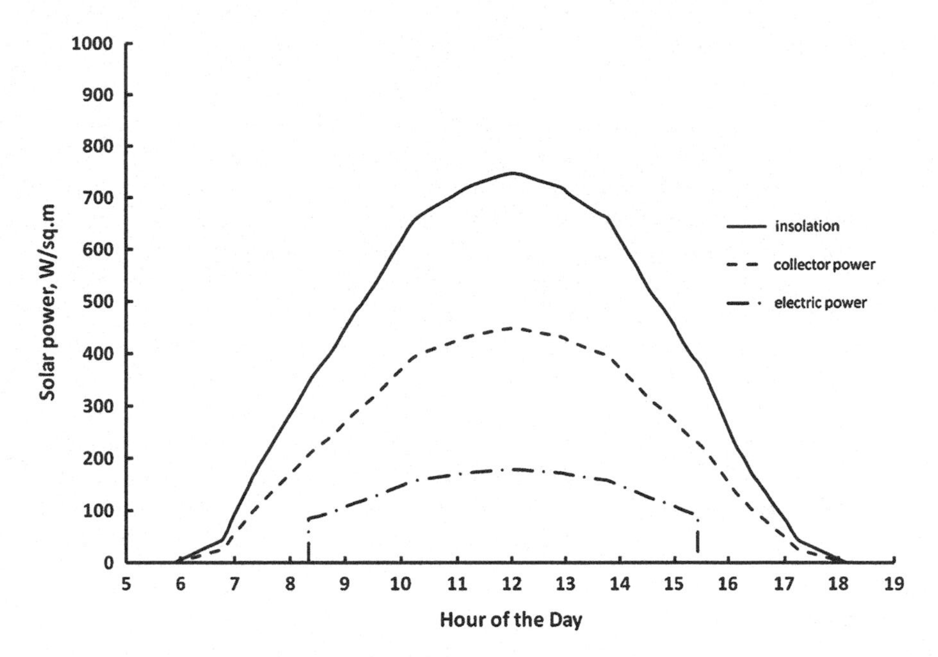

Fig. 7.9 Normal insolation, hours of operation of a solar thermal power plant, power losses attributed to the heliostats and power losses attributed to the receiver

It must be noted, however, that the thermal efficiency of a solar power plant—as well as the efficiency of most renewable energy power plants—does not have the same meaning as in the case of fossil fuel plants. In fossil fuel plants, the efficiency is a figure of merit that represents the benefit to cost ratio of the plant. The numerator is associated to the revenue that may be derived from the electric energy produced and the denominator in the efficiency expression of fossil power plants essentially represents the cost of the fossil fuel. Solar energy is free of any cost. The operators of solar plants do not pay for the heat received from the Sun, while they receive revenue or other benefits from the energy produced. A better figure of merit for solar and other renewable energy plants is the cost per kWh produced, usually expressed in $/kWh or $/MWh (the US$ would be substituted to the national currency in other countries). The thermal efficiency values may be used for the comparison of two cycles or plant designs or for engineering improvements of a given installation.

Thermal energy storage has been proposed in order to extend the operation of solar thermal power plants and to smooth their power output. Accordingly, some of the steam produced by the solar receiver during peak power would be stored in underground tanks to be used later. Since the steam is to be used within a few hours and the steam tanks can be well insulated, the thermal losses from such a system would be minimal. However, the storage system is still in the planning or feasibility stage of development and there are not any commercial solar thermal power plants that utilize thermal storage.

7.2.4 Solar Ponds

The concept of the solar ponds combines power production from solar energy and storage. A solar pond is essentially a pool of water, approximately 1 m deep. In a typical pool the water is heated by the Sun and rises to the surface, where it partly evaporates. The evaporation of the water vents a great deal of the pool's water as vapor; does not allow the accumulation of a great deal of energy in the pool; and keeps the average temperature of the water to low levels. Actually, the temperature of a typical pool in the summer is low enough for it to be used for recreational activities.

If the natural convection process that brings warmer fluid to the pool surface is suppressed, it is possible for the water temperature at the bottom of the pool to increase significantly by absorbing solar radiation. This may be accomplished in a salt-gradient solar pond, which uses natural salt (NaCl) as well as transparent partitions to suppress the natural convection in the pool. The depth of the solar pond is approximately 1 m and its bottom and walls are coated with a light absorbing material. The water in the solar pond is usually separated in three layers:

a) The bottom layer is saturated with salt and, therefore, its density is significantly greater than the density of pure water
b) The intermediate layer, which also contains salt at lower concentration, and
c) The upper layer, which is pure water without any salt.

Even though some solar ponds do not have partitions between the three layers, it is advisable to include transparent partitions in order to suppress molecular diffusion, which will disturb and annihilate the salt gradient.

When sunlight enters the solar pond, it is captured by the absorbing bottom and walls as well as the water. The temperature of the water close to the bottom rises, but the density of the salty water in this layer is high enough for natural convection to not occur. The transparent partitions between the three layers create a "greenhouse effect," thus trapping an additional quantity of heat in the two bottom layers. A great deal of the insolation absorbed remains at the bottom of the pond, where the temperature rises significantly. Temperatures as high as 95°C have been recorded in the bottom layer of well-designed solar ponds. Temperatures of this magnitude are sufficient to produce vapor in a suitable secondary fluid, such as pentane, butane or a refrigerant, which becomes the working fluid of a Rankine cycle operating with the solar pond as the boiler. Figure 7.10, depicts schematically the operation of a Rankine cycle powered by a solar pond. The thermodynamic states of this cycle and the equipment used are similar to those of the solar thermal power plant of Fig. 7.5 and that of the basic Rankine cycle, which was depicted of Figs. 3.7, 3.8a and 3.8b. The two main differences of the cycle, which is suitable with a solar pond, are:

a) The working fluid is a suitable fluid with lower boiling point and
b) The solar pond, which stores thermal energy, has substituted the boiler of the Rankine cycle.

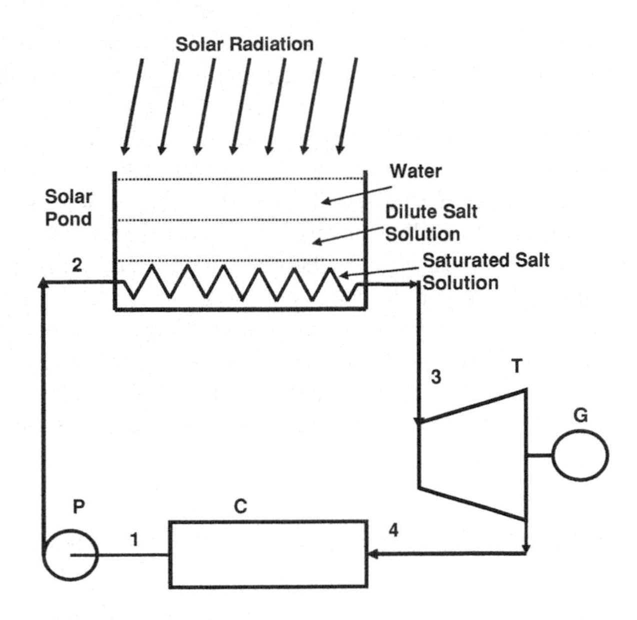

Fig. 7.10 A Rankine cycle operating with a solar pond as heater

Another characteristic of solar ponds is that, because their area is limited to a few hundred m^2, the corresponding power plants are very small, with power ratings on the order of a few hundred kW.

It has been recommended that the water from the uppermost layer of the solar pond be used for the cooling system of the small power plant. While this is possible, it is not advisable because: a) charging and discharging the upper layer of the pond would create waves that increase the reflectivity of the pond and also the water advection would accelerate the diffusion of the salt solution to the upper layers of the pond; and b) the uppermost water layer becomes warmer from conduction and, hence, has higher temperature than ambient water. Given that the upper temperature of the cycle associated with a solar pond is approximately 90°C, even a small increase of the cooling water temperature will have a significant and detrimental effect on the overall efficiency of the cycle. For this reason, if another source of water is available, it is highly recommended that it be used for cooling.

Because the water has very high heat capacity, a solar pond may store a significant amount of energy to be used when the Sun is low in the horizon. The production of electricity from a solar pond may continue even in the evening hours, when there is no natural insolation, thus providing a good alternative to produce power in the evening hours, when the demand is high because of air-conditioning needs. The main advantage of solar ponds is that they serve simultaneously as electric power generating plants as well as storage media.

The disadvantage of solar ponds is that they loose a significant amount of heat to the ground and to the atmospheric air. Because of this, the temperature of the lower zone becomes high enough to produce power for only a limited period of time during the year, typically for a few weeks during the summer. Even during these weeks, the solar pond temperature drops during the nights to levels that do not support power generation. Therefore the energy efficiency of solar ponds—defined as the electric energy produced during a year divided by the total incident solar energy—is expected to be very low. This energy efficiency was calculated to be 0.28% for a solar pond operating in San Antonio, Texas under the weather conditions prevailing in this City. The temperature of this solar pond was high enough to produce power between June 2 and August 31 of a typical year. The pond typically produced power for a few hours during these days from noon to early evening. The temperature and energy absorbed during the daylight hours were not high enough for power to be produced during the night, when stored energy is needed for electricity generation. Clearly, a solar thermal power plant is a better alternative for the production of electric power than a solar pond.

7.2.5 Passive Solar Heating: Solar Collectors

The most widespread use of solar energy is the passive heating of buildings and water. While passive heating does not produce electricity or any other form of energy, it helps avoid the use of fossil fuels or electric power for heating purposes and, thus, helps conserve the energy resources of the planet.

Passive solar heating in buildings is caused by an artificial *greenhouse effect:* Glass and several polymeric materials allow low-wavelength solar radiation to pass through them, but trap high-wavelength radiation. These materials effectively act as high-wavelength radiation filters. An enclosure that is covered by one of the greenhouse materials and faces the Sun "traps" a great deal of the solar energy. As a consequence, the temperature inside the enclosure will rise. Thermodynamic equilibrium between the enclosure and its surroundings will be established when the heat transfer from the enclosure—via conduction and advection—is equal to the difference of the radiation energy that enters and exits the enclosure. A simple piece of glass placed on the top of an otherwise open water tank, which is exposed to the Sun in the southern European countries, would raise the temperature of the water by 15–25°C. This is an adequate temperature for the water to be used in domestic applications. Well-insulated enclosures may reach temperatures 40–50°C higher than the ambient. Such enclosures are called *solar collectors* or simply *collectors*. The solar collectors are typically placed on the roofs of buildings, facing the prevailing direction of the Sun—south in the northern hemisphere and north in the southern hemisphere—and are used for water and space heating.

A schematic diagram of the cross-section of a solar collector is depicted in Fig. 7.11. Because a typical collector is placed on a roof, the solar collector depicted is at an angle with the horizontal. The inside material of the collector is coated with absorbing paint and usually has a dark color. Solar energy enters the top of the collector through the double glass and heats up the enclosed fluid, which is typically water or a solution. Depending on the design of the collector, the

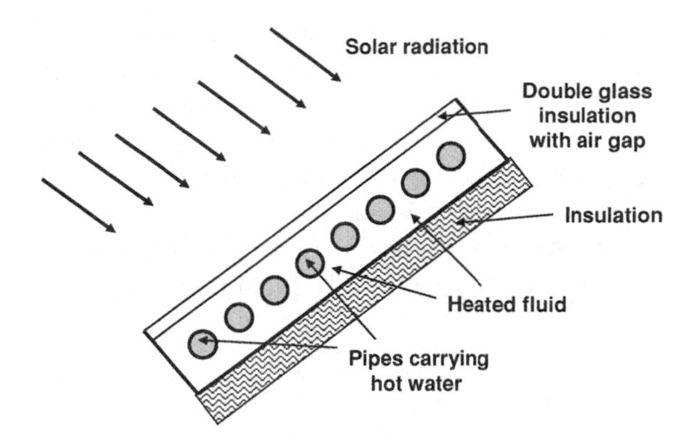

Fig. 7.11 A solar collector placed on a rooftop

temperature of this fluid may reach 70–85°C. Another, closed water circuit picks up the heat from the collector and dissipates it inside the building: the water that is circulated through the pipes of this circuit may supply the tank of a water heater and maintain it at the typically required temperature of 45–50°C, or it may circulate via space heaters to supply the heat required to maintain a comfortable condition in the building.

Solar collectors may trap direct or diffuse solar radiation. Because they also absorb diffused radiation, they may operate even in the cloudy autumn and winter days when most of the heating for the buildings is needed. Since the heated water may be stored for later use, the solar heating of buildings is not limited only to the daylight hours and may extend into the night hours. With the appropriate design of the heating system, solar heating may provide hot water and heat for buildings throughout the year and save a significant amount of primary energy resources.

Because of reflections at the outside surface of the collector and the incident angles, which are not 90°, the amount of radiation that enters the solar collector is less than the local solar insolation. This deficit in power entering the collector may be accounted by a radiation-loss factor, β_r. If the temperature of the enclosure is denoted by T_{en}, and that of the ambient is T_{amb}, and the useful heat extracted from the collector is $\dot{Q}$, the following energy balance equation applies at the steady-state operation of the solar collector:

$$\beta_r \dot{W}_{rad} = UA(T_{en} - T_{amb}) + \dot{Q}, \tag{7.10}$$

where U is the overall heat transfer coefficient of the collector and A its area. A well-designed collector has sufficient insulation to not lose any heat from the bottom and its sides. Therefore, the area A is the area exposed to the Sun.

The efficiency of the solar collectors is defined as the heat power removed by the water circuit to the insolation at the location of the collector:

$$\eta = \frac{\dot{Q}}{\dot{W}_{rad}} = \beta_r - \frac{U(T_{en} - T_{amb})}{(\dot{W}_{rad}/A)}. \tag{7.11}$$

In the southwest part of the USA typical values of the peak insolation are 900 W/m^2. With typical overall heat transfer coefficients $U = 4.5$ W/m^{2}K, radiation-loss factors, $\beta_r = 0.82$ and temperatures $T_{en} = 60$°C and T_{amb}=25°C, collector efficiencies may reach 65%. Because of the ambient temperature variations, solar collector efficiencies become lower during the winter months and higher during the summer.

The maximum temperature a collector may attain is when there is no heat removal, that is when $\dot{Q} \to 0$. From Eq. (7.10), the maximum temperature for the enclosure is:

$$T_{en}^{\max} = T_{amb} + \frac{\beta_r\left(\dot{W}_{rad}/A\right)}{U}. \tag{7.12}$$

Depending on the location of the collector and the type of insulation it has, the maximum temperature in the enclosure may reach 150–200°C. For example, with the values given in the previous paragraph for collectors in the southwestern part of the USA, the maximum temperature the collector may reach under typical weather conditions would be $T_{en}^{max} = 189$ °C. Since the collector enclosure without heat removal is a closed system, an internal temperature increase of this magnitude would result in steam formation, pressurization and damage to the enclosure. When designing solar collectors, it is important to incorporate safety systems for the enclosed fluid not to vaporize and for the internal pressure not to exceed values that would damage the collector.

Among the other uses of passive solar energy for households is solar cooking by *solar stoves*. This is typically accomplished with solar energy concentrators: when solar energy is concentrated in enclosures similar to those used as solar collectors, the temperature may rise significantly to reach the levels needed for the preparation of meals. The temperatures of solar stoves are regulated in a precise manner by allowing some of the internal fluid to evaporate or vent. Several of these solar stoves are being marketed successfully throughout the world.

An often neglected form of solar energy usage is clothes drying. Despite the fact that it is not continuously used, a clothes dryer consumes 5–6% of the total household energy in the USA. Drying the clothes under the Sun would save on the average 5–6% of the total energy used by households. However, in the USA there are several homeowners associations that have regulations against the solar drying of clothes for "aesthetic reasons." Similar regulations and behavioral norms exist in many other OECD countries. National, regional or local policies that ban such energy-wasting regulations would result in significant energy savings in the households by using a higher fraction of the free, passive solar heating.

It is apparent that most if not all of the heating needs of households may be provided by passive solar heating. Well-designed solar collector systems may supply heat and hot water to the households throughout the year. Cooking and clothes drying may also be accomplished with solar energy. The wider application of passive solar heating in households worldwide will reduce significantly the use of natural gas and other hydrocarbons and will save these natural energy resources for other uses.

7.3 Direct Solar-Electric Energy Conversion: Photovoltaics

Photovoltaic cells or *solar cells* convert the energy of the Sun directly to electricity. A product of the space race, the solar cell has been successfully used since the 1950s to power satellites, the International Space Station as well as remote terrestrial locations far from the electric grid. Because of the abundance of solar energy on the planet, the potential of photovoltaic devices to produce electric energy is immense: With maximum incident solar power density on the Earth's surface of approximately 1 kW/m^2, a few square meters of solar cells combined with energy storage for nighttime use, may provide the entire need of electricity for several households. Solar cells potentially may supply the entire energy needs of humanity, including thermal and electric energy. Since the Sun will always shine in the foreseeable future, photovoltaic devices with appropriate storage methods (see Chap. 12) have the potential to supply the energy demand of the Earth's population for millennia. Despite the advances in the manufacturing of solar cells, in the beginning of the twenty-first century, electricity from photovoltaic devices is still more expensive to produce than electricity from fossil fuel power plants.

Two developments in the early part of the twenty-first century may help shrink the price gap between photovoltaic and fossil fuel produced power: The fossil fuels are being depleted at a fast rate, causing their prices to increase and, at the same time new knowledge in materials science and technological advances are applied to photovoltaics, thus, causing the price of solar cells to decrease. With the two developments continuously occurring, there may be a point in the future when electricity produced by photovoltaic devices will be first competitive and then cheaper than electricity produced by other means. Because of the wide fluctuations and unpredictability of prices, we will examine the methods and technical merits of photovoltaic energy conversion without any reference to prices and economics.

7.3.1 Band Theory of Electrons

It is known from fundamental physics and chemistry that atomic nuclei are surrounded by electrons that move around the nuclei in well-defined orbits. The electrons have distinct energies and are subjected to the *Pauli Exclusion Principle,* which stipulates that two electrons may not exist in the same atom unless their states are different. Electrons in isolated atoms may exist in certain *levels of energy,* which are sharply defined. However, in groups of atoms, such as in crystals where atoms are arranged closely together and electrons interact with

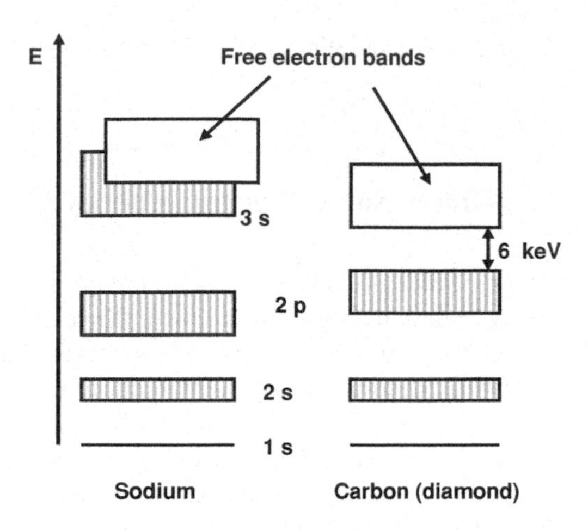

Fig. 7.12 Electron bands for a metal (sodium) and a non-metal (diamond)

several nuclei, these energy levels are not sharp and become energy bands, which are significantly wider for the outer electrons of an atom. The Pauli principle applies and excludes electrons from existing in the energy levels between the bands. These energy levels are sometimes called *forbidden bands.* At the upper end of electron energies there is a continuous band where electrons may exist as *free electrons.*

The free electrons are shared by the entire crystal and do not belong to specific atoms or nuclei. In metallic crystals the free electron band overlaps with the band, which is at the immediately lower level. In insulators, the free electron band is separated from its immediately lower band by a significantly high forbidden band. This is depicted in Fig. 7.12, where the bands of sodium and carbon (diamond) are shown in a schematic diagram. Because the free electron band in sodium overlaps with the immediately lower *3s band*, the outer electrons of sodium may easily become free electrons and be shared by the entire metallic crystal. From a microscopic point of view, the metallic crystal looks like atoms in a structured arrangement that exist in a sea of free electrons. An applied voltage would provide the motive force to support the movement of the free electrons of metals, thus, causing an electric current to pass through the metal. However, the forbidden gap of 6 eV in the outer electrons of the diamond is high enough not to allow a significant number of electrons to become free electrons. As a result, diamond and similar non-metallic elements do not conduct well electricity and heat. They are electric (and heat) insulators.

There is a type of materials, the *semiconductors,* where the energy gap between the free electron band and the immediately lower band is small enough, typically between 0.5 and 1.4 MeV. At low temperatures, semiconductors do not have a large number of free electrons and behave as insulators. However, at higher temperatures the thermal energy of the electrons is sufficiently high to enable a significant number of electrons to cross the last forbidden band, to become free electrons and to allow the material to conduct electricity and heat.

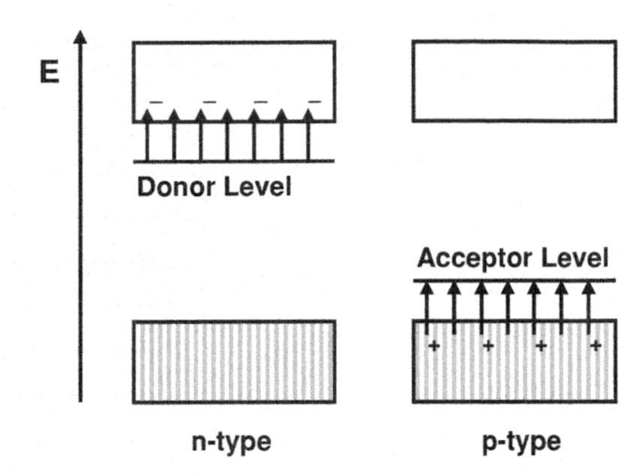

Fig. 7.13 Electron energy bands for n-type and p-type semiconductors

The transition of a semiconductor to a conducting material is facilitated by the addition of a small amount of an *impurity,* which creates an allowed energy level in the last forbidden band. This process is called *doping* and typical concentrations of the impurity are in the range 10^{-5}–10^{-6} (a few parts per million). Depending on the position of this allowed energy level, electrons from this energy level may "jump" into the free electron band. In this case, the conducting free electron band has a surplus of electrons, the semiconductor is called *n-type* and the new energy level is a *donor level.* Alternatively, electrons from the last valence band may "jump" onto the new energy level, which is called an *acceptor level.* The last valence band has a deficit of electrons or a surplus of "holes" for electrons and the semiconductor is called *p-type.* The electronic processes that create the n-type and p-type of semiconductors are depicted schematically in Fig. 7.13, where the symbols + and − denote the created electric charges that are caused by the jumping of electrons from one band or level to another.

With the addition of impurities, electrons may become excited and cross the small energy gap even at room temperatures. Crystalline silicon (Si) of very high purity is the typical semiconductor base material. Common doping/impurities materials are boron for the creation of the n-type and arsenic for the creation of the p-type semiconductors. In both types of semiconductors, the acceptance or donation of electrons by the impurity atoms creates charges in the valence level or free electron bands in the atoms of the semiconductor. This phenomenon is exploited in a *solar cell* or *photovoltaic cell (PV cell)* for the production of electric power.

7.3.2 Solar Cells and Direct Energy Conversion

When a layer of a p-type semiconductor comes in contact with a layer of a n-type semiconductor, the charged valence and free electron bands create a voltage difference across the combined material, which is called the *n-p junction.* If the

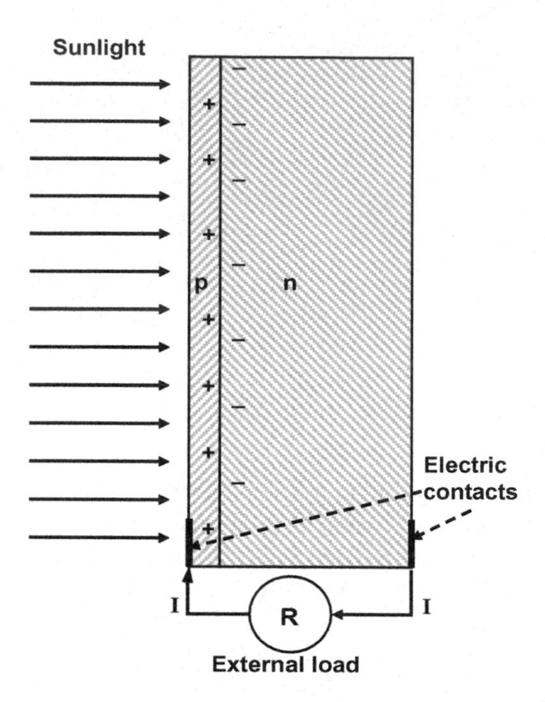

Fig. 7.14 The p-n junction and the production of electric power

opposite ends of the n-p junction are connected by an external electric circuit, electric current will flow until the electric balance of the valence and free electron bands is restored. In the absence of a mechanism or process that would continuously excite electrons from the donor or acceptor level to the valence or free electron bands respectively, the electric charge equilibrium will be soon established in the p-n junction and the electric current will stop.

The continuous excitation of electrons, which maintains the voltage difference across the p-n junction, is accomplished by the sunlight in the *solar or photovoltaic cell.* This is essentially a p-n junction and is shown in Fig. 7.14. The solar cell is composed of a very thin film of a p-type semiconductor joined with a significantly thicker layer of n-type semiconductor. Typically, the p-type film is on the order of 1 μm and the n-type layer is on the order of 1,000 μm. The incident sunlight penetrates the entire part of the p-type film, enters and partly penetrates the n-type material. The photons in the sunlight cause the excitation of electrons in both types of semiconductors: in the p-type film, electrons jump from the valence band to the acceptor level, while in the n-type layer, electrons jump from the donor level to the free electron band. Oftentimes, this process is described as the *creation of electron-hole pairs* close to the p-n junction.

The charge imbalance close to the p-n junction causes the diffusion of electrons from the n-type side to the p-type side of the solar cell. This creates a shortage of electrons in the n-type side and a surplus in the p-type side. The outer electric contact on the n-type side becomes a positive electrode and the outer electric contact on the p-type side becomes a negative electrode as shown in Fig. 7.14. This causes the current, I, to flow as shown in the figure, in a direction that is opposite to the flow of electrons. The continuous insolation on the surface of the p-n junction, creates continuously electron–hole pairs near the junction, maintains the voltage difference across the electrodes and, hence, provides continuous power to the external electric circuit. The higher the insolation power, the more electron-hole pairs are produced and the more power the solar cell produces.

It is apparent from Fig. 7.14 that the conversion of the solar radiation power to electric power occurs as a result of the electric material properties of the p-n junction. Furthermore, the conversion occurs directly from electromagnetic radiation to the electric potential created in the p-n junction, to electric power. Apparently, during the conversion process there is no significant thermal energy generated and, more importantly, there is no conversion of thermal energy to electric power. This process is called *Direct Energy Conversion (DEC).* A significant advantage of the DEC is that it is not subjected to the Carnot efficiency limitation (Eq. 3.22), which limits the efficiency of thermal engines. Because it is not limited, in principle the efficiency of a DEC engine/device may be as high as 100%. However, the current materials technology in the current and the projected generations of solar cells exhibit significantly lower conversion efficiencies for reasons that will be explained in the next section.

The voltage, current and power supplied to the electric load of Fig. 7.14 depend on the resistance of the external load, R. When R is very high and the circuit is almost open, the voltage developed by the cell is maximum, V_o. This maximum voltage, V_o, is a monotonically increasing function of the insolation, $\dot{W}_{rad}/A$, and reaches a plateau at high values of the insolation. At the other extreme, when R is very low, the cell is short-circuited, the current is a maximum at I_{ss} and the voltage developed by the cell almost vanishes. The short-circuit current, I_{ss}, is a linear function of the insolation. Figure 7.15a shows the typical variations of the open-circuit voltage and the short-circuit current for solar cells. The power produced by the solar cell is equal to the product of the voltage and current developed in the external circuit and depends on the external resistance R. By varying R between the two limits of open circuit and short-circuit, where the power produced is zero, one obtains the characteristic curves of a solar cell, which are depicted in Fig. 15b for two values of the insolation, $\dot{W}_{rad}/A$. It is easy to show from algebraic considerations that, under constant insolation, the maximum power is produced at the point where the slope of the tangent to the corresponding characteristic curve is equal to -1.

7.3.3 Efficiency of Solar Cells

As with all the electromagnetic waves and radiation, light may be considered as a wave or as a flux of individual particles. The light of different wavelengths is

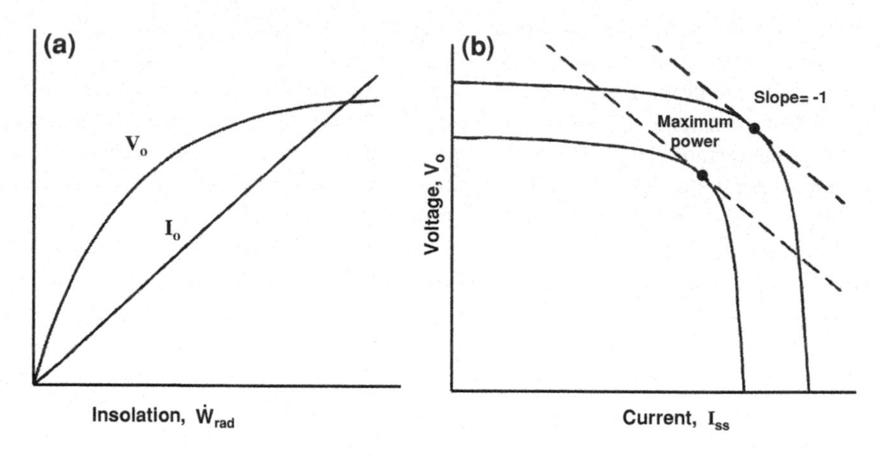

Fig. 7.15 **a** Typical variation of the open-circuit voltage and short-circuit current with insolation **b** Two characteristic curves and maximum power of a solar cell

considered to be transmitted by a stream of light-carrying particles, called *photons,* which possess momentum and energy. The energy carried by photons is a function of their frequency:

$$E = h\nu = h\frac{c}{\lambda}, \tag{7.13}$$

where h is Planck's constant, equal to $6.6256*10^{-34}$ Js; ν is the frequency of the light; λ is the wavelength of light; and c is the speed of light in vacuum, which is equal to $3*10^8$ m/s.

Of importance in calculations on photovoltaics is the number flux of photons, φ_p, which is incident on a given surface. Since the solar radiation is composed of a range of wavelengths, as shown in Fig. 7.2, there is also a range of photon energies that represent the solar spectrum. The energy of the average photon outside the Earth's atmosphere is approximately 1.48 eV. Given the value of the solar constant, $S = 1.353$ kW/m^2, the average number of photons that are incident on a surface of 1 m^2 at the outer end of the atmosphere is $5.8*10^{21}$ photons/(s-m^2). Therefore, a photovoltaic cell with area 45 cm^2, placed at the limit of the atmosphere and perpendicular to the rays of the Sun, would receive on average $2.6*10^{19}$ photons/s.

The photons interact with the structure of the photovoltaic p-n junction, and create the electron-hole pairs that are necessary for the development of the electric voltage. However, not all the photons have the energy or the capacity to create an electron–hole pair. Photons with low energies (high wavelengths) do not have enough energy to cause the needed "jump" of an electron from the donor level to the free electron band and are ineffective in producing electricity. On the other end of the spectrum, a photon may only create a single electron–hole pair, which entails a small amount of electric energy. Photons with very high energy, that is

Table 7.1 Photon characteristics and conversion to electric energy (data from El-Wakil 1984)

Frequency range (Hz)	Wavelength range (μm)	% of solar energy	Fraction converted	% Energy converted
$<2.72*10^{14}$	>1.1	22	0	0
$3.33–2.72*10^{14}$	0.9–1.1	13	0.91	12
$4.29–3.33*10^{14}$	0.7–0.9	20	0.73	15
$6.00–4.29*10^{14}$	0.5–0.7	28	0.55	15
$10.00–6.00*10^{14}$	0.3–0.5	17	0.36	5
$>10*10^{14}$	<0.3	0	0	0

high frequency or low wavelength, do not have the capacity to create more than one electron-hole pairs and, hence, a great deal of the energy of these photons is not used for the production of electric power. The excess energy of these photons is dissipated as heat on the solar cell. Table 7.1 shows the frequency, wavelength and energy carried by the photons, and the fraction of the energy that may be converted to electric energy in a semiconductor photovoltaic cell. One may conclude by adding the numbers of the last column of the Table that, under ideal conditions, only 47% of the incident sunlight radiation may be converted to electric energy.

The efficiency of a solar cell is defined as the ratio of the electric power it produces to the total insolation. It is apparent from this Table, that the reason for the low theoretical efficiency of semiconductor PV cells is the significant mismatch between the incident solar radiation and the capacity of the cell to convert the radiation to electricity. A "solar filter" that would convert the frequency of the entire solar radiation to an equal amount of energy with photons in the frequency range $4.29*10^{14}$ to $2.72*10^{14}$ Hz would be ideal. Such a filter would increase the theoretical conversion efficiency of photovoltaic solar cells to levels closer to 80%.

While crystalline silicon solar cells have been used for decades and have an excellent performance and reliability records, their production entails a great deal of energy input, because pure crystalline silicon requires a great deal of energy to be produced. Typically, the production of a PV cell requires four to five years of the energy it produces. This and the associated labor costs make photovoltaic energy appreciably more expensive than energy from conventional power units and are an impediment in the reduction of the production costs of silicon solar cells. More recent photovoltaic technologies have focused on the production of less expensive semiconductor materials regardless of the conversion efficiency. One such method is to deposit amorphous (e.g. non-crystalline) semiconductor materials, onto very thin film substrates of lower cost.

Thin film technology makes use of cadmium telluride (CdTe) or copper indium gallium selenide (CoInGaSe or CIGS), and amorphous silicon. The most recent designs of this technology strive to achieve the higher possible solar cell efficiency with thin films by using strategically chosen combinations of materials (multi-junction cells) and single materials (full spectrum cells) that respond to wider ranges of the available solar spectrum in a region, thereby producing more power.

A second method is to concentrate on the technology for the production of less expensive solar cells by using more common and less pure materials, regardless of the efficiency that is achieved. For example, thin film polymeric solar cells have been produced with gallium arsenide as the doping. These cells have lower efficiencies (6–8%) but significantly lower production cost. Such low efficiencies are acceptable because the input energy, the insolation, is free of charge. For the more widespread use of energy from photovoltaics, the technology will have to concentrate mainly on two factors:

a) Lower energy costs, which will have to be reduced to less than $0.1/kWh;
b) Storage methods for the energy to become available during the nighttime and the winter months.

7.3.4 A Futuristic Concept: The Space Solar Power Station

The two principal disadvantages of the terrestrial use of solar energy are:

1. The insolation at a given terrestrial location is highly variable, because of the Earth's rotation around the Sun and
2. Because of the atmospheric absorption there is significant power attenuation from the outer atmosphere to any given location as shown in Fig. 7.2.

Both of these terrestrial effects are absent outside the atmosphere of the Earth, where a solar receiver may be placed and its orbit designed so that the carrier satellite remains outside the shadow of the Earth. Such a receiver, which may be called a Space Solar Power Station (SSPS), will always face the Sun and will be capable to produce electricity continuously, because the Sun will always be in sight. Given the atmospheric attenuation losses and the periodic fluctuations of insolation, which are associated with the location of the terrestrial solar power plants, the SSPS will be capable to produce between 4 and 5 times more energy than an equivalent terrestrial power plant.

Of course, all the electricity produced in the space will have to be transported to the Earth in order to be used by the population. Microwave transmission systems have been proposed for this purpose. Accordingly, a SSPS would consist of very large solar panels that always face the Sun and a transmission antenna that continuously converts the electric power to microwave radiation and transmits it to a receiving station on the surface of the Earth. In this location the microwaves would be converted back to electric energy and will be distributed to the markets of electricity. In order to minimize atmospheric absorption, reflection and scattering a suitable microwave frequency, e.g. below 10 GHz, may be chosen for the transmission of the power. Clouds, rain, and hail may have some impact on the transmission by absorbing or scattering some of the energy transmitted. These effects may also be minimized by an optimum choice of the transmission frequency and the location of the receiving antenna.

The SSPS concept has received significant attention as a viable method for the utilization of solar energy, but an actual pilot plant has not been built yet. Some of the problems associated with the SSPS concept are:

1. The transportation of material to a space orbit, which requires significant amounts of energy by itself.
2. The effect of the transmitted microwaves on airplane traffic, birds and other animals in the path of the microwave radiation.
3. Interference with communication frequencies.
4. Biological effects on humans from the scattering of this radiation.
5. The significant cost of the producing and receiving stations and the unknowns associated with their operation, which implies a significant economic risk.

Given the inherent scientific and energy advantages of the SSPS concept, its potential to supply significant amounts of energy continuously to a fixed location (or locations) on Earth and the absence of storage requirements, this method of electricity production appears to be promising. While there are several serious impediments for its implementation at present, it is currently technologically feasible and may become economically feasible and advantageous in the future, when advances in materials science and propulsion will make the transportation of efficient photovoltaic cells to an orbit more commonplace and less expensive.

7.4 Environmental Issues of Solar Energy Utilization

Solar energy is one of the most benign forms of alternative energy. Its current utilization, either as solar-thermal or as photovoltaic conversion, causes very few environmental concerns. The principal environmental advantage of solar energy is that its utilization does not involve chemical or nuclear reactions. Thus, no chemical emissions (e.g. carbon dioxide, NO_x gases, etc.) and no radionuclides are emitted from a solar energy facility. This makes solar conversion an environmentally friendly energy source.

The most significant environmental impact of solar energy is associated with the production of the materials that comprise the photovoltaic cells. Silicon, germanium and phosphorous are produced and purified with the consumption of significant amounts of energy and involve the use of polluting chemicals such as sulfuric acid and cyanide. However, all the pollution associated with the production of solar cells is localized and contained at the production facility. The pollution at large production sites may be very well regulated and assurances may be given that all hazardous chemicals used for the production of solar cells will not be released to the environment. A more significant concern in the production of the photovoltaic cells is the high amount of energy consumed during their manufacturing process. Currently, the energy consumed for the manufacturing of a silicon-based solar cell is equivalent to the energy the cell will produce in approximately 4 years. Leaner manufacturing methods and different cells made of new materials may reduce this significant energy requirement.

Thermal pollution is associated with all solar energy operations. Solar thermal power plants reject a significant amount of heat in their condenser. Similarly, photovoltaic arrays reject all the incident radiation that is not converted to electricity. The heat power (waste heat) rejected by a plant, which produces electric power $\dot{W}$, with an overall efficiency η is given by the expression:

$$\dot{Q}_{rej} = \dot{W}\frac{1-\eta}{\eta}. \qquad (7.14)$$

A 100 MW thermal power plant operating with 40% thermal efficiency would reject 150 MW of heat and a 1 kW solar panel array with an efficiency of 20% would reject a total of 4 kW of heat. Because solar installations occupy a larger area than similar fossil fuel installations, the waste heat is dissipated over significantly larger areas and does not have the same adverse impact on the environment. Also, the fact that solar installations are, in general, located in remote, often deserted, areas, the thermal pollution caused by such installations seldom affects the human population and becomes unnoticed. Associated with the thermal pollution is the water use, if the solar plant is cooled by water. The amount of water used by such a solar thermal power plant is approximately the same as the water used by a thermal power plant, which rejects the same amount of heat power.

Land use is another environmental effect of solar energy utilization. Solar energy is diffuse and, hence, the production of significant amounts of it requires a large land area, which is significantly more than the area used by a thermal power plant, fossil or nuclear. For this reason, inexpensive, unutilized and usually deserted areas have been chosen for the location of solar energy facilities.

A notable beneficial environmental effect of solar energy that is utilized in urban environments, e.g. photovoltaic panels or passive solar panel heating on the sides and rooftops of buildings, is that some of the incident solar radiation is utilized inside the buildings and, therefore, it is not absorbed, but reflected back to the atmosphere or scattered. Considering the overall energy balance in the vicinity of the buildings, the solar energy systems produce a small but important cooling effect in the urban environment, which results in slightly lower ambient temperatures and reduces the need for building air-conditioning.

Problems

1. What is the mass rate of the Sun, in kg/s, which is converted into solar power? Hint: use the first statement in this chapter.
2. The solar collectors of a spacecraft that orbits the Earth have an area 42 m^2. What is the solar energy these collectors absorb in 3 h?

3. Using Eq. (7.1) prove *Wien's law*, that the wavelength at the maximum of the radiation spectrum of a black body is given by the relationship: $\lambda_{max}T = 0.0029$ mK.
4. Using Fig. 7.4 determine what is the annually averaged energy, in kWh, absorbed by a 10 m^2 solar collectors in the following locations: a) Hannover, Germany; b) London, England; c) Timbuktu, Mali; d) Calcutta, India; e) Buenos Aires, Argentina.
5. Photovoltaic cells with an overall average conversion efficiency of 10% and total area of 5 m^2 are placed in the following locations: a) Hannover, Germany; b) London, England; c) Timbuktu, Mali; d) Calcutta, India; e) Buenos Aires, Argentina and f) New York, USA. Determine the total amount of electricity, in kWh, produced annually in each location.
6. At a certain location, the average insolation on a flat plate collector is 7,200 kJ/(m^2*day). Such a solar collector is used to continuously provide 38,000 Btu/hr or heat to a process. If the collector efficiency is 42%, what should be the area of the collector?
7. It has been suggested that refrigerant R-134 be used instead of water for a new type of thermal solar collector. What is your opinion on this? Explain in detail your reasoning.
8. A thermal solar power plant uses 1,300 heliostats, each one with an area 25 m^2. The annually averaged peak insolation in the area is 700 W/m^2. The thermal solar energy collection efficiency of the plant is 56% and the efficiency of the cycle is 38%. Determine the annually averaged peak power produced by this power plant.
9. The drag force on a surface due to the wind flow is given by the expression:

$$F_D = \frac{1}{2}A\rho C_D V^2, \text{ with } C_D = \frac{24}{\text{Re}}\left(1 + 0.32\text{Re}^{0.687}\right) \text{ and Re} = \frac{\rho LV}{\mu} \quad (7.15)$$

where V is the velocity normal to the surface; C_D is the drag coefficient; Re is the Reynolds number; and L is a characteristic dimension, which is equal to the square root of the area of the collector. A is the area that is perpendicular to the wind direction. A circular solar collector is placed at an angle 30° to the horizontal and has a diameter $d = 7$ m. The wind flows in the horizontal direction. Make a diagram of the drag force on the collector versus the wind velocity, when the latter varies in the range 0–72 miles per hour. What do you observe?
10. A 20 MW (peak) thermal solar power plant is to be built in a region where the annually averaged peak insolation is 640 W. Because of high wind currents in the region, the area of the heliostats is not to exceed 12 m^2. If the overall thermal efficiency of the power plant is 18%, calculate the number of heliostats that must be constructed for the plant.
11. A 10 MW (peak) solar power plant is proposed for the Dallas, Texas region. At peak power, the overall efficiency of the plant is estimated to be 21%. Determine the minimum area of the heliostats that must be used for this plant.

12. Use typical values in your locality to calculate the solar collector efficiencies. Also, calculate what would be the maximum temperature the fluid in the collector may reach and suggest methods to keep the internal temperature below 90°C.
13. What is the energy of photons with wavelength 20, 50, 100, 200 and 800 nm?
14. A solar cell only converts the photons in the range $2.72*10^{14}$ to $4.29*10^{14}$ Hz. The efficiency of this photon conversion is uniform and equal to 58%. What is the maximum efficiency this solar cell will have?
15. "We have the know-how and the capability to produce a 60% efficient solar cell, but we lack the investment to produce it commercially." Comment on this statement by writing a 250–300 word essay.
16. A solar filter has been invented that would convert all the photons of the solar spectrum, which have frequency less than $2.72*10^{14}$ Hz, to photons of frequency $3.2*10^{14}$ Hz. The total number of photons will not change. Calculate the improvement on the maximum efficiency of the semiconductor solar cells this filter will cause.
17. A SSPS is rated to produce 20 GW of electric power and transmit it to the Earth via a microwave guide. The solar cells cover 85% of the total area of the SSPS. If the SSPS is of square shape and, if it faces the Sun directly, calculate the total area of the SSPS. You may assume that the solar cells in this station have maximum efficiency.
18. "The Sun is producing abundant and free energy that may be used to meet economically the entire energy need of this country, but the petroleum cartel undermines all efforts to do that." Comment on this statement by writing a 350–400 word essay.

Chapter 8
Wind Power

Abstract Since the time of the ancient sailboats and windmills, the power of the wind has been harnessed for ship propulsion and the performance of mechanical work. In the modern era, wind has been increasingly used for the production of electric power. In the first decade of the twenty-first century alone, the production of electricity from wind power worldwide has increased by a factor of eight. Similar to solar energy, wind is also a distributed, renewable source of energy. The energy density of the wind is low, but wind is available in all the geographical regions of the world and its geographical distribution is more uniform than that of solar energy. Wind turbines of different sizes and designs are currently used successfully for the production of electric power. The bigger and more efficient types of these engines have blades with lengths between 20 and 50 m, are located at the top of 50–140 m towers and are becoming ubiquitous in the landscape of several OECD countries. Wind is probably the most environmentally benign energy source. In theory, it has the capacity to satisfy the energy needs of entire countries and even that of the whole planet. However, it is also an intermittent source with availability and intensity much less predictable than any other source of alternative energy. This intermittency is a significant drawback and will limit the more widespread use of wind power, unless suitable energy storage systems are developed that would store part of the energy produced during windy periods.

8.1 Wind Patterns

It may be arguably said that wind power is a byproduct of solar energy: The uneven heating of different parts of the globe causes hot air to rise in regions that receive higher amounts of insolation. The rise of hot air creates a small pressure differential, which induces colder air from the surrounding regions to rush in. Thus, horizontal air currents and the wind patterns on the Earth's surface are

E. E. (Stathis) Michaelides, *Alternative Energy Sources*,
Green Energy and Technology, DOI: 10.1007/978-3-642-20951-2_8,

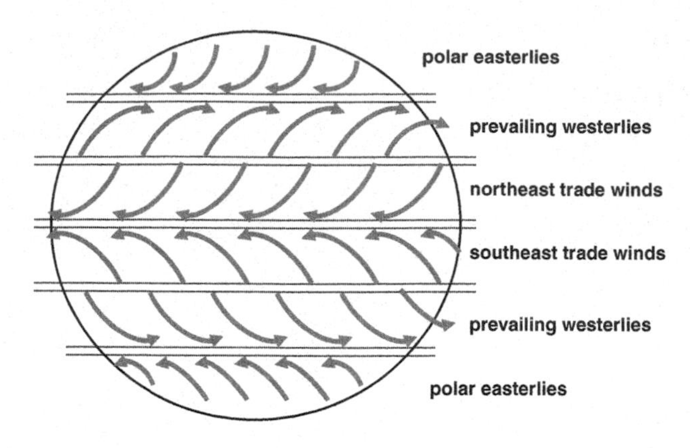

Fig. 8.1 The planetary winds on Earth's surface

created. Depending on the origin and effects of these currents, the currents are classified as *planetary* and *local.*

Planetary winds affect very large regions of the Earth, encompass large masses of air and are primarily caused by the higher amount of solar radiation received by the land masses near the equator. The hotter land masses near the equator cause the air at the tropical regions to rise and move towards the poles. This upward motion of the air masses is affected by the rotational motion of the Earth. The *Coriolis force* with its components on the horizontal plane is developed on the rising masses of air in both the northern and the southern hemisphere. As a result, the air rushing from the temperate zones to fill the relative vacuum at the equatorial zone develops a motion towards the west. This westerly motion causes the north–east *trade winds* in the northern hemisphere and the south–east *trade winds* in the southern hemisphere.[1] Early navigators, such as Columbus and Magellan mastered the effect (but not the causes) of these winds and planned their successful western journeys accordingly. At the same time, the rising warm air moves towards the poles, cools in the upper atmosphere and descends at approximately 30° latitude (in the nautical tradition these are the *horse latitudes*) in both the northern and the southern hemispheres. The downward motion at these latitudes in combination with the Coriolis force develops a general air motion in the temperate zones from the western to the eastern direction, which manifests itself as the *prevailing westerlies,* in both the northern and the southern hemispheres between 30 and 60° latitude. The pattern of the planetary winds is completed with the *polar easterlies,* which blow in both the south and the north Polar Regions. Figure 8.1 is a schematic diagram of the prevailing planetary winds.

Local winds have local effects and are usually caused by the uneven heating of neighboring masses of land or of land and sea. Well known among local winds is

[1] Winds are named from the direction they come from. Thus, a wind blowing from the west to the east is a westerly and a wind blowing from north to south is a northerly wind.

the *sea breeze*, which is caused by the higher temperatures of the land in comparison to a neighboring water surface: as part of the water evaporates, the surface of the water remains cooler than that of the surrounding land. As a result, the air close to the surface of the land rises, a pressure differential—or partial vacuum—is created and cooler air over the water surface rushes to fill this partial vacuum. Breezes are ubiquitous in the coastal areas as well as near large lakes. A lesser known mechanism of air current development is caused by the differential heating on the hills and mountain sides, which causes the air to rise during the day time when solar heating is high (anabatic flow) and to descend during the early evening hours when the mountain or hill sides cool faster than the neighboring areas (katabatic flow). In addition, mountains and hills create their own wind patterns by deflecting masses of air, which is already in motion. Other types of local winds are established according to the type of surface cover, the heating patterns, the climate, and the weather of a region.

Wind patterns are extremely important for the economy of a locality, a region or, a nation, because they are one of the prime determinants of local weather and, especially, of rainfall. Weather and rainfall determine the agricultural production of the area, and by extent, the regional economy. Local and regional wind patterns are intricately entwined into the social and economic fabric of modern societies and nations. Minor perturbations in the prevailing wind patterns may cause significant rainfall disruptions in a region, crop failures and subsequent economic instability. Global warming, which among its other effects is expected to cause significant disruptions in local climatic conditions and wind patterns, is also expected to produce rainfall pattern changes in several regions of the planet. Such changes may cause fresh water scarcity in regions where water is now abundant. This may lead to persistently lower agricultural production and significant economic hardships for the population. For this reason several governments have taken measures to control and limit the emissions of CO_2 and other greenhouse gases.

8.1.1 Early Types of Wind Utilization

Wind power was known and used extensively by ancient civilizations for ship propulsion. Historical records show that the Chinese, Indians, Egyptians and Hittites have been using sails to propel their boats before the second millennium BC. Greek ships sailed to Troy around 1,100 BC in an expedition that was recorded in the epic poems of *Iliad* and *Odyssey*.[2] Actually, propulsion by sail and manpower (rowing) were the only forms of seagoing power until the early nineteenth century and the advent of the steam engine.

---

[2] Wind energy is often called *Aeolic energy* named after *Aeolos*, the ancient Greek god of the winds.

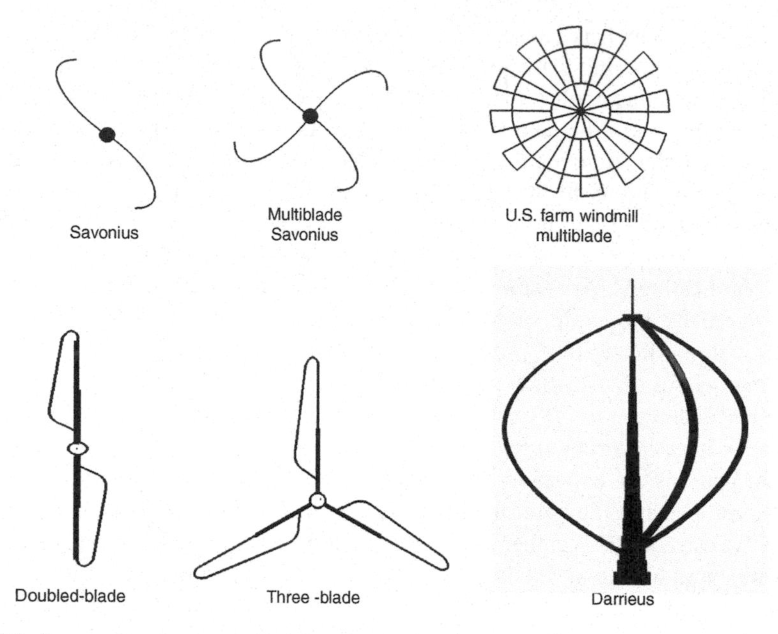

Fig. 8.2 Types of common wind turbines

Land use of wind power for the production of energy and the accomplishment of tasks came a great deal later. The use of windmills was recorded by Arab writers in the Asian regions that are now part of Iran and Afghanistan around 1,000 AD. Windmills were transported to Europe after 1,200 AD and became very common for the grinding of wheat and other food processing tasks. From Europe the use of windmills was transported to America. A type of windmills, the American multi-blade turbine, or *freetail,* was heavily used in the arid southwest part of the United States for pumping water from underground wells. This type of windmill is credited for the population growth and the economic development of the American Southwest.

Wind turbines can be very simple and their early designs demonstrate this simplicity: The *Savonius turbine* is made of an S-shaped metal sheet with a shaft running through its middle. The shaft of this turbine may be either horizontal or vertical but the vertical shaft configuration is the more common. The *Darieus turbine* is mounted on a vertical shaft and has two to four-blades. The blades of this turbine are bent and have an aerodynamics shape. This shape deflects the stream of air and produces a lift force on the blade, which acts in unison with the induced drag and assists in the harnessing of more wind power. A combination of the *Savonius* and the *Darieus* turbines may be constructed on a vertical shaft. Among the horizontal axis turbines are: the four-blade *Dutch windmill,* which was one of the most commonly, used turbines in the past, and the *US farm windmill,* which has as many as twenty-blades. Figure 8.2 depicts the schematic diagrams of these earlier turbine designs as well as a more modern three-blade wind turbine.

The early users of wind power were only concerned that the task the turbine was built would be done and that the turbine would be reliable. Turbine efficiency and optimization of the turbine operation was never a consideration in manufacturing the early engines. For this reason, none of the above devices is designed to convert wind power efficiently to electricity, or for that matter to perform any other task under the maximum possible efficiency. Maximum efficiency is one of the principal considerations of the construction of the engines that are currently built for the conversion of wind power to electricity. Typically, the modern turbines have two or three-blades and their design and operation have been optimized to produce maximum energy from the prevailing local wind conditions.

8.1.2 Wind Power Potential

Slightly more than 2%, of the total solar power received by the Earth, or $3.46*10^{12}$ kW, is converted to wind power. The total solar radiation energy converted into the mechanical energy of the wind is more than $1.1*10^{17}$ MJ per year. This is more than two million times more than the total energy that has been used by the human population in 2010. An immediate conclusion that may be drawn from these numbers is that wind power is an important alternative energy source, which, if properly harnessed, is capable to supply a significant fraction of the energy demand of the Earth's population. Of course, not all the wind power that is generated from the uneven heating of the sun may be harnessed for the production of electricity: Air currents over the sea or at high altitudes are too difficult, if not impossible, to be converted to electricity. However, a significant fraction of wind power is available near the surface of the Earth. This fraction is close to many populated regions and may be harnessed with conventional means. Even if a small fraction of the available power is converted to electricity, it would amount to significant contribution and would alleviate the need to use fossil or nuclear fuels for electric power generation.

The kinetic energy per unit volume of an air stream with velocity V and density ρ is:

$$K.E. = \frac{1}{2}\rho V^2. \tag{8.1}$$

Since the volumetric flow rate through an area A is: $\dot{V} = VA$, and the area swept by the blades of a turbine is a circle of diameter, D, the power of the air stream that is available to a wind turbine is:

$$\dot{W}_{av} = \frac{1}{2}A\rho V^3 = \frac{\pi}{8}D^2\rho V^3. \tag{8.2}$$

Thus, the wind power is proportional to the cubic power of the wind velocity and to the square of the diameter of the wind turbine. This is a significant

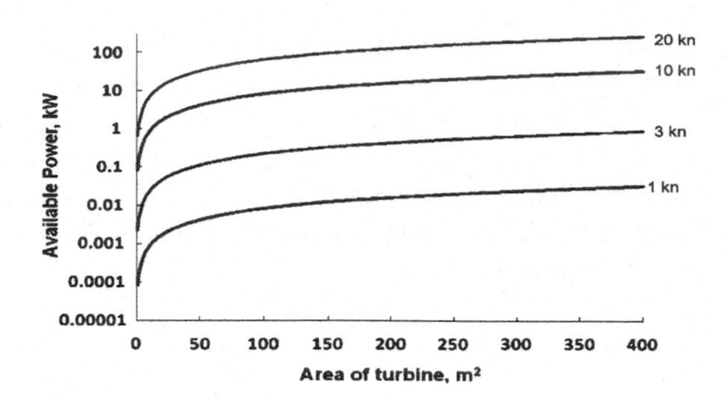

Fig. 8.3 Available power from the wind. 1 kn = 1.15 mph = 1.852 km/h = 0.514 m/s

conclusion for the harnessing of wind power: when the wind velocity is doubled, the power is increased by a factor of eight. Figure 8.3 shows the diagram of the power that is available by the wind as a function of the area swept by the wind turbine and with the wind velocity in knots [3] (kn) as a parameter. It is apparent that the power available from the wind is very high if wind turbines that cover very large areas are constructed and placed in operation at locations with high average wind velocities. A great deal of research and development effort, since the 1970s, has been devoted to the construction of wind turbines with longer blades that cover large areas.

Unlike solar radiation, which is a form of heat, the energy of the wind is kinetic energy, which is a form of mechanical energy. Therefore, the wind turbines, which convert the wind energy to electricity, are direct conversion devices and are not subject to the efficiency limitations (Carnot efficiency) of heat engines. The wind turbines that are currently used for the harnessing of wind power are by far simpler to construct and to operate than any thermal engine. An additional advantage of the wind turbines with respect to thermal engines is that their operation does not require the use of water for cooling purposes.

8.2 Principles of Wind Power

Large or small wind turbines harness the wind power for the production of electricity. The wind turbines are permanently located at a given location and are designed to generate maximum energy annually. Since the wind direction and velocity are subject to rapid and frequent changes, the wind turbines and their associated auxiliary machinery are designed to respond to both. For this reason, wind turbines are usually free to rotate about a vertical axis. The entire wind turbine system is pivoted on bearings and is free to swivel. A separate downwind vane or a simple rudder ensures that the wind turbine system is always

[3] 1 kn = 1.852 km/h. The distance 1.852 km is equal to one minute of arc latitude along any meridian on the surface of the Earth.

perpendicular to the wind. This type of motion is known as *yaw*. The characteristic time of yaw is typically significantly smaller than the characteristic time of wind direction variations. For this reason, in the following sections it will be assumed that the wind turbine is always in a position that the wind is perpendicular to the plane described by the rotation of the blades.

8.2.1 Spatial and Temporal Characteristics of Wind: The Boundary Layer and Exceedance Curves

It is well known from fundamental Fluid Dynamics that when a gas flows over a stationary solid surface, the velocity of the gas close to this surface diminishes and it is actually considered to be equal to zero at the surface. This is the so-called *no-slip condition* on stationary surfaces and is one of the basic principles on which the subjects of Fluid Dynamics and Aerodynamics are based. This no-slip condition applies to environmental flows and to the wind. Thus, the wind velocity approaches asymptotically the value zero on the surface of the Earth and increases with the height, $V(z)$. When a fluid stream flows over a surface its velocity vanishes close to the surface and increases with height to an almost uniform value. This uniform velocity is called the *free-stream velocity,* V_0. All the velocity variation takes place in a region close to the surface, which is called the *boundary layer.* Because the wind blows over large landscapes, the pertinent Reynolds numbers are very large and the flow is invariably turbulent. Hence, a *turbulent boundary layer* is developed over the terrestrial surface when the wind blows.

The boundary layer is the area of the flow where frictional forces and dissipation are of importance. Inertia is the dominant force outside the boundary layer. Two dimensionless parameters characterize the turbulent boundary layer:

a) The friction velocity, v^*, which is given in terms of the wind shear stress at the ground level, τ_w, and the density of air, ρ:

$$v^* = \sqrt{\frac{\tau_w}{\rho}}. \tag{8.3}$$

b) The wall coordinate, z^+, which is given in terms of the friction velocity and the kinematic or the dynamic viscosity of air, v and μ, respectively:

$$z^+ = \frac{zv^*}{v} = \frac{z\sqrt{\tau_w \rho}}{\mu}. \tag{8.4}$$

Very close to the ground and at heights less than 2 m there are two parts of the turbulent boundary layer that are called the *viscous region* and the *buffer layer,* where the viscous effects play an important role. Outside this thin layer, the viscous effects do not affect the velocity profile and the wind velocity may be approximated by the expression:

$$v^+ = \frac{V}{v^*} = \frac{1}{\kappa}\ln(z^+) + C. \tag{8.5}$$

where $\kappa = 041$ is the *von Karman constant,* and $C = 5.0$.

The shear stress, τ_w, is usually expressed in terms of the free-stream velocity, V_s, that is the velocity outside the boundary layer and a friction coefficient, f. The latter is an empirical parameter and is obtained from experimental studies:

$$\tau_w = \frac{1}{2} f \rho V_s^2. \tag{8.6}$$

Accordingly, the expressions for the friction velocity and the wall coordinate become:

$$v^* = V_s\sqrt{\frac{1}{2}f} \quad and \quad z^+ = \frac{0.7071\rho z V_s}{\mu} = 0.7071\mathrm{Re}_z. \tag{8.7}$$

It follows that the friction velocity is proportional to the free-stream velocity and that the wall coordinate is effectively proportional to a Reynolds number, which is defined by the height from the surface.

An approximate expression for the ratio of the wind velocities above the ground, at levels z_1 and z_2 is given by the so-called 1/7th power law. This is an approximation to a multitude of experimental data on turbulent boundary layers over solid surfaces. According to this law, the velocity in the boundary layer is proportional to the 1/7th power of the distance from the solid surface, that is $V(z) \approx kz^{1/7}$, where k is a constant. Therefore, the ratio of the velocities at two different heights is given by the simple algebraic expression:

$$\frac{V(z_2)}{V(z_1)} = \left(\frac{z_2}{z_1}\right)^{1/7}. \tag{8.8}$$

When wind velocities are measured in meteorological stations, by convention they are measured at a height of 30 ft (9.1 m). Hence, $V(z_1 = 9.1\text{ m})$ is known experimentally, and the last expression may be used to determine any other velocity within the boundary layer. For example, if the measured velocity is 10 kn, the wind velocity at 100 m above the ground is $10{*}(100/9.1)^{1/7} = 14.1$ kn.

Figure 8.4 shows the power law velocity profile in the first 100 m above the Earth's surface. Two trends are apparent from this figure and from the concept of the boundary layer:

a) Most of the variation in the wind velocity occurs in the very low part of the boundary layer, which is close to the ground; and
b) While the velocity profile flattens above 40 m, there is still wind variability in the upper part of the boundary layer.

For example, the wind velocity increases by 7.6% between the heights 30 and 50 m, where most of the wind turbines are located. Given the cubic dependence of

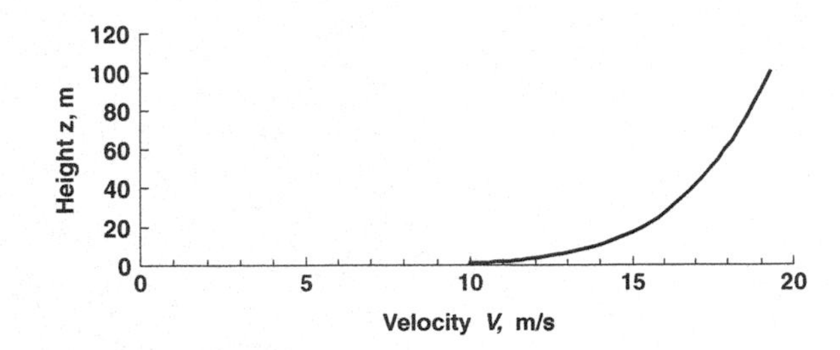

Fig. 8.4 Velocity distribution in the boundary layer

the power with respect to velocity, this implies that the available wind power increases by 25% between these two heights. The implication of this conclusion is that wind turbines should be placed as high as structural and material considerations allow.

The temporal characteristics of the wind emanate from the fact that the magnitude and direction of wind velocity vary significantly over a period of time. The mean wind velocity, V_m, and the mean power velocity, V_p, are two parameters that apply to the magnitude of the velocity and characterize the velocity variation and the available average power. These two parameters are defined as follows:

$$V_m = \frac{\int_{t_1}^{t_2} V(t)dt}{t_2 - t_1} \quad and \quad V_p = \frac{\left[\int_{t_1}^{t_2} [V(t)]^3 dt\right]^{1/3}}{t_2 - t_1}. \tag{8.9}$$

The mean power velocity, V_p, yields the average amount of available power for the wind during the time interval $t_1 < t < t_2$ and is derived from Eq. 8.2. Typically, the time interval $[t_1, t_2]$ is taken over one year and the two velocities signify the annually averaged wind velocity and the annually averaged available power.[4] Because, in general, $V(t) > 1$, it follows that $V_p > V_m$. A large number of empirical observations have shown that the two variables are correlated as: $V_p \approx 1.25\ V_m$. Since the available wind power is proportional to the cubic power of V_p, and since $1.25^3 = 1.953 \approx 2$, the expression for the available wind power becomes:

$$\dot{W}_{av} \approx A\rho V_m^3 = \frac{\pi}{4} D^2 \rho V_m^3 \approx \frac{\pi}{8} D^2 \rho V_p^3. \tag{8.10}$$

Wind velocities vary annually from zero, to values significantly higher than the mean wind velocity. The very high velocities however, occur only for short periods of time. Even though the extreme wind velocities may provide significantly higher power temporarily, they have very short duration. An engine designed to operate at the extreme velocities would provide large amounts of power only for a short period annually. When one makes decisions on the

[4] Usually the two integrals are computed from a large number, N, of velocity measurements. Hence, the two integrals are approximated by the sum of the respective integrants and the time interval by the total number of measurements, N.

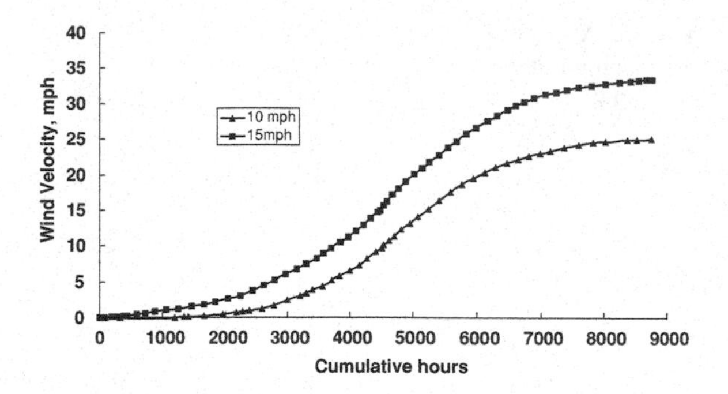

Fig. 8.5 Wind distribution curves for two locations with $V_m = 10$ and 15 mph

construction and operation of a wind turbine the magnitude of the wind velocities as well as the annual hours that certain ranges of velocities occur in the given location become important. For this reason, the *wind distribution curves,* which are also known as *exceedance curves, endurance curves,* and *duration curves,* are used for the construction of wind turbines. These are histograms that show how many hours annually the local wind velocity exceeds a certain value. Wind velocity is typically the abscissa of these diagrams and the hours of the year—0 to 8,760—is the ordinate. The wind distribution curves for two locations with average wind speeds, 10 mph (miles per hour) and 15 mph are shown in Fig. 8.5. These distribution curves provide valuable information on the duration of wind velocities and form the basis for the design and operation of wind turbines, as will be explained in more detail in Sect. 8.3.

8.2.2 Probability Distributions of Wind Speed and Wind Power

Rather than rely on the empirical data and the graphical form of the exceedance curves, engineers often develop from the experimental measurements probability distribution functions for the wind speed at a given location. The most suitable and common of these distributions is the *Rayleigh distribution,* which in its general form has the functional relationship:

$$p[x;\sigma^2] = \frac{x}{\sigma^2} exp\left[-\frac{x^2}{2\sigma^2}\right]. \tag{8.11}$$

The mean of the Rayleigh distribution is equal to $\sigma(\pi/2)^{1/2}$. When the latter is substituted for the average velocity, V_m, one obtains the following functional form for the wind velocity distribution:

$$p[V] = \frac{\pi V}{2V_m^2} exp\left[-\frac{\pi}{4}\left(\frac{V}{V_m}\right)^2\right]. \tag{8.12}$$

This probability distribution function satisfies the two essential conditions of any probability function:

$$\int_0^\infty p[V]dV \equiv 1 \quad and \quad \int_0^\infty Vp[V]dV \equiv V_m. \tag{8.13}$$

From differentiation of Eq. 8.11 it follows that the most probable wind speed at a location is $(2/\pi)V_m = 0.637V_m$.

According to the Rayleigh probability distribution, the average available wind power may be obtained by the following expression:

$$\frac{\dot{W}_m}{\rho A} = \int_0^\infty V^3 p[V]dV = \int_0^\infty \frac{\pi V^4}{2V_m^2} exp\left[-\frac{\pi}{4}\left(\frac{V}{V_m}\right)^2\right]dV = \frac{6}{\pi}V_m^3. \tag{8.14}$$

Since $6/\pi = 1.910$, the Rayleigh probability distribution implies that the average available power is almost equal to $1.25^3\ {V_m}^3\rho A$, which is the basis of the approximate Eq. 8.10. It may be also noted that the power probability function has a maximum at $V/V_m = (8/\pi)^{1/2} = 1.60$.

8.2.3 Fundamentals of Wind Power Generation

If one were to convert all the available power of the wind, as expressed by Eq. 8.2, to electricity, the wind would have to leave the turbine with vanishing kinetic energy and, hence, the wind will have to come to a stop downstream of the wind turbine. This would increase the downstream pressure and, naturally, the wind would stop and the turbine would not produce any power. It is apparent from the above that, for the continuous operation of a wind turbine, the air velocity downstream must be finite. Hence, the power extracted by the wind turbine would be less than that given by Eq. 8.2. The simple analysis of the air flow near the wind turbine, which follows, will help derive a realistic value for the maximum power a wind turbine would be expected to produce.

Let us consider an open thermodynamics system, or control volume, which surrounds the wind turbine. The cross-section of the control area at the turbine is equal to the area described by the rotating blades A_b $(= \pi D^2/4)$. A schematic diagram of the pressure and velocity distributions within this control volume is given in Fig. 8.6. The inlet and the outlet of the open system are far enough from the turbine so that the influence of the rotating blades on the static pressure is vanishingly small. Hence, the static pressure at both the inlet and the outlet is equal to the atmospheric pressure P_a. The static pressure variations in the entire control volume are small enough and it is assumed that they do not affect significantly the density of the air, which remains almost constant as air flows through the control volume. It is also assumed that the control volume is at a level significantly higher than the ground surface and that the velocity may be considered uniform and

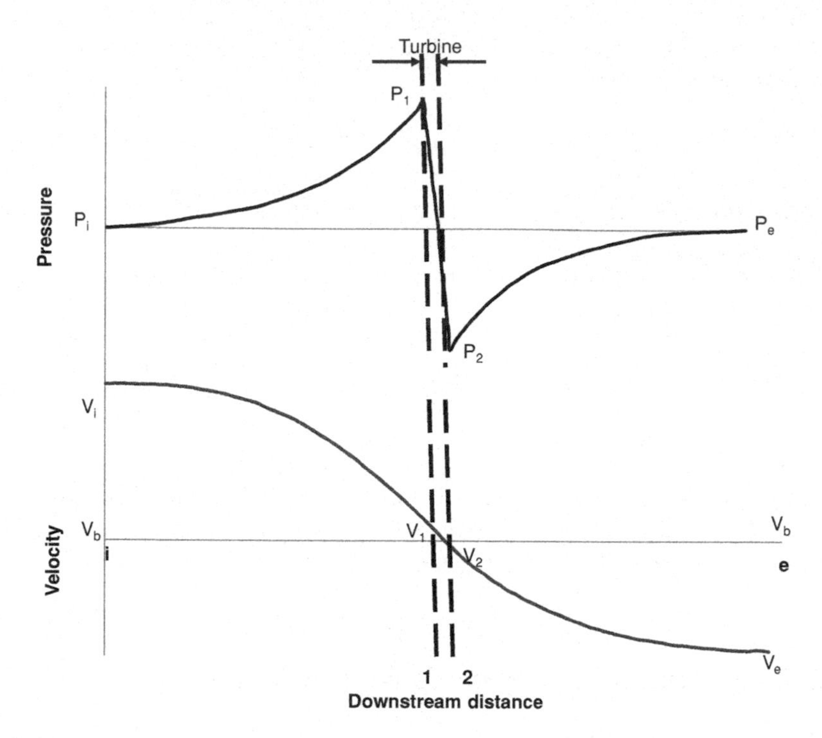

Fig. 8.6 Pressure and wind velocity variations close to the wind turbine

unidirectional. The inlet velocity of the air stream is V_i and the outlet velocity is V_e. It must be noted that the cross-sectional area of the control volume for this problem is not constant. The control volume area varies, in order to accommodate the changes in velocity and to ensure that all streamlines remain within the control volume. Hence, the mass flow rate,

$$\dot{m} = \rho A V. \tag{8.15}$$

is constant in the control volume and the mass continuity equation is satisfied in all cross-sections of the control volume. In the diagram of the Fig. 8.6 the wind turbine occupies the region 1–2, where, because of the extraction of mechanical power, it is expected that the static pressure will drop from P_1 to P_2 and the air stream velocity from V_1 to V_2. The control volume accommodates these changes.

Both the static pressure and the velocity vary continuously in the control volume. The first derivative of the static pressure is not necessarily continuous, because of the extraction of mechanical power at the turbine. The first derivative of the velocity is continuous, as depicted in Fig. 8.6, because the control volume is defined by the streamlines, no mass escapes or is added and the mass conservation equation is satisfied in the direction of the streamlines. Throughout the control volume there is not any source of significant friction and, hence, frictional losses are considered small enough to be negligible. Under these conditions, the momentum conservation equation in the control volume is the same as the Bernoulli equation and may be written as follows for the front section i to 1:

$$\frac{P_i}{\rho}+\frac{1}{2}V_i^2=\frac{P_1}{\rho}+\frac{1}{2}V_1^2. \tag{8.16}$$

Similarly, Bernoulli's equation is valid at the back of the control volume from 2 to e, where it may be written as follows:

$$\frac{P_2}{\rho}+\frac{1}{2}V_2^2=\frac{P_e}{\rho}+\frac{1}{2}V_e^2. \tag{8.17}$$

It must be pointed out that Bernoulli's equation is not valid in the region 1–2, because a significant amount of work is extracted by the blades of the turbine. This is a very thin part of the cross-section of the control volume, because the thickness of the blades is very small in comparison to the length of the control volume. One may subtract the last two equations, apply the condition: $P_i = P_e = P_a$ and derive the following expression for the difference of the static pressure, P_1-P_2, immediately before and after the turbine.

$$P_1-P_2=\frac{1}{2}\rho\left(V_i^2-V_1^2+V_2^2-V_e^2\right). \tag{8.18}$$

An inspection of the velocity variation in Fig. 8.6 and the fact that both the velocity as well as its first derivative are continuous functions, leads to the conclusion that, for thin turbine blades, points 1 and 2 are very close together and the velocities immediately before and after the wind turbine are approximately equal. If we define the velocity at the center of the wind turbine blade as V_b, we have the condition: $V_b \approx V_2 \approx V_1$. Therefore, the last equation yields the following expression for the static pressure:

$$P_1-P_2=\frac{1}{2}\rho\left(V_i^2-V_e^2\right). \tag{8.19}$$

The axial force that is developed on the blades of the wind turbine is equal to the product of the pressure difference and the area swept by the blades of the turbine, A_b:

$$F_x=A_b(P_1-P_2)=\frac{1}{2}A_b\rho\left(V_i^2-V_e^2\right). \tag{8.20}$$

In the absence of any other forces acting on the control volume, the axial force is also the net force developed on the entire control volume. From the momentum conservation principle, the force developed on this volume is equal to the momentum change of the mass flow rate, given by Eq. 8.15. Taking the turbine blade cross-section, where the area is A_b (= $\pi D^2/4$) and the velocity is V_b, as the reference section for the definition of the mass flow rate, the momentum conservation equation yields the following expression for the axial force:

$$F_x=(\dot{m}V)_i-(\dot{m}V)_e=\dot{m}(V_i-V_e)=\rho A_b V_b(V_i-V_e). \tag{8.21}$$

A comparison of the last two equations gives an expression for the unknown velocity at the turbine blade, V_b:

$$V_b = \frac{1}{2}(V_i + V_e). \quad (8.22)$$

Thus, the air velocity at the turbine blades is equal to the arithmetic average of the velocities at the inlet and the outlet of the control volume. Since the power developed by the turbine is equal to the product of the axial force, F_x, and the local velocity, V_b, an expression for the power developed by a turbine with diameter D may be derived as follows:

$$\dot{W} = \frac{\pi D^2}{16}\rho(V_i + V_e)\left(V_i^2 - V_e^2\right). \quad (8.23)$$

The last expression proves that, since V_e must be finite, the actual power obtained from a wind turbine is less than the available wind power, which is given by Eq. 8.2. In a typical wind turbine installation, the inlet velocity V_i is equal to the incoming wind velocity and the latter is determined by nature. The wind turbine and its operation may be designed in a way that the exit velocity, V_e, is optimized to yield maximum power. A simple way to perform this optimization is to differentiate the last expression with respect to the adjustable parameter V_e and ensure that the resulting expression vanishes:

$$\frac{\partial \dot{W}}{\partial V_e} = 0 \Rightarrow 3V_e^2 + 2V_eV_i - V_i^2 = 0 \Rightarrow V_e = \frac{1}{3}V_i. \quad (8.24)$$

Therefore, for the optimum operation of a given turbine, the downstream air velocity should be one-third of the incoming wind velocity. Under this condition the maximum power a wind turbine may produce at continuous operation is as follows:

$$\dot{W}_{max} = \frac{8}{27}\rho A V_i^3 = \frac{2\pi}{27}\rho D^2 V_i^3. \quad (8.25)$$

A figure of merit, or efficiency,[5] for the wind turbines is the ratio of the actual power they produce to the available power in the wind stream:

$$\eta = \frac{\dot{W}}{\dot{W}_{av}}. \quad (8.26)$$

A comparison of Eqs. 8.2 and 8.25 proves that the maximum efficiency a wind turbine may have is $\eta_{max} = 16/27 = 0.593$. Simply put, even an ideal wind turbine

[5] In some publications, the turbine efficiency is called *power coefficient* and is denoted by C_p.

will convert less than 60% of the available power in the wind current to electric power. This is sometimes referred to as *the Betz limit* or *Betz's law*.

8.2.4 Efficiency of Actual Wind Turbines

The steam and gas turbines that are currently used with power plants have isentropic efficiencies close to 80%. These engines are well designed, and also, operate with controlled conditions, such as: almost constant inlet pressure and temperature of the working fluid; several stages; optimum expansion and pressure drop between stages; relatively short blades with low range in the linear blade velocities, ωr, where r is the distance from the blade hub and ω the rotational speed of the turbine; no fluid loses or *spillage effects*; and optimum fluid velocity after each stage. In contrast, wind turbines have: only one stage; are designed to operate with a wide variety of inlet conditions; are open to the atmosphere with the implied fluid losses or *spillage effects*; have very long blades and, therefore, a very wide range of local blade velocities; and must be designed to operate within a wide range of inlet velocities. In addition, Betz's law limits the maximum efficiency of the wind turbines to less than 60%. As a result, wind turbines have significantly lower efficiencies than steam and gas turbines, typically in the range 20–45%.

The efficiency of a wind turbine would depend on several factors such as:

1. The ambient air velocity, which is variable.
2. The rotational design speed of the wind turbine, ω.
3. The type of wind turbine.
4. The size of the wind turbine, or blade diameter D.

A large number of empirical observations with several types of wind turbines have shown that a single parameter, *the tip-speed ratio*, defines very well the efficiency of a wind turbine. The tip-speed ratio (TSR) is the ratio of the velocity at the tip of the blade ($\omega D/2$) to the instantaneous velocity of the wind:

$$TSR = \frac{\omega D}{2V_i} = \frac{\pi n D}{60 V_i}, \tag{8.27}$$

where n is the rotational speed of the engine in revolutions per minute (rpm).

Figure 8.7 depicts the actual efficiencies of several common types of wind turbines as a function of TSR. It is observed that each type of wind turbine has a TSR where optimum operation occurs and that several types of engines, e.g. the Savonius and the American multi-blade types, have a narrow range of operation, outside which the engine does not produce power. The currently used two-blade and three-blade turbines, whose blades are long enough to cover very large areas, have the highest range of efficiencies, close to 40%.

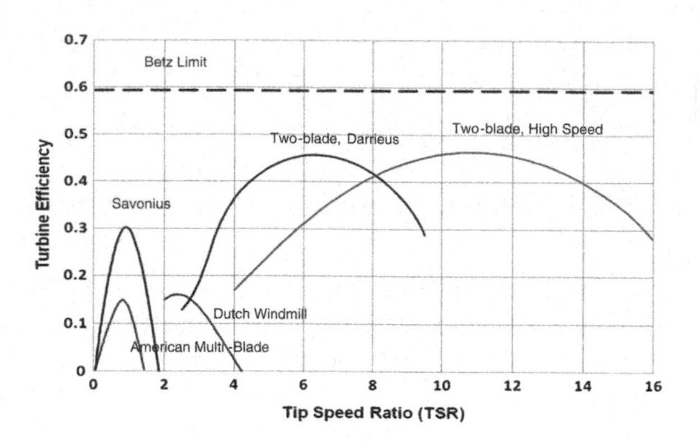

Fig. 8.7 Efficiencies of wind turbines as functions of TSR

8.3 Power Generation Systems: Parts of Common Wind Turbines

As shown in Fig. 8.7, the Savonius, Dutch, and American multi-blade turbines have significantly lower efficiencies than the two-blade engines. In addition, the first engines have high starting torque, which makes them suitable for mechanical work production applications, such as water pumping and grain grinding, but not for power generation. Two or three-blade wind turbines are now used almost exclusively for the production of electric power. The main parts of the systems that comprise these wind turbines are:

1. *The tower*: Since velocities close to the ground are very low and there must be good clearance between the lower part of the blades and the ground, the wind turbines are placed on top of a tower at a significant height above the ground. The height of the tower depends on the diameter of the blade and is of the order of magnitude of the blade diameter, D, allowing a clearance of $D/2$, between the ground and the lower part of the blade. Thus, towers are between 30 and 100 m high. The tower is a simple structural element, usually made of reinforced concrete, which is designed to withstand the axial force and resulting moment generated by the wind turbine. It is typically thicker at the lower part and is usually designed as a hollow structure to allow easy access to the top for engine repairs at the turbine hub. Some older (and shorter) towers were designed as trusses made of metal.
2. *The yaw bearings and yaw break*: Because the wind turbine must rotate to face the instantaneous direction of the wind, the entire electricity producing system is pivoted on strong bearings that allow the rotation of the system around a vertical axis. The drag force on a downstream rotating vane or a simple rudder provides the force for this rotation. In order to avoid overshooting in the rotation of the electricity generating system and unnecessary power fluctuations, the yaw break system slows the rotational motion by providing damping.

3. *The rotor blades*: They are the most important part of the generating system, where the wind energy is imparted to the engine. They are very long, typically 30–100 m in diameter. The rotor blades are designed aerodynamically with pitch angles that vary with the distance from the hub and they are made of low-weight and strong materials. Low density woven composites are now typically used for the turbine blades, which are typically hollow. The blades are connected to the hub, which extends to a horizontal metal shaft that becomes the prime mover of the engine. The shaft is supported by a series of bearings. In the more advanced and better optimized engines, a mechanism is put in place that changes the pitch of the blades to produce maximum power at the instantaneous wind velocity. These mechanisms are made of sensors and actuators, which measure the magnitude of the instantaneous wind velocity, adjust the position of the base of the blades inside the hub and, thus change the pitch of the entire blade. The actuator mechanisms are attached to the blades, rotate with them and are supported by their own *pitch-control bearings*.
4. *The gear box:* In order to minimize the centrifugal stresses, the rotational speed of the blades at operating conditions is fairly low, typically of the order of 100 rpm. A gearbox steps up the rotational speed of the prime mover to reach a range 2,000–3,000 rpm and transmits the power to a secondary high rpm shaft, which is connected to the generator. A small fraction of the blade power is dissipated in the gear box by friction. For this reason, larger wind power engines may require a cooling system for their gearbox.
5. *The generator*: Both permanent magnet generators and generators with electromagnets (exciters) are used for the conversion of the mechanical power to electricity. The generators of the more modern and larger engines are rated in MW (typically 1–3 MW) and include power electronics, such as *Variable Speed Constant Frequency* devices (VSCF), which convert the variable frequency of the secondary shaft to a constant frequency. Any power spikes in the system are usually absorbed by the inertia of the rotor.

One of the salient characteristics of wind power systems is that very high power fluctuations occur with relatively low wind velocity changes. For example, an increase of the wind velocity from 8 to 10 m/s (or 25%) would cause a power increase of almost 100%. Frequent power variations of this magnitude are undesirable because they are associated with high stresses on the blades, on the prime mover and gear as well as with strong power fluctuations on the electric grid. These types of problems are minimized by designing the wind turbines to produce almost constant power.

A glance at Figs. 8.5 and 8.7 proves that, if a wind turbine is designed to operate at the maximum range of the prevailing wind speeds, the turbine would produce maximum power, but for a very short time during the year. Therefore, the total energy that would be produced annually would be very small. For this reason, wind turbines are designed to operate and produce their rated power within a range of wind speeds that are significantly lower than the maximum prevailing wind

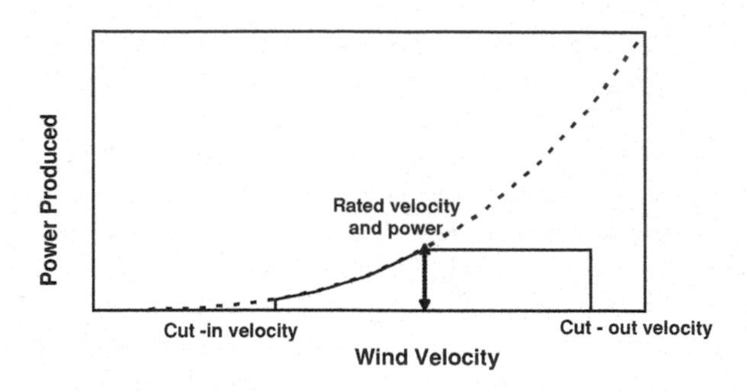

Fig. 8.8 The operation characteristics of a flat-rated wind turbine

velocity. Adjusting the pitch of the blades and the resistance of the prime mover shaft assists in producing constant power in the velocity range between the *rated velocity* and the *cut-out velocity,* which is the upper limit of the velocity range where the engine may operate safely. Such turbines are often called *flat-rated wind turbines.* At the cut-out wind velocity the engine is shut off for safety by *feathering* the blades and removing all the resistance from the prime mover. At this state, the blades may rotate but the engine does not produce any power. On the opposite end, where the wind velocity is very low, the wind turbine is designed to stop producing power at the *cut-in velocity* of the wind. Typically, the cut-in velocity is low and the turbine produces less power than the rated power. There is a range of velocities between the cut-in velocity and the rated velocity, where the power output of the turbine increases according to Eq. 8.2. Figure 8.8 shows the operation of a flat-rated wind turbine. This turbine produces power according to the cubic relationship of Eq. 8.2 between the cut-in and the rated velocities. The power the turbine produces is constant between the rated and the cut-out velocities. And, the turbine does not produce any power when the prevailing wind velocity is higher than the cut-out velocity. It is apparent that, when the wind velocity increases beyond the rated velocity, which is a design parameter, the turbine does not convert all the available power of the wind velocity. Actually, a large fraction of the available power from the wind is not harnessed when the turbine operates at its rated power. This limits significantly the average efficiency of the turbine, when this average is taken over the entire spectrum of wind velocities.

Table 8.1 lists key characteristics and velocity parameters of four types of wind turbines produced by two of the major wind turbine manufacturers. It is observed in this Table that the cut-in, the cut-out and the rated velocities are within narrow ranges in the four types of commercial turbines. The rated velocity of the turbines is in the range 27–30 m/s and the turbines do not produce any power when the wind velocity is more than 60 m/s.

Because of the very high variability of wind velocity, an important parameter in the operation of wind turbines is their capacity factor, CF, defined as the ratio of the average power produced annually to the rated power:

Table 8.1 Key parameters of four commercially available wind turbines (data obtained from www.gewindenergy.com and www.vestas.com)

Rotor diameter, m	104	100	80	70.4
Rated power, kW	3,600	2,500	2,000	1,500
Rated velocity, m/s (mph)	14(30.5)	12.5(27.2)	15(32.7)	13(28.3)
Cut-in velocity, m/s (mph)	3.5(7.6)	3.5(7.6)	4.0(8.7)	4.0(8.7)
Cut-out velocity, m/s (mph)	27(58.8)	25(54.5)	25(54.5)	25(54.5)
Rotor angular speed, rpm	8.5–15.3	N/a	9–19	12–22
TSR	3.3–6.0	N/a	2.5–5.3	3.4–6.2
Efficiency at rated power	0.26	0.27	0.20	0.29

$$CF = \frac{\dot{W}_m}{\dot{W}_r}. \tag{8.28}$$

An analytical expression for the capacity factor in terms of the ratio V_r/V_m, may be obtained using the Rayleigh distribution of Eq. 8.14 and expressing the velocity in terms of the rated velocity V_r, and the temporal pattern of the power produced as depicted in Fig. 8.8. It is noted that an actual flat-rated wind turbine will produce power in the range of wind velocities between the cut-in and cut-out velocities, $[V_{ci}, V_{co}]$. After performing the integration operation, one obtains the following expression for the capacity factor:

$$\begin{aligned} CF = {} & \frac{\pi^2}{16}\left[\exp\left[-\frac{\pi}{4}\left(\frac{V_r}{V_m}\right)^2\right] - \exp\left[-\frac{\pi}{4}\left(\frac{V_{co}}{V_r}\right)^2\right]\right] \\ & + \int_{V_{ci}}^{V_r}\left[\frac{\pi}{2}\left(\frac{V}{V_r}\right)^3\left(\frac{V}{V_m^2}\right)\exp\left(-\frac{\pi}{4}\frac{V^2}{V_m^2}\right)\right]dV \end{aligned}. \tag{8.29}$$

The first two terms of Eq. 8.29 stem from the range of velocities $[V_r, V_{co}]$, where the power produced is constant. Wind turbines are typically designed with rated velocities corresponding to the maximum of the power distribution function. From Eq. 8.14 the latter is: $V/V_m = (8/\pi)^{1/2} = 1.60$. Under this condition, the capacity factor is approximately equal to 0.35. Of this number, the first part of Eq. 8.29 contributes 0.13 and the second part 0.22. This implies that such a wind turbine operates at rated power 13% of the time and at less than rated power 87% of the time. However, because of the significantly lower power produced in the range under the rated power $[V_{ci}, V_r]$ when one looks at the total energy production of the turbine, 37% of the total energy (13/35) is actually produced during the 13% of the time, when the turbine operates at rated power.

8.3.1 Smaller Wind Turbines

While the power systems described in the last section pertain to relatively large power systems (1–3 MW), there are several smaller engines in the market that produce from a few kW to 100 kW. These are classified as small engines and are designed to produce power for local use, not for feeding the electric grid. Small engines are ideally suited for use in remote areas, which are not served by the existing electric grid. Farms are ideal locations for such smaller engines, whose output is used for domestic consumption as well as for pumping and irrigation. In 2010 there were 600,000 small wind turbines installed in the world. At an average power of 1.5 kW, this translates to 900 MW installed electric capacity, or close to the capacity of a large nuclear unit. In rural Texas alone, there were 20 MW of installed small wind turbine capacity. When these systems are also coupled with energy storage systems, they may supply without interruption the energy needs of buildings and small communities in rural areas.

Another type of system that has been marketed more recently is a small wind turbine, which is installed on roofs of houses in urban and suburban areas. Most of these designs have a vertical axis and are based on the Darieus concept. The number of blades on such systems varies from two to eight. These small engines are placed on the rooftops of houses and, depending on their size, may provide 0.5–5 kW of power. This is enough power to satisfy the needs of a household or a small business building. Since vertical axis engines operate independent of the direction of the wind, it is not necessary to pivot these engines for the production of optimum power. Several types of these small engines have aerodynamic blades and use both the drag and the lift components of the wind force for optimum power production. While it is possible to produce a considerable amount of power with the installation of many such small engines in suburban communities, where the average wind velocity at the rooftop level is adequate, the generated noise of the engines has become a significant drawback to their widespread implementation.

8.3.2 Other Wind Power Systems

Another wind engine concept is based on the *Magnus effect* or *lift effect:* When a rotating cylinder is in a crosswind a force is developed on the cylinder, which is perpendicular to the axis of rotation and the direction of the wind. A familiar manifestation of the Magnus effect is the *curveball* in baseball or tennis, where the ball changes direction because of its rotation (spin). Figure 8.9 shows the development and the direction of the *lift* or *Magnus* force on a rotating horizontal cylinder. If the cylinder rotates with angular velocity ω, the radius and length of the cylinder are R and L respectively, the air density is ρ and the wind velocity is V_i, then the Magnus force is given by the expression:

$$\vec{F}_M = 2\pi\rho R^2 L(\vec{\omega} \times \vec{V}_i). \tag{8.30}$$

where x denotes the vector product. Now, if the rotating cylinder moves in the direction of the Magnus force with a velocity V_c, it produces power, which is given by the scalar product of the force and last velocity vector:

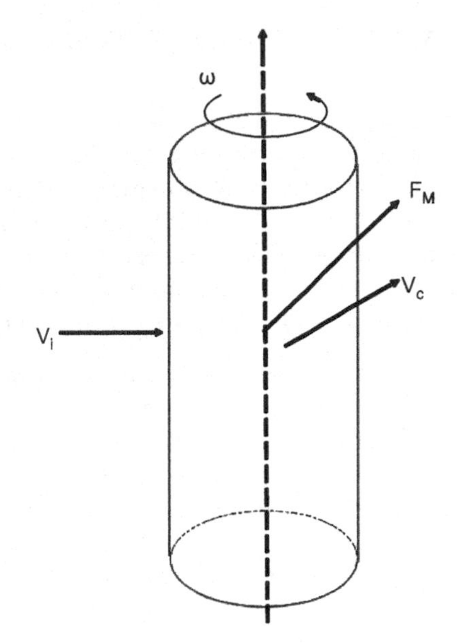

Fig. 8.9 Lift force developed on a rotating cylinder with crosswind

$$\dot{W} = \vec{F}_M \cdot \vec{V}_c = 2\pi\rho R^2 L(\vec{\omega} \times \vec{V}_i) \cdot \vec{V}_c. \tag{8.31}$$

With the appropriate combination of rotational speed and wind velocity, engines may be developed that would utilize the Magnus effect and produce a significant amount of power. One such engine, the *Madaras engine*, was proposed and developed in the 1930s in New Jersey, U.S.A. It consisted of several vertical cylinders of 30 m height, rotating at 120 rpm (2 rps) and placed on small, flat rail cars that were carried around a circular track. Small electric motors ensured the rotation of the cylinders. Because V_c changes direction as the cars move on their circular track, the direction of the rotation of the cylinders was reversed twice during a full trip around the track. This change in the rotational motion of the cylinders ensured that the developed Magnus force correlated well with the circular motion of the cars to produce power. Electrical generators were driven by the wheels of the cars to produce power, which was to be fed to the grid. The Madaras engine attracted a great deal of attention, but met an inglorious end: while it was in the development stage, it was destroyed by a storm and was not rebuilt.

The main advantage of engines, such as the Madaras engine, is that they produce power regardless of the direction of the wind and they are built on the ground. However, a significant disadvantage of such engines is that the lift force produced is a secondary and much weaker force than the drag force. Thus, the power that may be produced is significantly less than the available power of the wind. This is the main reason why efficient engines based entirely on the Magnus effect have not been built.

8.3.3 The Future of Wind Power

The first decade of the twenty-first century has seen a very significant expansion in harnessing power from the wind for the production of electricity. From approximately 14,500 MW worldwide installed electric capacity in 1999, the capacity in 2009 rose to 115,000 MW, an eightfold increase. Most of the electric capacity has been installed in the European Union. The worldwide installed electric capacity is expected to continue increasing at a high rate in the next decade. This trend is currently fueled by the following factors:

1. Wind power is abundant in several regions of the world and was previously relatively undeveloped.
2. The world population and governments, and especially those in Europe have become more aware of the adverse effects of the fossil fuels.
3. International agreements, such as the Kyoto agreement, mandate higher use from sources such as solar and wind power.
4. Governmental subsidies for clean or "green" power, which make the production of electricity from wind more competitive economically.
5. National or regional regulations and directives that call for a percentage of the electric power to be produced from renewable or "green" sources.

It is expected that, three of the current drawbacks for the utilization of wind power will be overcome, namely:

1. Wind power is diffused and is not harnessed in large power plants, which are the norm in the electricity production industry. A high capacity electric power plant (e.g. 1,000 MW) similar to the large fossil fuel or nuclear power plants, which utilize significant economies of scale, is impossible to construct. The national electric grids and distribution systems are designed for such large power plants and not for the smaller wind power engines. One may place together a large number of smaller wind turbines, 0.5–3 MW each, in *wind farms*, to produce larger amounts of power. In addition, electric grids may be adapted for the lower power inputs.
2. The prime locations for wind power generation are not necessarily in areas where electric transmission lines exist or where the transmission grid has the capacity to carry more power. New or more modern transmission lines may need to be built.
3. The technology of construction and installation of wind turbines is relatively new. This is not a significant disadvantage, because with longer experience, better and more cost-effective systems will be produced in the near future.

The main and most significant drawback, which limits the expansion of wind power, is that it is largely unpredictable and almost intermittent. The discipline of meteorology has advanced sufficiently so that, at present, we are able to predict with relative accuracy the wind speed and available wind power at a given location in the next few hours and, with lesser accuracy, in the next few days. As we move further into the future, e.g. weeks or months, the predictive methods for the strength of wind velocity and power become quantitatively unreliable. For example, science does not have an accurate answer to the question: "What will be the wind speed at the Isle of Jersey one month from today?"

Local governments and companies charged with satisfying the electricity demand of the population in a region or a nation simply cannot rely with any degree of certainty that wind turbines will produce sufficient electric power to satisfy the entire electricity demand at a given day or hour of the year. The available statistical methods are sufficient for reliable correlations to be obtained, but the correlation coefficients are very low for wind power to achieve the level of reliability required for the continuous and uninterrupted supply of the demand for electricity. An electricity generating company may not rely on wind power to supply a large fraction of the demand for electricity for its customers and must always have another option for backup. For this reason, wind power is and will be an alternative option to be harnessed whenever it is available, with the necessary backup of other electric power plants, which may produce electricity continuously during wind-less days or weeks. The absolute necessity of these backup units and the cost associated with them imposes the most significant drawback for the expansion of wind power. The development of new power storage methods and systems will help significantly with this issue. However, energy storage, especially over long time durations, is associated with significant energy losses, as may be seen in Chap. 12. This adds significantly to the cost of wind power, which is more abundant in the spring and fall, while the electric power demand peaks in the summer.

8.4 Environmental Effects

Wind power is one of the most benign and most environmentally friendly forms of electric energy production. It does not involve any chemicals and does not produce any harmful emissions or thermal pollution. The materials that make the towers and the components of the engines are commonly used structural and engineering materials. Hence, the construction and operation of the wind turbines does not impose any environmental threat. There are a few rather minor environmental issues associated with wind power, which are enumerated in the following paragraphs:

1. *Noise pollution*: A rotating engine always produces noise and wind turbines are no exception, especially the parts that operate at higher rpm's. However, large wind turbines are typically located in remote, rural areas with low population density. This mitigates the effect of the noise to human populations, but may have a significant effect on the wildlife in the nearby areas, which migrate because of the noise. Wind farms even if they are located in remote areas

disturb the balance of the local ecosystem. Noise pollution is also the limiting factor for the expansion and more widespread use of small engines in urban or suburban environments.

2. *Bird injuries and mortality*: Flying birds may be caught by the rotating blades and be killed. The overall motion of the blade and the pressure reduction that occurs immediately before the rotating blades, as shown in Fig. 8.6, detract the birds and often kill them. This may have a significant effect on migrating species of birds. Concerns for the bird populations that migrate to the shores of the Gulf of Mexico have put severe restrictions on the wind power development projects near the Gulf, where the sea breeze is always strong and the available wind power averages 500–600 W/m^2.
3. *Aesthetic pollution*: The picturesque landscape of remote, pristine areas is often disturbed by the placement of groups of high towers with wind turbines on top. The aesthetic pollution has raised significant opposition to the development of wind power, especially in areas with tourist interest.
4. *Radio and TV signal interference*: Given that many wind turbines are located near the top or the sides of hills and mountains, their operation interferes with the transmission of television and radio signals, especially with television re-transmission towers. This effect is mitigated by two trends: a) the location of the vast majority of wind turbines is in remote areas, where very few signals need to be transmitted, and b) the recent trend in communications is to transmit signals via fiber optic wires, at least in the most populated areas. This trend will become the norm in the future for all urban areas.

It is apparent that, while not insignificant, the environmental problems associated with wind power are much less harmful to the environment and the ecosystems than the effects of most other renewable energy sources. The further expansion of wind power and the substitution of conventional power sources by wind would be benevolent to the environment.

Problems

1. What is the power per unit area of a 10 knot uniform wind (in W/m^2)?
2. The drag force on a flat surface due to the wind flow is given by the expression:

$$F_D = \frac{1}{2} A \rho C_D V^2, \; with \; C_D = \frac{24}{\text{Re}}\left(1 + 0.32\text{Re}^{0.687}\right) \; and \; \text{Re} = \frac{\rho L V}{\mu}, \tag{8.32}$$

where C_D is the drag coefficient, Re is the Reynolds number and L is a characteristic dimension, which is equal to the square root of the area. The area A is the area that is perpendicular to the wind direction. A sail with 16.5 m^2 area faces the wind perpendicularly. What is the force on the sail when the wind velocity is 6, 10, 25 and 40 knots? What do you observe in these calculations?

3. The friction coefficient on the surface of the sea is $f = 0.21$. What is the shear stress produced by a wind of 12 knots?
4. A pitot tube at a height of 3 m, measures wind velocity 5.3 m/s. What is the wind velocity at a height of 50 m? If the center of a 45 m diameter wind turbine is placed at 50 m, what is the maximum power this turbine may produce?
5. The wind velocity at a height 3 m is measured to be 2.3 m/s. What is the wind velocity at 20 m? If the air density is 1.21 kg/m^3, what is the mass flow rate of the air that passes through a surface of 10 m width that extends from 2 to 30 m?
6. Starting from first principles, derive analytically *Betz's Law*. Justify every step in your analysis.
7. A 14 knot wind stream comes to a stop at the front of a tall building. What is the pressure increase (stagnation pressure) at the center of the surface of the building?
8. Determine the maximum power a wind turbine may produce at the following wind speeds: 1, 5 and 10 m/s.
9. What is the TSR of a 70 m diameter wind turbine that rotates at 80 rpm when the wind speed is 15 knots?
10. The cut-in velocity of a 30 m diameter flat-rated wind turbine is 1.5 m/s, the rated velocity is 8 m/s and the cut-out velocity 22 m/s. The turbine is placed in a location with the following wind velocity frequency:

$0 < V < 1.5$ m/s 22% of the time.
$1.5 < V < 3$ m/s 12% of the time.
$3 < V < 5$ m/s 12% of the time.
$5 < V < 8$ m/s 28% of the time.
$8 < V < 15$ m/s 12% of the time.
$15 < V < 22$ m/s 10% of the time.
$22 < V$ m/s 4% of the time.

What is the total energy (in kWh) you expect this turbine to produce during a year?

11. Modifications are performed on the turbine of problem 10 and its rated velocity is extended to 15 m/s. What is the annual total energy the turbine will produce?
12. During a 6 h period, the wind velocity varies linearly from 1.2 to 12 m/s. A 40 m diameter flat-rated wind turbine is subjected to this wind. The turbine has cut-in velocity 2 m/s, rated velocity 7 m/s and cut-out velocity 15 m/s. The conversion efficiency of the turbine is 70% of the maximum efficiency. Determine the amount of energy this turbine produces during the 6 h period.
13. Determine the exeedance curve corresponding to a Rayleigh probability distribution with $V_m = 14$ m/s.
14. Determine the most probable velocity and most probable power of a Rayleigh wind velocity distribution with $V_m = 28$ mph.

15. During a particular 24 h period, the wind velocity in Abilene, Texas could have been approximated with a half-sine function, $V = A\sin(\pi t/24)$, where t is measured in hours. The amplitude A was 16.5 m/s during that day. A 38 m diameter wind turbine with cut-in velocity 1.5 m/s, rated velocity 9 m/s and cut-off velocity 21 m/s was placed perpendicular to the wind. If the overall efficiency of the turbine-generator system is constant and equal to 0.32, what is the energy that was produced during that 24 h period?
16. It is apparent that a desirable attribute of wind turbines is to have high rated velocities. Under the conditions described in problem 13, what would have been the total energy produced if the rated velocity for the turbine were 12 m/s? How about if the rated velocity were 17 m/s?
17. A wind tower measures the velocity of wind at a height 30 ft (9.1 m) to be 12 m/s. Calculate the power density of the wind at this height. A large wind turbine with 48 m diameter is placed at this location. The hub of the turbine is 50 m from the ground. Determine the power density of the wind at the hub. Also determine the maximum power that can be delivered by this turbine at this wind velocity.
18. "We can extract so much power from the wind that it would be possible to satisfy the entire electricity demand of humanity using wind power alone." Comment in a 250–300 word essay.

Chapter 9
Geothermal Energy

Abstract Geothermal energy is the primordial energy of the Earth, produced in the interior of the Earth by nuclear reactions. It is a thermal form of energy that is conveyed by the magma to the crust of the planet. In the crust, some of the Earth's heat is advected by the water of deep aquifers to the surface. While, high-quality, high-temperature geothermal resources are met in regions that are close to the tectonic boundaries, other lower temperature resources are abundant in all geographic locations. Geothermal energy is primarily utilized by extracting dry steam or high-temperature liquid water from an aquifer and by using the steam or the vapor of another substance for the production of power with a turbine. Depending on the type and the characteristics of the geothermal resource, flashing or binary power plants are used for the production of steam and electricity. The geothermal electric power plants are simpler than fossil or nuclear power plants, i.e. they have a lesser number of components. Because the geothermal fluid emanates from the Earth's interior and carries other substances, including solids and non-condensable gases, the design of the equipment of a geothermal power plant poses several challenges, such as the avoidance of scale in the well and flashing chambers; and the removal of non-condensable gases from the condenser.

9.1 Introduction

Geothermal energy is the thermal energy convected from the interior of the Earth. Approximately 44 TW ($44*10^{12}$ W) of heat power is transferred from the interior to the surface of the Earth. Of this amount, 30 TW is generated by the radioactive decay of elements in the core of the Earth.[1] Given that the average power

[1] The implied power deficit of 14 TW (= 30–44 TW) is an indication that the earth cools continuously and its interior temperature decreases. Therefore, geothermal power will decrease in the future and, eventually, will cease. This is expected to take place in a geological time scale, of the order of billions of years.

E. E. (Stathis) Michaelides, *Alternative Energy Sources*,
Green Energy and Technology, DOI: 10.1007/978-3-642-20951-2_9,

consumed by the entire Earth's population is approximately 16 TW, it follows that geothermal power alone may satisfy the energy requirements of the humans. However, geothermal is thermal power that appears at the surface of the Earth at low temperatures, and its conversion to electricity is accomplished with significantly low efficiencies. In addition, a great deal of the geothermal power is dissipated in the oceans, which constitute more than 70% of the surface of the Earth. For this reason, geothermal energy may produce small amounts of electricity, but may not be reasonably relied upon to provide a high percentage of the total energy needs of the Earth's population.

Geothermal energy has been used since the ancient times. Among others, ancient Greeks and Romans used the high temperature water from hot springs and fumaroles for baths and even for the heating of public places. Later, therapeutic properties were attributed to these hot springs and a whole resort industry was developed around them, which peaked in the late nineteenth century. Hot geothermal water was used to fill individual or communal baths, to provide high moisture air (steam baths) and, in the winter months to provide heating for the resort.

The harnessing of geothermal energy for the production of electricity commenced in 1904 in the small town of Central Italy, Lardorelo. The local count of Lardorelo, apparently disenchanted with the unreliability and high cost of electricity in his area constructed a well that brought geothermal steam to the ground surface. He connected the steam to a turbine, which produced enough electric power to satisfy the demand of his household. In the 1920s the development of the Geysers field started, to the north of San Francisco, California, and continued in the 1950s and 1960s with the installation of more than 2,200 MW of electric power. Several other electric generation projects from geothermal energy around the world, for example in Wairakei, New Zealand; Tsukuba, Japan; and Reykjavik, Iceland, were completed after 1960. In 2010, there were a total of more than 9,000 MW of installed electric capacity from geothermal power plants in 24 countries. Another 16,000 MW of heat from geothermal origin were produced in 72 countries around the world and were used directly for a variety of purposes, including space heating, fresh water production from snow melting, aquaculture, and agriculture. With the emphasis on renewable energy sources that followed the 2007–2008 increase of petroleum prices, in the summer of 2010 an additional 4,000 MW of electric power installation and more than 20,000 MW of heat utilization projects were in the development and construction stages in the U.S. alone.

The *core* or central part of the Earth is in liquid state at temperatures that are estimated to be between 6,000 and 4,000°C and are maintained at this level from the radioactive decay of nuclear isotopes. The core is surrounded by the cooler layer, the *mantle*, which consists of the *magma,* a hot, semi-glassy, viscoplastic material, with temperatures ranging from 3,000 to 1,500°C. The upper layer of the Earth is a thin solid layer, the *crust,* of 8–10 km thickness, which is made up of six major *tectonic plates,* and several minor plates. The crust of the Earth consists of several solid plates that are floating on a pool of magma. The insulating characteristics of the crust help to maintain the lower temperatures at the surface of the Earth and to limit the heat transfer from the mantle to the atmosphere and the outer

Fig. 9.1 Regions of high geothermal activity are at the boundaries of tectonic plates

space. However, at the boundaries of the tectonic plates, which are referred to as *geological faults,* there are significant intrusions of hot magma into the crust layer. Heat from these pockets of hot magma is conducted to local aquifers, often called *geothermal reservoirs,* where water temperatures may rise to 200–300°C. Because the local static pressure is significantly high, water at these temperatures is below the corresponding saturation temperature and exists in the liquid state.

Figure 9.1 shows the regions of the Earth at the boundaries of tectonic plates, in dark, where magma intrusion and high geothermal activity has been observed. Among these regions is the western Pacific Rim, which encompasses the USA, Mexico, all the Central American countries, Ecuador, Peru, and Chile. The countries of Japan, Indonesia, New Zealand and the Philippines are in the East part of this Rim and have significant geothermal resources. In Europe the Mediterranean countries and Iceland have the majority of geothermal resources and geothermal power plants. The eastern African countries of Ethiopia, Eritrea, Kenya, and Tanzania also have significant geothermal resources and a few geothermal installations. The same regions are also characterized with higher volcanic activity and plate instabilities that generate earthquakes.

Figure 9.2 is a schematic diagram that shows how the intrusion of magma close to the Earth's surface creates a geothermal field in the local aquifer. Geothermal wells, which are drilled deep into the aquifer zone, bring the geothermal fluid to the surface of the Earth and supply a geothermal power plant that produces electric power. Oftentimes, the high pressure in the geothermal reservoir is augmented by the insertion of a *downhole pump,* which supplies a higher volume of geothermal fluid to the power plant.

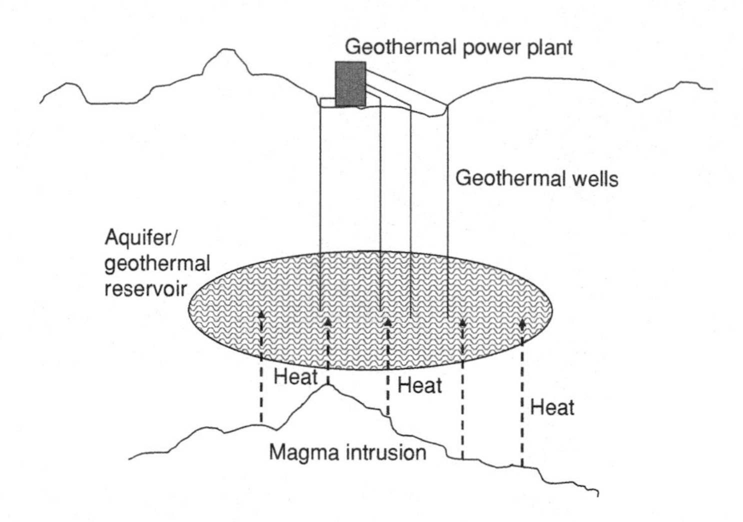

Fig. 9.2 A schematic diagram of geothermal activity

The *geothermal resource* has become almost synonymous with the *geothermal reservoir or aquifer*. Surface water that seeps through the ground and the permeable geological strata replenishes the water carried by the wells. In addition, water that has been used in the power plant is often injected back to the aquifer via *reinjection wells.* Because of the replenishment of the water and the very long time associated with the cooling of the Earth, geothermal resources are considered renewable energy sources. However, it has been observed that, in a period of a few decades, the pressure, temperature and volumetric flow rate of water produced by a specific geothermal well decline. When this occurs, the well may be shut and another well may be drilled at a different location, in the same aquifer. While the temperature of the geothermal reservoir depends to a high degree on the proximity of the reservoir to the heat source of magma, the static pressure of the reservoir is approximately equal to the hydrostatic pressure of a column of liquid water from the surface to the reservoir. This happens because rain or surface water finds its way toward the aquatic reservoir through the fissures of the rocks and other passages. Therefore, there is a continuous column of water from the surface of the Earth to the reservoir level at depth H. As a result, the reservoir pressure P_{res} may be approximately given in terms of the liquid water density, ρ, and the depth from the surface, z, as follows:

$$P_{res} = P_{at} + \int_0^H \rho g dz \approx P_{at} + \rho g H. \tag{9.1}$$

The geothermal fluid is essentially water. Depending on the location of the aquifer, the geothermal fluid contains several minerals, sometimes in significant quantities. For this reason the geothermal fluid is oftentimes referred to as *brine.* $NaCl$, KCl, and $CaCl_2$ are among the most common minerals of geothermal fluids.

The first two are soluble in water, while the last is essentially insoluble at lower temperatures and is carried in the fluid as small particles that may deposit in the piping. The geothermal fluids also contain a number of gases, among which are CO_2, CH_4, the malodorous H_2S, and oftentimes traces of the radioactive Ra, which is produced from the nuclear reactions in the core of the Earth. When modeling the flow and characteristics of the geothermal wells, the properties of pure water are typically used as an approximation. More accurate modeling would require the exact composition of the geothermal fluid as well as knowledge of the influence of all the constituents/impurities on the properties of the water substance.

9.1.1 Geothermal Resources

The type and quality of the geothermal resources depends on the type of fluid the wells produce and is intricately related to the specific exergy of the geothermal fluid. In general, geothermal resources are classified in the following types:

1. *Dry steam*: The water in these aquifers is at a significantly high temperature, typically in the range 200–280°C. As the water flows through the porous medium that constitutes the aquifer, its static pressure drops and a great deal of steam is produced inside the aquifer by local flashing. The steam, with a small fraction of droplets is carried in the geothermal well, where the pressure is further reduced and most or all of the droplets evaporate to produce a higher amount of steam. Thus, wells from a dry steam resource supply the power plant with saturated or superheated steam, which may be fed directly to a turbine for the production of power. The fields at Lardorelo and the Geysers are primarily dry steam geothermal fields. The specific exergy of the steam produced is significantly higher than the exergy of the other types of geothermal resources and, for this reason dry steam resources are considered to be of high quality.
2. *Liquid water*: When the reservoir temperature is lower, liquid hot water at high pressure is produced at the bottom of the well. As the water rises in the well, its static pressure decreases due to the gravitational and frictional losses in the pipe. The temperature of the geothermal fluid is also reduced because of heat losses to the surrounding rock. However, given the insulating properties of the rocks, the rate of temperature reduction is very low. The continuous reduction of pressure in the well may cause the static pressure of the fluid to become lower than the saturation pressure of the fluid at the prevailing temperature. At this level of the well, flashing of the water occurs and some steam is produced. As the fluid moves upwards to the wellhead, more steam is produced by flashing. A mixture of steam and liquid water is directed to the power plant, where additional steam may be produced by flashing. The dryness fraction (quality) of the fluid at the wellhead steam depends to a great extent on the characteristics (pressure and temperature) of the geothermal reservoir as well as on the depth of the well. From the thermodynamic properties of water and

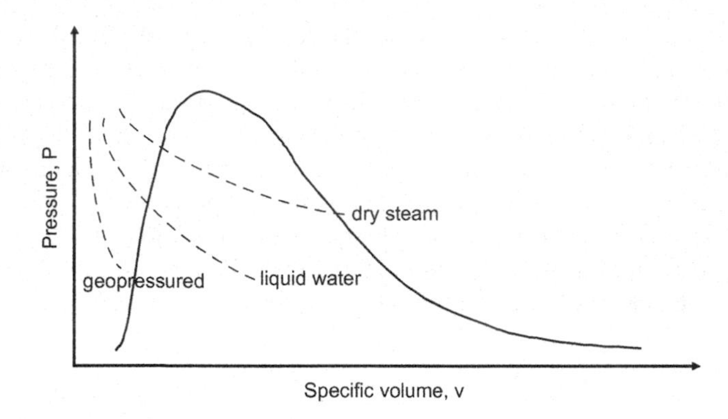

Fig. 9.3 Geothermal fluid processes on a P,v diagram

steam, one may easily deduce that the specific exergy of the fluid produced by liquid water resources is lower than that of the dry steam type.

3. *Geopressured*: Temperatures in the geopressured reservoirs are very low, typically in the range 120–160°C, which correspond to low saturation pressures. These reservoirs are typically located at greater depths than the dry steam and liquid water resources and, hence, the reservoir pressures are significantly higher. At all levels in the well, the pressure of the geothermal fluid is significantly higher than the saturation pressure. Thus, flashing does not occur and vapor is not produced. Liquid, hot water is produced and may be directed to the power plant. The exergy of the liquid water is significantly lower than that of steam and, hence, these resources are considered to be of significantly lower quality than the dry steam resources. The utilization of geopressured geothermal resources or the production of electricity is usually accomplished by a binary power conversion system, which will be described in Sect. 9.2.5. Figure 9.3 shows the pressure–volume diagram of the thermal processes that occur in the reservoirs and the wells of the dry steam, the liquid water and the geopressured resources. Since the amount of heat lost in the wells is very low, these processes are considered to be isenthalpic.
4. *Hot, dry rock*: As the name implies, these resources are hot rocks at a significant depth from the Earth's surface without a nearby aquifer to produce geothermal brine. Despite the fact that there is no local fluid to be carried to the surface, one may develop an engineering system to harness the thermal energy of the rock by drilling a number of wells, injecting a fluid to the wells, bringing this fluid to the surface and using the thermal energy of this fluid for the production of electric power. The working fluid is not necessarily water and may be another fluid with convenient thermodynamic properties, such as butane, pentane or a refrigerant. Figure 9.4 shows a map of the United States with the temperatures that are expected to be found at depths of 3,000 m. It is apparent that most of the high temperature resources are in the Rocky Mountains and the western part of the country. It is also apparent in this figure that, at

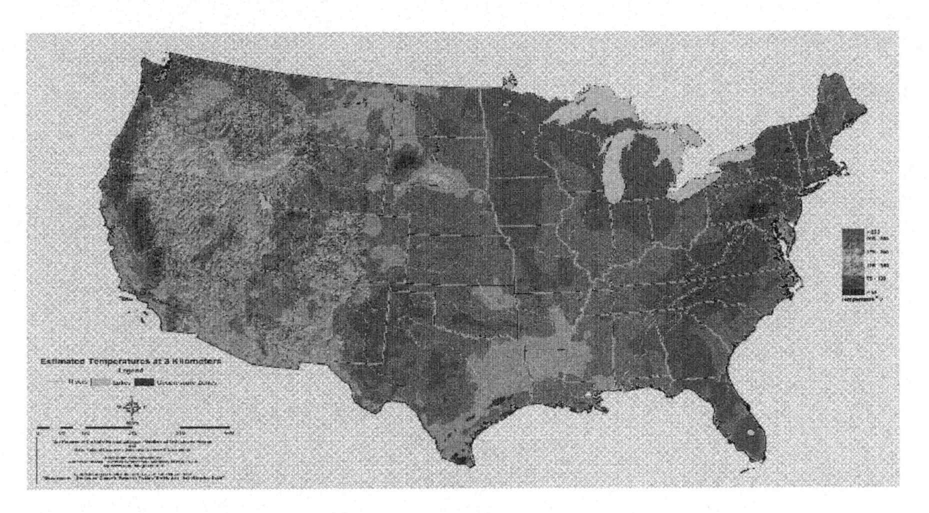

Fig. 9.4 A map of the United States, showing the expected temperatures at a depth of 3,000 m (Prepared by the Idaho National Laboratory Geothermal Technologies Program and Idaho National Laboratory Geospatial Science & Engineering)

a depth of 3,000 m there are several regions in the USA, where the temperature exceeds 150°C, Engineering systems may be designed and constructed in these areas to convert the Earth's thermal power to electricity.

While the dry steam type of geothermal resources are the most desirable for the production of electric power, these resources are scarce and most of them have already been utilized. Similarly, a high percentage of high temperature liquid water resources have been utilized economically and the few that remain will be developed in the near future. Geopressured resources and hot, dry rock resources are abundant on the planet. Despite the fact that the temperatures of these resources are lower, they represent the overwhelming majority of the geothermal resources on the planet, they have the capacity to produce economically a good amount of the electric power of several nations and they should be seriously considered as significant energy resources.

9.2 Geothermal Power Plants

All geothermal power plants utilize resources at moderate to low temperatures and rarely exceed 200°C. For this reason, they are built differently than fossil fuel and nuclear power plants, with an emphasis on higher second law efficiencies and high rate of exergy utilization. The type of geothermal resource that supplies the thermal energy dictates to a large extent the type of power plant that is built. A list of such plants from the simplest to the more complex ones is given in the following subsections.

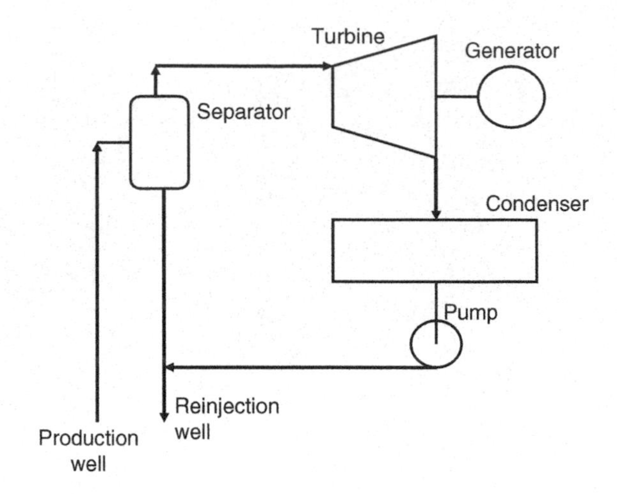

Fig. 9.5 Schematic diagram of a dry steam geothermal power plant

9.2.1 Dry Steam Units

These are the simplest geothermal power plants and utilize resources that produce dry steam. The original geothermal plant at Lardorelo and several of the units at the Geysers geothermal field in California are of this type. Essentially, the geothermal well produces dry (or almost dry) steam, which may be fed directly to a steam turbine that drives the electric generator. A schematic diagram of a dry steam power plant is depicted in Fig. 9.5. The steam from the well is fed to the *steam dryer* or *separator* where any water droplets that are present are removed. The steam is then directly fed to a turbine, which drives the electric generator, and finally exhausts in the condenser. The condensate, which is almost pure water, is then re-injected to the geothermal reservoir or is used locally as a water source. If the mass flow rate of the steam produced is denoted by $\dot{m}$ and the exergy of the steam by e_1, the maximum amount of power such an electric power unit may produce is:

$$\dot{W}_{\max} = \dot{m}e_1 = \dot{m}[h_1 - h_0 - T_0(s_1 - s_0)]. \tag{9.2}$$

The subscript 0 denotes properties at the *dead state*, which is characterized by the ambient temperature and the atmospheric pressure. In the actual operation of the plants, the turbine is not isentropic and the temperature in the condenser is slightly higher than the ambient, 35–45°C. Such thermodynamic irreversibilities would reduce the actual power produced by this power plant to:

$$\dot{W}_{act} = \dot{m}\eta_T(h_1 - h_{2s}) = \dot{m}(h_1 - h_2) \tag{9.3}$$

where 1 and 2 are the states of the steam at the entrance and exit of the turbine as shown in Fig. 9.5. For the calculation of the enthalpy at state 2 one uses the isentropic efficiency of the turbine, η_T, and the state 2s, which is the state that would be reached in an isentropic expansion 1–2*s*, that is, state 2*s* has equal

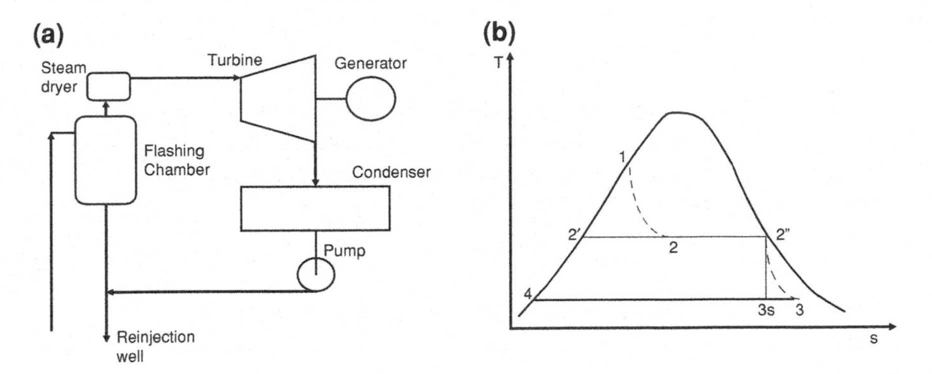

Fig. 9.6 **a** Schematic diagram of the single-flashing geothermal power plant. **b** Thermodynamic diagram of the single-flashing geothermal power plant

entropy to that of state 1 and its pressure is equal to the turbine exhaust pressure (Eq. 3.23).

Dry steam geothermal power plants are the simplest of all steam plants, require a modest amount of capital to construct and are very inexpensive to maintain. For this reason, the electric power produced by dry steam geothermal power plants is significantly less expensive than power produced by most of the other alternative energy sources.

9.2.2 Single-Flashing Units

A schematic diagram of a single-flashing geothermal power plant is shown in Fig. 9.6a. Geothermal fluid from the well at state 1, which may be liquid or two phase mixture (saturated steam and liquid water) enters a flashing chamber, where its pressure is reduced significantly. The flashing process takes place at constant enthalpy and, as a result a significant amount of steam vapor is produced. The vapor is separated from the droplets in the *steam drier* and then fed to the turbine where it expands to the pressure of the condenser, thus producing power. The condensate and the flashing chamber effluent are discarded or re-injected to the reservoir.

The operation of a single-flashing geothermal power plant is shown in the T-s diagram of Fig. 9.6b, where the state of the fluid produced by the well is assumed to be in the two phase region, state 1. At the flashing chamber, the prime symbol ($'$) denotes saturated liquid and the double-prime symbol ($''$) denotes saturated vapor. Flashing chambers are sufficiently insulated for the process 1–2 to be at constant enthalpy. Hence, the fraction of steam produced by this process is obtained by the expression:

$$h_1 = h_2 = (1 - x_2)h_{2'} + x_2 h_{2''} \Rightarrow x_2 = \frac{h_1 - h_{2'}}{h_{2''} - h_{2'}}. \tag{9.4}$$

Typically, 10–25% of the geothermal fluid is converted to steam in the flashing chamber. If the total mass flow rate of the geothermal fluid from the well is denoted by $\dot{m}$, the amount of steam fed to the turbine is $\dot{m}x_2$ and the total power produced is calculated from the expression:

$$\dot{W}_{act} = \dot{m}x_2\eta_T(h_{2''} - h_{3s}) = \dot{m}x_2(h_{2''} - h_3). \tag{9.5}$$

The pressure at state 2 is a parameter in the operation of the single-flashing geothermal units and may be optimized in order to yield maximum power. When the pressure P_2 is very close to the pressure P_1 the steam produced has higher exergy but the fraction of steam produced, x_2, is low. Since the power produced is proportional to this fraction, the total power produced would be low. On the other hand, when the pressure P_2 becomes very low, the exergy of the steam produced (or the isentropic enthalpy difference h_2-h_{3s}) is very low and the total power produced would be, again, very low. Between these two extremes there is an optimum value where the total power produced, as given by Eq. (9.5), is maximized. This optimum value may be obtained by a parametric study, in which the pressure P_2 or equivalently the saturation temperature $T_2 = T_{sat}(P_2)$ is the optimization parameter. From such an optimization study, it was concluded that, when the well produces saturated liquid or compressed liquid at temperature T_1 and the condenser temperature is T_3, then the optimum temperature and pressure for the flashing chamber are given by the expressions:

$$T_{2opt} = \frac{T_1 + T_3}{2} \quad \text{and} \quad P_{2opt} = P_{sat}(T_{2opt}), \tag{9.6}$$

that is, the flashing temperature separates the available range of temperatures in two equal parts.

With typical geothermal resource temperatures near 200°C and condenser temperatures of 40°C, the optimum temperature of the flashing chamber is typically 120°C and the pressure of the flashing chamber is close to 2 bar. Because only 15–25% of the geothermal fluid is converted to steam, 75–85% of the mass of this fluid is discarded or reinjected at this high temperature, thus, wasting a significant amount of the exergy of the geothermal fluid. Clearly this high-temperature fluid may be utilized to produce more steam and, consequently, more power. This is accomplished in the dual-flashing geothermal power plants and the binary plants.

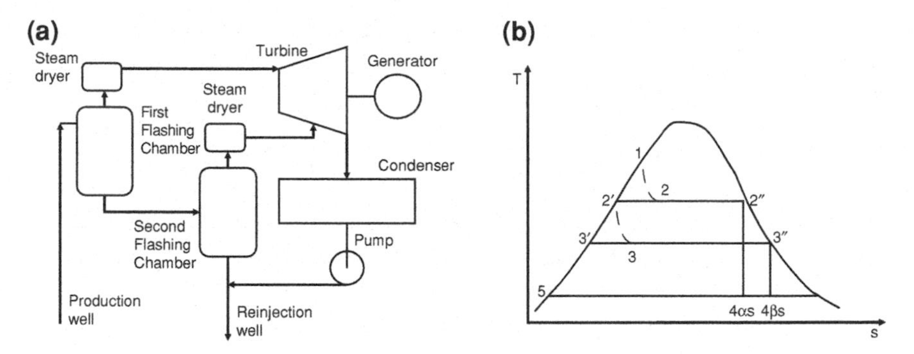

Fig. 9.7 **a** Schematic diagram of the dual-flashing geothermal power plant. **b** Thermodynamic diagram of the dual-flashing geothermal power plant

9.2.3 Dual Flashing Units

The dual flashing geothermal power plants utilize a second flashing chamber, which admits the liquid effluent from the first flashing chamber. The reduced pressure and temperature of the second flashing chamber causes a second, additional quantity of steam to be produced, at the reduced pressure P_3. This steam is also passed through a *steam drier/separator* and subsequently directed to the low-pressure part of the turbine, where it produces an additional amount of power. The schematic diagram of the dual flashing geothermal power plant is shown in Fig. 9.7a and the thermodynamic diagram of the power plant in Fig. 9.7b. As in the case of the single flashing unit, the quantity of steam, x_2, produced in the first flashing chamber is calculated by an energy balance and is given by Eq. (9.4). Following a similar analysis in the second flashing chamber, the steam fraction at state 3 may be calculated from the expression:

$$h_{2'} = h_3 = (1 - x_3)h_{3'} + x_3 h_{3''} \Rightarrow x_3 = \frac{h_{2'} - h_{3'}}{h_{3''} - h_{3''}}. \tag{9.7}$$

It must be noted that the mass flow rate of the effluent from the first flashing chamber is not $\dot{m}$, but $\dot{m}(1 - x_2)$. Hence, the vapor mass flow rate from the second flashing chamber to the turbine is: $\dot{m}(1 - x_2)x_3$. The total power produced by the dual-flashing geothermal power plants emanates from the two streams with dryness fractions x_2 and x_3 and is given by the expression:

$$\begin{aligned} \dot{W}_{act} = \dot{m}\eta_T\left[x_2(h_{2''} - h_{4\alpha s}) + (1 - x_2)x_3\left(h_{3''} - h_{4\beta s}\right)\right] = \\ \dot{m}\left[x_2(h_{2''} - h_{4\alpha}) + (1 - x_2)x_3\left(h_{3''} - h_{4\beta}\right)\right] \end{aligned}, \tag{9.8}$$

where the symbols α and β denote the different states at the end of the expansion processes of the two streams of steam that are fed into the turbine.

As in the case of the single flashing power plants, the pressures and temperatures of the two flashing chambers are free parameters that may be optimized.

When the geothermal well produces compressed liquid or saturated liquid water the optimum temperatures for the flashing chambers are given approximately by the following expressions:

$$T_2 = T_4 + \frac{2(T_1 - T_4)}{3} \quad and \quad T_3 = T_4 + \frac{T_1 - T_4}{3}, \tag{9.9}$$

In analogy with the single-flash units, the flashing temperatures of the dual-flash units separate the available range of temperatures in three equal parts. The corresponding pressures of the two flashing chambers are the saturation pressures of the last two temperatures: $P_2 = P_{sat}(T_2)$ and $P_3 = P_{sat}(T_3)$.

9.2.4 Several Flashing Processes: A Useful Theoretical Exercise

While the dual flashing power plants utilize a higher fraction of the exergy of the geothermal fluid than the single flashing units, still the effluent of the second flashing chamber is liquid water at an elevated temperature. In principle, a third (and then a fourth) flashing chamber may be employed to extract a higher amount of the available power in the geothermal fluid. The isenthalpic flashing process is a highly irreversible process that produces a significant amount of entropy. The inclusion of several flashing processes and the production of small amounts of vapor that are fed to the turbines for expansion, reduces the amount of entropy generated and also reduces the total irreversibility of the power producing processes. The addition of more flashing chambers would decrease the amount of energy destroyed and, as a consequence, would increase both the first and second law efficiencies of geothermal power plants.

An interesting exercise in the theory of thermo-mechanical energy conversion is to consider the multi-flash system, which is shown in Fig. 9.8: The vapor from the kth flashing process is fed to the kth turbine and the liquid is fed to the $(k + 1)$th flashing chamber. The vapor from this chamber is fed to the $(k + 1)$th turbine, the liquid to the $(k + 2)$th flashing chamber etc. If a total of n flashing chambers is used, the total power produced by the geothermal power plant is:

$$\dot{W}_{tot} = \sum_{k=1}^{n} \dot{m}_{k-1} (\Delta h)_k \eta_k, \tag{9.10}$$

where $\dot{m}_{k-1}$ is the amount of vapor produced in the $(k-1)$th flashing chamber, $(\Delta h)_k$ is the isentropic enthalpy difference in the kth turbine and η_k is the efficiency of the kth turbine. Since steam is produced by an isenthalpic process, the mass flow rate of the steam produced in the $(k-1)$th chamber is given by the expression:

$$\dot{m}'_{k-1} = \dot{m}'_{k-2} \frac{h_{(k-1)'} - h_{k'}}{h_{fg}} = \dot{m}(1 - x_1)(1 - x_2)\ldots(1 - x_{k-2})x_{k-1}. \tag{9.11}$$

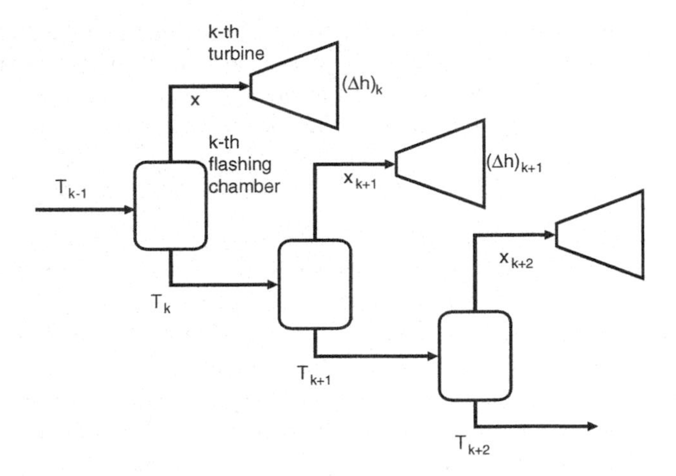

Fig. 9.8 Schematic diagram of three consecutive stages of the multi-flashing geothermal power plant

In Eq. (9.11) $\dot{m}$ is the mass flow rate supplied by the geothermal wells to the first flashing chamber; the prime symbol (′) denotes the saturated liquid state and the double prime (″) the saturated vapor phase; h_{fg} is the latent heat of vaporization, which is almost constant over small ranges of water/steam temperature. Hence, the total power from the entire series of flashing and expansions may be written as follows:

$$\dot{W}_{tot} = \dot{m} \sum_{k=1}^{n} \left[\prod_{i=1}^{k-2} (1 - x_i) \right] x_{k-1} (\Delta h)_k \eta_k. \tag{9.12}$$

When the number of the flashing processes is very large, only a very small amount of steam is produced in each flashing chamber and, hence $(1-x_i) \approx 1$. In addition, the isentropic enthalpy drop and the liquid to liquid enthalpy drop shown in Eq. (9.11) may be approximated in terms of the specific heats for the saturated water vapor, $c_{p''}$, and the saturated liquid, $c_{p'}$, as follows:

$$(\Delta h)_k \approx c_{p''} (T_k - T_0) \quad and \quad h_{(k-1)'} - h_{k'} = c_{p'} (T_{k-1} - T_k), \tag{9.13}$$

where T_0 is the condenser temperature. Assuming that the properties h_{fg}, $c_{p'}$, and $c_{p''}$ are constant, one may derive the following approximation for the total power produced by the multiple flashing systems, solely in terms of the flashing temperatures, $T_1, T_2, T_3 \ldots T_k$:

$$\dot{W}_{tot} \approx \frac{\dot{m} c_{p'} c_{p''}}{h_{fg}} \sum_{k=1}^{n} (T_{k-1} - T_k)(T_k - T_0) \eta_k \tag{9.14}$$

One may use the theory of Lagrange undetermined multipliers to maximize the total power produced, when the geothermal water enters the plant at temperature

T and the condenser temperature is at T_O. As with the single- and the dual-flashing units, the optimization process will yield that the available range of temperatures be divided in n equal parts. The optimization process yields the following expressions that determine all the intermediate temperatures, T_1, T_2, T_3...T_k, ...T_n:

$$T - T_1 = T_1 - T_2 = T_2 - T_3 = \ldots = T_{k-1} - T_k = \ldots T_{n-1} - T_n = T_n - T_0, \tag{9.15}$$

Equations (9.6) and (9.9), may be considered as special cases of this general expression, derived for $n = 1$ and $n = 2$ respectively.

At the limit, $n \to \infty$, which is practically approximated by introducing a very large number of flashing chambers, the total amount of power produced by the geothermal power plant becomes equal to the theoretical maximum power. The latter is expressed by the product of the mass flow rate and the exergy of the geothermal fluid with respect to an environment at temperature T_O:

$$\lim_{n \to \infty} \left(\dot{W}_{tot} \right) = \dot{W}_{\max} = \dot{m}[e(T) - e(T_0)]. \tag{9.16}$$

In principle, the maximum amount of work may be extracted by an isentropic *two-phase turbine*, which makes possible the expansion of hot liquid fluids. Although there is considerable engineering research going on in this area, such a turbine is far from becoming reality in the near future. Common turbines operate with vapor expansion and, for this reason, geothermal energy is currently utilized by producing vapor in a flashing or a binary power plant.

In practice, there are diminishing returns in the marginal amount of power that may be extracted by having more than two flashing chambers. The added cost of such installations does not justify the additional energy that may be produced. For this reason, there are not any known geothermal power plants, which utilize more than two flashing chambers. Instead, binary power plants are used, which typically utilize a higher fraction of the exergy of the geothermal fluid and, as a consequence, have higher Second Law efficiencies.

Another practical consideration in the design of flashing geothermal power plants is the pressure in the flashing chamber: Flashing chambers have significantly high volume and, over time, develop cracks that allow air to leak from the outside, if the pressure is sub-atmospheric. Air leakage to a flashing chamber would increase the percentage of non-condensable gases to the turbine and the condenser, thus, reducing by a higher percentage the net power of the plant. On the other hand, a limited amount of steam leakage from the flashing chamber to the environment would not decrease by a great deal the net power produced. For this reason, single and dual-flashing power plants are designed with the flashing chambers to be above atmospheric pressure. For example, if the optimum temperature of a dual flashing unit, T_3, is calculated to be less than 100°C, the second flashing chamber is designed in the temperature range 102–105°C. This choice maintains a positive pressure inside the second flashing chamber, $P_3 > P_{atm}$, and prevents the costly leakage of atmospheric air into the second flashing chamber.

9.2.5 Binary Units

The construction of binary units is recommended when the geothermal fluid at the wellhead is at low temperature. Given this low resource temperature and the practical consideration to keep the pressure and temperature of the flashing chambers higher than 1 atm and 100°C respectively, the production of steam by flashing would be very low and a flashing-type power plant would produce very low power. In such cases, it is advantageous to use a heat exchanger for the transfer of the thermal energy from the geothermal fluid to a *secondary* or *working* fluid, which evaporates and undergoes expansion in a turbine and subsequent condensation. Effectively, the heat exchanger becomes the boiler of a simple Rankine cycle that uses the secondary fluid for the production of power. A schematic diagram of a binary geothermal power plant is shown in Fig. 9.9. The geothermal fluid enters the main heat exchanger from the production well and exits at the re-injection well. In the heat exchanger heat is transferred to the secondary fluid, which is the working fluid of the simple Rankine cycle. The secondary fluid exits the heat exchanger as a saturated or superheated vapor at sufficiently high pressure and temperature to power the turbine, which drives the electric generator. The Rankine cycle is completed via the condenser and the condensate pump of the secondary fluid.

Suitable choices for the secondary fluid are substances with high saturation pressures and low corresponding saturation temperatures. Candidate fluids are most of the refrigerants, including ammonia, and several hydrocarbons, such as propane, butane, iso-butane and pentane. Figure 9.10 shows the operation of the heat exchanger in a Length–Temperature (L–T) diagram: The primary geothermal fluid supplies with heat the secondary fluid and, thus, cools between states 1–4. The secondary fluid preheats during the process 4–3, evaporates during the process 3–2 and becomes superheated during the process 2–1. The three processes are separated by the dotted lines in Fig. 9.10. If one denotes by $\dot{m}_p$ the mass flow rate of the primary fluid and by $\dot{m}_s$ the mass flow rate of the secondary fluid, then the energy conservation principle (first law of thermodynamics) in the three parts of the heat exchanger yields the following equations:

$$\begin{aligned} \dot{m}_p\left(h_{p1} - h_{p2}\right) &= \dot{m}_s\left(h_{s1} - h_{s2}\right) \\ \dot{m}_p\left(h_{p2} - h_{p3}\right) &= \dot{m}_s\left(h_{s2} - h_{s3}\right). \\ \dot{m}_p\left(h_{p3} - h_{p4}\right) &= \dot{m}_s\left(h_{s3} - h_{s4}\right) \end{aligned} \tag{9.17}$$

The primary fluid remains in the liquid state and, hence, its enthalpy difference may be approximated using its specific heat capacity in all three processes. The same approximation may be used in the preheating process, 4–3, of the secondary fluid. Also, during the evaporation process, 3–2, the enthalpy difference of the secondary fluid is the latent heat h_{fg}. Hence, the set of the above equations may be

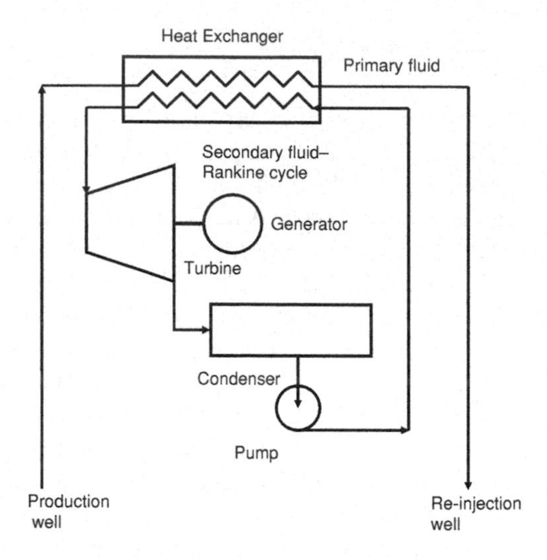

Fig. 9.9 Schematic diagram of a binary geothermal power plant

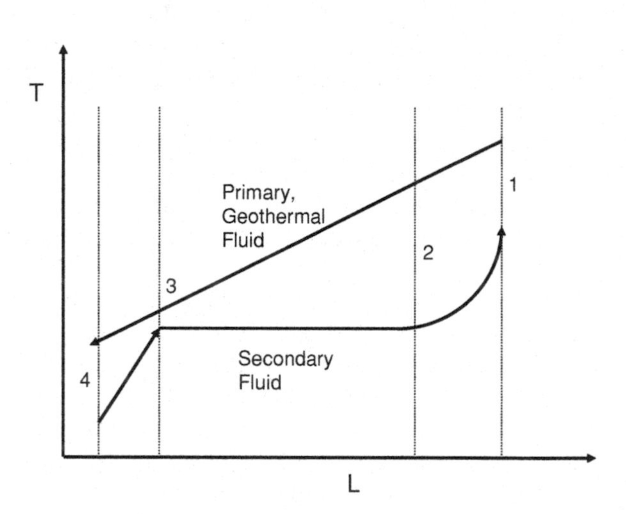

Fig. 9.10 L-T diagram that elucidates the three parts of the main heat exchanger

written in terms of the corresponding temperatures at the primary and secondary side as follows:

$$\begin{aligned} \dot{m}_p c_{pp}\left(T_{p1} - T_{p2}\right) &= \dot{m}_s(h_{s1} - h_{s2}) \\ \dot{m}_p c_{pp}\left(T_{p2} - T_{p3}\right) &= \dot{m}_s h_{fg} \\ \dot{m}_p c_{pp}\left(T_{p3} - T_{p4}\right) &= \dot{m}_s c_{ps}(T_{s3} - T_{s4}) \end{aligned} \tag{9.18}$$

where c_{pp} and c_{ps} are the specific heat capacities of the primary and the secondary fluid, respectively.

The mass flow rate, $\dot{m}_p$and the primary fluid inlet temperature, T_{p1}, of a typical geothermal binary power plant are imposed by the thermodynamic state of the produced geothermal fluid. Also, the evaporation temperature of the secondary fluid, T_{s2}, is decided from design optimization considerations. In addition, heat exchanger design consideration impose a 3–7°C difference at the *pinch point* of the heat exchanger at state 3, that is the difference $T_{p3}-T_{s3}$ is defined to be between 3 and 7°C. These conditions and the three Eq. (9.18) may be used to determine the mass flow rate of the secondary fluid, $\dot{m}_s$, the exiting temperature of the secondary fluid, T_{s1} and the exiting temperature of the primary fluid, T_{p4}. The equations define all the temperatures in the heat exchanger and make its design feasible.

In general, binary geothermal power plants utilize a higher percentage of the exergy of the geothermal fluid than flashing units. Another advantage of the binary power plants is that the suitable choice of a secondary fluid will result in smaller turbines and heat exchangers. In addition, the absence of flashing eliminates any non-condensable in the condenser, which cause a significant decrease of the power, as described in Sect. 9.3. Among the disadvantages of the binary geothermal power plants are:

a) The high cost of the heat exchanger
b) The higher cost of the secondary fluid turbine and
c) The lower temperatures in the primary fluid side may cause significant scaling in the heat exchanger pipes and this may interrupt the operation of the power plant. This may be avoided by the occasional scrubbing and cleaning of the primary side of the heat exchanger.

9.2.6 Hybrid Geothermal-Fossil Power Units

One of the main thermodynamic disadvantages of the geothermal power plants is that the geothermal resources are at moderate or lower temperatures, typically less than 220°C. As a result, the steam produced in the flashing units or the vapor produced in the binary units is at even lower temperatures. For this reason, the first-law efficiency of all the geothermal power plants is lower than the efficiency of fossil fuel and nuclear power plants. A method to increase the temperature and the power produced as well as to improve the efficiency of the geothermal power plants is to use fossil fuels in order to superheat the steam produced by the flashing units, or the vapor of the secondary fluid produced by the binary units. The superheated steam or vapor has higher energy and may produce a higher amount of power in the turbine. A schematic diagram of a single-flash-fossil hybrid power plant is depicted in Fig. 9.11. It is observed that the hybrid plant has all the components/equipment of the single-flashing unit, with the only addition of the steam superheater. Hybrid geothermal power plants have in general higher first- and second-law efficiencies but require the consumption of fossil fuels.

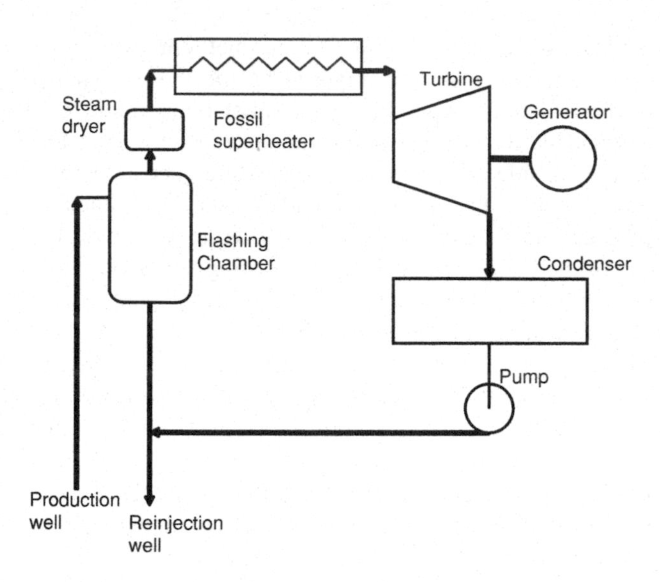

Fig. 9.11 Schematic diagram of a geothermal flashing hybrid power plant

The amount of fossil fuel, $\dot{m}_f$, to be used in such a hybrid plant is determined by the amount of steam vapor produced $\dot{m}_v$ heat balance in the fossil superheater and the superheat temperature:

$$\dot{m}_f \Delta h_f = \dot{m}_v (h_e - h_i), \tag{9.19}$$

where Δh_f is the heat of combustion (heating value) of the fuel and the parenthesis in the right hand side of the equation represents the enthalpy rise of the vapor in the superheater.

9.3 Effects of Impurities in the Geothermal Fluid

Because the geothermal fluid emanates from natural aquifers in the crust of the Earth, it carries several substances other than pure water. These substances are characterized as:

a) *Solids*, which are typically dissolved in the form of ionic salts and primarily consist of NaCl, KCl, Ca_2Cl, Ca_2HCO_3, and $CaCO_3$ and
b) *Non condensable gases*, primarily CO_2, H_2S, CH_4, NH_3, N_2 and C_2H_6. Among the gases CO_2, is the most common constituent and accounts for at least 85% of the mass of the non-condensable gases.

The main effect of dissolved solids on the operation of geothermal power plants is that at lower pressures and temperatures their solubility decreases and they come out of solution. The kinetic considerations of the solution process favor the separation of solids, at the level where steam is initially produced. This often occurs inside the geothermal well, at a level where the static pressure becomes less than the saturation pressure. At this level, the so-called *flashing point* of the geothermal well, steam is produced in the form of bubbles. The evaporation process induces some of the solids to come out of the water solution. Naturally, a fraction of the solids will be deposited on the walls of the pipe. If left untreated, the continuous deposition of salts in the pipe forms scales, which over time will result in pipe restriction and eventual well "clogging." Well clogging may be avoided by treating the affected part of the well with an acid solution or by regularly "scrubbing" the well with metallic brushes. Other parts of the power plant, where scales may be formed from salt deposition are the flashing chambers, where a large amount of steam separates from the geothermal fluid, the remaining water solution becomes supersaturated and solids come out of the solution.

Despite the absence of flashing processes and steam separation, scaling may also become a problem in the heat exchangers of the binary power plants, because ionic solubility is a strong and monotonic function of temperature. Between the entrance and the exit of the binary unit heat exchanger there is always a significant drop of the fluid temperature, typically in the range of 120–150°C. This implies a significant decrease of the salt solubility along the heat exchanger, which leads to the separation of part of the solids from the solution. A high percentage of the solids is carried by the flow, but a small part is driven to the pipe wall by thermophoresis and is deposited. Pipe fouling and "clogging" may occur with the continuous operation of the equipment. This is detrimental in the transfer of heat and the overall operation of the heat exchanger. The solids may be removed and the heat transfer coefficient may be improved by the periodic scrubbing of the heat exchanger pipes or treatment with a mild acid (e.g. HCl) solution.

Non-condensable gases, which emanate from the Earth's magma, account for up to 10% in dry steam installations and up to 2% by weight of the liquid brine resources. The latter, when used with a flashing unit may produce steam that contains up to 35% non-condensables. It has been observed that these gases are flashed out in the initial stages of operation of the geothermal reservoir and, hence, the amount of non-condensable gases in the geothermal wells decreases significantly after the first few months of operation. CO_2 is the main constituent of these gases. Traces of the malodorous hydrogen sulfide (H_2S) are often present and easily detected in the vicinity of geothermal power plants.

The presence of the non-condensable gases has significant effects in the net work the power plant produces and the design of the condenser. As their name implies, these gases do not condense in the condenser and, hence, may not be removed by the condensate pump, which only carries liquid. If left in the condenser, the gases will accumulate over time and will cause a significant increase of the condenser pressure. Since the condenser pressure determines the back-pressure of the turbine, this would imply a significant decrease of the power produced by the turbine. Therefore, the non-condensable gases must be continuously removed from the condenser at the expense of a high amount of power. Among the equipment used for the removal of the gases from the condenser are steam ejectors, pressurized water ejectors, rotary compressors or systems of steam ejectors, and radial blowers (ERR systems).

For a thermodynamic analysis of the gas removal from the condenser, we must recall that the gases and water vapor form a homogeneous mixture in the condenser under a total pressure P_t and a common temperature T. Since there are two vapor constituents in the homogeneous mixture, steam and gases, the total pressure of the mixture, P_t, is equal to the sum of the saturation pressure of the steam, $P_{sat}(T)$, and the partial pressure of the non-condensable gases, P_g:

$$P_t = P_{sat} + P_g. \tag{9.20}$$

The condensation process liquefies most of the water vapor and the condensate is removed by the condensate pump. The remaining steam becomes part of the homogeneous mixture with the non-condensable gases and must be removed by the mechanical gas removal equipment. Let us consider a volume V of the steam-gas homogeneous mixture in the condenser. The partial pressure of the gas, $P_g = P_t - P_{sat}(T)$, is low enough for the gas to be considered as an ideal gas. Therefore, if the apparent molecular weight of the non-condensable gas is denoted by M_g, the mass of the gas in the volume V would be:

$$m_g = \frac{(P_t - P_{sat})VM_g}{\mathbf{R}T}, \tag{9.21}$$

where $\mathbf{R}$ is the universal gas constant, 8.314 kJ/kmol K, and the apparent molecular weight of the gases is often assumed to be equal to the molecular weight of CO_2, 44 kg/kmol.

For better accuracy, the thermodynamic properties of the steam may be obtained from steam tables. Thus, the mass of steam in the same volume, V, would be equal to:

$$m_s = V\rho_s = V/v_s, \tag{9.22}$$

where v_s is the specific volume of the saturated steam at the temperature T (usually denoted as v_g in most steam tables). Hence, the ratio of steam to gas masses in the homogeneous mixture that occupies the volume V is:

$$\frac{m_s}{m_g} = \frac{\mathbf{R}T}{(P_t - P_{sat})M_g v_s}. \tag{9.23}$$

In order to avoid the accumulation of the non-condensable gases the entire mass flow rate of them, $\dot{m}_g$, must be removed. Because this mass is in a homogeneous mixture with the steam, a corresponding mass flow rate of steam must be also removed continuously from the condenser. From Eq. (9.22) the mass flow rate of steam that must be removed with the gases is [1]:

$$\dot{m}_s = \dot{m}_g \frac{\mathbf{R}T}{(P_t - P_{sat})M_g v_s}, \tag{9.24}$$

The corresponding volumetric flow rate the gas extraction equipment must remove from the condenser is:

$$\dot{V} = \dot{m}_g \frac{\mathbf{R}T}{(P_t - P_{sat})M_g} = \dot{m}_s v_s. \tag{9.25}$$

A glance at the last two equations proves that the condensation process in the presence of non-condensable gases is not isothermal at the saturation temperature $T_{sat}(P_t)$ that corresponds to the turbine back pressure, P_t. If $P_t \approx P_{sat}(T)$, a very large amount of steam must be extracted in the vapor form at the expense of a large amount of power. The amount of steam is reduced significantly by designing the condenser temperature, T, to be a few degrees lower than $T_{sat}(P_t)$. Usually, a difference of 3–6°C is sufficient for the condensation to proceed smoothly. This practice effectively increases the total condenser pressure, P_t, to values higher than those the cooling system would otherwise support. In addition to the power used for the gas extraction equipment, the pressure rise in the condenser implies that the back pressure of the turbine, P_t, must be higher than the corresponding back pressure in the absence of the non-condensable gases. This causes a reduction of the isentropic work as well as the actual work the turbine produces.

The presence of non-condensable steam fed to the turbine and the need for the operation of gas extraction equipment uses a significant fraction of the power produced by the plant. Hence, the main effect of the presence of the non-condensable is a significant reduction of the net power produced by the geothermal power plant. Table 9.1 shows the net work produced under ideal conditions when the amount of CO_2 present in the steam fed to the turbine is in the range 0–40%. The actual percentage of work reduction will be higher because of irreversibilities in the equipment.

It is apparent from Table 9.1 that the non-condensable gases may consume a very high percentage of the available geothermal power because they have to be extracted. When the amount of non-condensable gases from the flashing chamber exceeds 8–10%, a good engineering practice is to feed them to a turbine that exhausts to the atmosphere and does not have a condenser. The power forgone by the atmospheric expansion compensates for the power needed to extract the gases from the condenser. Since almost all the non-condensable gases are released at the first flashing chamber, dual flashing plants may use an atmospheric turbine for the expansion of the vapor and gases from the first flashing chamber and a condensing turbine for the expansion of the vapor from the second flashing chamber.

In addition to the power lost for their extraction, a second drawback of the presence of non-condensable in a condenser is the significant drop of the overall heat transfer coefficient. Figure 9.12 shows schematically how the reduction of the heat transfer coefficient occurs. The turbine exhaust, which is composed of steam and non-condensable gases, is fed directly to the cooling tubes of the condenser.

Table 9.1 Power reduction in a flashing unit because of the presence of non-condensable gases

Percent CO_2	Net work (kJ/kg)	Percent reduction
0	620	0
10	485	21.8
20	417	32.7
30	362	41.6
40	324	47.7

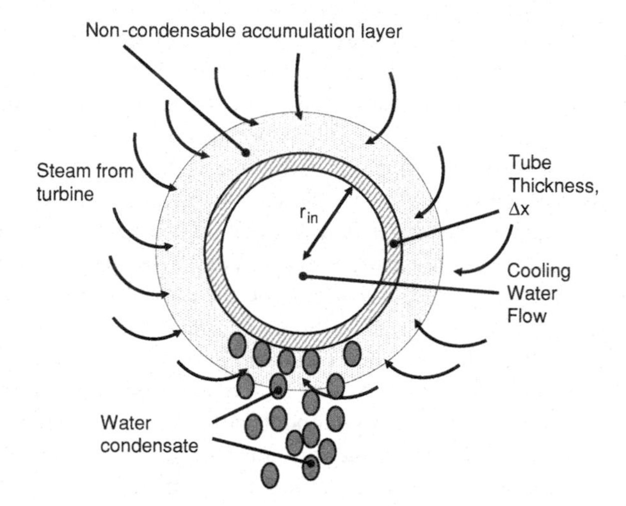

Fig. 9.12 Formation of the gaseous layer at the outer surface of condenser tubes, that causes the reduction of the outside heat transfer coefficient h_{out}

Steam condenses at the outer surface of the tubes but the non-condensable do not condense and, thus, form a gaseous layer at the external surface of the tubes. Even though the non-condensable are continuously removed from the condenser, the gaseous layer surrounding the external surface of each tube becomes a permanent presence. Steam that is directed to the external surface of the tubes must diffuse through this outer layer before it cools sufficiently to condense. The diffusion process is very slow and impedes the heat transfer to the cooling fluid. This causes a large drop of the outside heat transfer coefficient, h_{out}, of the condenser tubes. It has been observed experimentally that a 2% fraction of non-condensable gases would reduce the outside heat transfer coefficient of a tube by more than 50%. This effect causes a corresponding reduction of the overall heat transfer coefficient of the tube, U, which is given by the following expression:

$$\frac{1}{U} = \frac{1}{h_{in}} \frac{r_{in} + \Delta r}{r_{in}} + \frac{\Delta r}{k} + \frac{1}{h_{out}}. \tag{9.26}$$

In this expression h_{in} is the inside heat transfer coefficient, which pertains to the cooling water and is unaffected by the presence of non-condensable; k and Δr are the conductivity and thickness of the tube respectively; and r_{in} is the inside radius of the tube. Accordingly, the rate of heat transfer for the heat exchanger tubes is given by the expression:

$$\dot{Q} = UA(LMDT). \tag{9.27}$$

LMDT is the logarithmic mean temperature difference. The consequence of the significant reduction of h_{out} and U because of the presence of the non-condensable gases is that condensers, which are typically designed to make possible the steam condensation with non-condensable gases, must have a significantly higher surface area, A, than condensers that operate with steam vapor alone. This practice increases significantly the capital cost of the power plant.

9.4 Cooling Systems

Fossil fuel and nuclear power plants use energy sources (fuel) that have very high energy density (in kJ/kg) and may be readily transported to the location of the power plant. As a result, nuclear and fossil fuel plants are located close to the regions of electric power demand and close to sources of available cooling water, usually rivers or lakes. On the contrary, the energy density of geothermal resources is rather low and transportation of the geothermal fluid far from the point of production is neither economical nor thermodynamically sound, because the resource will cool during the transportation. One of the characteristics of geothermal power plants is that they are built very close to the geothermal resource in order to minimize energy losses upon transportation. More often than not, the area where the geothermal resource is located does not have a source of cooling water, such as a river or a lake. For this reason, alternative cooling methods must be used to supply the condenser of a geothermal power plant with cooling water.

Artificial lakes or ponds are probably the best and most economical way to supply the condenser of the geothermal power plants with cooling water. The pond is constructed on location and its water is supplied from the geothermal wells or from local water wells. Since the condensate is not reused or re-injected, it may be discharged to the artificial pond in order to replenish the losses from natural evaporation in the pond.

Mechanical or natural draft wet cooling towers are also used with geothermal power plants. The water needed for the operation of the wet cooling towers may be supplied by the condensate and, hence, no other source of water is needed. If the condensate water is used for another purpose, some of the effluent from the last flashing chamber may also be used after a suitable treatment.

Dry cooling towers, either mechanical or natural have also been used, where the condensate may not be used or is consumed as a source of water for other reasons. However, the large overall temperature difference needed for the operation of the dry towers results in higher condenser temperature and, hence, lower efficiency and lower power production. Because, in general, geothermal power plants have

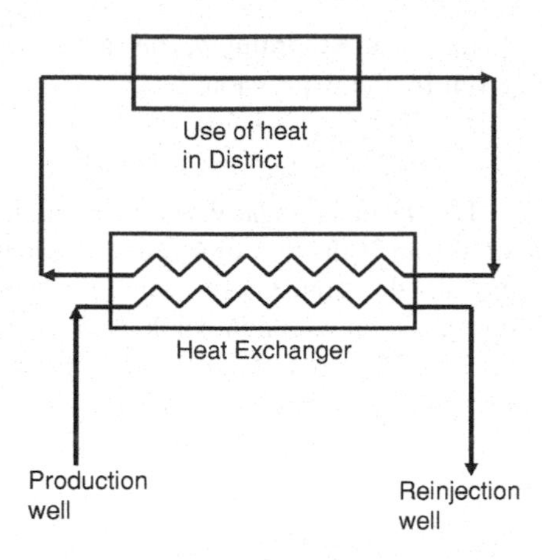

Fig. 9.13 District heating system

lower efficiencies than fossil fuel plants, this additional reduction of the overall power plant efficiency makes the dry cooling towers the least desired option for the production of geothermal power.

9.5 Geothermal District Heating: An Example of Exergy Savings and Environmental Benefit

The earliest use of geothermal energy was to provide hot water for baths and houses. When geothermal resources are available near the population centers, this practice continues today with the heating of entire districts of houses and businesses. This is called *district heating* and has been successfully practiced in several locations in Iceland, Japan, New Zealand as well as in Oregon, USA. The engineering system for district heating is very simple and is shown in Fig. 9.13. The geothermal water from the production well passes through a heat exchanger, where it transfers its enthalpy to a cleaner secondary fluid, which is often clean water. The latter is distributed via well-insulated pipes to the district, where it transfers its enthalpy to raise the interior temperature of the group of houses and commercial establishments. Fluid valves at the entrance of the main heat exchanger control the amount of geothermal water used and balance the demand for heating with the supply. Since the geothermal fluid does not exit the heat exchanger, dissolved gases or solids are re-injected and, hence, do not pose any environmental problem. Some of the carried solids may come out of solution and deposit at the pipe walls of the heat exchanger. For this reason, periodic cleaning of the main heat exchanger—e.g. during the summer season when heating is not needed—is highly recommended.

The use of geothermal water for district heating is an excellent example of how exergy and primary energy sources are saved by the use of alternative energy sources. Let us consider two alternatives for the heating of a group of houses, which require a heating load of 10 MW-t. The first is burning a hydrocarbon fuel, such as methane and the second is the use of a low temperature geothermal resource that provides water at 90°C and may be cooled in the heat exchanger to 45°C. If the heating load is met by the geothermal water, a mass flow rate $\dot{m}_w =$ 52.9kg/s would be used according to the energy balance equation:

$$\dot{Q} = \dot{m}_w c_p (T_{in} - T_{out}). \tag{9.28}$$

If geothermal water were not available and the heating load were met solely by the combustion of methane in conventional household burners with 70% combustion efficiency and given that the heat of combustion of methane, $-\Delta H$, is 50,020 kJ/kg, we would need for the same task of space heating 0.29 kg/s of methane. Thus, the use of geothermal district heating results in the conservation of a significant quantity of methane.

An exergy analysis of the two alternatives also gives an informative example of the benefits of geothermal heat utilization: assuming an environment at 25°C (298 K), the rate of exergy content in the 90°C (363 K) water, which is needed for this energy consuming task is:

$$\dot{E} = \dot{m}_w (e_{90} - e_{25}) \approx \dot{m}_w \left[c_p (363 - 298) - 298 c_p \ln \frac{363}{298} \right], \tag{9.29}$$

or 11,331 kW. Given that the exergy loss (change in the Gibbs free energy) of methane is -$\Delta G = 51{,}978$ kJ/kg, the exergy usage from the combustion of 0.29 kg/s methane is 15,074 kW. This is significantly more than the exergy use of the geothermal fluid.

It is apparent that, even though the geothermal water is only partially cooled to 50°C and not to the environmental temperature of 25°C, the use of the geothermal water for district heating still has a thermodynamic advantage in comparison to the use of fossil fuels, such as methane or other hydrocarbons. Because fossil fuels may produce very high temperatures upon combustion, it is best to use them with high-temperature Rankine cycles with first-law efficiencies in the range 35–40%, and not for the production of relatively low-temperature water for space heating or household water use. In addition to the thermodynamic advantages, there is always the environmental advantage that the combustion of 0.29 kg/s of methane would have continuously released approximately 0.80 kg/s of CO_2 in the atmosphere during the heating season. Therefore, the substitution of the fossil fuels with geothermal water for district heating saves precious natural resources (methane in this case) and prevents environmental pollution.

9.6 Environmental Effects

Geothermal energy is a clean source of energy. The operation of the geothermal power plants causes minimal environmental impact. The main source of atmospheric pollution from geothermal units is the discharge of the non-condensable gases, primarily CO_2. The quantity of CO_2 released to the atmosphere is by far less than that produced by an equivalent fossil fuel plant. Geothermal power plants produce 1,000–5,000 times less CO_2 per kWh produced than fossil fuel plants. Hydrogen sulfide, H_2S, is another gas that is released from the condenser of a geothermal power plant. At very small concentrations, less than 1 ppm, H_2S is malodorous (it smells like rotten eggs) and leaves a distinct odor near geothermal power plants. At higher concentrations, H_2S desensitizes the olephatic nerves and may not be detected by its odor. Sulfur abatement methods may be used to eliminate the release of H_2S. The Geysers field, north of San Francisco, California uses these methods to eliminate a very high percentage of H_2S and to comply with the air standards of the state of California, which are among the strictest in the world. Other gases that are released by a geothermal power plant, such as NH_3, are in traces and do not pose environmental problems.

Soil subsidence may become a problem in the vicinity of geothermal power plants: As steam or water is removed from the geothermal aquifers, open cracks shrink and the permeability of the aquifers decreases. The surrounding rocks and soil are displaced to fill all major voids and the soil surface subsides, but not always in a uniform manner. Uneven soil subsidence on the surface may pose problems to structures and buildings. Water re-injection and rainwater seepage mitigates to a large degree the soil subsidence. Since most of the geothermal resources and power plants are located far from the major population centers, soil subsidence does not pose a significant problem to large populations. A beneficial effect of this geological volume reduction is that mechanical stresses that are caused from geological plate movements are released. The stress release in the interior of the Earth's crust alleviates earthquakes or significantly mitigates their strength and their effects on the structures.

Thermal pollution, which is caused by the heat released to the environment as a result of the operation of the cooling system, is another environmental effect. Because, geothermal power plants operate at lower temperatures, their first-law efficiency (W/Q_{in}) is significantly lower than that of fossil fuel or nuclear power plants. Subsequently the thermal pollution per unit energy produced (kJ/kWh) is higher (about 1.7–2.5 times more) than that of the fossil fuel plants.

Finally, noise, which is always a problem with all power plants, is also an environmental concern for the geothermal units. Ejectors of non-condensable gases, especially steam ejectors, add significantly to the noise pollution. However, since the vast majority of geothermal power plants are isolated and far from population centers, noise pollution is not a significant environmental concern and only affects the wildlife in a way that is similar to the noise effect of wind turbines.

Problems[2]

1*. A geothermal well produces 60 kg/s of saturated liquid water at 210°C. What is the maximum power (in kW) this well may produce? The atmospheric temperature is 27°C.

2*. A geothermal well has a diameter 25 cm and produces dry steam at an average velocity of 9 m/s. The steam is at 5 bar pressure and 200°C. The steam is supplied to a turbine with an isentropic efficiency $\eta = 0.78$ and exhausts in a condenser at an average temperature 36°C. Calculate: a) the mass flow rate of steam; b) the power produced by the turbine; c) the total amount of kWh the plant produces annually; and d) the annual revenue to the operator, if the average sale price of energy is $0.087/kWh.

3*. Typical dry steam wells are 25 cm in diameter and produce continuously at a typical average steam velocity of 10 m/s at the wellhead. The wellhead steam conditions at a particular location are T = 200°C and P = 7 bar. It is proposed to build a 60 MW geothermal power plant in this location. The turbines to be used have an isentropic efficiency $\eta = 0.80$ and enough cooling water is available in this location to maintain the condenser of a power plant at 38°C. What is the power that a single well may produce and how many wells must be drilled to supply sufficient steam to this geothermal power plant?

4*. Five geothermal wells produce together 310 kg/s of a mixture of water and steam at 180°C with a dryness fraction 12%. What is the maximum power a power plant may produce from the five wells? The atmospheric temperature is 27°C.

5*. A geothermal well produces 68 kg/s of saturated liquid water at 230°C. The water is used in a single-flash unit and the isentropic turbine efficiency is 80%. What is the total power produced and what are the first and second-law efficiencies of this power plant?

6*. A geothermal well in Warakei, New Zealand produces an average of 56 kg/s of a two-phase mixture of steam and water at 190°C. The dryness fraction at the well head is x = 0.15. The steam is separated at the wellhead and fed to a turbine with isentropic efficiency $\eta = 0.82$. The rest, which is saturated liquid water is flashed and the steam produced is fed to a low pressure turbine with $\eta = 0.80$ and back pressure 8 kPa. Determine: a) the amount of power produced by the two turbines; and b) the first and second law efficiencies of the power plant.

7*. A geothermal well produces 65 kg/s of saturated liquid water at 230°C. The water is used in a dual-flash unit. What is the total power produced and what are the first and second-law efficiencies of this plant?

---

[2] Steam tables or other water properties are needed for the solution of the problems marked with an asterisk (*).

8*. A geothermal resource/aquifer has been tested and confirmed to continuously produce saturated liquid water at 230°C. It is desired to construct a 60 MW geothermal power plant, which will utilize this resource. Two designs are proposed: a) a single flash unit and b) a dual-flash unit. A nearby cooling pond may supply sufficient cooling water to keep the condenser of the plant at 40°C. For the two types of units under consideration calculate:

a) The total amount of steam that needs to be produced.
b) The total amount of water the wells must provide.
c) If each well produces 60 kg/s of water, what is the number of wells that must be drilled?

9*. A geothermal well produces 60 kg/s of liquid water at 20 bar and 160°C. What is the maximum power you can get from this resource? Design a binary cycle with R-134 or another refrigerant to produce power from this resource.

10*. Two geothermal wells produce 110 kg/s of water at 170°C, 15 bar. A binary plant, operating with refrigerant-134, is to be designed to extract power from this resource. Ample cooling water is available to maintain the condenser temperature at 40°C. A temperature difference of 5°C must be maintained at all points of the main heat exchanger. What is the optimum power this cycle may produce?

Hint: you will need to optimize using an iteration process for the boiling temperature in the secondary cycle.

11*. The average apartment in Paris, France needs 1.2 kW of heat power in the winter months. It is proposed that geothermal district heating be used in the 15th District (15th Arrondisement) for a number of buildings with a total of 1,260 apartments. How much heat power is required? If the geothermal water enters the district heating plant at 70°C and is re-injected at 42°C, how much water volumetric flow rate, in m^3/s, is required for this plant?

12*. A geothermal district project in Iceland has a continuous water supply of 7,200 tons of water per hour. The water enters the district at 73°C and is re-injected at 40°C. This geothermal heat district operates for 285 days per year. Determine: a) the heat power in kW, this project delivers; b) if the only other heating alternative is propane, the annual amount of propane that is saved because of this type of heating; and c) the amount of CO_2 emission avoidance annually because of this project.

13. "Because the interior of the Earth is cooling fast, geothermal energy will not be available. Therefore, there is absolutely no reason to invest in geothermal projects." Comment in a short essay of 250–300 words.

14. "$44*10^{12}$ Watts of power are continuously produced from geothermal resources and this represents three times the entire energy consumption by humans. Therefore, if we concentrate on developing our geothermal resources, we will not need any other form of energy." Comment in a short essay of 250–300 words.

Reference

1. Michaelides EE (1982) The influence of noncondensable gases on the net work produced by the geothermal steam power plants. Geothermics 11(3):163

Chapter 10
Biomass

Abstract Wood, agricultural crops, agricultural residues, herbaceous grasses, algae, sea plants, municipal solid waste, sewage, and animal waste are all forms of biomass that are being used in order to satisfy the energy demand of modern society. Biomass is still the principal fuel used for cooking and domestic heating in most agrarian societies and many developing nations, while in the industrialized nations it is primarily used for the production of other fuels, the *biofuels*, or it is burned for the production of electricity. The production of agricultural crops necessitates large areas of arable land and significant amounts of fresh water for irrigation, two scarce resources on Earth. The consumption of these crops to satisfy the energy demand of the population is fast becoming a controversial socioeconomic issue because of the direct competition of energy production with food production and food availability to the poorer segments of the society and poorer nations on the planet. This "*to eat or to burn*" dilemma in conjunction with the increasing population of Earth, is expected to introduce limitations on the use of agricultural crops and to increase the use of waste products for energy production. This chapter describes the types of biomass that are available as well as their energy content. It presents several useful calculations and facts on the conversion of biomass to transportation fuels and computes, as a special case, the energy balance related to the conversion of corn to ethanol. It appears that in the conversion of corn to ethanol, more energy goes into the materials and processes needed than the energy content of the produced ethanol. The chapter also investigates the environmental effects, social, economic and regulatory issues that will play an important role to the increased utilization of the various types of biomass as energy sources.

E. E. (Stathis) Michaelides, *Alternative Energy Sources*,
Green Energy and Technology, DOI: 10.1007/978-3-642-20951-2_10,

10.1 Biomass

The term *biomass* encompasses all organic plant matter as well as organic waste derived from plants, humans, animals, and aquatic or marine life. The term is broad enough to include: fire wood; methanol derived from wood; alcohol made from grains; methane derived from the anaerobic decomposition of waste; aquatic plants, such as seaweed and kelp; algae; animal waste; municipal solid waste; sewage; and liquid fuel derived from the re-processing of spent engine oil or kitchen oil. Biomass, most notably wood, was the predominant fuel at the beginning of the industrial revolution when households used wood for domestic heating and cooking purposes. As the per capita energy consumption rose and the population of the Earth dramatically increased, burning of wood was insufficient to satisfy the energy needs of humans and was replaced first by coal and then oil and gas. The few areas that have not secured other energy resources to satisfy the growing energy needs of their populations, such as the Central Asian plateaus and parts of Africa, have suffered severe biomass depletion and deforestation, which has significant adverse environmental and micro-climatic implications.

In the beginning of the twenty-first century the use of biomass varies widely among countries: while in most of the OECD countries the use of biomass for energy production is less than 3%, several developing nations sill derive more than 30% of their energy needs from forestry or agricultural products. For example, Cuba produces more than 35% of its electricity from sugar cane; and Brazil uses sugar cane and corn for the production of biofuels and electricity to satisfy more than 25% of its total energy needs. In the OECD nations the most important forms of biomass that are used for the production of energy are:

a) wood for domestic heating and cooking;
b) alcohol, which is mixed with gasoline as transportation fuel;
c) diesel fuel derived from wasted liquid fuels, lubricants and cooking oil;
d) methane and combustible gases derived from the anaerobic decomposition of municipal and animal waste; and
e) forestry and lumber industry byproducts for the production of process heat.

Biomass is a natural method of solar energy storage in the form of chemical energy. Plants use the chlorophyll in their leaves as catalysts and large quantities of solar energy to convert atmospheric carbon dioxide and water into the complex molecules of glucose and fructose as, for example, in the following chemical reaction for the formation of glucose during the photosynthesis process:

$$6CO_2 + 6H_2O \rightarrow C_6H_{12}O_6 + 6O_2 \tag{10.1}$$

This reaction has many intermediate stages, where chlorophyll plays an important role. The overall reaction is highly endothermic with ΔG^0= 480,000 kJ per kmol of formed glucose. From glucose and fructose, plants form the more complex organic compounds, such as sucrose, starch and other complex organic

Table 10.1 Time for the formation of biomass in years

Biomass form	Time to form or regenerate (years)
Forest/timber, northern temperate zone	91
Forest/timber, southern temperate zone	25
Fast growing timber, Sycamore	2–3
Switch grass, corn and sugarcane	0.5
1 ton solid waste in USA per person	1.2
1 ton solid waste in the European Union per person	4.3

molecules with high chemical energy content. In general, plants form carbohydrates according to the general reaction:

$$xCO_2 + xH_2O \rightarrow carbohydrate\ molecule + xO_2 \tag{10.2}$$

From the thermodynamic point of view, plants may be considered as thermal engines that receive solar energy and store it as chemical energy in the form of complex organic molecules. Their conversion efficiency is in the range of 0.1–1.5%, which is significantly lower than that of photovoltaic solar cells or man-made power plants. However, the plants may be planted and grown with low capital cost, low technology, and, in producing carbohydrates they remove carbon dioxide from the atmosphere and release oxygen. For this reason, the production of biomass is beneficial to the environment because it removes CO_2 molecules. The combustion of biomass is neutral to the environment from the carbon footprint point of view, because it returns back to the atmosphere the same number of CO_2 molecules, which were removed during the production stages of the biomass.

Biomass is considered a renewable energy source because the plants and crops grow at short timescales in comparison to the human timescales. As a renewable source, the time of reproducibility of the various biomass forms is important. Table 10.1 shows the time, in years, for several types of biomass to reproduce themselves and regenerate. It is apparent that, while agricultural crops, wastes, and fast growing timber may be considered as renewable energy sources, forests and especially those grown in the northern latitudes do not regenerate fast enough and, thus, their classification as renewable energy sources is questionable.

It must be noted that fossil fuels, such as coal and petroleum, are also products of plants and animals. However, the time to produce or to regenerate the fossil fuels is of the order of millions of years, which is by far greater than the human lifetime or planning timeframe. For this reason, fossil fuels are considered depletable and not renewable energy sources. A glance at Table 10.1 proves that it is not prudent to use timber excessively for the production of energy, especially in the northern zones. Excessive use of timber from forests leads to deforestation, with all the adverse environmental and micro-climate effects that follow it. On the other hand, utilization of municipal or agricultural waste is not only beneficial for the production of energy, but also offers an alternative pathway for the reduction of the volume of the waste and effectively removes the produced methane, which is a potent greenhouse gas (GHG).

The widespread use of biomass for energy production has several disadvantages, which may be summarized as follows:

A. *Periodic production with uncertain yield.* There are one or two harvests of crops every year, primarily in the warmer seasons, and the yield of the land depends significantly on the weather conditions of a particular season. Also, crops are only available during a 2–3 month period, during which all the material must be harvested and used, or put into storage to avoid spoilage. Another, auxiliary supply of fuel will be needed for the continuous operation of a power plant, which is designed to burn biomass.
B. *Diffuse nature of the energy source*: With a range of conversion efficiencies between 0.1 and 1.5%, a glance at Fig. 7.4 proves that the annually-averaged energy yield of the land is on the order of 1 W/m^2. While the total amount of energy available is large, the energy density is very low. This implies that large tracts of land must be used for the production of significant energy from biomass and that there are large costs for the collection, transportation, processing and delivery of the energy produced.
C. *Further processing is needed*: Energy in biomass is in the form of chemical energy, which must typically be processed, by fermentation or combustion, for its final use as transportation fuel or as fuel for power plants. In contrast, the yield of solar energy in W/m^2 is 20–30 times higher. In addition, solar energy is produced and may be used directly as electricity, while biomass must be processed to another fuel or burned to produce electricity. Direct conversion bypasses the Carnot limitations in the conversion of thermal energy.
D. *Competition with food production*: Several food crops, such as corn, sugar cane, and oily seeds are typically included in the biomass and are used for the production of energy. Socioeconomic considerations preclude the use of large areas of arable land for the production of energy when this practice would have the effect of rising food prices or making food scarce in several regions of the planet.
E. *Highly variable energy per unit mass or volume*: The energy content of biomass varies widely according to the type and origin of biomass and its moisture content. Dry agricultural products, such as rice hull and dry bagasse[1] have significantly higher energy density (in kJ/kg or in kJ/m^3), while municipal waste and sewage have lower energy density.

Two figures of merit that characterize a fuel are the high heating value (HHV) and the low heating value (LHV). Both are measured in terms of the mass of the fuel (kJ/kg). The former represents the maximum heat that may be extracted from

[1] Bagasse is the residual from sugar cane production, after the stalk is ground and pressed to release its recoverable sugar content.

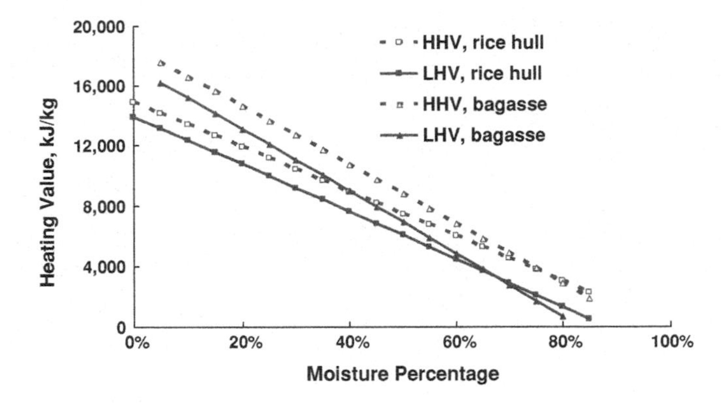

Fig. 10.1 Heating values of rice hulls and bagasse as functions of moisture

the combustion of a fuel when the water vapor from the combustion products condenses, while the latter is the heat released during the combustion when the water exits the combustion chamber as vapor. Essentially, the HHV is equal to $-\Delta h$ of the reaction with water as liquid and the LHV is equal to $-\Delta h$ when the water is vapor. It is easy to deduce that HHV > LHV. Figure 10.1 depicts the HHV and LHV of rice hulls and bagasse fuels as a function of their moisture. It is apparent from this figure that drying the biomass before its use increases significantly its energy density and its value as a fuel. Drying the biomass also reduces its volume and its weight and is highly recommended before biomass is transported over long distances. Table 10.2 shows the low heating value (LHV) of several types of biomass, most of which may be burned untreated. The pertinent numbers for anthracite are also given for comparison. While there is no standard convention on the term, oftentimes, the LHV, either in kJ/kg or Btu/lb, is referred to as the *energy density* of the fuel and this term will be used in the rest of this chapter.

10.1.1 Biomass Production, World Potential

As a renewable energy resource, the potential of biomass to contribute to the world energy challenge may be measured in terms of annual production rates. The current production of renewable biomass includes primarily agricultural byproducts, forestry byproducts, animal waste, and urban waste, including sewage. These may be called the currently *recoverable residuals* of biomass. In order to calculate the full potential of biomass one may include *energy crops*, which are crops grown for the production of biomass to be used solely for energy production. The energy crops include sorghum, energy cane, sugar cane, switchgrass, eucalyptus trees, miscanthus, giant reed, and leauceana lucacephala. Since a great deal of the Earth's land is not utilized, one may add to the total potential for biomass the

Table 10.2 Energy density (LHV) of several types of biomass

Biomass	Energy density (kJ/kg)	Energy density (Btu/lb)
Rice hull, 0% moisture	13,905	5,970
Rice hull, 30% moisture	9,175	3,940
Bagasse, 0% moisture	17,266	7,414
Bagasse, 30% moisture	11,038	4,740
Dry wood (average)	18,500	7,934
Municipal solid waste	4,000–8,000	1,715–3,430
Sewage/animal waste	1,164–1,863	500–800
Anthracite	31,952	13,720

Table 10.3 Potential annual production of biomass in several regions of the planet in Quads

Region/country	Recoverable residuals	Potential from energy crops	Total potential
North America	5.9	34.8	40.7
Europe	3.8	11.4	15.2
Africa	2.6	52.9	55.5
China	3.4	16.3	19.7
Japan	0.3	0.9	1.2
Total Earth[a]	31.2	267	298.2

[a] Estimate with all currently unutilized arable land plowed for energy crop production. It does not include the energy input for the growth of the crops.

energy crops these regions may produce. Table 10.3 shows the annual energy value of the recoverable residuals as well as the annual amount of energy that may be produced in the continents from land utilization for the production of energy. The third column in the Table, *potential from energy crops*, assumes that all arable areas, currently not used for the production of food are used for the production of energy crops. The latter, does not take into account the availability of water for agriculture.

The total global energy consumption in the beginning of the twenty-first century is approximately $480*10^{15}$ kJ/yr, of which $105*10^{15}$ kJ/yr ($\approx$100 Quads) are consumed in the USA. It appears from Table 10.3 that biomass has the potential to supply more than 50% of the energy needs of the Earth's population and this has become an argument of the biomass proponents. However, this conclusion is misleading because the numbers in Table 10.3 do not include the amount of energy, which is needed for the seeding, fertilization, growth, harvesting, and processing of the energy crops. It will be seen in the section on the production of ethanol, that this energy input is significant and comparable to the energy content of the final product. Also, these numbers do not take into account the fresh water availability, which is necessary for the production of the energy crops. For example, the production of a single gallon of ethanol from corn requires the use of approximately 1,000 gallons of freshwater. In addition, about 90% of the total potential to produce energy ($267*10^{15}$ kJ/yr) stems from the rather doubtful assumption that, all the currently available land may be harvested to produce energy crops. These lands are more valuable because they are needed to feed the growing population of the planet.

It is also apparent that several regions of the world, such as Africa, where the climate is favorable and large parts of the continent are underutilized, have very high potential to produce biomass for energy conversion. In other regions or countries, such as Japan, Belgium, and China, where most of the land is already used for agriculture and urban development, the potential to produce biomass energy is very limited. It must also be noted that, in estimating the potential for biomass, it is very difficult to assess the potential of certain regions, without considering impending future technological advances in agriculture. For example, how does one compare the biomass potential of the desert in the Middle East to the potential of the forested land in Indochina?

A significant amount of the recoverable residuals, approximately 10–20% in the OECD countries, stems from animal waste and urban waste. The latter includes sewage as well as solid waste. While, in terms of mass, the amount of waste is significant, its energy density is very low. Human and animal waste, if properly and efficiently processed, may provide a very small percentage of the urban energy needs. The real value of urban waste processing locally lies in the reduction of the volume of the waste, so that less volume/weight needs to be transported at the expense of more energy and cost. In addition, waste processing enables the containment and utilization of the methane, which is produced during waste decomposition. If untreated, urban waste products will produce more of this GHG, with all its adverse climatic consequences.

10.1.2 Methods of Biomass Utilization

The ways biomass is used for the production of electricity and biofuels have been adapted from the methods used for the combustion of fossil fuels as well as from methods for the chemical or physical transformation of fossil fuels to other forms of transportation fuels, e.g. coal gasification or coal liquefaction. The following are some of the ways of biomass utilization:

1. *Direct Combustion*: is used for cooking; space heating, e.g. in fire places or boilers; process heat in industrial plants; and steam generation for the production of electricity. Because the biomass contains a high percentage of humidity, temperatures achieved in biomass-only burners are significantly lower than conventional fossil fuel burners. When one considers the energy spent in the transportation of biomass, the efficiency of power plants that only use biomass is in the range 15–20%, roughly half of the conventional fossil fuel plants.
2. *Co-firing*: is the combustion of a small percentage (2–10%) of locally produced biomass and coal. Because coal is the principal fuel in a co-firing burner, steam

may be produced at high temperatures in the boiler and the efficiency of the power plant does not suffer.

3. *Gasification in air*: the product of gasification is typically a mixture of hydrocarbons, which is used for process heat, e.g. crop drying; running stationary engines in the immediate vicinity of the gasifier; space and water heating; and small electric power producing plants.
4. *Pyrolysis*: it produces combustible gaseous and liquid products, such as methane, ethane and methanol, which may be used in gas turbines, space and water heating as well as in transportation fuels.
5. *Hydrolysis and fermentation*: produce liquid fuels, such as methanol and ethanol. They are commonly used in mobile and stationary engines, gas turbines, generation of steam, and as gasoline or diesel additives.
6. *Anaerobic decomposition*: occurs naturally in municipal and agricultural waste sites, in sewage treatment plants and in animal waste plants. The decomposition produces a gas that is rich in methane, carbon monoxide and carbon dioxide. This artificial combustible gas has a low heating value, but may, nevertheless, be mixed with other fuels to be used in gas turbines or steam power plants. Vacuum pumps extract the mixture of gases from the heaps of waste in the disposal sites. The gas is transported to power plants via pipelines for combustion.

The co-firing of a coal power plant with agricultural or municipal solid waste is sometimes used as a way to reduce the amount of fossil fuel input to the power plant, thus reducing the cost of the fuel as well as the amount of CO_2 produced from coal. Some NO_x reduction is also achieved by co-firing. The co-firing of a small percentage, typically under 5% of the heat input, in a coal-fired power plant may be accomplished without any modification to the fuel supply system. The waste product, e.g. rice hull, bagasse, or tree clippings, may be directly fed to the coal pulverizer, where they are mixed with the coal and subsequently injected to the boiler. Such mixtures of coal and biomass produce the high temperatures required by the Rankine steam cycle. If the size of the biomass parts is more than 1 cm (0.4 inches) a hammer mill system is recommended to be installed. This system reduces the size of the biomass parts to less than 1 cm and prevents the "clogging" of the feed lines. For the co-firing of municipal solid waste, modifications have to be done to the feeding system of the power plants because of the significantly low heating value of the municipal waste. For the municipal waste, a second feed system is usually installed, which feeds the waste products directly into the boiler.

Table 10.4 shows the savings in coal and prevention of CO_2 pollution achieved by the co-firing of a 1,450 MW power plant with bagasse, which contains 46% moisture and has a heating value of 6,886 kJ/kg. The average rate of heat input in this power plant is 4,527 MW and its average overall efficiency is 32.03%.[2] The

[2] The numbers pertain to the Independence Steam Electric Station (ISES) coal-fired power plant in Arkansas, USA, which is owned and operated by Entergy Inc.

Table 10.4 Fuel savings and CO_2 production prevention by the use of biomass fuel in co-firing

Heat from biomass (%)	Heat from biomass (MW)	Coal input (kg/day)	Biomass input (kg/day)	Coal savings (kg/day)	CO_2 emission savings (kg/day)
0.00	0	16,439,805	0	0	0
0.50	23	16,344,361	284,035	95,445	179,692
1.00	45	16,248,916	568,070	190,889	359,383
1.50	68	16,153,471	852,105	286,334	539,075
2.00	91	16,058,027	1,136,139	381,779	718,766
3.00	136	15,867,137	1,704,209	572,668	1,078,150
4.00	181	15,676,248	2,272,279	763,557	1,437,533
5.00	226	15,485,359	2,840,349	954,447	1,796,916

Table proves that the use of biomass, which would have been otherwise wasted, not only saves fuel, but also reduces the net CO_2 emissions of the power plant.

10.1.3 Aquatic Biomass

Sea and surface waters cover almost 75% of the Earth's surface. Plants that grow in these aquatic areas include seaweed, kelp, and several types of algae. Because the amount of potable freshwater on the planet is only 0.3% of the total and because fresh water is becoming a scarce resource in several regions, it is not considered good environmental stewardship to devote large quantities of freshwater for energy plants or algae. However, the salt-water and brackish/marsh-water areas, which comprise 97.3% of the total water supply of the planet[3] is readily available for the production of biomass and biofuels. Harvesting marine organisms for the production of energy is a very sensible way to produce liquid fuel for transportation or solid fuel for the production of electricity because:

a) Aquatic biomass does not use any useful parts of land, which is a scarce resource.
b) No irrigation is necessary for the production of aquatic biomass.
c) Nutrient trace elements, such as K and P are abundant in the sea and, hence, no fertilizers are necessary.
d) Oftentimes, the food of aquatic bacteria and algae is sewage. This helps in the reduction of liquid waste and will lead to sewage reduction and possible elimination by feeding it to energy-rich algae.

Biologists have produced a number of synthetic bacteria and have discovered algae with high content of fatty oils, which thrive in aquatic environments. When these organisms are dried up, the fatty oils are processed to produce a fuel similar

[3] The remaining 2.7% of water is in the form of glaciers.

to diesel, which is often called *biodiesel*. Similarly, other sea plants may be dried to produce liquid or solid fuel.

One of the advantages of algae is that they may grow in very short times. For example, micro-algae can regenerate in 48–72 h and cyanobacteria can regenerate in 5–20 h. The short growth and harvesting times lead to better utilization of the solar energy in comparison to land crops and result in higher overall conversion efficiencies. For example, the production of biodiesel from algae in gallons per hectare per year is 10–20 times higher than that of palm oil or jastropha and 50 times higher than that of oilseed.

A significant disadvantage of the use of aquatic biomass for the production of energy is its water content. Drying the aquatic biomass is necessary before it is processed to produce biofuels or be burned in a boiler for the production of electricity. The high energy requirement of the drying process often exceeds the energy that may be obtained from the aquatic biomass itself. When solar energy is used for drying, the process is energy neutral but it is significantly slower and rather unreliable, especially during a wet, rainy season. Because of this, the contribution of aquatic biomass to the energy demand in the first decade of the twenty-first century is almost zero. Technological breakthroughs in the drying of algae, the extraction of oils and lipids from the algae cells and better engineered biological organisms are needed for aquatic biomass to contribute significantly to the energy challenge of the world.

10.2 Biofuels

The term *biofuels* refers to liquid fuels derived from biomass, which are used for transportation. It encompasses all types of liquid fuels derived from plants or animals, including methanol, ethanol and diesel fuel. Methanol, CH_3OH or wood spirit, has been produced for centuries from the distillation of wood products, natural gas or from coal. It is a liquid fuel, with boiling point 65°C. Its energy density of 21,000 kJ/kg is approximately 50% lower than that of gasoline. Because of this, when used in cars, the tailpipe NOx emissions are significantly lower. Small amounts of methanol, up to 10%, mixed with regular gasoline, may be used in the current fleet of vehicles without any significant modifications.

Ethanol, C_2H_5OH or alcohol has a boiling point of 78°C, is a colorless fluid and burns clean to produce water and CO_2. There is significant experience in the production of ethanol primarily as the main product of wheat or grape juice fermentation for the production of alcoholic beverages and distilled spirits. Ethanol as a fuel is produced primarily from corn or sugar cane. Figure 10.2 shows diagrammatically the physicochemical processes in the production of ethanol from corn. Yeast is a catalyst in the production of ethanol and accelerates the fermentation process. It must also be noted that the fermentation process releases some CO_2, which needs to be accounted for, when evaluating the environmental advantages of using ethanol. During fermentation, 1 bushel or 26 kg of corn yields

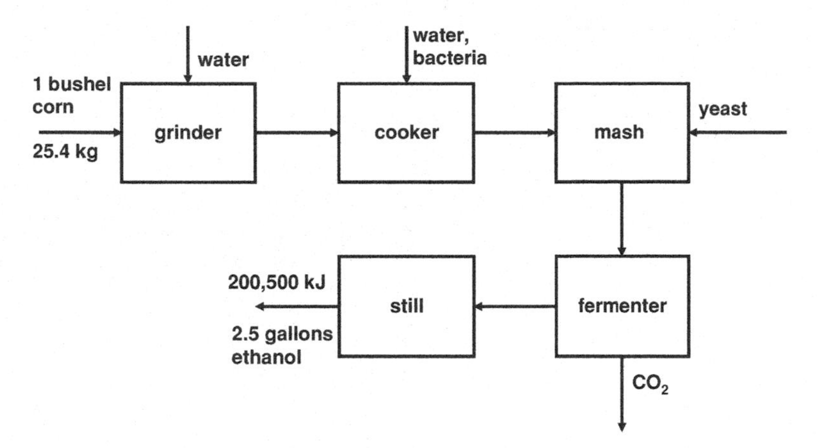

Fig. 10.2 The production stages of ethanol

approximately 2.5 gallons of ethanol, which has an energy value of 200,500 kJ (190,000 Btu). However, as it will be seen in the next section, growth of corn and the process of ethanol production also consumes a great deal of energy resources.

A mixture of 10% ethanol and 90% gasoline by volume, or E10, is called *gasohol* and is used as a transportation fuel in many states in the USA as well as in other parts of the world. This mixture may be used in all spark-ignition engines without any engine modification. E20 has been introduced in several states and countries, also without any significant engine modification. The use of higher percentage of ethanol in the fuel mixture requires modifications in the fuel supply system of the engine. A much richer in ethanol mixture, E85, is composed of 85% denatured[4] ethanol. It has been used in Brazil and was also introduced in the European Union, where it is primarily used in Sweden. The so called *flex-fueled engines* for cars and trucks that use this mixture have their engines modified to adjust electronically the air-to-fuel ratio and the ignition timing in order to use the E85 fuel. In general, the addition of ethanol in gasoline reduces the specific energy of the fuel, because the energy by volume of ethanol, e.g. in kJ/gal or kJ/l, is approximately 67% of the specific energy of common gasoline. Thus, a car running on E15 would achieve approximately 90% miles per gallon or 90% km per liter than if it were running on pure gasoline. On the positive side, this car would produce much less atmospheric pollution because ethanol burns cleaner, producing significantly lower amounts of NO_x and almost zero SO_x than gasoline.

Finally, biodiesel is a generic name given to liquid transportation fuels with a mixture of heavier hydrocarbons and alcohols, such as $C_{10}H_{22}$, $C_5H_{11}OH$, and $C_8H_{17}OH$. Biodiesel is produced from oily seeds grown in nature, such as cotton seeds, olive oil residue and pumpkin seeds. It may also be produced artificially,

[4] Denatured alcohol is ethanol that has additives to make it undrinkable. It may be toxic if consumed by humans or animals.

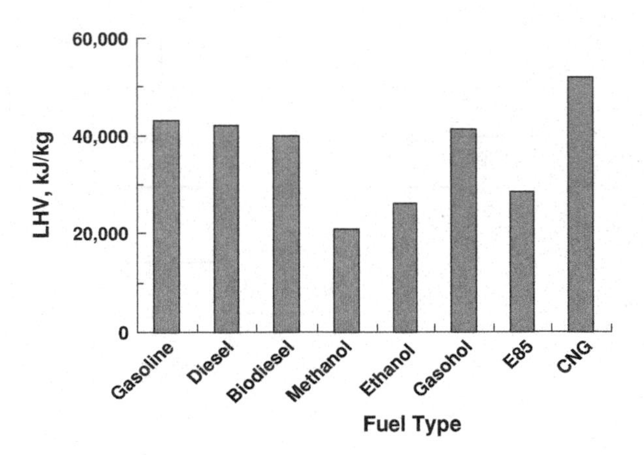

Fig. 10.3 LHV, in kJ/kg, of common fuels and biofuels

using waste products from industrial processes or from the food industry. For example, spent cooking oil and used engine oil may be refined and used as biodiesel for cars and trucks. Because biodiesel has almost the same composition and very similar transport and combustion properties as common diesel fuel, standard diesel engines do not need to be modified to run on biodiesel. Figure 10.3 shows the LHV of several common biofuels as well as common liquid fossil fuels used in the transportation sector. It is observed in this figure that the fuels, which are most commonly used in today's internal combustion engines, have LHV's close to 40,000 kJ/kg.

Another class of biofuels may be derived directly from agricultural products, especially oily seeds, such as cotton, peanut, flax, soybean, sunflower, safflower, sesame, palm, jastropha, and Chinese tallow. These seeds grow in both warm and cold areas of the planet, which means that almost all countries may produce one or more of these crops for the production of transportation fuels. However, because most of these crops have traditionally provided food sources for the population, their widespread use for the production of biofuel will become problematic if it is accompanied with a rise in food prices or food shortages.

10.2.1 Ethanol Production from Corn

The production, processing and use of biomass entail several energy-intensive processes. The energy consumption during these processes must come into the calculations of the net energy yield and the overall contribution of biomass in solving the energy challenge. For example, if corn is to be used for the production of ethanol, some of the processes needed are: plowing, seeding, fertilizing and harvesting of the fields; drying and transportation of corn; fermentation, removal of solids and purification; and transportation of the produced fuel. A complete energetic analysis of these processes is presented in the following paragraphs. The

Table 10.5 Preprocessing energy consumption for the growth of corn

Study	N_2	P	K	Lime	Herbicide	Insecticide	Total energy
Pimentel (2003)	9.388	0.924	0.785	0.930	0.886	0.063	12.976
Patzek (2004)	8.099	0.431	0.636	0.583	0.663	0.290	10.702
Shapouri et al. (2002)	6.120	0.257	0.740	0.469	1.235	0.059	8.880
Wang et al. (1997)	7.506	0.638	0.350	0.469	0.729	0.053	9.745
Average	7.778	0.563	0.628	0.613	0.878	0.116	10.576
% std. deviation	17	51	31	36	29	100	17

Energy inputs associated with the0 agricultural fertilizers and pesticides needed in GJ/ha
Piementel, *Natural Resources Res. Vol. 12, No. 2, 2003.*
Patzek, T., *Critical Reviews in Plant Sci. Vol. 2, No. 6*, 2004.
Shapouri, Duffield and Wang, *US Dept of Agriculture, Agricultural Economic Report, 2002.*
Wang, Saricks and Wu, *Argonne National Lab. Report, 1997.*

analysis is based on four pertinent studies conducted in four U.S.A. laboratories and gives the results on the energy inputs and outputs related to the production of ethanol from corn. The average and the percent standard deviation are also calculated and shown in the tables that follow.

Table 10.5 shows the energy, in GJ per hectare, which enters the production of the fertilizers (N_2, P, K, $CaCO_3$), pesticides and insecticides that are necessary for the production of corn. 1 hectare (ha) is approximately equal to 2.5 acres. It is observed that these energy inputs are very significant and average approximately 10,500 MJ/ha. The disagreement on the total energy between the studies is rather small as signified by the low value of the percent standard deviation, in all values except the one for the insecticides. This is not significant, however, because insecticides account for only 0.5–2% of the total energy at this stage of ethanol production. It is also observed that most of the energy at this stage is in the input of nitrogen, which represents a high percentage of the mass of the fertilizers.

Table 10.6 shows the energy inputs on the field. The field inputs include the liquid fuels that are consumed by the machinery used in the plowing, irrigation and harvesting processes. The term "other energy" includes natural gas and electricity, which is oftentimes used for field irrigation. One observed discrepancy between the four studies is that the first two take into account the energy that goes into the repair and manufacture of the equipment used in the production of corn, while the last two studies do not take this item into account. Proper accounting of all energy inputs should also include the energetic cost of producing and repairing of the heavy machinery used for the production and transportation of corn such as trucks, tractors, ploughs, cranes, railroad cars, locomotives, barges, etc. It must be noted that the "total pre-processing" column includes the sums of the total energy used in the field from Table 10.5.

Once the corn is produced, it must be transported to the ethanol plant for processing. Based on the calculated production of corn, Table 10.7 includes the yield of ethanol in liters per hectare (l/ha); the transportation energy input for the corn and the processing energy input for the production of ethanol in MJ/l of ethanol produced. At this stage, the most consuming energy processes are the

Table 10.6 Corn seeding, plowing, irrigation and harvesting energy needs including machinery utilization and repair GJ/ha

Study	Diesel	Gasoline	Seeds	Machinery	Other energy	Total preprocessing
Pimentel (2003)	3.8	2.3	2.6	8.9	3.2	33.8
Patzek (2004)	3.1	0.9	2.6	8.5	3.4	29.2
Shapouri et al. (2002)	3.7	1.2	2.6	0.0	5.9	22.3
Wang et al. (1997)	2.9	1.0	2.6	0.0	3.6	19.8
Average	3.4	1.4	3	4.4	4.0	26.3
Percent standard deviation	13	48	0	116	32	24

Table 10.7 Ethanol yield and energy inputs during corn transportation and ethanol processing

Study	Yield (l/ha)	Transport (MJ/l)	Fuel (MJ/l)	Total inputs (MJ/l)	Total processing (GJ/ha)
Pimentel (2003)	3,201	1.3	14.74	16.07	51.4
Patzek (2004)	3,175	1.7	14.45	16.19	51.4
Shapouri et al. (2002)	3,137	0.44	14.45	14.89	46.7
Wang et al. (1997)	3,189	0.43	14	14.39	45.9
Average	3,176	1	14	15	49
Percent standard deviation	1	66	2	6	6

milling of the corn and the distillation of the fermentation product. The energy content of byproducts of ethanol production is included in the credits column of this Table and given in MJ per litter of ethanol produced. The total processing energy consumption in GJ per hectare is given in the final column of Table 10.7. The very low value of the fractional standard deviation for the total processing energy signifies that, even though some of the partial numbers of the four studies do not agree well with each other, all four studies agree very well in the total energy used for the processing of corn.

The energetic study for the production of ethanol from corn is completed by adding all the energy inputs associated with the production of the final product and by comparing them with the energy output of the final product. The LHV and the HHV of ethanol, when expressed in terms of volume are: 21.1 and 23.3 MJ/l respectively. From these numbers, the LHV and HHV per hectare are calculated to be 67,000 and 74,000 MJ/ha, respectively. Table 10.8 shows the total energy inputs associated with the production of corn and the transportation and processing of ethanol. The sum of the energy inputs for the production of ethanol is given in the fourth column of the Table. The last two columns are the figures of merit for the production of ethanol from corn, which are defined as follows:

Table 10.8 Total energy inputs and output input ratio based on the LHV of 67,000 MJ/ha and HHV of 74,000 MJ/ha

Study	Total preprocessing	Total processing	Total energy consumption	LHV ratio	HHV ratio
Pimentel (2003)	33.8	51.4	85.2	0.79	0.87
Patzek (2004)	29.2	51.4	80.6	0.83	0.92
Shapouri et al. (2002)	22.3	46.7	69	0.97	1.07
Wang et al. (1997)	19.8	45.9	65.7	1.02	1.13
Average	26.28	48.85	75.13	0.90	1.00
Percent standard deviation	24	6	12	12	12

$$LHV\ ratio = \frac{67,000}{Total\ Energy\ Input\ in\ MJ/ha}\ and$$
$$HHV\ ratio = \frac{74,000}{Total\ Energy\ Input\ in\ MJ/ha}. \quad (10.3)$$

Obviously, if a figure of merit is less than 1, the corresponding study indicates that the ethanol from corn transformation consumes more energy than it produces. When the ratio is higher than 1, the ethanol produces more energy than what is required in its production. Of the two columns, the LHV ratio is of more relevance to the production of ethanol, because ethanol is primarily used as transportation fuel and the exhausts of vehicles do not condense the produced water vapor to recover the HHV. A remarkable observation on Table 10.8 is that three of the four studies indicate that the LHV ratio is less than one and the result of the fourth (1.02) is practically equal to one. Even if the HHV ratio were used, the resulting figures of merit are very close to 1. This demonstrates that, in the production of ethanol from corn and similar sources, which are based on starch, by the conventional method of fermentation, we usually consume more or an equal amount of energy than we actually obtain from the final product. Based on these studies, the production of ethanol from corn for its use as a transportation fuel is highly questionable as a viable energy saving or energy production method as well as a method to reduced the GHG emissions. Because this method of producing transportation fuels consumes more energy than it delivers, it should not be considered as a viable alternative to fuel production.

While the production of ethanol from starch is clearly an unacceptable option, another method of the production of ethanol, from cellulosic materials,[5] has a more favorable energetic analysis and may be a better alternative in the production of biofuels as explained in the study by Farrell et al. [1]. The favorable energetic analysis relies on the complete utilization of all the byproducts for the production of ethanol, which contain significant amounts of energy. However, the utilization

[5] Cellulosic materials consist of the fibers that are commonly called "wood." Their scientific name is lignocellulose, which is principally composed of cellulose, hemicellulose and lignin.

of these byproducts is not accomplished at present by ethanol producing installations because of the additional equipment needed, transportation costs, and the significant capital expense that is required.

10.3 Environmental Effects

In the early twenty-first century, the increased use of biomass has been advocated as a viable method to reduce pollution and liquid hydrocarbon independence in several OECD countries. Governmental programs and regulations have been adopted for the widespread use of the E10 fuel and economic subsidies have been offered for the production of biofuels. However, the widespread use of crops for energy becomes highly questionable when one considers the environmental and social effects of biomass use. The following sections examine critically these effects.

10.3.1 Land Use

The most important consequence of biomass production for energy is the use of agricultural land, which is a scarce resource in the planet. Agricultural crops and forests use a very large area and the development of energy crops would be very costly in the use of land. On the other hand, agricultural waste, such as bagasse, rice husk, or forestry waste is a very welcome source of energy because it is normally considered unwelcome waste and, in most cases, their producers will even pay for their removal. The widespread planting of switchgrass and eucalyptus trees for combustion or corn for the production of ethanol fuel entails the utilization of agricultural land that could have better been used for the production of additional food, which is still sorely needed in several countries. When in the years 2005–2008 a great deal of the US corn production was diverted to ethanol production, the price of feeding the livestock increased significantly. As a result, the price of milk doubled in the U.S.A. and several food items, including meat and cheese, became significantly more expensive.[6] The same cause in Mexico resulted in the tripling of the price of corn tortillas, which led to social unrest and obliged the Mexican government to impose price controls. Several countries, including most of East Asia and Africa, simply cannot afford to divert scarce agricultural land for energy crop production, because almost every parcel of land is currently needed to feed their populations.

---

[6] Other factors, such as higher energy and fertilizer costs as well as commodity speculation contributed to this increase. However, in all studies conducted on this topic, the use of corn for the production of ethanol was the primary variable that caused the rapid food price increase.

A wiser and much more effective use of land for energy would be the installation of photovoltaic cells to produce electricity. This energy may be stored in the form of hydrogen (see Sect. 12.5) and then exported in exchange for food. This exchange may be called the *food for energy trade*, and is based on the fact that photovoltaic energy conversion has much higher conversion efficiency (close to 20%) than natural plants (close to 0.5%). For example, Brazil has vast areas of arable land and, through its vast network of rivers, a great deal of freshwater for irrigation. This country has the capability to produce significantly more food than it does now. On the other hand, the countries in the Saharan West Africa (e.g. Chad, Mali, Niger and Mauritania) have very small areas of arable land, but are located in the best region for the production of solar power as may be seen in Fig. 7.4. The electric energy from the photovoltaic cells may be stored as hydrogen fuel and subsequently transported.[7] If food were produced in excess in Brazil and hydrogen in the Sahara, one may envision a trade route across the mid-Atlantic ocean carrying food from Brazil and bringing back the excess hydrogen fuel from the Saharan nations. Similarly, hydrogen fuel produced in the highlands of Ethiopia by solar energy may be traded for food grown in Mozambique or Tanzania. Even, when one calculates the externalities of this type of trade, such as the production, storage and transportation costs, the *food for energy trade* would result in a much better use of arable and non-arable land areas than, simply, the growth of energy crops. In addition, a stable food for energy trade would alleviate outbreaks of famine in the Saharan countries and other areas of the Earth.

10.3.2 Fresh Water Requirements

A significant environmental effect of the growth of biomass for energy is the high requirements of fresh water for irrigation and processing. For example, the production of a single gallon of ethanol from corn requires approximately 1,000 gallons of water. If all the available land were to be used for the growth of energy crops, this would require very high quantities of fresh water, which are not available in all regions. Unless the energy crops are planted close to large river systems with unused or underutilized agricultural land, such as the Amazon, the Mississippi, or the Siberian Rivers, the competition for the existing water resources would result in a significant strain to local populations. Planting energy crops in the Southwestern part of the USA is not feasible because of the scarcity of the water. Similarly, planting additional crops to be used for energy production along the Rhine in Germany, the Nile in Africa or the Mekong in Southeast Asia is impractical, because all of the agricultural land along these river systems is

[7] Of course this large scale production of electricity and hydrogen would entail significant capital cost and investment in the Saharan countries. This may be viewed as an investment that will in the future contribute to the economic development of this region and the employment of their populations.

currently utilized in the, more socially acceptable way, of food production. Simply, the irrigation requirements of energy crops lead to the conclusion that very few regions on the planet are suitable for the more widespread production of such energy crops.

The situation is entirely different if sea water were to be used for the production of energy crops: there are vast areas of land close to the ocean, which are not currently used for the production of food. A technological breakthrough that would allow irrigation of energy crops with sea water will change this situation and may make feasible and more socially acceptable the widespread production of energy crops in several coastal regions of the world.

10.3.3 Use of Fertilizers and Pesticides

The expected globally widespread use of fertilizers and pesticides is another significant environmental effect of more biomass production. The known energy crops are very fast-growing plants and need the input of large amounts of fertilizers, pesticides and insecticide chemicals. These chemicals contain phosphorous, sulfur, nitrates, arsenic and trace metals such as zinc, lead, and manganese. Many of these elements are toxic to humans or harmful to the environment. The widespread combustion of biomass will become another pathway that introduces these toxic elements to the atmosphere and to the living organisms.

In addition, all the chemical compounds in the fertilizers, pesticides and insecticides invariably find their way to the hydrosphere—rivers, lakes, estuaries, and oceans—following the runoff waters from rainfall. The addition of these chemicals in the hydrosphere alters the chemical composition of the freshwater and seawater. This is especially significant for the fragile ecosystems in lakes, estuaries, and bays with low water flow or circulation. The periodically observed *hypoxia*[8] in the northern Gulf of Mexico is caused by the high nutrient content in the waters of the Mississippi River, which carries fertilizers and other nutrients from 43% of the land mass in the USA. The water effluent alters locally the chemistry of the northern Gulf of Mexico, off the Louisiana coast and disturbs the ecological balance of the species. Similarly, lake *eutrophication*[9] in Europe and North America causes the depletion of oxygen by algae, which endangers other more complex species, such as fish and shellfish.

---

[8] Very low oxygen concentration that kills the larger fish.

[9] This is the uncontrolled growth of the algae population because of the high concentration of nutrients. The large population of algae consumes the dissolved oxygen at a fast rate and causes the death of fish and other large aquatic animals.

10.3.4 Unintended Production of Methane

A fourth environmental effect of biomass is the potential accelerated production of methane. If left untreated for periods of a few weeks, biomass decomposes naturally and produces carbon dioxide and methane. The latter is sixteen times more potent GHG than carbon dioxide. The anaerobic decomposition of biomass, e.g. when it is immersed in water or buried underground, always produces methane gas, which diffuses into the atmosphere and contributes significantly to global warming. If biomass is not used promptly and is left to decompose, either in the field or in storage facilities, it would cause significantly greater damage than the environmental benefit of removing some of the carbon dioxide from the atmosphere.

The Brazilian experience during the 2002–2008 period of uncontrolled and high demand for biomass products is a warning to what may happen to the land if only economic incentives for the growth of biomass are offered: Because of the high demand for biofuels that are produced from sugar cane and the high price of sugar cane, Brazilian farmers converted large parcels of land along the Amazon to biofuel producing crops. In addition, they cleared very large tracts of tropical forest near the Amazon River to establish farms for the production of sugar cane, corn and other high energy yielding crops. The result was that large trees were uprooted and were burned or, even worse, were left decomposing to produce methane. Bearing in mind that 30–40% of the mass of the tree is in the underground roots, this organized deforestation and conversion to arable land brought into the atmosphere a great deal more GHG's than what the subsequent biomass production removed.

10.3.5 Other Effects

Other, less significant environmental effects of expanded biomass production and production of energy crops include:

- Soil erosion and depletion of soil nutrients.
- Loss of biodiversity.
- Partial or total deforestation, which may lead to desertification.
- Growth of monocultures, which are highly vulnerable to agricultural diseases and bacteria.
- Higher river silt concentration and enhanced siltation rates that may lead to river or estuary eutrophication.
- Changes in land use and irrigation patterns, which may change the micro-climate of the region.
- Dust production from plowing and harvesting.

It must be noted that almost all the adverse environmental effects of extensive biomass production are associated with the energy crops and not with the treatment of agricultural, human and animal waste. The expanded production of energy from any type of waste and the reduction of the volume of waste on a global scale has beneficial environmental and ecological effects, in addition to producing needed energy.

10.4 Social, Economic and Other Issues for Biomass Utilization

The widespread use of energy crops is highly controversial. Unlike the other renewable energy sources, which may only be utilized for energy production, the use of biomass is intricately connected to food production, world poverty and malnutrition, high rates of freshwater use, high rates of arable land use, and agricultural land deterioration. Increased reliance of the world population on biomass-derived energy would certainly have adverse effects on the food and fresh water supply, with all the social consequences this would imply. For this reason, there are several key issues, which must be resolved before national or regional policies are formulated that encourage and further promote the higher utilization of energy crops. Some of these issues were identified earlier in this chapter, but are repeated here for clarity.

A. *Economic subsidies*: Food production and several agricultural products enjoy substantial subsidies from most national and regional governments. Farming subsidies are among the few items that are exempted from the international treaties brokered by the World Trade Organization (WTO). Frequently, the subsidies are in the form of tax credits. For example, in the U.S.A. there is a federal tax credit of $0.51 per gallon of ethanol that is used as a transportation fuel. Since biomass is essentially a farming product it enjoys very generous subsidies in most countries including: tax credits; direct grants; low-interest loans for its production and processing; exemption from taxation of the fuels used for its production; high depreciation rates and even subsidies on the equipment used for its production and processing.
   In the 1990s the real cost for the production of one gallon of ethanol in the U.S.A. and Europe was close to $11, yet because of the several subsidies, ethanol was mixed with gasoline to be sold at significantly lower prices. While this practice is a prime example of economic inefficiency, several of the agricultural subsidies may be justified socially when directed to small farmers, who in most countries live close to poverty levels and produce other staple foods for the benefit of the society at large. However, since the 1990s large enterprises and several multinational conglomerates have entered the biomass production industry and have derived very high profits from subsidized biomass production. In the latter case, the social, economic, and national interest in continuing and extending these subsidies become highly questionable.

Clearly, any current or future subsidies for energy crops must be justified on social, economic, and environmental grounds. While economic subsidies for the production of corn and wheat for food are justified, subsidies for the large scale production of these staples for their conversion to biofuels are highly questionable. The regulatory institutions that govern the production of biomass must ensure that any economic subsidies benefit the small farmers and not the large corporations, maintain affordable prices for the foodstuffs and do not result in environmental deterioration.

B. *Food scarcity*: The high expansion of ethanol production in the period 2001–2008, especially ethanol derived from corn, has been accompanied by a significant rise in commodity prices worldwide between 2003 and 2009. Even though economic studies on the effect of corn used for ethanol production on the increased food prices vary significantly, these studies attribute 15–60% of the increased food prices in the production of corn ethanol. This is an indisputable fact that the diversion of agricultural products for the production of energy drives the prices of food products higher. Simple demand and supply considerations dictate that the removal of corn from the marketplace for the production of ethanol will always increase its price as food. Concerns over food availability, especially in the famine-prone regions of the world, have the potential to stifle any use of food products for energy or to at least cause the removal of all subsidies.

The popular press has reported that one billion people on Earth, or one out of seven Earth inhabitants, suffer from malnutrition and that in every six seconds, one child dies of malnutrition or outright starvation.[10] The rising food prices have affected citizens living at or near the poverty level even in the wealthier OECD nations. Under this light, the increased use of subsidized corn or other agricultural products to feed the engines of *Hummers*, *BMW's*, *Cadillacs*, and *SUV's* is neither socially responsible nor economically justifiable. The United Nations has issued a warning against the use of food crops for energy and several national governments are expected to follow by removing subsidies or placing restrictions on which agricultural products may be used for energy production. Because the viability of the biomass industry depends to a large extent on the levels of national and regional subsidies and grants, the regulatory environment and not the free markets will finally determine the expansion of several types of biomass for the use of energy. An international policy along the line "*do not burn what you can eat*" may justifiably become the cornerstone of the future of biomass use.

C. *Poverty levels worldwide*: Some of the concerns related to world-poverty over the price of food are partially offset by a worldwide increase in the agricultural incomes in the first decade of the twenty-first century. Because a large percentage of the population living in poverty levels derives its income from

[10] Associated Press, October 15, 2009.

agricultural employment, the increase in their incomes has been a welcome effect of biomass production and a rationale for several national and regional governments to continue with the agricultural subsidies. However, as explained in section A, in considering economic subsidies for the production of biomass, a government has to ensure that larger, for-profit corporations do not take unfair advantage of such subsidies, which are intended for the small farmers.

D. *Stability of energy prices*: The interest in biomass and biofuels is not new. The *energy crisis* of the 1970s and the oil embargo in Europe and North America caused a dramatic rise in the production of biomass, especially biofuels, that continued into the early 1980s. The interest in the conversion of biomass waned in the mid 1980s when energy prices dropped. Several biomass conversion plans were abandoned at that time and a great deal of investment capital was lost. It has become apparent that high energy prices are needed to stimulate higher biomass utilization. The OPEC production trends and the resumption of the higher oil demand in Asia in 2010 are indications that the energy prices will not tumble in the second decade of the twenty-first century as they did in the 1980s and 1990s. These indications point to rising biomass utilization in the future, especially biomass derived from waste products.

E. *Greenhouse gas policy*: GHG emission concerns are the primary reason for the increased use of all renewable energy sources, including biomass. Well-intentioned but injudicious regional and national regulations may require a higher use of biomass or biofuels. For example, the regulatory change from the E10 fuel (gasohol), which is currently used in most of the states in U.S.A., to E20 will almost double the demand for ethanol. Actually, a provision was included in the Energy Policy Act of 2005, which mandates that 7.5 billion gallons of renewable fuels be used in gasoline by 2012. Similar regulations in the European Union and other nations will further increase the production of ethanol and other biofuels. Given the energy balance for the production of ethanol from starch, such regulations will not always achieve their goal for the reduction of GHG emissions, unless it is also stipulated in the regulatory or legislative framework that ethanol may not be derived from the starch of foodstuff but from cellulosic plant materials, with LHV ratios significantly higher than 1.

F. *Technological advances*: These pertain to the growth as well as the processing of the biomass. Examples include: the engineering of algae and bacteria with higher fat content; the faster and more efficient drying of the aquatic biomass; higher crop yields of the land; energy crops that may grow with salt water; ethanol from cellulosic processes; catalytic pyrolysis at lower temperatures; and bacteria that process solid and liquid waste at a faster rate.

G. *Global and regional climate change*: It is rather ironic that climate change, the very cause for the increased use of biomass, will have a significant impact on the production of biomass and biofuels. Potential regional climate changes are a key issue in the crop yield of the land and the production of biomass. Warmer weather would increase the crop yield and the amount of biomass if it is accompanied by sufficient rainfall. On the other hand, drier weather may spell the end of the use of biomass for energy because sufficient food supply will always have a priority over energy supply.

10.5 The Future of Biomass for Energy Production

It is apparent that the use of food crops for energy production, whether this is electricity or biofuels, is detrimental to the environment, it places a strain on water supply, and is economically inefficient and socially undesirable. Providing high subsidies for the production of energy crops in the industrialized countries is economically inefficient. In addition, complete and reliable energy balances, such as the ones in Sect. 10.2.1 for the production of ethanol from corn, indicate that the increased use of energy crops for energy production may actually increase, not reduce GHG emissions. The use of vast tracts of land for the production of energy crops is also a very inefficient as well as a detrimental use of the land area, especially in the industrialized and economically advanced countries, where capital for high-technology investments for energy production is available. For example, given the availability of capital for the investment, the same land may be covered with photovoltaic solar cells and be used at 20–30 times higher efficiency for the production of electricity.

Biomass derived from human, animal or agricultural waste and aquatic biomass are totally different. Wastes do not have an economic value, they must be disposed and, in general, they harm the environment. If left unattended they decompose and produce methane, a potent GHG. The reduction of the mass or the volume of the waste by extracting biofuels or directly burning them is environmentally beneficial and economically efficient. Similarly, the use of municipal sewage for the growth of algae that may be used for the production of biofuels and the use of agricultural and forestry wastes in a co-firing process makes good sense and, in most cases is economically justified. For this reason, it is believed that in the future, the production of biomass from energy crops will be curtailed, while the biomass from waste products will increase significantly. Technological advances that will lead to increased production of aquatic biomass and especially of marine biomass will materialize when faster and less energy consuming drying processes are invented, or when organisms, which are more suitable for energy or fuel production, are engineered.

Problems

1. In your opinion, what are the primary reasons for the large disparity of the solid waste production per capita in the U.S.A. and in Europe?
2. Use of biomass from fast growing trees, such as sycamore, is often advocated as an alternative to burning coal. Grown sycamore threes weigh 430 kg on

average and the heating value of their wood is 15,300 kJ/kg. Find out how much heat power is consumed annually by a 400-MW coal power plant with an overall efficiency 40%. If the entire amount of this heat input were to be supplied from harvested sycamore trees, how many grown sycamore trees would the power plant need to consume annually? You may assume that the energy for the planting, harvesting and transportation of the sycamore trees to the power plants is negligible.

3. The country of Estonia consumed 8.42 TWh of electricity in 2009 at an average overall thermal efficiency 32%. It has been suggested that Estonia produces all its electricity from locally grown pine trees. The average pine tree requires 11 m^2 to grow, becomes 400 kg in 8 years and has a heating value 13,700 kJ/kg. How much area is required for this solution to Estonia's energy problem? Is this alternative feasible? Assume that the energy for the planting, harvesting and transportation of the pine trees to the power plants is negligible.
4. How much electric energy (in Terajoules, TJ) was produced in your country? If this energy was produced by power plants with an average overall efficiency 32%, how much heat energy (in TJ) was consumed? Assume that the entire area of your country were planted with native trees that grow to 500 kg in 7 years, have heating value 14,000 kJ/kg and each requires an area 10 m^2 to grow. How much area for tree planting would your country need? Is this a feasible solution for the production of electricity in your country?
5. Rice-hull residue with 30% humidity is used in the boiler of a small thermal power plant. The plant produces 60-MW of electricity and has an average efficiency 34%. How many tons of rice-hull the power plant consumes every year? If the average distance for the transportation of the fuel is 58 miles, the transportation trucks carry 12 tons per trip and consume 5 miles/gallon of diesel on average, how many gallons of diesel are needed annually for this plant to operate?
6. A 600-MW coal electric power plant substitutes 8% of its fuel with "tree clippings," which are tree branches that are cut to clear roads as well as the access lanes (rights of way) of power lines. The tree clippings are fed in the boiler together with the coal. How much coal is saved annually by this practice and how much CO_2 emissions are prevented?
7. The New York metropolitan area has a population of approximately 10,000,000. If the New York population behaves like the average U.S. person, how many tons of solid waste do they produce every year? What is the range of the heating value of the solid waste for the region per year? If it were possible to use this waste in thermal power plants with 35% overall efficiency, how much power (in MW) would this waste produce?
8. Cuba produces 38% of its electric power from bagasse. In 2008, Cuba produced 14.67 TWh of electricity in power plants with overall efficiency 32%. How much bagasse was consumed?

9. What do you think will be the main parts and processes of a biodiesel plant that uses marine algae? Make a conceptual design of a plant that produces biofuels from marine organisms.
10. "We have enough land in this great country to produce enough corn and corn-based biofuels for us to become completely independent of foreign oil. We only need the will and a small amount of investment." Comment on this statement by writing a 250–300 word essay.
11. How many gallons of ethanol may be produced annually from a farm of 100 hectares (250 acres)? What are the inputs of fertilizers, pesticides and total energy that are required for such a farm?
12. "At a time when 50% of the children in the world are under-nourished, it is unconscionable of the leaders of the richer nations to divert food resources for the production of fuel for their high-mileage cars." Comment on this statement by writing a 250–300 word essay.

Reference

1. Farrell AE, Plevin RJ, TurnerBT, Jones AD, O'Hare M, Kammen DM (2006) Ethanol can contribute to energy and environmental goals. Science 311:506–508

Chapter 11
Power from the Water

Abstract More than two-thirds of the planet's surface is covered by water, and power may be obtained from both the surface water and the deep oceans. Hydroelectric, tidal, ocean current, wave, and OTEC power installations are among those that may convert the power of the water directly to electricity. Unlike wind power, these renewable energy sources are either continuous or predictably variable. The naturally occurring water cycle and the resulting rainfall transfer millions of tons of water annually to high elevations with significant potential energy. River flows on the planet carry the very high amounts of potential and kinetic energy, which currently turn the turbines of several hydroelectric power plants and have the capability to provide at least 25% of the total electric energy demand of the planet. The 240 MW tidal power plant in La Rance, France, has proven that there are reliable systems in operation to convert the tidal energy of the sea to electric power. Tidal power plants in prime locations in England, Norway, USA, Canada and other parts of the world have the capability to produce between 25 and 100% of the electricity demand in several coastal countries. The harnessing of the energy in ocean currents, such as the Gulf Stream, is at present a rather futuristic idea, which has the potential to cover the entire global demand for electricity. Several wave power systems and devices have been invented in the last half of the twentieth century, which are now in the testing and pilot-plant phases. If successful, harnessing the waves and the widespread use of wave power have the potential to produce more than 50% of the electric demand in island and coastal countries, such as Portugal, Norway, Great Britain, Japan, South Africa. and Australia. The Ocean Thermal Energy Conversion (OTEC) cycle also has the potential to supply almost unlimited electric power to coastal communities, but needs significantly more research and development efforts to reliably and economically meet this promise. All these electric power sources associated with the flow of water are examined in this chapter in a scientific manner. Engineering systems and cycles for the utilization of these resources are described. The environmental and ecological impacts of these systems are also examined critically.

E. E. (Stathis) Michaelides, *Alternative Energy Sources*,
Green Energy and Technology, DOI: 10.1007/978-3-642-20951-2_11,

11.1 Hydroelectric Power

The use of the potential and kinetic energy of water as a source of power has been known for thousands of years. Watermills were used in Asia and Europe to grind or mill the grain and produce flour. Today, similarly constructed plants transform the potential and kinetic energy of water to electricity. Figure 11.1 shows schematically how the potential and kinetic energy of the water may be transmitted to the prime mover, which is the shaft of the water wheel. In this case, if the torque on the water wheel is denoted by T and the angular velocity by ω, the total power recovered from the river flow is:

$$\dot{W} = T\omega. \tag{11.1}$$

When the water flow is not constant and predictable, or when relatively constant power is needed, a dam is built to restrict the flow of water upstream and create a water reservoir that maintains the water flow to the water wheel at a controlled rate. Damming the stream and the creation of the reservoir makes the water flow through the water wheel more predictable and reliable throughout the several seasons of the year. The dam also raises locally the level of the water surface, thus increasing the potential energy that may be converted to electric power.

The harnessing of hydroelectric power in contemporary power plants relies on the difference of the potential energy of the water and the mass flow rate of the water flow. The maximum amount of power that may be produced by a stream of water flowing at mass flow rate $\dot{m}$, when it falls over a height by Δz is:

$$\dot{W}_{\max} = \dot{m}g\Delta z. \tag{11.2}$$

A schematic diagram of a typical hydroelectric power plant is shown in Fig. 11.2. A river with relatively high mass flow rate is restricted by a dam, close to a point where a significant drop in elevation occurs. Locations close to a waterfall or near water rapids are excellent candidates for the building of dams. The dam creates an artificial water reservoir upstream, which may store several billions or trillions of tons of water.[1] Because of their large size, the capacity of the artificial reservoirs is measured in acre-feet. One acre-foot is equal to 326,000 gallons or 1,239 m^3. Water from the artificial reservoir enters a large pipe, the *penstock*, and is directed to a system of hydraulic turbines, which produces electric power. The water discharge from the turbine is directed to the same river, downstream. Frictional and other losses in the penstock are typically very small. The actual work produced by the hydraulic turbine is lower than the maximum work because of the finite efficiency of the turbine, which is typically in the range 72–84%:

$$\dot{W} = \eta\dot{m}g\Delta z \tag{11.3}$$

[1] One cubic meter of water contains approximately 1 t, or 1,000 kg, of mass.

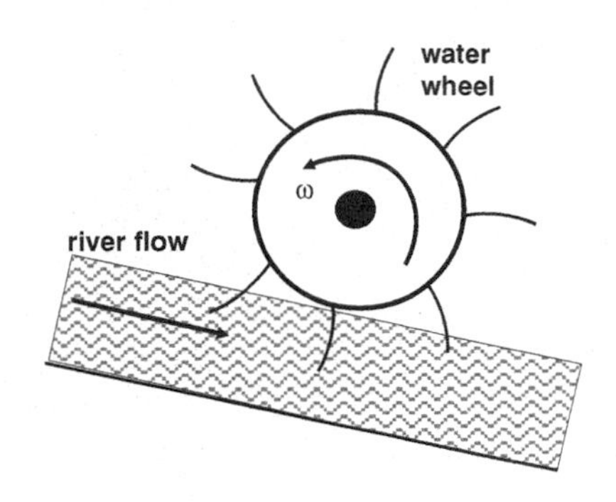

Fig. 11.1 Operation of a water wheel

A large hydroelectric power plant has several, typically 5–30, penstocks that lead to a number of turbines and operate in parallel to produce power. The river supplies the artificial reservoir and has sufficient water to maintain an approximately constant head, Δz.

From elementary thermodynamics it is known that the potential energy of fluids is very low (as specific energy in kJ/kg) in comparison to the thermal energy. One kg of water at 30 m height difference has specific energy ($g\Delta z$) equal to 0.294 kJ/kg. In comparison, one kg of steam has specific internal energy close to 3,000 kJ/kg, which is 10,000 times higher. Even when one considers the Carnot limitations of the thermal energy conversion, e.g. 30–40% practical overall conversion efficiency, the specific thermal energy of the steam is by far higher for the production of electric power. The very high power of the hydroelectric power plants is provided by the vast amounts of water, which are available in a river and pass through the several hydraulic turbines. Typical flow rates in large rivers are in the range 10^4–10^5 m^3/s. Such flow rates bring the equivalent mass flow rates of 10^7–10^8 kg/s to the power plants. Simple calculations show that these hydroelectric plants with 75% turbine efficiencies will are the capacity to produce between 2,200 and 22,000 MW of electric power. For comparison, a typical nuclear power plant produces approximately 1,000 MW of electric power. Of course, smaller rivers with lower flow rates would produce a lower amount of electric power. Table 11.1 shows several of the largest hydroelectric power plants in the world and their power generation capacity. It must be noted that, with the exception of the "Three Gorges" plant in China, most of the other hydroelectric power plants are older installations, with some, e.g. Krasnoyarsk and Hoover Dam, operating continuously since the 1930s.

11.1.1 Global Hydroelectric Energy Production

The year 2009 statistics of the world total hydroelectric power installations are approximately 770,000 MW. The existing hydroelectric power plants have collectively the capacity to produce 6,745 TWh of electricity (6,745*10^9 kWh) or

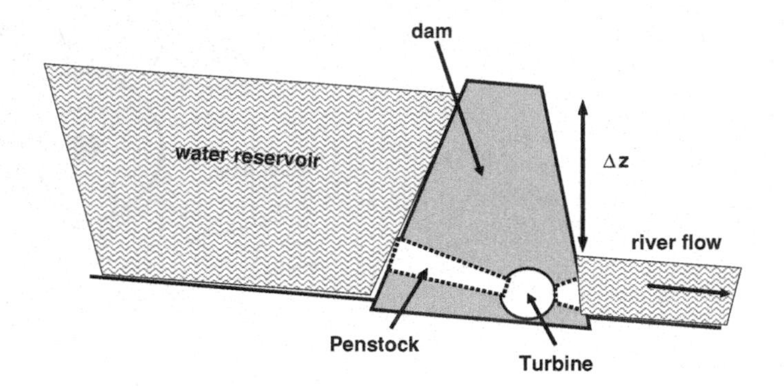

Fig. 11.2 A schematic diagram of a hydroelectric power plant

Table 11.1 Some of the largest hydroelectric plants in the world

Name of Plant	River/country	Power (MW)
Hoover dam	Colorado/USA	1,500
Niagara falls	Niagara/USA	1,950
Grand coulee	Columbia/USA	6,500
La grande	St Lawrence/Canada	10,000
Itaipu	Parana/Paraguay-Brazil	12,600
Three gorges*	Yangzee/China	18,200
Krasnoyarsk	Yenisei/Russia	6,400
Guri	Caroni/Venezuela	10,300

* Installed in 2009 and planned to increase to 22,500 MW by 2012

one third of the world's electric energy demand, which is approximately 20,000 TWh. In addition, several hydroelectric power plants have been constructed to provide electric power to a particular industry, typically an aluminum smelter, which requires very high amounts of electric energy for its operation. For example, the Brokopondo power plant in Surinam was constructed to provide electricity for the local Alcoa aluminum plant and the Manapouri Power Station in New Zealand was constructed to supply electricity to an aluminum smelter at Tiwai Point.

By reducing the flow to the turbines or by shutting down the flow through penstocks, the power produced by a hydroelectric power plant may be easily controlled and adjusted at short times. For this reason, hydroelectric power plants are used as intermediate and peak demand units (see Chap. 12) and many do not produce power continuously at their *rated power*. In 2009 approximately 3,200 TWh of electricity were produced from hydroelectric power plants. This implies that the average *plant capacity factor* of hydroelectric units was 47%. This is approximately 16% of the global electric energy demand and represents more

Table 11.2 Leading countries in hydroelectric power production

Country	Installed capacity, GW	Energy produced, TWh	Percentage of total electric energy
Brazil	69	364	86
Canada	89	370	61
P.R. China	172	585	17
India	34	116	16
Japan	27	69	7
Norway	28	140	98
Russia	45	167	18
Sweden	26	65	44
USA	80	251	6

than 90% of the electric energy from renewable energy sources. Table 11.2 shows the countries with the largest total installed hydroelectric capacity. Smaller countries, such as Norway have the capacity to produce their entire demand for electricity from hydroelectric power. However, this is not the case with larger electricity consumer countries, such as USA and the PRC, which produce 6 and 17% of their electricity from hydroelectric energy respectively.

Hydroelectric power plants may operate for a very long time if they are well maintained. The main mechanical parts, the turbines, may be frequently maintained, refurbished and replaced if necessary. Penstocks, valves, and water gates do not contain any moving parts and last much longer. What may limit the life of a hydroelectric power plant is the silting of the reservoir upstream the dam. There is a stagnation region in front of the dam, where water comes to a standstill, turbulence is significantly reduced and the small particles that constitute the silt fall to the bottom. Oftentimes the sedimentation process is assisted by particle flocculation, which produces coarser grains. The sedimentation/siltation process reduces the water storage capacity of the reservoir and, if untreated, may shut down the reservoir. Excessive siltation in hydroelectric reservoirs may be avoided by:

a) Diverting part of the water close to the bottom of the river to cause sediment bypassing
b) By special weirs
c) By appropriate projects upstream that reduce soil erosion and particulate entrainment
d) By periodic dredging of the reservoir.

A well-maintained water reservoir will last for more than 100 years and the corresponding hydroelectric power plant for an equal amount of time. This longevity makes the hydroelectric power plants one of the least expensive systems for the production of electricity.

11.1.2 Planned Hydroelectric Installations and Future Expansion

Except for the construction of an often costly dam, hydroelectric power plants are relatively simple in operation, inexpensive to construct and, most important, do not require fuel for the production of electricity. Our accumulated experience of the plants built early in the twentieth century proves that the large hydroelectric power plants will operate for more than 80 years with relatively low maintenance. For this reason, most of the large-scale hydroelectric power plants produce electricity at significantly lower cost than thermal power plants. Also, most of the prime, large-capacity hydroelectric resources in the OECD nations, e.g. Japan, USA, Norway, and Canada, have been utilized and power plants in such locations have been in operation for at least 40–50 years. In the future, almost all of the immediate expansion of hydroelectric power will occur in developing countries and especially in the Peoples Republic of China and the continents of South America and Africa. In addition to the "Three Gorges" power plant, which is in the final stages of completion, there are currently several other hydroelectric power plant projects in China totaling 70,000 MW. Planned hydroelectric capacity in Brazil and Argentina is close to 10,000 MW and the proposed project of "Grand Inga" in the Democratic Republic of Congo alone is expected to produce 40,000 MW.

In countries that have already utilized most of their large-scale resources, the construction of *small hydro* (less than 30 MW) or *micro hydro* (less than 1 MW) is the way to utilize a higher percentage of the hydroelectric resources and to produce more renewable electric power. These plants are regional and may serve a small community or a single industrial plant. A glance at Eq. (11.3) proves that a smaller river with volumetric flow rate 10^3 m^3/s and elevation drop of 4 m has the potential to produce 39.2 MW and will actually produce 29.4 MW with typical turbine-generator pairs that have 75% efficiencies. Smaller units at rivers with volumetric rates 100–500 m^3/s would produce electric power in the range 3–15 MW. There is a multitude of locations in Europe, and North America, where such rivers/resources exist and where small hydroelectric power plants may be constructed to produce a higher amount of electric energy from this clean and renewable source. These small hydroelectric resources are abundant and their increased utilization for the production of electricity may quadruple the hydroelectric energy produced in the OECD countries, where all the other, large-scale hydroelectric resources have already been utilized. An additional advantage of such *small or micro hydro* units is that, because the units are typically very small, it is not necessary to construct large dams and, hence, their environmental and ecological impacts are minimal. Smaller hydraulic turbines are available to produce lower amounts of power in these units. Oftentimes, these smaller power plants may be combined with flood control, irrigation and recreation projects to enhance the utility and benefits of the entire project to the local community.

11.1.3 Environmental Impacts and Safety Concerns

Hydroelectric power is essentially a clean, renewable energy source. However, the construction of the hydroelectric power plants, especially those of very large scale, poses a few environmental problems: At first is the construction of the dam and the reservoir behind it. Dams often exceed 50 m in height and 20 m width and require a very high volume of material, primarily concrete, to construct them. For example, the construction of the Hoover Dam in Nevada, U.S.A. required as much concrete as a 3,200 km, four-lane highway. Secondly, the dam causes the flooding of a very large area behind it, which becomes the water reservoir. The flooding of the area upstream requires the displacement of communities, the abandonment of whole towns, and the relocation of inhabitants and businesses. The construction of the "Three Gorges" dam in China was for a long time a very controversial project because it resulted in the displacement of 1,240,000 persons and the abandonment of several towns. It also caused extensive ecological change and when it was built, it caused the flooding of several important cultural and archaeological monuments. Thirdly, the construction of the dams prevents fish migration and restricts the waterways. Several studies have shown that hydroelectric dams along the Atlantic and Pacific coasts of North America have reduced salmon populations by restricting the fish access to spawning grounds upstream the rivers. This ecological impact is mitigated by the installation of artificial "fish ladders," where the fish are able to swim upstream in a bypass waterway. Because the turbines harm the salmon spawn, in several reservoirs along the Columbia River, young salmon is transported downstream by barges during the spawning season.

The generation of electric power alters the downstream water environment. Because of sedimentation and the passage of water through sieves, water exiting a turbine usually contains a very small amount of suspended sediment. This prevents the river bottom nourishment downstream with fresh sediment and may lead to the scouring of the river beds and loss of riverbanks. Because the power generation in the power plant fluctuates, the turbine gates are opened and closed several times a day. This results in frequent fluctuations of the river flow, which contributes significantly to the erosion of the river downstream. The dissolved oxygen content of the water is also reduced because of the rapid swirling flow in the turbines. This, in addition to water warming as it passes through the turbine systems, may endanger the fish population and the ecological systems downstream.

The environmental impacts of the hydroelectric dams and power plants are relatively few and benign in comparison to the impacts of the fossil fuel power plants. However, a significant safety concern is dam failures, which have resulted in several of the largest man-made disasters in the history of electric power production. The Banqiao Dam failure in Southern China resulted in a great flood, the deaths of 171,000 people and left millions homeless. The failure of the Vajont Dam in Italy, which was built in a geologically unstable region, caused the deaths of 2,000 people. In this respect, the smaller dams and small hydroelectric power plants, which are touted for the future, create a lesser risk, because they contain a much smaller volume of water and they affect a smaller fraction of the population. However, even these small plants are not immune from failures and disasters and must be maintained meticulously. For example the Kelly Barnes small hydro-electric dam in the USA failed in 1967 and resulted in 39 deaths ten years after its power plant was decommissioned. Good design, correct construction in the right location and meticulous maintenance will, in general, help avoid dam failure and disasters. It must be noted though that this is not an unfailing guarantee of safety, because large dams are tempting industrial targets for terrorism and wartime sabotage. In the early twenty-first century, appropriate security measures to thwart terrorist activities and sabotage are becoming part of the maintenance operations in all the large dams in the OECD as well as in many developing countries.

11.2 Tidal Power

Tides are created because of the gravitational and kinematic effects due to the position and the combined motion of the Earth, the Moon and the Sun. The Moon, even though it has much smaller mass than the Sun, plays a more important role in the creation of tides because it is significantly closer than the Sun. The mass of the Moon "pulls" the ocean water masses in its direction and creates a "bulge" on the surface of the ocean. The effect of the Moon's pull is modified by the effect of the Sun, as shown in Fig. 11.3. When the Earth, the Moon and the sun are aligned, the tides are amplified and are called *spring tides*. When the three bodies form a right-angle triangle the tidal effect is modulated and we have the *neap tides*. There are two periods of spring tides and two of neap tides during every Moon cycle, which lasts for approximately 29.5 days.

A simple analysis of the effects of the gravitational potentials of the Earth and the Moon on the surface of the Earth's ocean may be made with the help of Fig. 11.4, which shows the combined effect of the Moon's gravity on the surface of the ocean and the rotation of the Earth around its axis. When we consider an equipotential surface around the surface of the Earth and perform the force balance resulting from the gravitational forces of the Moon and the Earth on a unit mass of water, at the surface of the ocean, we will obtain the following approximate expression for the height of the ocean "bulge" at any point on the surface of the ocean:

$$h(\theta) = \frac{r^4 m}{M d^3}\left(\frac{3}{2}\cos^2\theta - \frac{1}{2}\right), \tag{11.4}$$

where θ is the angle formed between the line that connects the centers of the Earth and the Moon and the line that connects the center of the Earth with the point under consideration.; M is the mass of the Earth; m is the mass of the Moon; r is the average radius of the Earth, which is approximately 6,387 km; and d is the

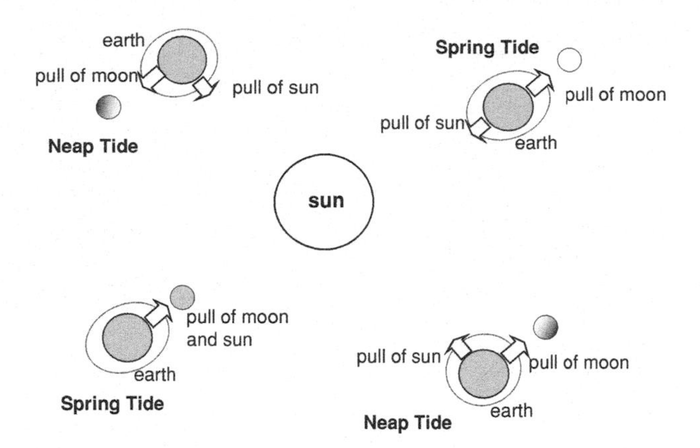

Fig. 11.3 Spring tides and neap tides

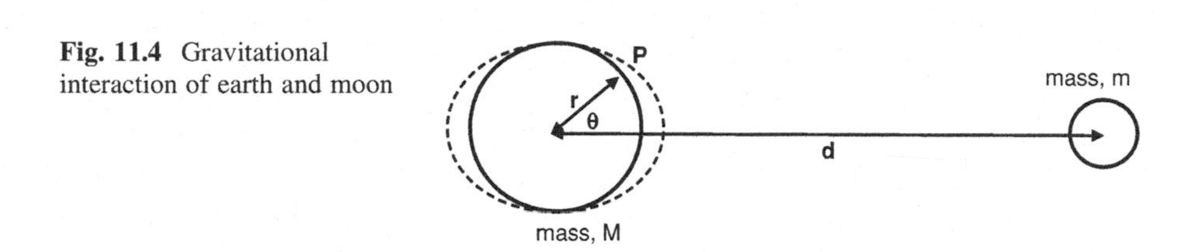

Fig. 11.4 Gravitational interaction of earth and moon

Moon to Earth distance, which is approximately 384,420 km. Given that $m/M = 0.0123$, Eq. (11.4) reveals that the maximum bulge on the surface of the oceans is 0.36 m. This value is close to the globally-averaged height of the tides on the oceans, which is measured to be approximately 0.5 m. The *range* of the tides, H, is the difference between the maximum and the minimum heights of the water. The average range of the surface of the oceans is twice the maximum height, or approximately 1 m.

The bulges on the tidal deformations on the surface of the ocean form the *tidal waves*, which are shallow, long wavelength waves that move on the surface of the Earth's ocean. At any time, there are two highs and two lows on the surface of the ocean.

The Earth rotates around its axis with a period of 24 h, and the apparent motion of the moon is a rotation around the Earth with a period of 24 h and 50 min. This implies that the tidal waves propagate on the surface of the oceans with a period, which is half the apparent period of rotation of the Moon. Therefore, the period of the ocean tides is 12 h 25 min or approximately 45,000 s.

According to elementary wave theory, typical speeds of propagation of the tidal waves are close to 200 m/s. From the wave equation:

$$c = \lambda v = \frac{\lambda}{T}, \tag{11.5}$$

Table 11.3 Locations with high potential for tidal energy conversion

Location	Country	Range, m	Average Power (MW)	Annual energy, (GWh)
Cobequid	Canada	10.7	20,000	175,200
Passmaquoddy	Canada	5.5	1,800	15,768
Severn river	England	9.8	1,680	14,717
La Rance	France	8.4	349	3,956
Mont St. Michel	France	8.4	9,700	84,972
Kimberlay	Australia	6.4	630	5,519
Mesen (Mezen)	Russia	6.6	14,000	122,640
Khambat	India	7.0	7,000	61,320
Golfo Nuevo	Argentina	3.7	6,600	57,816

These numbers were calculated using the "one pool" tidal system according to Eq. (11.7)

one may conclude that the tidal wavelengths are approximately 9,000 km. This distance is greater than the width of the Atlantic Ocean and of the same order of magnitude as the maximum width of the Pacific Ocean.

The presence of the continental shelves, the variable ocean depth and the irregularity of the coastlines interfere significantly with tidal waves and cause local distortions and resonances. The resonances create tidal waves with different amplitudes locally, which are significantly higher than the global average range of 1 m. These are most common in bays, gulfs and estuaries. Tidal ranges between 5 and 10 m have been observed in several narrow bays and estuaries. Several of these high ranges are listed in Table 11.3. These ranges reveal a few of the locations on the Earth with high potential for electricity production from tidal power.

Table 11.3 shows that the amount of tidal power, which is available for conversion to electricity, is very high. The average power available in several of the locations listed in this Table is equal to that produced by tens of conventional fossil fuel plants. For example, the power available in the Cobequid area alone is equivalent to 31% of the total electricity that was consumed in Canada in 2009. Similarly, the full utilization of the tidal power resources in France and England has the capacity to provide between 25 and 30% of the electricity consumption in the two countries. Clearly, tidal power if developed and well utilized may provide a high percentage of the electric energy demand and will prove to be an excellent alternative to fossil fuels.

11.2.1 Systems for Tidal Power Utilization

It is apparent from the above that, in every coastal location on the planet, the sea level is expected to rise and fall with a period of approximately 12.5 h or 45,000 s. Therefore, tidal power is a predictably variable, renewable energy source. Electric

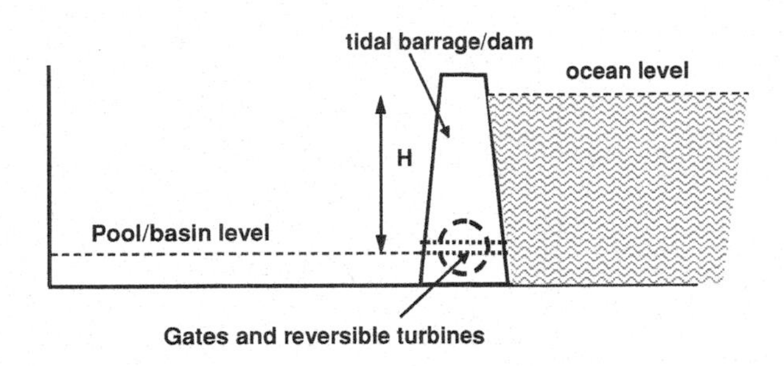

Fig. 11.5 Single pool tidal system or tidal *barrage*

power may always be produced from tides in a predictable way. However, because this does not necessarily coincide with the variation of the electric energy demand tidal energy systems must be designed to store some or all of the energy that is available in tides.

A simple system to utilize the tidal power is the *tidal barrage* or *single pool tidal system* that is shown schematically in Fig. 11.5. The tidal barrage is essentially a dam that separates the ocean from a basin, which is typically a gulf or an estuary. Gates are located close to the bottom of the barrage and lead to hydraulic turbines. The latter operate at low pressure heads and are reversible. This implies that they produce power when the water flows in either direction. The gates are closed with the pool empty until the tide reaches the highest level, which is equal to the tidal range, H. At this instant the gates open, the turbines operate producing electricity and the pool fills with water. The total amount of energy produced during the filling of the pool from level 0 to H is:

$$W = \int \eta g h dm = \int_0^H \eta g h A \rho dh = \frac{1}{2} A \eta \rho g H^2, \tag{11.6}$$

where η is the average efficiency of the turbine, A is the cross sectional area of the pool, assumed to be uniform, and ρ the water density, approximately 1,000 kg/m^3. When the pool fills with water, the gates close, the level of the pool is at H and the plant does not produce electric energy. The level of the ocean drops until the low tide point is reached, when the pool is at level H above the ocean. At this instant, the gates open again and the reciprocating turbines produce power for the second time within a single tidal wave cycle. Hence, the single pool system produces twice the amount of energy shown in Eq. (11.6) during the approximately 12.5 hours of the period of the tidal wave. Therefore, the average power produced is:

$$\dot{W} = \frac{A \eta \rho g H^2}{T}, \tag{11.7}$$

where $T \approx 12.5$ h $= 45{,}000$ s.

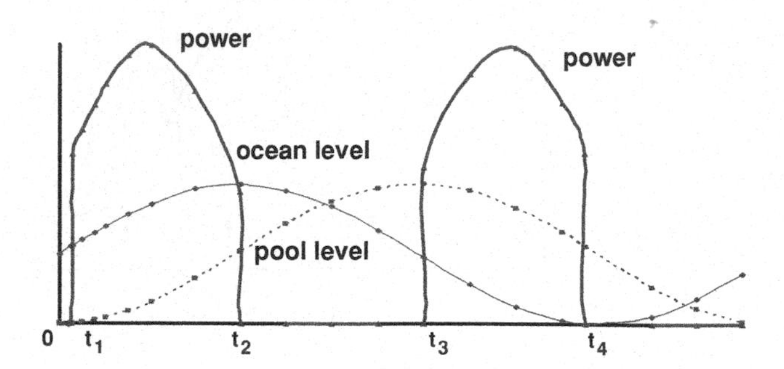

Fig. 11.6 Power production, ocean and pool levels in the single pool tidal system

It is apparent from this analysis that the single pool system does not produce power continuously. Rather, its operation produces this amount of power in two "bursts," during the fast filling and emptying processes, which last approximately 20–30 min. The actual power produced during the short time of the power plant's operation is significantly higher than that calculated from Eq. (11.7). For example, if the discharge occurs within 20 min (1,200 s), the actual power would be 45,000/(2*1,200) or 18.75 times higher than the average power indicated by Eq. (11.7). The system remains idle during the rest of time. This type of operation for the utilization of tidal power has several disadvantages including the following:

1. Very high power is only produced over very short periods of time, when the demand may not exist. In the absence of high electricity demand, the power level of other facilities will have to be reduced to accommodate the produced power. This may involve significant and undesirable power fluctuations in the other power plants that supply electricity to the electricity grid.
2. The system must employ very high power turbines to accommodate the "bursts" of power produced. The high-power turbines are significantly more expensive and add to the capital cost of the power plant and the final cost of electric energy.
3. Switching on and off the turbines, exciters, and generators adds significantly to the wear of the equipment and shortens their useful lives.

An alternative operation of the single pool system is to operate the turbines for longer periods of time and to shut them down only when the power produced falls below a minimum, predetermined value. The disadvantage of this type of operation is that the pool fills up and the difference in the level of the water may be significantly lower than H over the periods when power is produced. The turbines produce a lower amount of power and energy during the tidal cycle. The main advantage of this type of operation is that the tidal power plant uses smaller and less expensive turbines. In addition, other power installations are not unduly burdened by its operation. Figure 11.6 shows the operation of the plant as a function of time and the power produced by it. The upper part of the figure shows the water levels of the ocean, $h_o(t)$, and that of the pool, $h_p(t)$. The instantaneous amount of work produced under this mode of operation is:

$$W(t) = A\rho\eta g\left|h_o^2(t) - h_p^2(t)\right|. \tag{11.8}$$

The total work produced during a tidal cycle is:

$$W = A\rho\eta g\left[\int_{t_1}^{t_2}\left|h_o(t) - h_p(t)\right|\frac{d\left|h_o(t) - h_p(t)\right|}{dt}dt + \int_{t_3}^{t_4}\left|h_o(t) - h_p(t)\right|\frac{d\left|h_o(t) - h_p(t)\right|}{dt}dt\right], \tag{11.9}$$

and the average power of the power plant is:

$$\dot{W} = \frac{2A\rho\eta g}{45{,}000}\left[\int_{t_1}^{t_2}\left|h_o(t) - h_p(t)\right|\frac{d\left|h_o(t) - h_p(t)\right|}{dt}dt + \int_{t_3}^{t_4}\left|h_o(t) - h_p(t)\right|\frac{d\left|h_o(t) - h_p(t)\right|}{dt}dt\right]. \tag{11.10}$$

The two functions of the water height in the ocean and the pool, $h_o(t)$ and $h_p(t)$, may be approximated as sinusoidal functions. Hence, the total energy produced and the average power may be evaluated analytically or numerically. One of the characteristics of the single pool tidal system is that the choice of the time periods t_1 to t_2 and t_3 to t_4, when the system is in the power-producing mode, may be appropriately chosen and optimized to suit electricity production objectives, such as maximum power generation or power production during peak demand.

The *two-pool tidal system* includes another controllable parameter for the optimization of power generation objectives by making use of an additional water pool where water may be discharged. The two-pool tidal system utilizes two water pools, which are filled and emptied during each tidal cycle. The system may be optimized in one of the following three ways:

a) To produce the maximum amount of energy during a tidal cycle
b) To produce maximum power during periods of peak demand
c) To produce some power but also store energy for use during peak demand.

Figure 11.7 is a schematic diagram of the two pool system. At high tide, pool A is filled by the ocean water, thus producing some power, while pool B is at its lowest level and closed. When pool A is filled, it starts discharging to the low-level pool B, thus producing additional power. By the time the water level in the two pools A and B equalizes, the level of the ocean drops and both pools A and B may discharge their water to the ocean. Power is produced during this discharge phase. The capacities of the two pools may be chosen so that the effects of the gravitational head variability are minimized. This type of operation of the two-pool system results in continuous and almost uniform power generation.

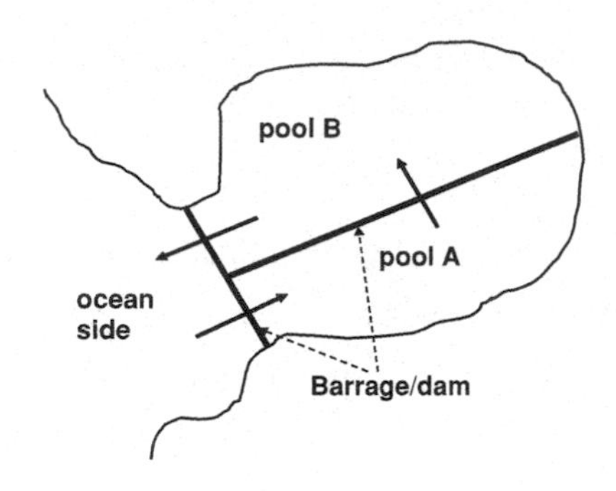

Fig. 11.7 Operation of the two pool tidal system

It must be noted that, for the functioning of the two-pool system, it is not necessary to employ additional turbines than the single-pool system. Turbines are more expensive and add significantly to the cost of the installation. Instead of installing more turbines, one may construct a suitable network of pipelines, penstocks, inlets, outlets and diverting gate valves to always direct the flow from the higher water level to the lowest level, while passing the water through the same hydraulic turbines.

11.2.2 Environmental Effects of Tidal Systems

It is apparent from Table 11.3 that tidal power may produce a significant amount of electricity, thus replacing fossil-fuel power plants and avoiding the pollution effects associated with them. Tidal power is renewable, clean and does not emit any pollutants. There are some adverse environmental effects of tidal systems, which are enumerated here:

1. For the building of the pool systems, significant construction is required, which uses a great deal of materials, especially cement.
2. Dams and barrages pose an environmental hazard because they obstruct the passages of fish and other marine life.
3. Because the tidal systems are constructed in coastal environments, they invariably interfere with fish spawning and have an adverse effect on the populations of the fish and other aquatic life.
4. Because of the large water flows involved, there will be daily fluctuations in the turbidity and chemical composition of the water, which may have adverse effects on the aquatic life of bays and estuaries.
5. Local navigation of boats and recreational watercraft is adversely affected by the construction of barrages, dams and locks.
6. Additional sedimentation in the "pools" will necessitate frequent dredging and disposal of the, sometimes, contaminated silt.

On the positive side, the construction of a tidal system may have multiple purposes to add to the quality life in the surrounding communities. For example:

1. The upper surface of a barrage that connects two sides of an estuary may be paved and used as a road that connects communities.
2. Recreational activities, such as sailing, boating and swimming may be promoted within the enclosed estuary.
3. The additional construction may be used for flood control and avoidance of sea erosion.

11.3 Ocean Currents

Permanent ocean currents occur in several parts of the sea and are generated by a variety of forces including: the Coriolis force; the predominant surface wind; temperature and salinity gradients; the predominant tides. Variability in depth contours and shoreline configurations affect the ocean currents' direction and power potential. High intensity ocean currents are frequently met in the waterways between islands, at the tip of large peninsulas and capes, for example off Cape Hateras in North Carolina, U.S.A. or at waterways that connect oceans, as for example the Straight of Magellan at the southernmost point of South America.

An ocean current is in all respects similar to a wind current. As with the wind power, the available power of ocean currents is given by the expression:

$$\dot{W}_{av} = \frac{1}{2} A \rho V^3 = \frac{\pi}{8} D^2 \rho V^3. \tag{11.11}$$

The power production limitation on wind power, or Betz's Law is also applicable to the ocean currents. The maximum power that may be produced from a water turbine, which is placed within the current, would be:

$$\dot{W}_{\max} = \frac{8}{27} \rho A V_i^3 = \frac{2\pi}{27} \rho D^2 V_i^3. \tag{11.12}$$

A significant difference between wind and ocean currents is that the density of the water is approximately 800 times higher than that of air. This implies that significantly more power may be produced by water currents. For example a small ocean current turbine with 3 m blade diameter would produce a maximum of almost 152 kW of electric power. Such an engine would only produce 0.20 kW from the wind. As a result, sea current turbines may be significantly smaller and more compact than their wind counterparts.

Because water is dense and the volume of the water that circulates in the oceans is very large, if suitably utilized, ocean currents have a very high potential for the production of electric energy. For example, the *Gulf Stream* alone, which flows with a cross-sectional area of 100 km (62 miles) wide and 800 m (2,600 ft) to 1,200 m (3,900 ft) deep with average velocity 2 m/s, carries approximately

400,000 MW of power in the form of kinetic energy and may produce maximum power of 237,000 MW. If utilized continuously, the Gulf Stream energy would be sufficient to provide 50% of the electric energy consumption in the USA or 12% of the electricity consumed in the entire planet in 2009. It is evident that, tapping even a small fraction of this enormous potential resource to produce electricity would provide a significant amount of energy and replace the use of fossil fuels. In addition, because ocean currents are continuous, this form of renewable energy is not intermittent or predictably variable and ocean current turbines may be used for the continuous production of power.

While the use of ocean currents for the production of electricity is feasible and may prove to have significant payoffs, an ocean current power plant has not been built yet, even at a pilot scale. The principal impediments for such a project are:

1. Ocean storms and high currents that may damage underwater installations.
2. Transmission of electricity onshore and feeding into the national grid.
3. Lack of experience and pertinent research on underwater turbines and large-scale electricity generating systems, which must operate in a hostile, underwater environment.
4. Most strong water currents are in international waters. Lack of international treaties and governing laws increase the uncertainty and risk of investment.
5. Sabotage and terrorism concerns for systems that, by their very nature, are inherently built a large distance offshore.
6. Ecology, reproduction and migration of marine life.

11.4 Wave Power

The ocean waves are directly caused by the wind and, indirectly, from solar energy. Similar to the tide power, and because the oceans cover more than 70% of the Earth's surface, wave power is abundant and, if appropriately harnessed, it has the potential to provide a great deal of electric power.

11.4.1 Wave Mechanics and Wave Power

All waves are characterized by their: amplitude, α, which is half of the vertical distance between crest and trough; wavelength, λ, which is the horizontal distance between two successive crests; period, τ, which is the time at a point on the surface between two successive crests; and frequency $f = 1/\tau$. The wave velocity, c, is equal to λ/τ. When one follows the wave motion at a point on the water surface one sees that the physical fluid particles perform circular oscillations with period τ as shown in Fig. 11.8, with the wave velocity c superimposed on the physical velocities at the crest and the trough. From these considerations, we have the following expressions for the wave velocities at the crest and the trough:

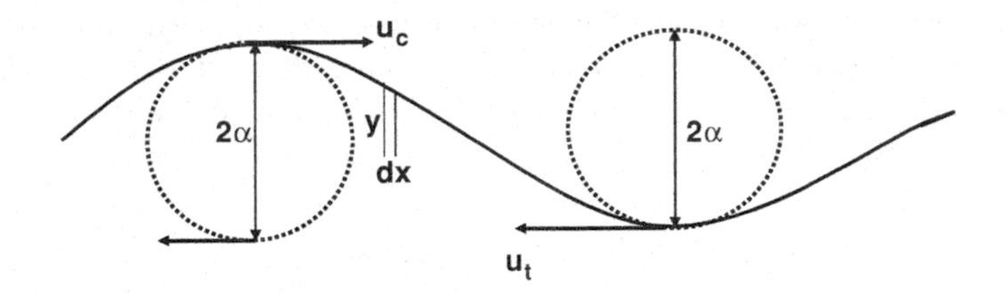

Fig. 11.8 Surface wave velocities

$$u_c = \frac{2\pi\alpha}{\tau} - c \quad and \quad u_t = -\frac{2\pi\alpha}{\tau} - c. \tag{11.13}$$

The pressure at the surface of the water is equal to the atmospheric pressure. When we apply Bernoulli's equation between the crest and the trough, we obtain the following expression:

$$u_t^2 - u_c^2 - 2g\alpha = 0. \tag{11.14}$$

The system of the last two equations together with the wave equation, $\lambda = c\tau$, yields the following expressions for the wave velocity, period and frequency:

$$c = \sqrt{\frac{g\lambda}{2\pi}}, \quad \tau = \sqrt{\frac{2\pi\lambda}{g}} \quad and \quad f = \sqrt{\frac{g}{2\pi\lambda}}. \tag{11.15}$$

The motion of the waves in the Cartesian coordinates may be expressed by a sinusoidal function, which has the form:

$$y = \alpha \sin\left(\frac{2\pi}{\lambda}x - \frac{2\pi}{\tau}t\right). \tag{11.16}$$

For the calculation of the potential energy of the wave, let us consider a length of the wave, L, in a direction perpendicular to the direction of wave propagation. Referring to Fig. 11.8, the infinitesimally small volume, defined by the width dx, height y and length L has a potential energy equal to:

$$dE_P = \rho g \frac{y}{2} dV = \frac{1}{2}\rho g L y^2 dx. \tag{11.17}$$

One may substitute for the height y from Eq. (11.15) and integrate this expression over the entire wavelength of the wave, λ, to obtain the total potential energy in one whole wavelength as follows:

$$E_P = \int_0^\lambda dE_P = \frac{1}{2}\rho g L \int_0^\lambda \alpha^2 \sin^2\left(\frac{2\pi}{\lambda}x - \frac{2\pi}{\tau}t\right) dx = \frac{1}{4}\rho g L \lambda \alpha^2. \tag{11.18}$$

It is known from wave theory that the total wave energy is equally partitioned between the potential and kinetic energy of the wave, that is $E_P = E_K$. Therefore, the total wave energy is twice the amount of the potential energy, which is expressed by Eq. (11.18):

$$E_T = E_P + E_K = \frac{1}{2}\rho g L \lambda \alpha^2. \tag{11.19}$$

The total wave power is equal to the product of the total energy and the wave frequency. Thus, the total power may be written as follows with the help of the expressions in Eq. (11.15) as:

$$\dot{W} = E_T f = \frac{E_T}{\tau} = \frac{\rho L \alpha^2 \lambda^{1/2} g^{3/2}}{8^{1/2}\pi^{1/2}} = \frac{\rho L \alpha^2 g^2 \tau}{4\pi}. \tag{11.20}$$

Oftentimes, the power per unit width and the power per unit area, or power density, are used for the computation of the total wave power. Since the width is L and the surface area is $L\lambda$, these two parameters may be calculated are as follows:

$$\begin{aligned} \frac{\dot{W}}{L} &= \frac{\rho L \alpha^2 \lambda^{1/2} g^{3/2}}{8^{1/2}\pi^{1/2} L} = \frac{\rho \alpha^2 \lambda^{1/2} g^{3/2}}{8^{1/2}\pi^{1/2}} = \frac{\rho \alpha^2 g \sqrt{g\lambda}}{\sqrt{8\pi}} \\ \frac{\dot{W}}{A} &= \frac{\rho L \alpha^2 \lambda^{1/2} g^{3/2}}{8^{1/2}\pi^{1/2} L\lambda} = \frac{\rho \alpha^2 g^{3/2}}{8^{1/2}\pi^{1/2}\lambda^{1/2}} = \frac{\rho \alpha^2 g \sqrt{g}}{\sqrt{8\pi\lambda}}. \end{aligned} \tag{11.21}$$

Typical power densities per unit width of waves near the shores of the Atlantic and Pacific oceans are in the range 20–80 kW/m. Since the wave amplitudes are increased significantly during storms, the wave power varies from these levels to much higher values. For this reason, any system that would convert the wave energy to electricity must also be able to withstand the high forces and high power levels associated with high storm waves.

11.4.2 Systems for Wave Power Utilization

Everyone who has observed the waves on the sea or a lake has noted that the waves on the surface waters are actually formed by the superposition of several types of waves with different wavelengths, frequencies and speeds. As a result, the appearance of the sea surface is significantly more irregular than that of a sinusoidal wave. One can also see that most of the wave energy is contained in the long-wavelength, high-amplitude waves. While devices designed for the utilization of the sea waves should respond to and be able to absorb the energy from all the constituent wavelengths, for maximum energy production, these systems must be able to absorb efficiently the energy of the more energetic, high wavelength and high-amplitude waves. Several devices have been developed in the last half

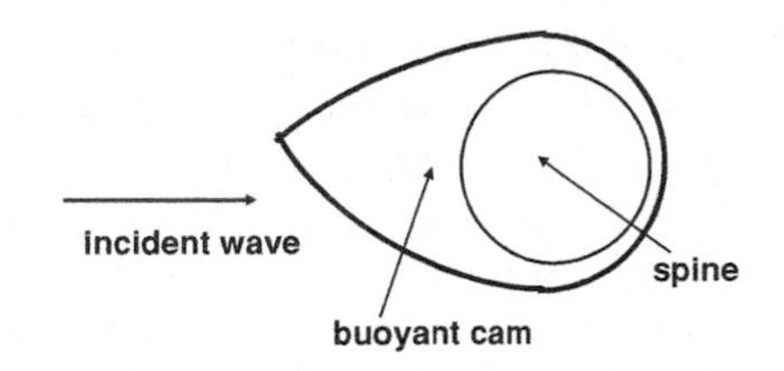

Fig. 11.9 Schematic diagram of the *salter duck*

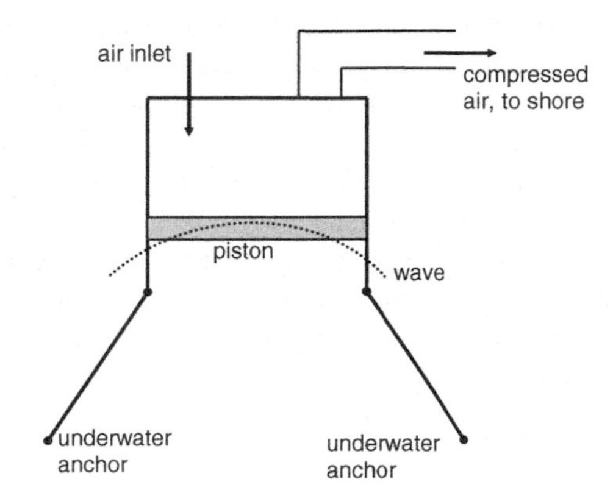

Fig. 11.10 Inverted piston-in-cylinder wave engine

century for the utilization of the sea wave power with various degrees of success. Among these devices are the following.

The *Salter Duck* was developed in the 1960's and took its name because its up and down motion resembles that of a duck in the water. It consists of a central cylindrical spine, around which is positioned a buoyant cam with a conical prismatic shape, as shown in Fig. 11.9. The cam is positioned so that its leading edge is in the principal direction of the incident waves. The shape of the cam is such that it responds to the motion of the incident waves, but it does not disturb the flow behind. The cam absorbs the energy of the incoming waves without creating any other waves on the lee side and converts the wave motion to its own motion, which in turn is transmitted to the prime mover of an electric generator.

Another, simple mechanism to harness the up-and-down motion of the waves is to use the inverted cylinder and piston assembly, which is depicted in Fig. 11.10. The wave action moves the buoyant piston in a vertical motion. When the piston moves downwards, a valve opens at the top and allows air to flow into the cylinder. When the motion of the piston is reversed, the inlet valve closes and the confined air is compressed until a predefined pressure is reached. At this moment the outlet valve opens and the compressed air is directed to a duct, which carries the compressed air to a power producing turbine, usually located onshore.

Another wave system utilizes the splashing effect of the waves on shores, which, by raising the level of the water periodically, results in the flooding of narrow parts of the coastline. This effect, essentially converts the vertical motion of the wave to an almost horizontal motion of the water on the shore, thus creating a low-level, high-frequency tide. The horizontal motion may be converted to power by a waterwheel or a low-head water turbine. The velocity of the water is amplified by the use of a tapered channel, which directs the sea water of the wave to the low-head turbine. The TAPCHAN (tapered channel) system that was developed in Norway results in a 3–5 m elevation of the water level. This is a sufficiently high head to operate a small Kaplan turbine.

A fourth system, which has been proposed for ocean waves far from the shore, is the *Dam-Atoll* system, which is named from the combination of the actions of dams and atolls. The latter are small islands of volcanic origin. As the long-wavelength waves approach an atoll, they wrap around the small island and create a vortex with a significant spiral motion. The water is directed to the center of the Dam-Atoll system and its spiral motion drives a turbine, which is located at the center of the system. Because the Dam-Atoll systems float on the water, they may withstand strong sea storms. It has been estimated that an 80 m diameter, 20 m high Dam-Atoll system would produce approximately 1–1.5 MW power. For maximum power production, the Dam-Atoll systems must operate far from the shore. This presents the technical problem of power transmission to the shore.

11.4.3 Environmental Effects of Wave Power and Other Considerations

Wave power is clean, renewable, almost continuous, and does not pollute the atmosphere. Because the wave systems are built in the open seas, they do not affect the environment and the ecology of the coastal areas as much as tidal systems. The following are the most important environmental effects of the conversion of wave power to electricity:

1. Because the wave power density is relatively low, the wave conversion systems are massive and require large quantities of materials for their construction.
2. Most wave power systems are lengthy. Therefore, the navigation of ships will be obstructed.
3. All wave power systems include moving parts, which would kill fish and other sea animals if they are trapped in the systems.

Mechanical systems for wave power conversion must operate in adverse conditions: First, the systems must operate and maintain their structural integrity under heavy and low seas, under storms, and calm weather. This imposes a large constraint on the size, the strength of the materials and the design of the systems, which must withstand a very high range of forces. Because of the very high density of water, the forces that sea-wave systems must endure are by far higher than those of the wind-power systems. Secondly, the salt in the sea water is corrosive to most materials, and any wave energy system must be designed around this constraint.

Thirdly, the anchoring systems are not fail-safe. Often, the anchors fail or drift and this may cause the failure or damage of the power system on the water surface. Fourthly, some systems, such as the Dam-Atoll systems, must be located far from the shore. This makes the transmission of the electric power produced problematic. Such systems are also more susceptible to damage from the strong storms and high waves that occur in the high seas. These are some of the reasons why wave power systems are expensive and there are not many systems in operation, other than experimental or pilot systems.

An advantage of future wave power systems is the high correlation between the predominant wind and high energy waves. A promising wave power site is usually a good candidate for the construction of towers that produce wind power simultaneously. The construction of both types of electric production units at the same location would bring advantages from economies of scale as, for example, common generators, transmission lines, and maintenance crews.

11.5 Ocean Thermal Energy Conversion (OTEC)

The concept of Ocean Thermal Energy Conversion is almost as old as the theory of heat engines. The OTEC concept is included in most elementary Thermodynamics textbooks as a demonstration of the consequences of the Second Law of Thermodynamics: In tropical latitudes, water at the surface of the sea may reach temperatures in the range 27–32°C. At the same time, and because the sunlight does not penetrate at depths below 30 m, water in the bottom of deep bays is colder and may be in the range 5–8°C. These conditions create the two heat reservoirs that are necessary for the operation of the elementary heat engine, which is depicted in Fig. 3.7. The warm sea water at the surface and the cold sea water at the bottom constitute the two heat reservoirs that are necessary for the operation of a heat engine. All that is needed is the engineering system that would make this heat engine operational.

From the outset it must be noted that the efficiency of the OTEC heat engine would be very small. Even if we take the extreme temperatures mentioned in the previous paragraph, 32 and 5°C, the Carnot efficiency of the OTEC cycle would be less than 9%. In contrast, the Carnot efficiencies of gas cycles, where the higher temperature is in excess of 1,500°C, are higher than 80% and those of the Rankine cycles are close to 70%. The inherently very low Carnot efficiency of the OTEC cycle implies that the actual efficiency of OTEC systems would be even lower. Thus, the expected range of actual OTEC power plants is in the range 1–2%. However, the amount of heat that may be extracted from the water of the oceans is almost inexhaustible and free. Despite the low expected efficiency, a successfully designed OTEC power system would be capable to provide significant amounts of electric power in an environmentally friendly way.

11.5.1 Two Systems for OTEC

One way to utilize the temperature difference in the oceans is to use a simple Rankine cycle with a refrigerant substance as the working fluid. A schematic diagram of the components of the Rankine cycle is shown in Fig. 3.8 and the thermodynamic, *T- s* and *P-v*, diagrams of the cycle are shown in Figs. 3.9a and b respectively. A pump is used to pressurize liquid refrigerant (process 2–3 in the diagrams). The fluid is then directed to a heat exchanger at the surface of the sea, which acts as the boiler, and the refrigerant evaporates (process 2–3). The high pressure vapor is directed to a turbine, which produces power (process 3–4). Finally the vapor from the exit of the turbine passes through another heat exchanger, at the bottom of the sea, which acts as the condenser of the system (process 4–1). The condenser exhausts at the pump inlet and the cycle is repeated. A variation of this operation is to have the condenser close to the surface or onshore and to pump cold water from the bottom of the sea to the condenser, instead of directing the vapor to the bottom and then pumping the refrigerant liquid to the surface.

While the Rankine cycle is a simple way to convert the temperature difference of the sea water to electricity, having the two heat exchangers, each of which requires a finite temperature difference to operate effectively, is a significant drawback. When one allows for a reasonable temperature difference in each heat exchanger, typically 5–10°C, the range of temperature that becomes available for the operation of the working fluid is very narrow and the Rankine cycle is not practical to operate. Also the two heat exchangers typically have very large areas because a great deal of heat must be transferred through them in comparison to the power produced. Assuming an average cycle efficiency of 2%, an optimistic figure given all the constraints of the cycle, the boiler of a 1 MW OTEC power plant must exchange 50 MW of heat with the sea water. Similarly, the condenser must reject 49 MW of heat, which implies that the two heat exchangers would have very large surface areas. In addition, any surface fouling of the heat exchangers would adversely affect the rate of heat transferred to the detriment of the electric power produced.

The *Claude cycle* was developed by Georges Claude, who actually built an OTEC power plant in Matanzas Bay, Cuba in 1929. This cycle does not use any heat exchangers and uses as the working fluid water vapor, which is produced from sea water. It is an open cycle similar to the *single flash cycle* used with the geothermal power plants. A schematic diagram of this cycle is shown in Fig. 11.11 and the corresponding *T*-s diagram of the cycle is shown in Fig. 11.12. Warmer water at state 1 enters the low-pressure evaporator where it flashes and releases vapor at state 2. The produced vapor, at state 2”, is separated by gravity from the liquid water, at state 2’, and is directed to the turbine, where it expands to the lower pressure of the condenser and reaches state 3. The expanded steam enters a direct-contact condenser, where it condenses to state 4 and is pumped back to the

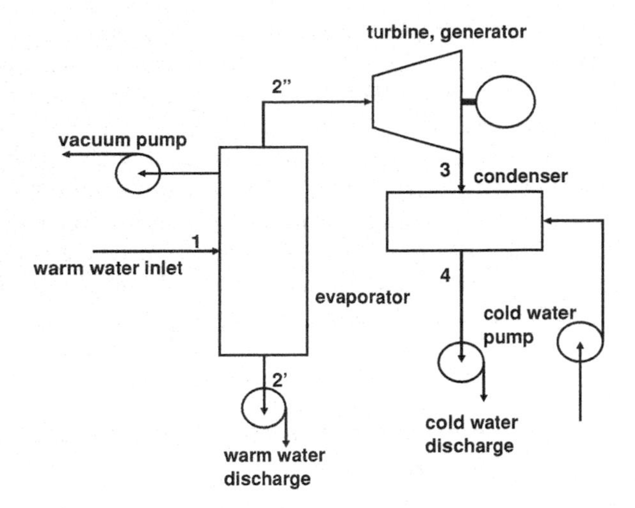

Fig. 11.11 Schematic diagram of the Claude cycle for OTEC

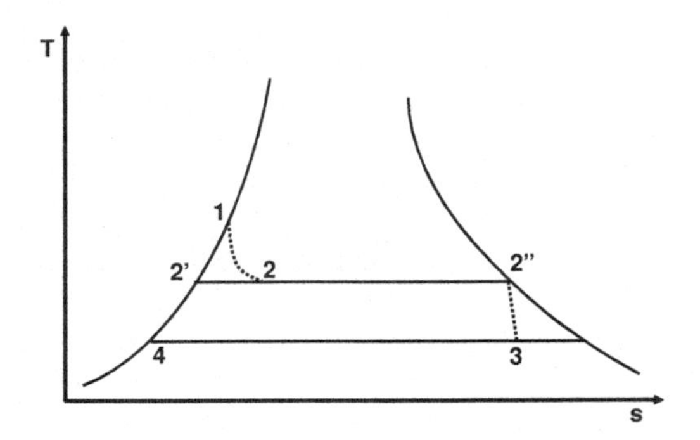

Fig. 11.12 Thermodynamic T-s diagram of the Claude cycle. The states of the working fluid correspond to the states in Fig. 11.11

sea. The remaining warm liquid water at state 2' is also pumped out of the evaporator.

The evaporator, where the flashing occurs, may be modeled as an adiabatic system, because any heat transfer from this equipment is negligible in comparison to the enthalpy of the warm water that enters. Applying the First Law of Thermodynamics to the evaporator, we may calculate the amount of vapor produced per unit mass of the water flow rate:

$$h_1 = (1 - x_2)h_2' + x_2 h_2'' \Rightarrow x_2 = \frac{h_1 - h_2'}{h_2'' - h_2'}. \tag{11.22}$$

Furthermore, and since the specific work of the turbine is h_2"-h_3, the total power produced by the Claude cycle is:

$$\dot{W} = \dot{m} x_2 (h_2'' - h_3), \tag{11.23}$$

where $\dot{m}$ is the entire mass flow rate that enters the evaporator at state 1. The flashing temperature, T_2 may be determined by maximizing the amount of power produced by the OTEC cycle. As with the geothermal flashing cycles, in an optimized Claude cycle, T_2 is approximately equal to the average of T_1 and T_3.

The Claude cycle necessitates the use of several pumps that require a significant amount of auxiliary power. Theoretical efficiencies of this cycle may be as high as 4%, but the practically achieved efficiencies are not expected to be more than 1–2%. One of the advantages of the Claude cycle is that the machinery does not need to be anchored and may operate aboard a ship. The produced power must then be transmitted onshore by a cable.

11.5.2 Environmental Effects of OTEC and Other Considerations

OTEC is another form of clean and renewable energy, which does not cause atmospheric pollution. Because the relative temperature differences between surface and sea-bottom water are present throughout the year, OTEC may become a continuous and rather reliable source of electric power. There are very low-impact and rather insignificant environmental and ecological effects associated with electricity production from OTEC, namely the local disturbance of aquatic life and, possibly, small impediments in the local ship navigation. Because the warm OTEC resources exist primarily in smaller bays, all environmental and ecological effects are localized in the vicinity of the OTEC power plant.

All OTEC cycles suffer from the inherent disadvantage of the low temperature differences and the implied low thermodynamic efficiencies. For this reason, all processes must be optimized and the equipment must be maintained in good condition. This becomes problematic in the ocean/coastal environment: Salty water causes corrosion and deterioration of the equipment, while the growth of algae tends to foul the fast heat exchanger surfaces, pipelines, and submerged water pumps. These are the main reasons why the few OTEC pilot plants that were built on shore or on barges have failed and were decommissioned after a few months of operation.

11.6 Types of Water Power Turbines

Water turbines are the most important energy conversion equipment in hydroelectric, tidal and ocean current power plants. Water power could not have been utilized at its present rate if it were not for the modern water turbine. The characteristics of water power are such that the water turbines must be versatile and must operate at low as well as at high heads ($g\Delta z$) as well as at low and high power

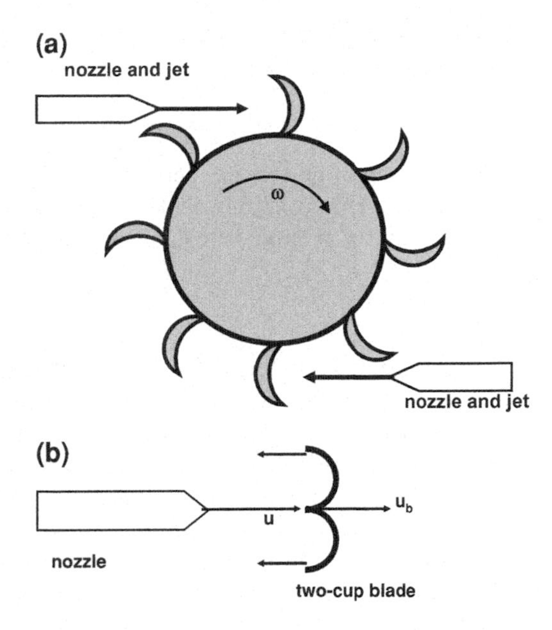

Fig. 11.13 **a** Schematic diagram of the Pelton wheel. **b** Schematic diagram of the two-cup blades of the Pelton wheel

outputs. Two basic types of water turbines the *impulse turbines* and the *reaction turbines* are employed for water power conversion. The impulse turbines utilize solely the kinetic energy of the incident fluid for the production of power, while the reaction turbines utilize primarily the higher static pressure of the fluid.

The *Pelton wheel* is a typical impulse turbine. Figure 11.13a shows a schematic diagram of the Pelton wheel, which for all practical purposes is a waterwheel with specially designed water caps. Two high velocity water jets ensuing from diametrically opposed nozzles strike the blades of the Pelton wheel in opposite directions. The velocity of the water jets is controlled by varying the exit area of the nozzles with spear valves. The blades of the Pelton wheel consist of two cups that are joined at one end and are designed to deflect the water by 180° and, thus, achieve the maximum possible change of the momentum of the incident water jet. A more detailed schematic diagram of the blades, which includes the water velocities, is also shown in Fig. 11.13b.

Let us consider the water jet with a volumetric flow rate $\dot{V}$ ensuing from the nozzle and moving toward the two-cup blade of the Pelton wheel. The absolute velocity of the jet is u. The blade also moves with absolute velocity u_b and, hence the velocity of the water relative to the blade is $(u\text{-}u_b)$. From mass conservation, the relative velocity of the water, as it leaves the blade is $-(u - u_b)$. Therefore, the total change of the water velocity relative to the blade is $-2(u-u_b)$ and the total change of momentum is $-2\rho\dot{V}(u - u_b)$. From the momentum conservation principle the latter must be equal and opposite to the force acting on the blade, that is:

$$F_b = 2\rho\dot{V}(u - u_b) = 2\dot{m}(u - u_b). \tag{11.24}$$

Since the blade moves with absolute velocity u_b, the power generated by the blade because of the jet action is:

$$\dot{W}_b = 2\rho\dot{V}(u - u_b)u_b. \tag{11.25}$$

For maximum power, the derivative of the last expression with respect to u_b must vanish. This condition yields $u_b = 0.5u$ and, hence, the maximum power generated by the jet may be calculated from Eq. (11.25) to be:

$$\dot{W}_{max} = \frac{1}{2}\rho\dot{V}u^2. \tag{11.26}$$

A glance at the last equation proves that the maximum power that may be generated by a Pelton wheel is equal to the rate of kinetic energy carried by the water jet. The actual power generated by a Pelton wheel will be lower because of irreversibilities (primarily friction) associated with its operation. In practical operations, these irreversibilities are accounted for by the efficiency of Pelton wheels, which is in the range 75–80%.

Of the reaction turbines, the *Kaplan turbine* and the *Francis turbine* are not fully reaction turbines because they utilize both the pressure and the velocity of the incident fluid for the production of power. The impeller or runner of the Kaplan turbine resembles a propeller, while that of the Francis turbine is a spiral annulus, which diverts the water flow. The *degree of reaction*, *R*, defines the ratio of the specific energy of water that is converted into power to the total specific energy, *e*, converted[2]:

$$R = \frac{P_i - P_o}{\rho e} = 1 - \frac{u_i^2 - u_o^2}{2e}, \tag{11.27}$$

where the subscripts, *i* and *o*, denote the inlet and outlet of the turbine. The second part of Eq. (11.27) is a consequence of the application of Bernoulli's equation between the inlet and the outlet. One may also write the last part of Eq. (11.27) in the following form, which gives directly the specific energy of the water, $e = e_o - e_i$, which is converted to power in terms of the turbine inlet and outlet conditions.

$$\frac{P_i}{\rho} + \frac{u_i^2}{2} = e + \frac{P_o}{\rho} + \frac{u_o^2}{2}. \tag{11.28}$$

Regarding the choice of water turbines for specific applications, in general the Kaplan turbines are preferred in installations with high volumetric flow rates and low hydrostatic heads. Hydroelectric power plants with high volumetric flow rates and heads less than 50 m, would use Kaplan turbines. The Pelton wheel is ideal for

[2] The fluid energy in this case is actually equal to the exergy of the fluid and the difference, $e = e_o - e_i$, is the difference in exergy. For this reason, the symbol *e* is used for both quantities.

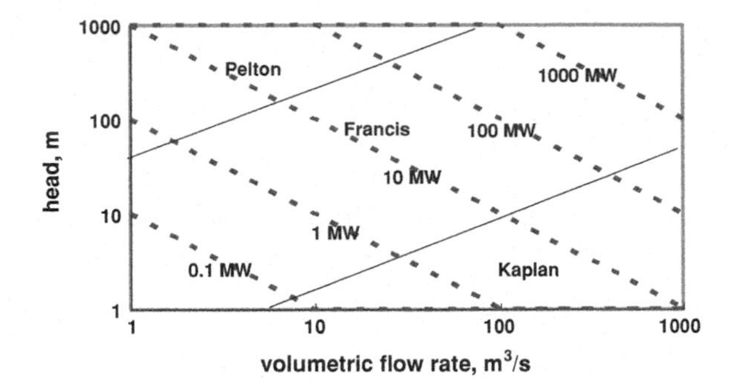

Fig. 11.14 Diagram for the choice of water turbines according to the volumetric flow rate and predominant water head

applications that involve high heads, up to 1,000 m, and low flow rates, less than 100 m^3/s. For all conditions between these two limits, the Francis turbines are used, which can handle low volumetric flow rates at relatively low heads as well as high volumetric flow rates at relatively high heads. Figure 11.14, depicts a diagram for the typical uses of the three types of water turbines.

11.7 Concluding Remarks on Water Power

It becomes apparent, when one considers the several forms of water power, that this is a significant but underutilized energy resource. Despite the clear environmental advantages and the enormous potential of surface water to produce a significant percentage of the electricity demand, of the several types of water power plants that may be put in operation, only the hydroelectric power plants have been extensively used. There is significant technological expertise with hydroelectric power plants that extends to more than 120 years. The technology of these plants is considered proven and reliable. For this reason, the construction of large-scale hydroelectric power plants is expected to continue in the developing countries, including the emerging economies of India, Brazil, and the Peoples Republic of China. In the OECD countries, where most large-scale hydroelectric resources have been utilized, the next expansion of water power plants will take place with the small-scale and micro-scale plants, which have the capacity to contribute significantly to the electric demand of most countries.

However, the current situation and prospects are entirely different with the other types of water power conversion plants: Several prototype and pilot projects that utilize tidal, wave, and OTEC power have been built, but very few power plants that utilize these resources are currently in operation. Regarding ocean current plants, there have been several studies with good results but no actual projects. There are several economic, social, and technological reasons for this experience, some of which are summarized in the following paragraphs.

The primary reasons for the lack of more water power projects are economic and financial. Water power projects have long lives, very low variable costs (no fuel cost and low maintenance cost) but very high capital cost. The high capital cost associated with the construction of the massive dams, barrages, water turbines, wave engines, and OTEC plants are very high and must be amortized over the long life of the projects. In most of the cases, there have been no loan guarantees for the construction of these types of power plants and the owner or operator had to assume all the risks of the project. Therefore, any private corporation that undertakes the construction of a water power plant must place a great deal of capital and borrowing capacity at risk. In addition, lack of wide technological experience and available operational data with water power plants other than hydroelectric, makes the tidal, OTEC, wave, and ocean current plants high risk ventures, which are subject to very high insurance premiums. The high initial capital cost and insurance in general result in the high cost of the electricity produced by most water power plants, except from hydroelectric plants. It has become apparent with all water power projects that if individual pilot projects are not built first in the OECD countries, which have the capital structure and technology to build and put them successfully in operation, the very high potential of all water power will not be utilized in the entire world.

Tidal electric power production has had a significant success with the 240 MW power plant in La Rance, France. However, the low fossil fuel prices of the 1980s and 1990s have hampered the construction of more tidal power plants in other countries. Another tidal project across the English Channel in the Severn estuary, England, did not progress beyond a few feasibility studies. Despite the great promise, which is apparent in all the feasibility studies, there has not been a state or private entity that would undertake the construction of this project. The economic and financial reasons presented in the previous paragraph are the main reasons for the lack of more activity in tidal power.

The utilization of ocean water currents has not progressed beyond a few academic and feasibility studies. Lack of any operational expertise; the essentially unknown environment where ocean current sites are most energetic; and the risk-averse attitude of energy entrepreneurs contribute to the lack of any projects. Clearly, a great deal of research and development is needed in this technological area, which must come from state investment. One of the development areas is the construction of reliable underwater turbine-generator systems. Because of the very high potential of ocean current power on several sites worldwide, these ventures may prove very successful investments in the future, if their expected operational data are verified and validated at the feasibility stage.

Wave power has been produced by several experimental and pilot power plants at very small scale. A major deterrent of the further expansion of this type of power is the structural reliability of the equipment used and their survivability during major storms. A great deal of research and development is needed in the characterization and design of materials and equipment that would be used with wave power plants. There are currently several wave power engine designs and a few pilot projects on the coastal areas of England, Norway, and Ireland. If these pilot projects prove to be successful and reliable in the near future, we will see a significant development of wave power before 2030.

OTEC power has received significant attention especially in USA and Japan. However, the few pilot plants that were built in the past have not proven to be reliable and were not successful in delivering their rated power continuously and for a long time. The main problem with the OTEC plants is equipment deterioration resulting from biological and chemical fouling and the structural reliability of the equipment with sea storms. In addition, the inherent very low power plant efficiencies limit significantly the power that may be produced from an OTEC installation. Despite the very high global potential of OTEC power and the fact that it is a continuous and very reliable energy source, expansion of OTEC activities are not expected to happen unless a significant technological breakthrough is achieved with the heat exchangers that are needed.

In conclusion, in the near future hydroelectric power is expected to play a more important role in the generation of electricity from renewable sources and will continue to be a viable and economic substitute for fossil fuels. Of the other water power sources, a few tidal power plants may be built on prime locations followed by several low-scale wave power pilot projects and demonstration plants. However, and despite their very high potential, it is rather unlikely that many ocean current and OTEC power plants will be built in the first part of the twenty-first century or that they will contribute significantly to the future global energy demand.

Problems

1. A small river is to be dammed to create a hydroelectric power plant. After the completion of the dam the river level will rise by 28 m. The river has a cross-sectional area 340 m^2 and the average water velocity is 0.6 m/s. If the combined turbine-generator efficiency of the power plant is 0.75, what is the power this river may produce?
2. The maximum water-level difference in the Three Gorges dam is 110 m. How much water flow rate (in kg/s and m^3/s) is required from the river to produce the rated power of the plant if the combined turbine-generator efficiency of the power plant is 0.76?
3. How much annual CO_2 emissions (in tons per year) would be avoided if the USA doubles its hydroelectric capacity and reduces the coal-fired power plants at the same time? The average overall thermal efficiency of the coal-fired power plants is 36% and the heating value of carbon is 32,700 kJ/kg.
4. Small waterfalls of 2–5 m exist in several locations on the planet. Typical water flow rates in these waterfalls are 2-100 m^3. If you were to divert 50% of the water for electricity production, what would be the upper and lower range of the power that would be produced from such waterfalls? A typical efficiency of the turbine-generator pair is 75%.

5. If the proposed tidal power plant in Cobequid, Canada is to produce 20,000 MW of average power what would be the cross-sectional area of the dam to be constructed?
6. The average range of the tides in the Gulf of Mexico is 0.6 m (2 ft). It is proposed to construct a 1.5 MW tidal power plant for the process of fish and shrimp. If the height of the barrage is not to exceed 0.7 m, what would be the length of the barrage?
7. In several places the Gulf current is equal to or exceeds 4 knots. If a small water turbine with blade diameter 3.2 m were placed in these, what would be the maximum power this turbine would produce? How many of these water turbines are required to produce the equivalent of one typical nuclear power unit (1 GW)?
8. It is proposed to place 5,000 small water turbines with blade diameters 3.5 m at the bottom of the ocean and inside the Gulf stream, where the average velocity is 2 m/s. What is the maximum power these turbines would produce? What type of engineering and technological difficulties do you expect this scheme may have in the short- and the long-run?
9. The waves in the North Sea have average amplitude 3 m and wavelengths 30 m. What is the total power of these waves, the power per unit width and the power per unit area?
10. Waves of 30 m height and 60 m wavelength have been observed during adverse weather conditions in the Gulf of Mexico (e.g. during hurricanes and tropical storms). What are the total power and the power per unit area for these waves? What are the engineering problems that would be encountered in harnessing these waves?
11. What is the maximum power that may be produced from an OTEC cycle when the upper temperature is 34°C, the lower temperature is 6°C and the cycle receives 180 kW of heat? How much heat is rejected from this cycle?
12. Refrigerant 134a is used as the working fluid in an OTEC power plant, which utilizes a simple Rankine cycle without superheat. The condenser is at 12°C and the boiling of R-134a occurs at 28°C. The efficiency of the turbine is 76%. What is the thermal efficiency of this cycle and how much heat input is required for the cycle to produce 0.5 MW?
13. What is the actual power that may be produced from an OTEC Claude cycle when the high water temperature is 34°C, the lower temperature is 10°C and the cycle receives 200 kW of heat?
14. "The oceans comprise more than 70% of the surface of the Earth and may provide an unlimited amount of power from ocean current and thermal-ocean energy. We simply have to invest in this vast potential and soon we will have energy independence." Comment with an essay of 300–400 words.

Chapter 12
Energy Storage

Abstract In the beginning of the twenty-first century, our society mostly uses energy that has been accumulated for millennia: Fossil fuels have stored in their chemical compounds vast amounts of energy, which we now use for our primary energy needs, unfortunately, at a very high rate that cannot be sustained. When fossil-fuels are exhausted or when their use is curtailed because of environmental factors, such as the production of greenhouse gases, humans will have to rely more on alternative energy sources among which are nuclear, wind, and solar power. While the production of power by nuclear energy may be controlled, wind and solar power are intermittent or periodic sources, which may produce power when there is limited demand and may not produce power at all, when demand is high. For example, during the night hours of July 17th, when there is high demand for air-conditioning and electricity in the northern hemisphere, there is no Sun and the average wind power is significantly lower than average. If the society relies on alternative energy sources and stored energy is not available during this night, humans will experience a significant shortage of energy with the inconvenience this entails. As the human society moves more to the direction of alternative energy sources, the need for energy storage becomes apparent and more acute in order to satisfy the fluctuating energy demands. This chapter provides an exposition to the several energy storage methods that are currently used to smoothen the electric power demand and those that may become feasible and popular in the future. First, the electric energy demand patterns in a contemporary society that makes significant use of air-conditioning in the summer are explained. Secondly the various methods for energy storage, such as electromechanical, thermal—in both latent and sensible heat form—and chemical are explained with the devices/ systems that make energy storage possible. The alternative method of storing "coolness" with materials that exhibit a temperature hysteresis in their melting/ solidification processes is elucidated within the context of thermal storage methods. Finally, details are given on the possible switch to hydrogen as a clean energy storage medium, the long advocated "hydrogen economy" and the inherent

E. E. (Stathis) Michaelides, *Alternative Energy Sources*,
Green Energy and Technology, DOI: 10.1007/978-3-642-20951-2_12,

advantages of Fuel Cells as devices for the direct conversion of chemical energy to electricity.

12.1 The Demand for Electricity: The Need to Store Energy

It is apparent from our daily activities that the societal demand for electricity is not constant, but is subjected to diurnal, weekly and seasonal fluctuations: In general, electricity consumption is higher during the day and falls off during the night, when the majority of the population retires. On a given day, there are demand peaks that coincide with several daily activities of the population, such as the preparation of the daily meals, etc. In a typical week, there is higher demand during the working days and lower during Saturdays, when many of the offices and businesses close. The lowest demand is on Sundays, when most of the businesses close and most of the population does not work. In the United States and the OECD countries, the widespread use of air-conditioning has increased significantly the use of electricity during the hot summer season. This has resulted in a significant peak of electricity demand in the early afternoon hours, when ambient temperatures are higher and the need for air-conditioning is heightened. Figures 12.1 and 12.2 depict the hourly electric power demand during two typical winter and summer days in San Antonio, Texas. The figures include weekday days and weekends. The data were supplied by CPS Energy

One may draw the following conclusions from these two figures:

1. On a given working day, for both winter and summer, the peak daily electric power demand is approximately twice the minimum demand.
2. During the summer, the peak work-day demand is approximately three times more than the minimum demand on Sundays.
3. On a given day, the summer power demand is more than 60% higher than the winter demand on a similar day.
4. During the early hours of the morning, the electric power demand may almost double within a three-hour period (6:00–9:00 am). Similarly, in the late evening the electric power demand drops by almost 50% within a 4–6 h period.

Since electricity cannot be stored, all electric power demand must be met almost immediately on the supply side by the electric power plants, which are connected to the electricity grid. As a result of the frequent and wide fluctuations in the demand for power, the power plants must be able to meet satisfactorily the demand fluctuations and provide power to the consumers. For this reason, all the power plants do not operate continuously and, when they operate they often operate at reduced and not at full power. Several parameters that characterize the capability of a power plant to meet the electric power demand are:

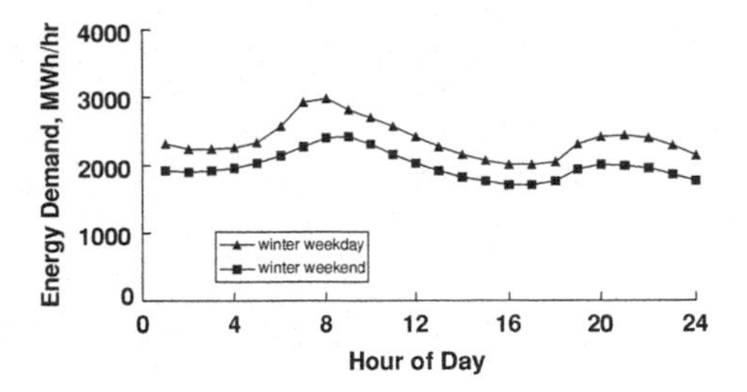

Fig. 12.1 Typical electric power demand during the winter in San Antonio, Texas

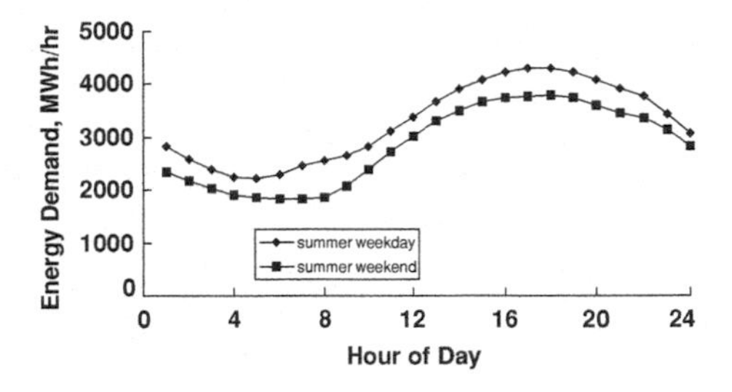

Fig. 12.2 Typical electric power demand during the summer in San Antonio, Texas

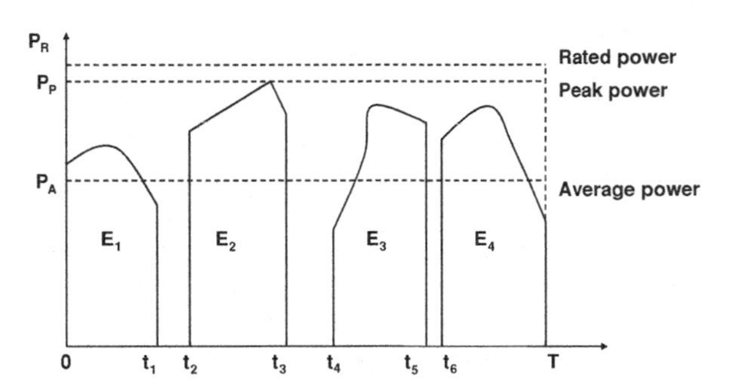

Fig. 12.3 Variable power parameters of a power plant

a) The *rated power* is the nominal power a plant may produce.
b) The *peak power* is close to the rated power and, depending on the type of the plant and the availability of cooling water, may be a few percentage points higher or lower than the rated power.
c) The *average power*, an average usually taken over a period of a year.
d) The *availability factor* is the fraction of time that the plant is on line and may contribute to the electric power demand, regardless of the level it contributes.
e) The *power operating factor* or *plant capacity factor* is the ratio of the total electric energy generated to the total capacity to generate energy during a time period. This takes into account the fluctuations of the energy produced during a given period.

Figure 12.3 shows schematically these parameters for the operation of a hypothetical power plant during the time interval [0, T], which is typically one year. The power plant is on line and produces power during the time intervals

[0, t_1], [t_2, t_3], [t_4, t_5]. and [t_6, T] and is off-line (e.g. for maintenance or because power is not needed) during the time intervals [t_1, t_2], [t_3, t_4], and [t_5, t_6]. The rated, peak and average power are shown in the figure. The availability factor of the power plant, AF, is defined by the expression:

$$AF = \frac{t_1 + (t_3 - t_2) + (t_5 - t_4) + (T - t_6)}{T}. \quad (12.1)$$

The areas in the polygons and the symbols E_1, E_2, E_3, and E_4 denote the total energy produced by the plant during the time intervals [0, t_1], [t_2, t_3], [t_4, t_5], and [t_6, T] respectively. The availability factor is a measure of the faction of time the plant contributes to the production of electricity. The power operating factor (or capacity factor), *POF*, is a measure of the average power produced by the plant throughout a year and is defined in terms of the rated power, $\dot{W}_{rat}$, as:

$$POF = \frac{E_1 + E_2 + E_3 + E_4}{\dot{W}_{rat} T}. \quad (12.2)$$

It is apparent that the POF is different than, simply, the average power divided by the peak power or the rated power of the power plant. It is also apparent that the total energy produced by this power plant is significantly different than the total energy the plant *could have produced* if it operated continuously at its rated power during the time interval [0, T].

The availability factor of the power plants depends significantly on their type and design: Large steam power plants operating with a steam cycle, such as nuclear and coal power plants may be damaged if they are started and stopped often (taken *out of line*) and must remain in operation for long periods of time. Some of these power plants may operate at reduced rather than peak power for long intervals. Smaller power plants, especially those operating with a gas Brayton cycle—the so-called gas turbines—may be taken in and out of line often, and are suitable to be used to meet the high frequency fluctuations of the electric power demand.

The availability of solar, wind, and wave power plants depends greatly on the availability of the respective resource: a solar unit does not produce any power during the night, a wind turbine is motionless when there is no wind and an engine powered by waves, does not produce any power during calm seas. A second factor that affects the availability of a power plant is the *operational cost* of the plant, which primarily depends on the cost of the fuel. Units that are expensive to produce electricity are taken off line more often, while units that produce power cheaper operate for as long as it is feasible. In terms of energy generated per kg of fuel (e.g. kWh/kg of fuel) nuclear fuel and coal are less expensive than other fuels. Thus, nuclear and large coal power plants are typically used in the production of electricity. Their availability is higher than the availability of gas power plants, which use a more expensive fuel in terms of kWh of electric energy produced per unit currency ($) of fuel cost. The often quoted rate, $/kWh or ¢/kWh (cents per kWh), is one of the determining factors of the availability of a power plant. Based on these considerations, the power plants are divided in the following categories:

1. *Base-load plants,* which are large, very efficient units operating with Rankine cycles and, in general, utilize nuclear fuel or coal. These units operate almost continuously, except for periods of maintenance. During periods of lower demand (e.g. nights) their load is reduced in the range of 50–70% of rated power, but the plants are not taken off the system. Typical POF's for these plants are close to 75%.
2. *Intermediate or cycling plants,* which are usually older, less efficient steam units or gas-cycle units specifically designed to be taken in and off line relatively often, have a POF in the range 25–50% and typically come in line and supply power during the day-time.
3. *Peak power or peaking plants,* are designed to come on line and off line often, sometimes two or three times during a single day. They are usually Brayton cycle units operating with liquid or gaseous fuel, diesel or natural gas. Electric energy produced from peaking units is more expensive than that from nuclear or coal units. The peaking units produce a relatively small amount of electric energy at higher average cost, but they are very important because they enable the utility to meet the peak demand of its customers. Peaking units may be in operation for as little as two hours during a week and their POF is in the range of 5–15%.

The main objectives of the electricity production companies (utilities) are:

a) To satisfy the demand, and
b) To produce electric power at minimum cost.

Because peaking units produce electric energy more expensively, these corporations follow energy management methods that minimize the production of power from the more expensive peaking units. Among the methods presently used for the management of the power produced are the following:

1. Incorporate into the electric grid several geographic areas where peak demand occurs at different times.
2. Charge the consumer higher prices for peak power and supply the households with *smart meters*, which reveal the price charged to the consumer. Naturally, the majority of the consumers curtail their demand for power during periods when the price of energy is high, by conserving energy or switching off appliances that may be used at another time. For example, during a high-price interval, a home owner may raise the temperature of the air-conditioning thermostat or postpone the use of the dish-washer, the washing, and dryer machines for the late evening, when demand is less and the electricity price is lower.

3. Shifting the demand by 10–15 min, by delaying the starting of the operation of devices that consume significant amounts of power. This is achieved when the electric power production corporation remotely controls high powered units (e.g. air-conditioning units, industrial refrigeration units, etc.) and arranges for their staggered operation in a way that power demand is smoothened across the area supplied with electricity. When the shift is short (10–15 min) the effect of this practice to the consumer is almost imperceptible.
4. Use of units that have lower operational cost all the time as base load units and allow more costly units to run only when necessary.
5. Construct smaller, low capital cost units as peaking units, even though the fuel cost of these power plants is more expensive.
6. Use energy storage systems to produce power at low cost and use the power later, when demand peaks.

Since the 1990s energy storage systems are becoming more important in the modern society. The realization of the adverse effects of carbon dioxide accumulation in the atmosphere has put significant pressure on governments, corporations and individuals to "*become greener*" by using a higher percentage of non-polluting, renewable energy sources in their mix for power production. Among the most popular choices are wind power and solar energy. However, during a windless summer night, neither a solar nor a wind power plant is capable of providing the necessary energy to run the air-conditioning systems and other household appliances our society uses. Total or even highly increased reliance on these two sources would result in prolonged energy shortages that are unacceptable to the society. Therefore, a significantly increased solar and wind power capacity by constructing more solar and wind units by itself, would not solve any of today's energy problems, unless it were accompanied by the construction of systems that store this energy for later use. With energy storage systems, some of the energy produced during a windy evening in the spring or during a sunny summer day may be stored to be used during the windless summer night. Thus, energy storage and energy storage systems are becoming very important in the increased utilization of the intermittent and periodic alternative energy sources, such as wind, solar, tidal and wave energy. The most frequently used systems for the storage of energy—electromechanical, thermal and chemical systems—will be presented in a systematic way in the next few sections. Hydrogen energy storage is essentially chemical, but will be treated in a separate section because of its enormous potential to store and supply very large quantities of energy and because of the transformational implications of hydrogen as the fuel of the future.

12.2 Electromechanical Storage

Electromechanical energy storage systems store energy as potential energy, kinetic energy, or electric energy. Chief among them are above and below ground water pumping, compressed air, mechanical springs, flywheels, superconducting coils, and ultra-capacitors.

12.2.1 Pumped Water

Water is abundant, inexpensive, easily transportable in pipelines and, among the fluids, has relatively high density. One cubic meter of water has a mass of approximately 1 ton. At a height of 360 m, the potential energy of this volume of water is 3.6 million J or 1 kWh. A natural or artificial lake or even a smaller artificial pond may store millions of cubic meters of water for very long periods of time. This makes the transformation of the potential energy of water an ideal medium for energy storage. Typical power plants with pumped-water energy storage units are located in valleys, close to high hills or mountains. If a natural lake does not exist, an artificial lake or pond is constructed, at or near the top of the hill or mountain. During periods of low demand, e.g. nights, the pumps transport the water to the lake or pond. This mass of water, which has significantly high potential energy may be used in the future, when peak power is demanded by the consumers. Oftentimes, the pump-motor systems that are used for the pumping of water may be reversed to operate as turbine-generator systems that produce the electric power. Thus, separate systems are not needed for the storage and the production of power. This reduces significantly the capital costs of energy storage and utilization. A schematic diagram of a typical pumped water system is shown in Fig. 12.4. Excess energy produced in the power plant is transmitted to the pumping/generating station. Water from a nearby river or lake is pumped via a simple pipeline to the artificial lake at a significantly higher elevation. The pumped water has high potential energy, which may be converted to electric power in the generating station by reversing the flow in the pipeline.

A 100×100 m^2 artificial lake with an average depth of 5 m, may store 50,000 cubic meters of water. At an elevation of 800 m, the potential energy stored in such a lake is $4*10^8$ kJ or 110,000 kWh. Typically, there will be minor energy losses during the transmission of the water. The most significant energy losses are at the hydraulic turbine, which is used in the generating station. Since typical turbine/generator efficiencies are in the range 70–72%, the energy that may be produced from such an artificial lake is closer to 77,000 kWh. However, these are not the only energy losses for the pumped water storage system: the pumping of the water is achieved by motor/pump systems, which have efficiencies also in the range of 70–72%. For the pumping of the 50,000 cubic meters of water to the higher elevation of the artificial lake, the energy consumption is approximately

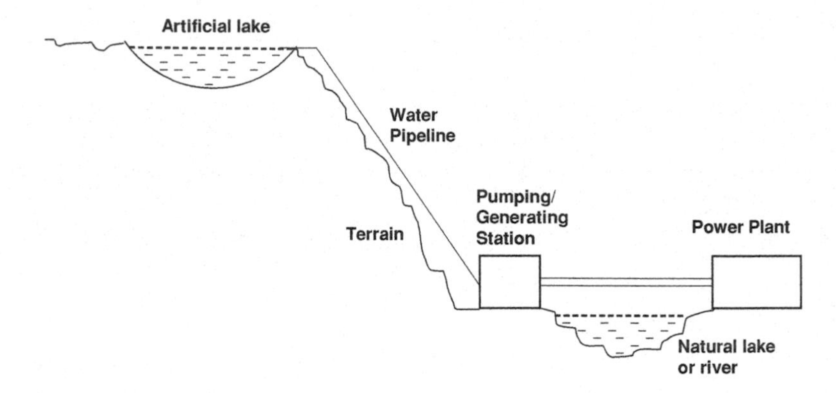

Fig. 12.4 Above ground pumped water system

110,000/0.7 or 157,000 kWh. Therefore, the electric energy recouped from this storage system is close to 50% of the energy used to produce it. Thus, the typical *storage efficiency* of the pumped water system is close to 50%, and of course this means that only half of the energy that is stored is finally recovered.

Despite the rather low storage efficiency, pumped water systems present a viable alternative to energy storage because:

a) Pumped systems may store a significant amount of energy (millions of kWh) at relatively low capital cost, and
b) The energy may be stored for long time periods.

Except for a small amount of water that may be lost to evaporation, there are no other losses. Oftentimes the evaporated water is recouped and even enhanced from rainfall water.

An alternative concept to storing water above ground is to store water in underground caverns. In underground storage, the pumping/generating station is located deep in a cavern and water flows to it from the surface gravitationally. When the power plant produces electric power in excess, the pumping station retrieves water from the cavern and stores it at the Earth's surface, close to the power plant. Underground energy storage has the same advantages and the same storage efficiency as the above ground storage. Some disadvantages of underground storage are:

a) Underground caverns sometimes leak and a lot of stored water is lost in the ground fissures
b) The storage capacity is limited to the size of the cavern
c) Power must be transmitted from the power plant to the underground pumping station, and
d) Access and maintenance of the underground pumping/generating equipment is difficult and more expensive.

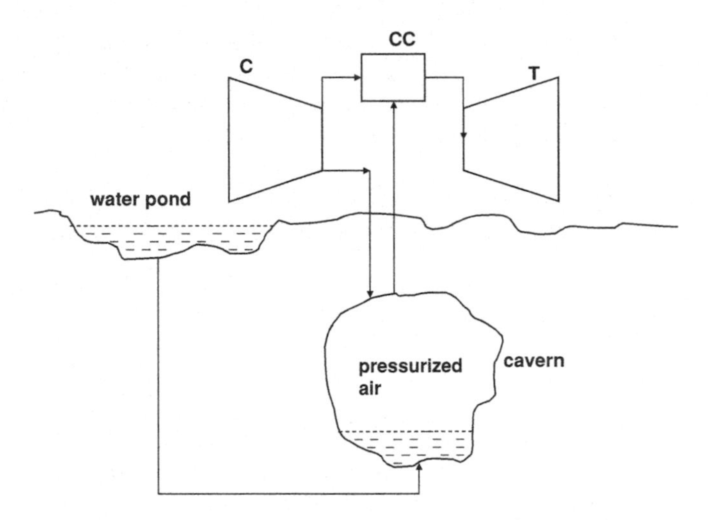

Fig. 12.5 Energy storage with compressed air

12.2.2 Compressed Air

The concept of compressed air storage working in conjunction with a Brayton power cycle and with a water reservoir as pressure regulator is demonstrated in Fig. 12.5. When there is no demand for its power, the gas turbine (T) simply supplies enough power to drive the compressor (C). In this mode of operation the function of the compressor is to pressurize and transport the air in the cavern. Any displaced water from the cavern is stored in the above-ground water pond. The displaced water helps maintain the higher pressure in the cavern, but is not essential for the energy storage. During hours of high power demand, the pressurized air is extracted from the cavern, is fed to the combustion chamber (CC) and finally to the turbine, thus augmenting the power produced by the Brayton cycle. A variation of this method of energy storage is to feed the compressed air directly to the turbine without heating in the combustion chamber.

The maximum amount of work that may be extracted from the combination of the pressurized air and water system may be obtained from the exergy of the air–water system. Writing the exergy of this system in terms of the volume of the cavern, V_c, the total exergy of the air–water system is:

$$E - E_0 = PV_c\left[\frac{c_v}{R}\left(1 - \frac{T_0}{T}\right) + \left(\frac{P_0}{P} - \frac{T_0}{T}\right) - \frac{T_0 c_p}{RT} ln\frac{T}{T_0} + \frac{T_0}{T} ln\frac{P}{P_0}\right] + V_c \rho_w g H, \tag{12.3}$$

where H is the average depth of the cavity, measured from the surface of the water reservoir; T, P, and ρ_a are the temperature, pressure, and density of the air respectively under the conditions prevailing in the cavity; T_0, and P_0 are the ambient temperature and pressure conditions, R is the gas constant; c_v is the specific heat of air at constant volume, which is 0.72 kJ/kgK and ρ_w the density of water. The last term in Eq. (12.3) represents the work that may be extracted when the water level is lowered to the level of the cavern with the use of hydraulic turbine at the cavity level.[1]

Salt caverns, hard-rock caverns, aquifers, and artificial reservoirs have been suggested as reservoirs for air storage. An optimum storage medium would have the following characteristics:

a) Low thermal conductivity to maintain the higher air temperature
b) High permeability to charge and discharge with very low drop in pressure
c) Absence of cracks and fissures to contain the air without significant air mass loss and
d) High wall elasticity to avoid crack generation under frequent loading and unloading. Salt caverns and hard rock caverns have most of these characteristics and appear to be promising media for air storage.

Outside of possible losses in ground fissures, another disadvantage of compressed air storage is that the exergy per unit volume of compressed air is low. Hence, very large volumes and masses are required for the storage of significant amounts of energy. Air in the cavern typically assumes quickly a temperature that is close to the ambient temperature, T_O, and its exergy is given by the expression:

$$E - E_0 = PV_c\left[\frac{P_0}{P} - 1 + ln\frac{P}{P_0}\right] + V_c\rho_w gH, \tag{12.4}$$

where the pressure P is not the original pressure of the hot air, P_T, but the pressure developed after the cooling process in the reservoir, $P = P_T{*}(T_O/T)$. It is apparent that very large caverns are required for significant energy storage. In addition, there have been no extensive tests on crack formation and propagation under the cyclic loading of caverns at high pressures. Considerable research and testing must be done on this subject, before either salt caverns or hard-rock caverns are extensively used for energy storage.

Two compressed air energy storage (CAES) power plants have been built. The first one is a 290 MW unit in Huntorf, Germany, which was built in 1978; the second is the 110 MW unit of the Alabama Electric Corporation in McIntosh, Alabama, USA, which was commissioned in 1991. Both power plants use large underground caverns to store the compressed air. The two caverns of the Huntorf unit have a total capacity of 310,000 m^3 and operate between 43 and 70 bar pressure. These two power plants have demonstrated the technical soundness and reliability of CAES technology in storing and supplying significant power to the electric grid. However, the energy losses in the storage/production stages due to equipment efficiencies have been significant in both power plants. Because such

[1] During the storage mode, an exergy storage system is a closed system and for this reason, the expressions used for the exergy function are those pertinent to closed systems.

losses increase the cost of the electric power, there are no more CAES plants that are currently operational although there are a few in the planning stages.

12.2.3 Springs, Torsion Bars and Flywheels

All three are simple mechanical energy storage devices. The energy stored in a spring with constant, k, is proportional to the square of the displacement from the equilibrium position, x_0:

$$E = \frac{1}{2}k(x - x_0)^2. \tag{12.5}$$

Typical values of spring constants for large steel springs are of the order of 1 MN/m. With typical displacements on the order of 0.2 m, the energy that may be stored in such springs is of the order of 20 kJ (0.0056 kWh), which is a very small number in term of typical electric power units. Therefore, even large steel springs cannot be used for the storage of significant amounts of electric energy.

Torsion rods are very similar. The energy stored in a cylindrical torsion rod of length L and radius R, with shear modulus G and angular displacement Ω from its equilibrium position Ω_0 is:

$$E = \frac{1}{2}(\pi R^2 L)G(\Omega - \Omega_0)^2. \tag{12.6}$$

Steel has a shear modulus $G = 81*10^7$ N/m^2. A 2 m long cylinder of steel with a radius 0.1 m and subjected to an angular displacement $\pi/12$ (15°) will store 1.7 MJ or 0.48 kWh. Even though this stored energy is higher than that in springs, it is very low in comparison to typical electric energy needs. As with the springs, the torsion bars are not good candidates for the storage of significant amounts of energy that would provide electric power. These two devices may store mechanical energy for long periods of time and may be used in highly specialized low energy applications, such as mechanical clocks and weight balancing devices.

The energy stored in a flywheel with mass m and radius R, which is rotating at n revolutions per second ($\omega = 2\pi n$) is:

$$E = \frac{4\pi^2}{2}mR^2n^2. \tag{12.7}$$

A metal flywheel with 100 kg of mass uniformly distributed at a radius of 1 m and $n = 10$ revolutions per second, stores 197.4 kJ (0.055 kWh) of energy. Flywheels, like the other mechanical energy storage devices, do not have the capability to store significantly higher amounts of energy than springs and torsion bars for the production of electricity. Flywheels are occasionally used with electric motors of elevators to store small amounts of mechanical energy during the descent of the elevator. Also, they are used with subway cars in some urban transportation systems. They store energy during braking and release it during the acceleration process. Flywheels have also been extensively used with reciprocating engines, where they are attached to the crankshaft. During the power pulse, the flywheel absorbs some of the energy and releases it in the latter part of the stroke, when the power of the engine is reduced. Thus, the function of the flywheel is to make uniform the power delivered by the reciprocating engines and to smoothen their operation, but not to store significant amounts of energy.

The amount of stored energy is not significant enough for flywheels to be used extensively as energy storage devices in the power industry. Combinations of very large masses and rapid rotation would be required for the storage of significant quantities of energy. For example, a large flywheel of 3 m radius, which stores 100 kWh, while rotating at 20 revolutions per second, must have a mass of 5,066 kg. In addition, flywheels could not store energy for long periods of time: air friction and friction on their axis with its supports reduces the rotational speed, thus resulting in high losses of energy even in timescales of the order of minutes.

Because the flywheel material is subjected to high centrifugal forces, suitable materials for flywheel design must have high strength, high resistance to crack growth, and low cost. A type of steel, called the *maraging steel,* is suitable material for flywheels because of its high strength and density. Fiber reinforced composite materials are also suitable for use in flywheels, primarily because of their high-strength and relatively low cost. Because the material strength of the flywheel is one of the limiting factors in its design, the specific energy (energy stored per unit mass) is limited by the maximum strength of the material. The theoretical maximum specific energy is given by the expression [1]:

$$\left(\frac{E}{m}\right)_{\max} = 3.77 * 10^{-7} k_m \frac{\sigma}{\rho}, \tag{12.8}$$

where ρ is the density of the flywheel material, σ is the allowable stress and k_m is a dimensionless *mass efficiency factor.* Under optimum design conditions, the latter may reach the value 1.0 for isotropic materials (steel alloys) or 0.5 for anisotropic materials (composites).

The use of smaller flywheels has been advocated as an energy storage device in automobiles. A flywheel may absorb some of the engine power during braking and release it during the acceleration of the automobile. While this would improve the fuel consumption of a typical automobile, the weight of the flywheel and its mechanism would add significantly to the weight of the automobile. In addition, a fast rotating flywheel would reduce safety and reliability: gyroscopic forces on the axis of the flywheel are developed during turns and other changes in direction. Their effect would be to decrease the stability of the automobile, and this will increase the frequency of road accidents.

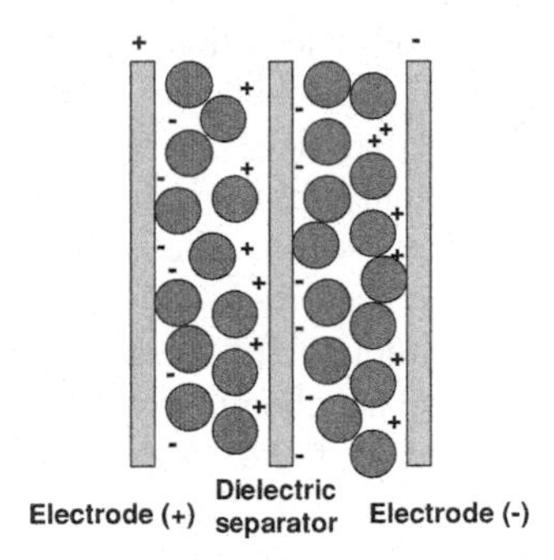

Fig. 12.6 The electrical double layer

12.2.4 Capacitors, Ultra capacitors, and Superconducting Coils

Capacitors and induction coils store electric or magnetic energy. When this energy is released it may be directly and almost instantaneously converted to electric energy at efficiencies close to 100%. Therefore, the storage of energy in capacitors or coils does not suffer from significant conversion penalties.

The electric energy stored in a capacitor is given by the expression:

$$E = \frac{1}{2}CV^2, \tag{12.9}$$

where C is the capacitance of the material, which depends greatly on the size of the capacitor, and V the applied voltage. Conventional capacitors are formed by two metal plates, or a sphere and a surrounding spherical shell. The two have equal and opposite charges. Dielectric materials inserted between the charged plates counteract the attractive forces; maintain the gap between the plates and cause different voltages and different amounts of energies to be stored. Optimizing the properties of these dielectric materials leads to higher energy densities for any given size of capacitor. The charge separation creates a voltage between the two plates, and the energy stored in the gap may be used in an external circuit that connects the two plates. This energy may be harnessed in an external circuit.

Conventional capacitors store very modest amounts of energy in a given volume, primarily because the dielectric material between the plates is bulky. Recently the *ultra capacitors* have been developed, which do not use as much volume for the dielectric material. The ultra capacitors utilize the concept of the double electric layer and use the polarization of an electrolytic solution to store energy electrostatically. The concept of the electric double layer is illustrated in Fig. 12.6, which depicts a typical ultra capacitor cell. Two porous plates, or collectors, are suspended within an electrolyte solution, with a voltage potential applied across the collectors. The applied electric potential on the positive electrode attracts the negative ions in the electrolyte, while the potential on the negative electrode attracts the positive ions. A dielectric separator between the two electrodes prevents the charge from moving between the electrodes. Since the electric double layer is formed on a very thin liquid electrolyte layer of the order of molecular size, the separation distance between the two electrodes may be made extremely thin, on the order of nanometers. This allows the packing of several "plates" with much larger surface area into a very small size, resulting in extraordinarily high capacitances in packages of relatively low volume.

Although the ultra capacitors are electrochemical devices similar to the batteries, no chemical reactions are involved in their energy storage mechanism. Because of this, there is only a very small irreversibility associated with their operation and, thus, they may be charged and discharged hundreds of thousands of times, with no significant energy dissipation. In contrast with traditional capacitors, electric double-layer capacitors do not have a conventional dielectric. Rather than two separate plates, which are separated by an intervening substance, these capacitors may use as "plates" two layers of the same substrate. When their electrical properties are optimized, the electrical double layers that are formed cause the effective separation of the charges and significant storage of electric energy, despite their very thin size and small volume. This gives very high volumetric energy storage capacity for the ultra capacitors.

The material between the plates of an ultra capacitor is a porous material that is typically made of activated charcoal. This is a carbon powder that consists of fine particles with rough surfaces, which in bulk form a low-density volume of particles with inter-particle holes and passages that replicate the structure of a sponge. The overall surface area between the electrodes of these materials is many times greater than a traditional porous material, such as aluminum. The higher surface area allows a great deal more ions and radicals from the electrolyte, which are the charge carrying agents, to be stored in any given volume. This implies that electric double-layer capacitors using charcoal are limited to low potentials on the order of 2–3 V, a very low voltage for most practical applications. The electric double-layer capacitors, which would be used at higher voltages, must be made of matched capacitors of individual electric double-layer capacitors connected in series, in a way that is similar to the cells in higher-voltage batteries.

Activated charcoal is not an ideal material for ultra capacitors, because the charge-carrying ions and free radicals are often larger than the holes and passages allowed in the random packing of the charcoal particles. Recent research on improved materials has revealed other materials of higher usable surface areas. Carbon nanotubes, which are made of only carbon atoms arranged in long cylindrical matrices, have similar charge storage capability as charcoal, but are arranged in a more regular pattern that exposes a significantly higher surface area of carbon to the electrolyte and its charges. Other materials that are promising for energy storage and are under consideration are activated polypyrrole and even nanotube-impregnated common paper.

The two main advantages of the ultra capacitors are:

a) They may store a high amount of energy per unit mass and
b) They may be charged and discharged in fractions of a second.

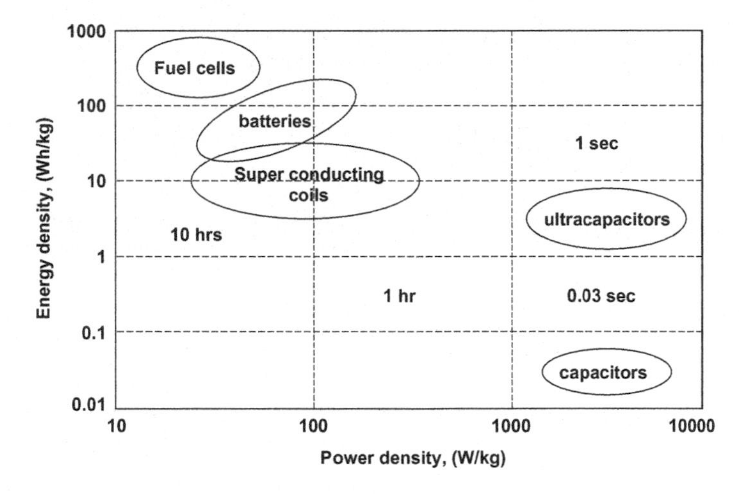

Fig. 12.7 Storage capacity of various electrochemical devices

In contrast, batteries need several minutes or hours to fully charge. Figure 12.7 shows a *Ragone chart* or "bubble chart" of the energy and power densities of ultra capacitors as compared to the other electrochemical energy storage devices. The energy and power densities are defined as the ratios of the total energy stored per unit mass of the device and the nominal power the device may produce per unit mass. It is apparent in this figure that, because of the short duration of charging and discharging, capacitors and ultra capacitors are ideal devices for storing small amounts of energy over short periods of time. When larger amounts of energy storage are required, and when the storage is to exceed a few seconds, the conventional batteries or fuel cells, which store energy in chemical form, are by far more suitable storage devices.

The energy stored in an electric coil or solenoid is given by the relationship:

$$E = \frac{1}{2} LI^2, \tag{12.10}$$

where I is the current flowing and L is the inductance of the coil. The latter is proportional to the radius of the coil, r, and to the square of the total number of turns of the coil. When made with the usual type of materials, a coil would exhibit significant ohmic energy losses (I^2R), because of the finite electric resistance, R, and would not be a good medium for energy storage. The discovery and recent use of superconducting materials at relatively high temperatures allows very high currents to pass through superconducting coils with minimum energy loss ($R \to 0$). Current density values on the order of 10^8 Amperes/m^2 and currents on the order of 10,000 Amperes in cables are routinely achieved with superconducting coils. With typical inductance values of 500 Henry (Henry = Volt*second/Ampere), Eq. (12.10) shows that a typical superconducting coil would store $2.5*10^{10}$ J or approximately 7,000 kWh, which is a significant amount of energy. Therefore, coils made of superconducting materials have been recommended as energy storage devices, because: a) they may store very high quantities of energy; b) the charging/storing time is very short and c) the released energy may be controlled very well. What prevents their more widespread use is that they are expensive to manufacture and operate.

12.3 Thermal Storage

Thermal storage may be accomplished for two purposes:

a) To store energy in the form of sensible or latent heat for later use, and
b) To store "coolness," that is to produce materials at lower temperatures, which will be used at a later time for refrigeration or air-conditioning.

12.3.1 Sensible and Latent Heat Storage

During the *sensible heat storage*, the temperature of a material, which may be pressurized water, an organic fluid, or a solid (including solid beds), is raised for the energy stored to be extracted at a later time as heat. The material that stores heat is typically well insulated. Its energy is released by the convective flow of air in its interior, when this becomes desirable. Typical sensible energy storage units are space heaters that have been used in several European countries. These are made of packed bricks or other materials with high specific heat capacity. Electric resistance units and air channels run between the bricks. These units are supplied with electric power during the night and early morning hours from the base-load power plants, when the demand for electricity is very low. During the day, when there is high demand for electricity, the heating units gradually release their energy by circulating a controlled amount of air in their interior channels, thus, keeping the space temperature constant.

One may achieve better temperature control and heat flow with *latent heat storage*. This is accomplished with pressurized steam, another pressurized vapor, or a molten metal. Typically, the vapor is stored in a well-insulated, pressurized vessel. The energy stored in the vapor or the molten metal may be used for space heating, process heating or conversion to electric power. An advantage of the latent heat storage method is that, if the pressure of the storage material is kept constant, the energy released is at constant temperature. This is very important for some industrial processes as well as for power production, where constant temperatures are desired. All other factors being equal, stored latent heat is significantly higher than sensible stored heat. For example, 1 kg of vapor water—steam—at 1 atm may supply approximately 2,550 kJ[2] at 100°C for space or process heating, while one kg of liquid water at the same temperature may only supply approximately 300 kJ for space heating. Thus, latent heat storage is ideal for storing large quantities of thermal energy for heating and for the production of electricity.

Latent heat storage has been proposed for transforming base-load power plants to peaking units, by storing very high amounts of energy in steam or a molten metal. When a steam power plant does not run at full load, some of the steam produced by the boiler or reactor may be diverted from the turbine and be stored at high pressure in "steam accumulators." At high-demand periods, this steam may be fed to auxiliary turbines, where it produces additional power that is needed during the peak hours. Nuclear power plants must operate almost continuously and are essentially base-load units. Power plants operating with high-temperature cycles, such as HTAGR or HTGR power plants, which are depicted in Fig. 5.10, are excellent candidates for conversion to an intermediate or peaking unit by storing some or all of the heat produced during off-peak hours in a pool of molten salt at significantly high temperature. The stages of the production of electricity are as follows:

a) During off-peak hours, all the heat produced by the reactor is transferred by the circulating helium to the pool of the salt, which melts and heats up. The pool temperature may rise to 650–700°C, or even higher if ceramic storage materials of very high melting points are used.
b) The stored energy may be removed, either by circulating a molten metal of lower boiling point, such as lead, through the molten salt and then using a heat exchanger to produce steam at high pressure and temperature, or by circulating water through the molten salt and, thus, raise steam directly.
c) The steam, which is produced during the day, is directed to a turbine and is used solely for the production of electricity. During this operation, the molten salt may cool to a lower temperature and, depending on the design of the unit, some or all of it may solidify. Because of the high temperatures that are achieved in the storage medium, the temperature of the steam may be in the range 550–600°C, which is in the higher range of temperatures achieved by modern fossil fuel plants. This implies that the steam component of this power plant would be as efficient as any modern fossil-fuel power plant.

A variation of this scheme would allow the HTAGR power plant to produce a small amount of base-load electric power continuously and store the rest. Such a power plant would continuously store a percentage of the produced heat in the molten salt pool and will produce additional steam and electric power during the periods of higher electricity demand.

Whether a system uses latent or sensible heat for the storage of thermal energy, the maximum amount of work that may be obtained from the system is given by its exergy. It must be recalled that a storage system is a closed system. When the specific properties of the system at its elevated temperature and pressure (T, P) are

[2] Upon condensation the steam will release 2,257 kJ/kg and will become liquid water at 100°C.

denoted by u, v, and s and the properties of the same system at atmospheric conditions (T_O, P_O) by u_O, v_O, and s_O, the difference of the exergy between the state at the high temperature, T, and the dead state is:

$$W_{\max} = m(e - e_0) = m[u - u_0 + P_0(v - v_0) + T_0(s - s_0)]. \tag{12.11}$$

In the case of steam, or another material undergoing phase change, these properties may be obtained from the pertinent tables, such as steam tables. If the storage system is incompressible and there is no latent heat involved in the storage process, as in the heating of heavy oils or solid rocks, the maximum amount of work that may be produced is given approximately by the simpler expression, in terms of the specific heat capacity, c, as follows:

$$W_{\max} = m(e - e_0) = m\left[c(T - T_0) + cT_0 \ln\frac{T}{T_0}\right]. \tag{12.12}$$

12.3.2 Heat Losses in Thermal Storage Systems

The storage material (water, steam, molten salt, molten metal or solid rock) is at a temperature that is significantly higher than that of the environment. Because there is no material which is a perfect insulator, heat would be continuously lost from the thermal storage medium at a rate which is proportional to the storage temperature. Given enough time, and if the stored energy is not used, the energy dissipates in the environment and is eventually lost. For this reason, thermal storage is a good way to store thermal energy for short to intermediate periods of time. The following example illustrates the heat and temperature loss from a storage system:

Let us assume that sensible heat (e.g. superheated steam) is stored in a cylindrical vessel of diameter D and length L, that the storage material has density ρ, specific heat capacity c, and the ambient temperature is T_{amb}. The initial temperature of the vessel is T_o and the temperature at a given instant is denoted as T. The rate of heat transfer from the cylindrical vessel depends on the overall heat transfer coefficient U. At a given time, the total rate of heat lost to the surroundings is:

$$\dot{Q} = U\left(2\frac{\pi}{4}D^2 + \pi DL\right)(T - T_{amb}). \tag{12.13}$$

Assuming a uniform temperature in the vessel, the heat lost from the cylindrical vessel would reduce the temperature of the material by a rate, dT/dt, which is given by the following heat transfer equation:

$$\frac{1}{4}\pi D^2 L\rho c\frac{dT}{dt} = -\dot{Q} = -U\left(2\frac{\pi}{4}D^2 + \pi DL\right)(T - T_{amb}) \tag{12.14}$$

Equation (12.14) is an ordinary differential equation with respect to the instantaneous (uniform) temperature of the storage tank, T. Under any conditions this equation may be solved numerically to yield the function $T(t)$. A good approximation to the solution may be derived by assuming that the properties of the storage medium, as well as the overall heat transfer coefficient, U, are constant. Under these conditions, the solution to this differential equation is:

$$\frac{T - T(0)}{T_{amb} - T(0)} = \exp(-t/\tau_{st}), \tag{12.15}$$

where τ_{st} is the characteristic thermal storage time, which is given by the coefficients of Eq. (12.14) as follows:

$$\tau_{st} = \frac{DL\rho c}{2(D + 2L)U}. \tag{12.16}$$

The last three equations demonstrate that the temperature of a sensible heat storage system drops exponentially and finally reaches the ambient temperature. A well insulated vessel (very low U) would have a significantly lower heat loss to the surroundings and would delay significantly the temperature drop. At any rate, Eq. (12.15) demonstrates that thermal storage systems may be best used for short to intermediate energy storage situations. In the long term, all the thermal energy is lost to the environment. Currently used insulation materials have high enough overall heat transfer coefficients, U, so that they may store significant quantities of heat for periods of 10–12 h without an appreciable temperature drop. The state of the art in insulation materials presently allows for the development of systems with diurnal cycles (e.g. storage of heat during the night and use during the day) but not for seasonal cycles (storage of heat during the spring/winter months to be used during the peak demand of the summer). The "coolness" storage system, which is described in the following section and operates on a diurnal cycle, is also feasible with commonly used insulation materials.

12.3.3 Storage of "Coolness" to Offset the Peak Power Demand

In the U.S.A. as well as many other OECD countries, the southern migration of the population and incessant need for air-conditioning has shifted the peak demand for electric power from the winter to the summer. This is demonstrated in Figs. 12.1 and 12.2, where it is apparent that the peak demand for electricity during the summer is 35% higher than the peak power demanded during the winter, for both weekends and weekdays. Although these figures were obtained from the San Antonio market, a southwestern city in the USA with very high cooling demand in the hotter months of the year, similar trends, with a lesser percentage difference between summer and

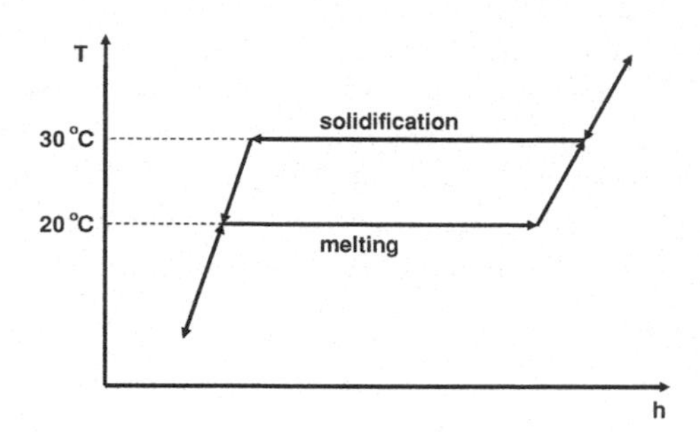

Fig. 12.8 The melting/solidification curve of a eutectic material

winter are observed in all the major cities of the United States and of most of the OECD countries, where summer air-conditioning is common.

Since the peak power demand is spent on cooling, it makes sense to store "coolness," that is to produce materials at temperatures lower than the internal temperatures of buildings. "Coolness" or low temperature materials may be stored as latent heat/coolness, as sensible heat, or a combination of the two. As with the storage of heat materials, systems and materials working primarily with latent storage have an advantage because they would store significantly higher amounts of "coolness" per unit mass. A "coolness" storage material is cooled during periods of off-peak hours. The material receives heat from the hotter air of a building during peak electricity demand, cools this air in a heat exchanger and supplies the cooler air back to the building. The air-conditioning system uses electric power during off-peak hours and avoids the use of significant electric power during the peak afternoon hours.[3] The coolness system has two significant advantages:

1. It shifts the electric power demand from peak hours to off-peak hours, when electricity is supplied by base load plants. This implies that utilities will not have to have available more power plants in order to meet the peak demand.
2. Because base-load and intermediate power plants are significantly more efficient than peaking units, the electric power used in air-conditioning is produced at higher overall plant efficiency. This implies that a lesser amount of energy resources is used for the satisfaction of the air-conditioning needs of the population.

A system for storing "coolness" during the summer months may also be used to store heat during the winter months. For this purpose, the materials used must store high quantities of energy at temperatures higher than the typical space temperature in winter (e.g. close to 80°F or 27°C). At the same time this material must also store high quantities of "coolness" at temperatures lower than the typical temperature in the summer (e.g. close to 74°F or 24°C). Excellent materials for cooling during the summer and heating during the winter are materials that exhibit a temperature hysteresis in their melting and solidification processes, by a few

[3] A low amount of electric power is still needed for the circulation of air.

degrees. These materials melt at lower temperature than their solidification temperature as depicted in the temperature-enthalpy diagram, which is shown in Fig. 12.8. Ideally, such materials melt in the range 65–75°F and solidify in the range 80–90°F.

During the summer period the material stores "coolness:" The solidification of the salt occurs at 30°C, during off-peak hours and the temperature of the material may be dropped to values that are lower than the melting/solidification range. During the peak demand hours, the material melts and absorbs heat, which is stored as enthalpy, from the cooled space at 20°C. At this temperature the material cools the space air to the lower air-conditioning temperatures, thus keeping the temperature of the cooled space at the desired low levels. During the winter, the operation of the air-conditioning medium is reversed. When the material solidifies it supplies heat to the heated space at a temperature of 30°C, which is above the typical space temperature, thus, maintaining the space warmer. At off-peak hours the material absorbs most of its heat/enthalpy at 20°C during its melting process and then its temperature rises above 30°C.

The so-called *eutectic salts*, such as the sodium sulfite decahydrate or Glaubert's salt ($Na_2SO_4 \cdot 10H_2O$), exhibit this type of temperature hysteresis and are good materials for cooling and heating systems. Because of the finite temperature differences during the heat transfer processes, these materials operate better if their range of operating temperatures becomes wider than the 10°C, which are shown in Fig. 12.8. A wider use of heat storage based on eutectic salts or similar materials for cooling and heating purposes has the potential to shift significantly the peak electricity demand during the summer months and to reduce the peak heating requirements during the winter months. This will enable electricity production corporations to use their more efficient power plants for the satisfaction of the demand for electricity.

12.4 Chemical Storage: Batteries

12.4.1 The Electrochemical Cell

Most of the energy we use today for the production of electricity and propulsion is derived from chemical energy stored in fossil fuels: coal, oil, and natural gas consist of chemical compounds that have been formed millions of years ago and are now being used at an alarmingly fast rate. Chemical energy has very high energy density (in kJ/kg) and is easy to be used by combustion or direct conversion to electricity. For example, the chemical energy that may be released from the combustion of solid graphite ($-\Delta H^o$) is 32,770 kJ/kg. For octane this number is 49,010 kJ/kg; and for methane, 55,510 kJ/kg.

The energy density of most stable chemical bonds is significantly high and, hence, chemical storage is a very desirable method for energy storage. Apart from

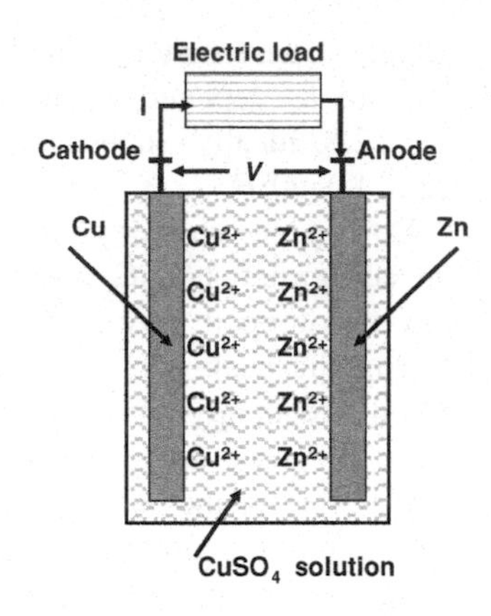

Fig. 12.9 Electric power equal to $\dot{W} = VI$ is produced from the operation of the Daniel cell

having high storage density, an energy storage device with desirable properties must charge and discharge quickly, to provide not only large quantities of energy, but also high power when it is needed. Two very important parameters in chemical energy storage are:

a) The characteristic time of storage, which is the time it takes for the chemical compounds to be formed (or formation time) and
b) The characteristic time of discharge, in this case the combustion time.

The formation time of fossil fuels is on the order of hundreds of millennia and, hence, chemical storage in the form of fossil fuels is impractical, even though the discharge time, via a combustion process, is of the order of minutes or seconds.

Other chemical compounds, such as acids, bases, and metallic salts have shorter formation times and are frequently used in *electric accumulators*. These are energy storage devices that are commonly referred to as "batteries" or "electrochemical cells." Common batteries consist of several individual cells in series. When electric power is desired, a chemical reaction, which is controlled by an external circuit, takes place inside the cells and, simultaneously, a voltage difference appears in the electrodes. The "Volta Cell" is the first electrochemical cell that was invented by Alexandro Volta in the late eighteenth century and comprises of copper and zinc electrodes with copper sulfate as the electrolyte. Its successor, the "Daniel Cell" was invented in 1836 and makes use of a semipermeable porous medium that prevents the deposition of copper on the zinc anode. Figure 12.9 depicts the operation of the Daniel cell: The cell is made with solid copper and zinc *electrodes*, which are called the *cathode* and the *anode*. The electrodes are surrounded by water solution of copper sulfate ($CuSO_4$) and zinc sulfate respectively. These solutions constitute the *electrolyte*. A porous material, originally made of clay, separates the two types of the electrolyte, allows the SO_4^{2-} ions to pass through its pores, but inhibits the passage of the Cu^{2+} and Zn^{2+} ions and, thus, effectively keeps them in the vicinity of their respective electrodes. At the anode, a solid zinc atom deposits two electrons and enters the solution as a positive ion according to the reaction:

$$Zn \rightarrow Zn^{2+} + 2e^{-}. \tag{12.17}$$

At the cathode, a copper ion (Cu^{2+}) absorbs two electrons and is converted to a solid copper atom, which is deposited on the cathode:

$$Cu^{2+} + 2e^{-} \rightarrow Cu. \tag{12.18}$$

Therefore, the overall reaction that takes place in this electrochemical cell is:

$$Zn + Cu^{2+} \rightarrow Cu + Zn^{2+}. \tag{12.19}$$

The anode of the Daniel Cell has a surplus of electrons and, hence, a negative charge, while the cathode has a deficiency of electrons, and a positive charge. This creates a potential difference (voltage) between the anode and the cathode, which may be utilized in an outside circuit for the production of electric power: $\dot{W} = VI$. At the same time, the cathode has a surplus of sulfuric ions (SO_4^{2-}) and the cathode has a deficiency of SO_4^{2-}. These ions move through the electrolyte and the porous material from the anode to the cathode and maintain the electric current in the electrolyte. The released electrons travel from the anode to the cathode (by convention, "current" flows in the opposite direction to the electrons) and deliver electric power to the circuit.

The reactions stop and the cell stops producing power and voltage difference when all the zinc material in the anode has eroded and has been converted to ions. At this point the cell may be *regenerated* or *recharged* if an external voltage is supplied to the electrodes with the positive pole connected to the cathode and the negative to the anode of the cell. During the recharge process, energy is supplied to the cell and the reactions described by Eqs. (12.17) and (12.18) are reversed: Zn comes out of the electrolyte solution and forms the cathodic material; Cu enters the electrolyte solution as Cu^{2+} ions; the sulfate ions, SO_4^{2-}, move from the cathode to the anode; and the electrons move in the outside circuit from the cathode to the anode. The recharging time of a Daniel cell and other similar cells is typically on the order of hours. This makes such cells cumbersome to use with applications that require rapid charging and discharging.

In contrast to common uses of chemical energy in fossil fuels, which is by combustion and thermal energy release, electrochemical cells convert the energy stored in the chemical bonds directly into electricity, without producing heat or thermal energy as an intermediate stage of the energy conversion process. Because of this, electrochemical cells are not subjected to the Carnot limitations, which are described in Chap. 3. The total chemical energy that may be converted to electric energy is equal to the exergy of the electrode materials. According to (Eq. 3.34), the exergy is equal to the change of the Gibbs free energy at atmospheric conditions, $-\Delta G^o$, of the overall chemical reaction in the cell. In the absence of thermodynamic irreversibilities this energy is converted to electric work (voltage times electric charge transferred). If one kmol of a substance reacts (e.g. 1kmol of Zn in the Cu–Zn cell) the charge transferred would be equal to ζF, where F is the *Faraday constant* (=96,500,000 Coulomb/kmol) and ζ is the valence number of the reaction, or the number of electrons transferred per reacting atom or ion. Hence, the electromotive force of the cell during a reversible operation, or the maximum voltage the cell may produce is given as follows:

$$-\Delta G^o = \zeta F V_{max} \Rightarrow V_{max} = \frac{-\Delta G^o}{\zeta F}. \tag{12.20}$$

In the Zn–Cu reaction of Eq. (12.19) $\Delta G^o = 216{,}160$ kJ/kmol and $\zeta = 2$. Hence, the maximum voltage the Zn–Cu cell may produce is approximately, 1.1 V. This is a typical value for most electrochemical cells. Several electrochemical cells must be connected in series to develop the voltage of 10–20 V, which is needed in many applications. In practice, internal irreversibilities related to the passage of ions in the electrolyte and external irreversibilities caused by the flowing current, would reduce the voltage that is actually developed in a cell typically by 10–25% from the maximum value given by Eq. (12.20).

12.4.2 Commonly Used Battery Types

One of the most common types of cells for batteries is the lead cell, which is invariably used in the automotive industry. The lead cell consists of an anode, made of solid lead (Pb); a cathode made of solid lead oxide (PbO_2); and an electrolyte between the two electrodes, typically made of a sulfuric acid solution (H_2SO_4). The acid in the electrolyte solution separates into hydrogen (H^+) and sulfuric (SO_4^{2-}) ions. The chemical activity in the electrolyte causes chemical reactions to occur at the surface of the electrodes, which release or absorb electrons. This in turn causes the erosion of the material of the two electrodes and its dissolution in the electrolyte as described by the following reactions:

A. Anode:

$$Pb + SO_4^{2-} \rightarrow PbSO_4 + 2e^-, \tag{12.21}$$

B. Cathode:

$$PbO_2 + 4H^+ + 2e^- \rightarrow 2H_2O + Pb^{2+}, \text{ and} \tag{12.22}$$

C. Electrolyte:

$$2H_2SO_4 \rightarrow 4H^+ + 2SO_4^{2-}; \quad Pb^{2+} + SO_4^{2-} \rightarrow PbSO_4. \tag{12.23}$$

Table 12.1 Voltage, charge capacity and energy density for several types of Li-ion cells

Cathode	Voltage (V)	Capacity (A-h/kg)	Energy density(kJ/kg)
$LiCoO_2$	3.7	140	1,865
$LiMnO_2$	4.0	100	1,440
$LiFePO_4$	3.3	120	1,426
Li_2FePO_4F	3.6	115	1,490

It is apparent from these reactions that the lead in the anode and the lead oxide in the cathode are eroded by the hydrogen cations and the sulfuric anions, which are formed by the ionization of the sulfuric acid. The excess of electrons in the anode and the consumption of electrons in the cathode results in the polarity and the development of the voltage difference of the cell. The overall reaction that occurs in the entire cell may be obtained from the sum of Eqs. (12.21) through (12.23) as follows:

$$Pb + PbO_2 + 2H_2SO_4 \rightarrow 2PbSO_4 + 2H_2O. \tag{12.24}$$

The Gibbs free energy change of the last expression, ΔG^o, would determine the maximum voltage developed from this cell according to Eq. (12.20).

More recently, the lithium batteries have been developed for use in electronic devices and several light vehicles. The term "lithium battery," or "Li-ion battery" refers to different types of batteries comprising several types of cathodes, anodes and electrolytes. The most common type of lithium batteries for consumer applications uses metallic lithium as anode and manganese dioxide as cathode, with a lithium salt dissolved in an organic solvent. Another type of Li-ion battery is the lithium-thionyl chloride cell, a liquid solution of thionyl chloride ($SOCl_2$) and lithium tetrachloroaluminate ($LiAlCl_4$) acts as the cathode and electrolyte respectively. A porous carbon material serves as the cathode of the cell. This type of batteries are well suited to be used in very low-current applications, where reliability and long life is necessary, such as wireless alarm systems. Because of their complexity, lithium-thionyl chloride cells are used only in specialized industrial applications, or are installed into devices where no consumer replacement may be performed and, generally, they are not sold in the consumer market.

Table 12.1 shows several types of Li-ion cells, the typical voltage they produce, in Volts, the electric charge capacity they have, in Ampere-hours per kg of weight and the total energy that may be stored in such cells, in kJ/kg. It is apparent that Li-ion batteries may store a significant amount of energy per unit weight and for this reason they have been recommended as an alternative to gasoline-powered vehicles. However, the recoverable energy density (kJ/kg) stored in fossil fuels, such as gasoline is approximately thirty times more than the energy density of the Li-ion cells. When taking under consideration the finite conversion efficiencies of the components, an electric vehicle must carry approximately 1 ton of battery weight if it were to have the same range as a vehicle with an 18-gallon gasoline tank. In addition, the recharge time for conventional batteries and Li-ion batteries is on the order of hours, which implies that a quick recharge, similar to a tank filling at a gas station, is impossible.

As any car operator knows, a battery has a finite life: after several months and a few thousand charge-recharge cycles (currently on the order of 1,000) batteries do not recharge and must be chemically regenerated. In an economy heavily dependent on batteries for energy storage, the extensive transportation and chemical regeneration of batteries that contain large quantities of heavy metals will cause significant environmental and public health problems. Almost all the solid battery types, including lithium batteries, contain to a small or large degree heavy metals (Pb, Ni, Co, Mn, Cd, etc.) which have been proven to be harmful to humans. Since the useful life of a solid battery is finite, the battery must be replaced or refurbished after approximately 1,000 full charge–discharge cycles. Let us assume that our economic system uses these solid batteries for the storage of significant amounts of energy, which is produced from renewable sources, in households. This type of economic system will necessitate the use of the equivalent of 20–25 car batteries in every household in the USA, which will be typically charged and discharged every day. These batteries will need to be replaced and regenerated every 24–48 months. Hence, every household must have the equivalent of six to ten batteries per year sent to a central facility for refurbishment or replacement. The shear volume of this battery traffic is simply very high for the society to sustain. The environmental and public health problems that will be created from the use and transportation of such a storage system are prohibitive. A contemporary society simply cannot rely solely on metal oxide batteries for energy storage without risking significant environmental damage and future public health problems. Even if the life-cycle of the solid batteries improves by a factor of two, a rather unlikely number if the cost of batteries is to be kept at affordable levels, the environmental risk by misplaced or discarded batteries during their transportation would still be enormous.

Another disadvantage of all batteries is the imperfect electrical insulation of the battery cells, which causes a *current drift* or a very weak internal current, even when the poles of the battery are not connected to an external circuit. A car battery that is not operated and charged for a few weeks is completely discharged and becomes a "dead" battery. The drift is continuous and results in the gradual loss of energy. The current types of batteries are not good energy storage systems for long periods of time.

Even though since the 1990s there has been significant progress in the manufacturing of durable, reliable and high capacity Li-ion batteries, it is apparent that at present they are not ready to replace the gasoline tanks of the vehicles. For this replacement to happen, a significant amount of research and several technological advances must first occur that will: a) improve the battery energy density; b) reduce the recharge time; c) lengthen significantly the life cycles of batteries; and d) eliminate the current drift.

12.5 Hydrogen Storage: The Hydrogen Economy

Hydrogen has been advocated as an environmentally friendly and physically powerful energy storage medium. The advantages for the use of hydrogen as an energy storage medium may be summarized as follows:

1. It is the lightest element and the specific chemical energy stored in hydrogen (142,700 kJ/kg at ambient conditions vs. 49,500 kJ/kg for octane) is higher than that of most other materials. On a volumetric basis, as a compressed gas hydrogen may store many more times the energy stored in compressed air: At 20 bar pressure, 1 m^3 of H_2 stores 229 MJ of energy in chemical form, while 1 m^3 of air at the same pressure would store merely 6.0 MJ of energy as exergy of compressed air. At 700 bar pressure, 1 m^3 of hydrogen will store 8,010 MJ and the compressed air only 459 MJ.
2. It is a stable molecular compound and does not change form if stored. This implies that it may be used after a long time.
3. It combines readily with oxygen to form water, which is harmless.
4. It is abundant on the surface of the Earth and it may be relatively easily and economically produced by the electrolysis of water.
5. It is not harmful to the environment if released and does not pose any health threat to humans.
6. It may be readily used in fuel cells (Sect. 12.6) with high conversion efficiencies.
7. It has been used for a long time as an industrial material and as propulsion fuel in the space shuttle and several types of rockets. There is significant expertise in storing and handling it.

Hydrogen has also several disadvantages as an energy storage medium: At first it is flammable and explosive, and, hence, it must be kept in special containers and be transported under well-controlled conditions. Because of its high specific energy content and its high diffusivity, hydrogen forms a powerfully explosive mixture with air. For this reason, hydrogen containers or pipelines must be well-designed to avoid significant leakage and mixing with air. Secondly, since hydrogen is the lightest element, its density is very low under atmospheric conditions (0.08 kg/m^3 vs. 1.2 kg/m^3 for air). High pressures or very large containers must be used for the storage of a significant mass of this gas. Thirdly, because the hydrogen molecule is very light, hydrogen may diffuse through the matrix of several metals including steel. When hydrogen diffuses through metal matrices it causes *hydrogen embrittlement* and *decarburization*.

Hydrogen embrittlement (or hydrogen grooving) occurs when hydrogen atoms diffuse through the metal matrix and accumulate in very small cracks. The accumulation of the gas in these, so called "voids" of the metal matrix creates high internal pressure, lower ductility and lower strength for the metal. Cracks in the metal matrix may easily propagate and open under relatively low internal pressure to release the stored hydrogen. Decarburization of steel is another consequence of the hydrogen diffusion: hydrogen atoms combine with the carbon in the grain boundaries and voids of the steel matrix to produce methane. This gas does not diffuse through the metal matrix. It accumulates in cracks and induces a significantly higher local internal pressure, which causes the cracks to propagate and the metal to break. A similar process occurs with copper alloys that contain oxygen: The diffusing hydrogen combines with oxygen and forms water (steam) which accumulates in the metal matrix and induces a higher pressure. Oftentimes, this process is called steam embrittlement and leads to a significant reduction of the strength of a copper container.

Despite the disadvantages of hydrogen as an energy storage medium, the high energy storage capacity of this chemical in combination with its abundance on the surface of the Earth and the relative easy way to produce it by electrolysis, make this gas a very viable energy storage medium. Improved materials and metal coatings in combination with improved methods of storage and transportation will evolve in the future. These would enable the extensive use of hydrogen as an energy storage medium. Several scientists have suggested that the widespread use of hydrogen as energy fuel will transform the economy of the planet into the *hydrogen economy.*

The term hydrogen economy was coined in the 1970s. In the context of the hydrogen economy, hydrogen is not a primary energy source but an energy carrier, similar to what electricity is in the beginning of the twenty-first century. Under the hydrogen economy concept, hydrogen will be generated by the electrolysis of water using the energy harnessed by the abundant solar or wind power, or by a chemical method using the heat generated by nuclear power plants. Assuming that the current problems of hydrogen storage and transportation are solved and suitable materials for the storage of hydrogen become readily available, the hydrogen produced will be chemically stable and may be easily stored and transported. The use of the stored and transported hydrogen would virtually eliminate the emission of carbon dioxide and, therefore, will become a method to alleviate the greenhouse effect and global warming. A hydrogen economy may become a panacea for developing countries without fossil fuels: since renewable energy sources, such as wind and solar, are widely and uniformly distributed on the planet, all countries will be capable to achieve energy independence and avoid expensive fossil fuel imports, by using a combination of renewable energy and hydrogen.

Whether a hydrogen economy will evolve in the near or far future would depend greatly on the technological advances related to the storage and transportation of hydrogen as well as on the technological progress that will be made in other methods for energy storage. Proponents of a world-scale hydrogen economy argue that hydrogen is the cleanest source of energy known to end-users, particularly in transportation applications, where it does not release particulate matter and greenhouse gases. Critics of the transition to a hydrogen economy contend that the cost of switching to a national or a global hydrogen distribution system may be prohibitive and an intermediate step may become economically more viable: for example, synthetic fuels from locally-produced hydrogen and atmospheric CO_2, such as ethanol and methanol, might accomplish the same goals of a hydrogen economy at significantly lower investment. Since the CO_2 input will be from the atmosphere, this scheme will not contribute to the growth of the GHG emissions.

At the end of the first decade of the twenty-first century one may see small but persuasive signs that point to a future transition to a hydrogen economy:

a) Several European communities from Iceland to Greece have adopted public buses that make use of hydrogen fuel cells.
b) Several hospitals in the OECD countries have installed combined units that accomplish the electrolysis of water and storage of hydrogen to be used with fuel cell units for emergency power locally. These systems are advantageous for emergency use because of their low maintenance requirement and non pollutant emissions. Such devices may be located at any place, as opposed to internal combustion driven generators, which must have adequate exhaust space.
c) Countries such as Portugal, Iceland, Norway, Denmark, Germany, Japan, and Canada as well as several states in the U.S.A., such as California, Oregon, Minnesota, and Texas have started investing in hydrogen distribution network systems. Even though these systems have proven to be initially costly, technological breakthroughs and improved methods of hydrogen transport may become lucrative in a future society dominated by hydrogen.
d) Hydrogen, fuel-cell pilot programs have started in all of the OECD countries as well as in Russia, China, India, and the many of the countries of the Middle East.

The concept of a hydrogen economy has become a futuristic concept that has drawn a great deal of criticism and debate, primarily stemming from the expense of the hydrogen fuel cells and the expense of a hydrogen distribution infrastructure. Our society is dominated at present by fossil fuels to the point that it is almost impossible to think of a mode of transportation outside the framework of liquid fossil fuels, such as gasoline and diesel. However, one must not forget that both of these fuels were entirely novel and very little used until the end of the nineteenth century. Hydrogen is to the humans of 2011, what gasoline was to the humans of 1880. During the twentieth century, a century that is characterized by the widespread use, the extensive exploration and the rapid depletion of fossil fuels, a planet-wide infrastructure was developed for the mining, transportation and distribution of liquid fossil fuels. The petrol/gasoline station, which is ubiquitous in the modern society, was not always present in almost every corner of urban developments. To the citizen of 1880s, today's infrastructure for the transportation and distribution of liquid fossil fuels would have appeared prohibitedly expensive and out of reach. A future with an extensive hydrogen infrastructure appears the same way to the citizens of the early twenty-first century.

It is an undisputable fact that gasoline and diesel will be exhausted at some point in the near or far future. There will be a point in the future of the humanity, when the use of liquid fossil fuels will be significantly curtailed and will finally

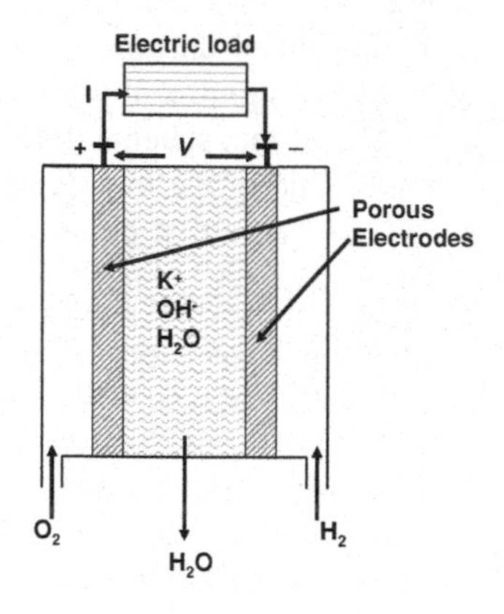

Fig. 12.10 Schematic diagram of the hydrogen–oxygen fuel cell

cease. Another fuel will inevitably take the place of the currently used liquid fossil fuels, and hydrogen is a good candidate to become this fuel. This inevitable evolution will be followed by considerable private and public investment in the production, transportation, and distribution of hydrogen. The hydrogen filling station may become in the future as ubiquitous as the gasoline/petrol station is in the beginning of the twenty-first century. The difference between the two fuels is that hydrogen may be continuously produced by renewable or other reliable and environmentally friendly sources of energy in all the countries and all human communities. As a result, the hydrogen distribution station and the hydrogen economy is sustainable and will last for a much longer period of time than the liquid fossil fuels.

12.6 Fuel Cells

Fuel cells are an integral part of the transition to a hydrogen economy, because they offer a direct conversion of chemical energy to electric energy that is both convenient and has very high efficiency. Fuel cells are similar to batteries, but are open thermodynamic systems, which may operate continuously. They are continuously supplied with fluid fuels and oxidants, their electrodes are not part of the reaction process and, hence, do not need to be regenerated or recharged. Although fuel cells were invented almost one century before the internal combustion engine, in 1802 by Sir Humphrey Davy, and were used for the operation of a tractor as early as 1839 by Sir William Grove, they have not been widely used in practical operations. This, despite the distinct thermodynamic and environmental advantages they present for the conversion of energy.

A schematic diagram of a fuel cell operating with hydrogen as the fuel and oxygen as the oxidant is shown in Fig. 12.10. Hydrogen is supplied to the cell on the side of the cathode, and oxygen is supplied on the side of the anode. The two

Table 12.2 Several types of fuel cells and corresponding maximum voltage at 298 K

Fuel	Reaction	$-\Delta G^o$ (kJ/kmol)	V_{max} (V)
H_2	$H_2+1/2O_2 \rightarrow H_2O$	236,100	1.22
CO	$CO+1/2O_2 \rightarrow CO_2$	275,100	1.43
CH_4	$CH_4+2O_2 \rightarrow CO_2+2H_2O$	831,650	1.08
CH_3OH	$CH_3OH+3/2O_2 \rightarrow CO_2+2H_2O$	718,000	1.24
C_2H_5OH	$C_2H_5OH+3O_2 \rightarrow 2CO_2+3H_2O$	1,357,700	1.17

electrodes are composed of a porous material and enclose a strong basic solution of potassium hydroxide (KOH), which is dissociated into K^+ and OH^- ions. Hydrogen diffuses into the pores of the cathode, combines with the hydroxyl ions (OH^-), forms water and releases two electrons according to the reaction:

$$H_2 + 2HO^- \rightarrow 2H_2O + 2e^-. \tag{12.25}$$

On the side of the anode, oxygen also diffuses into the pores and combines with water molecules and electrons to form hydroxyl ions (OH^-):

$$1/2O_2 + H_2O + 2e^- \rightarrow 2HO^-. \tag{12.26}$$

Overall, the water forming reaction from oxygen and hydrogen takes place in the fuel cell with the potassium ions being the catalyst. As in the case of the batteries, the anode has a deficiency of electrons and, hence, a positive electric charge, while the cathode has a surplus of electrons and a negative charge. The potential difference created between the anode and the cathode induces an electric current to pass through an external circuit. Hydrogen and oxygen may be continuously supplied to their respective chambers, from where they are diffusing in the pores of the cathode and the anode respectively. The water produced may also be continuously removed from the fuel cell. In practice, instead of water, a weak solution of potassium hydroxide (KOH) is removed from the cell. In order to restore the chemical balance, a hydrogen–oxygen fuel cell must be supplied periodically with a concentrated solution of KOH.

In general, fuel cells are supplied with a fuel and an oxidant, which is normally air at atmospheric pressure. Several types of fuels, other than hydrogen, have been proposed to be used. Table 12.2 lists some of these fuels; the corresponding overall oxidation reactions; the Gibbs free energy of these reactions; and the electromotive force or maximum voltage that may be obtained according to Eq. (12.20). It is apparent from this table that all these practical fuel cells produce a maximum voltage of the order of 1 V. Therefore, several fuel cells in series would be required to produce a significant voltage difference and this is one of the main disadvantages of fuel cell systems: Operating too many fuel cells in series or having very high currents would increase significantly the electrical losses (irreversibilities) of the fuel cell systems.

12.6.1 High-Temperature Fuel Cells

Higher values of the produced voltage with fuel cells are achieved by having several fuel cells in series, or "stacks." Typically, fuel cells are in stacks of 20–30 units, which provide operational voltages close to 30 V. The fuel cells are joined by the *interconnect*, which is either a metallic or ceramic layer that connects the individual cells. The function of the interconnect is to connect each fuel cell in series, so that the voltage each cell generates may be added. Because the interconnect is exposed to both the oxidizing and the reducing side of the cell at high temperatures, it must be an extremely stable material. Ceramic materials as well as metals have been used as interconnect materials.

One of the ways to increase the voltage of an individual fuel cell is to operate the cell at a temperature higher than the ambient: ΔG^o is a strong function of temperature and there are several reactions, for which the maximum voltage at elevated temperatures is significantly higher than the voltage at ambient conditions. An added advantage of such fuel cells is that the rates of all chemical reactions are faster at elevated temperatures. This implies that the power produced by the high-temperature cells may be significantly higher than the power produced by the same cells at ambient temperature. High-temperature fuel cells have been put in service where the operating temperature is in the range of 600–1,000°C and the maximum voltage is in the range of 3–7 V. Examples of such high-temperature fuel cells are the Solid Oxide Fuel Cells (SOFCs).

Because of the high temperatures, SOFCs do not use water as the electrolyte, but a dense layer of ceramic that conducts oxygen and fuel ions. Both cathode and anode must be good electric conductors and of high porosity to allow the diffusion of ions. The lanthanium-strontium-manganite (LSM) is one of the cathode materials that are frequently used for high temperature fuel cells and the yttria stabilized zirconia mixed with nickel metal for high conductivity is an anode material that is commonly used. The oxidation reaction that occurs between the fuel and the oxygen ions produces both water and electricity. If the fuel is hydrogen, water is the only product of the reaction. However, if the fuel is a hydrocarbon a high-temperature fuel cell may be used for the production of hydrogen.

While the high operating temperatures of SOFCs allow the kinetics of oxygen ion transport to be sufficiently fast for the good performance of the cell and the production of sufficient power, when the operating temperature approaches the lower limit for the SOFCs operation, which is around 600°C, the electrolyte begins to exhibit large ionic transport resistance. This lowers the voltage generated and adversely affects the overall performance of the high temperature fuel cells [2].

12.6.2 Thermodynamic Losses and Fuel Cell Efficiency

Similar to the other electrochemical cells, fuel cells are direct energy conversion devices (DEC) and are not subject to the Carnot limitations of heat engines. In principle, the entire chemical energy of a fuel cell, $-\Delta G^o$, may be converted to electric energy ($-\Delta G^o = VIt$) in a fuel cell that operates reversibly. In practice, thermodynamic losses occur, fuel cells do not operate reversibly, a fraction of the chemical energy is converted to heat and, as a result, the actual voltage obtained is lower than the maximum voltage given by Eq. (12.20). The following are the most common irreversibilities that result in power losses in fuel cells:

1. *Polarization losses*: Polarization or over-potential losses are due to the imperfections of the materials, of the microstructure, and of the design of the fuel cell. Polarization occurs from the ohmic resistance of oxygen ions as they are transported through the electrolyte; the electrochemical activation barriers at the anode and the cathode (this is almost absent in high-temperature fuel cells, where activation rates are high); and the concentration polarization, which is due to the very slow diffusion of ions through the porous anode and cathode.
2. *Ohmic polarization losses*: Ohmic losses emanate from the ionic conductivity through the electrolyte, which is a material property of the electrolyte and the ions involved. Ionic conductivity may be enhanced by: a) operating at higher temperatures, which decreases significantly the ohmic losses; b) doping methods that optimize the crystal structure of the electrolyte and control concentrations of material defects in the electrolyte matrix; and c) decrease of the thickness of the electrolyte layer.
3. *Concentration polarization losses*: These losses are the result of the finite fuel and oxidant diffusion processes that govern the movement of gases in the overall electrochemical reaction. Since the rate of mass transport of gases is subject to the Fick's diffusion law, the maximum rate of gas diffusion, which is directly related to the maximum current density that may be obtained, is achieved when the concentration of the fuel in the electrochemically active area is zero. The measure of the concentration polarization is the potential difference between two modes of operation of the fuel cell: a) when the current is flowing and b) when the external circuit is open and current is not flowing at all.
4. *Activation polarization losses*: These losses are the result of the reaction kinetics involved with the overall electrochemical reactions. Each chemical reaction has an activation barrier, which must be overcome for the reaction to proceed and the voltage to be developed. This barrier leads to additional polarization. The activation barrier is the result of the many electrochemical reaction steps, where the rate limiting step is responsible for the polarization. Good design and optimization of the microstructure of the electrolyte and the electrodes of the fuel cell may reduce significantly the activation polarization losses. This is in general accomplished by increasing the *triple phase boundary length* (*tpbl*), which is the electrochemically active part of a fuel cell.

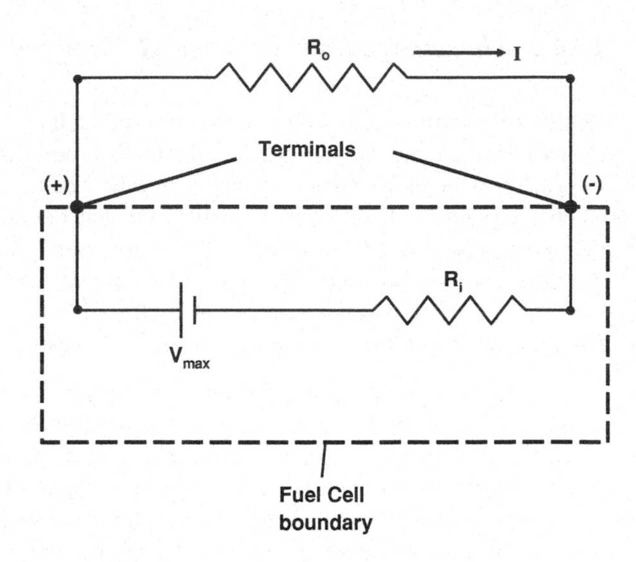

Fig. 12.11 The electric circuit diagram of a fuel cell

All the irreversibilities associated with the operation of a fuel cell may be lumped together in the concept of the *internal resistance* of the cell.[4] The internal resistance is the cause of energy dissipation that would have been produced by the cell under ideal conditions. Figure 12.11 shows the electric circuit model of the fuel cell with the internal resistance R_i. While the cell produces its full electromotive force or maximum voltage, $V_{\max}$, because of the internal losses, R_iI, the voltage available to the external circuit, V, and which produces the power delivered is lower:

$$V = V_{max} - IR_i \tag{12.27}$$

The power produced by the fuel cell is the power that is available to the circuit outside the cell:

$$\dot{W} = VI = V_{\max}I - I^2R_i. \tag{12.28}$$

Under the normal operation of fuel cells and batteries an amount of power equal to I^2R_i is dissipated inside the cell as heat and should be removed. As with all the power producing devices, fuel cells produce their own waste heat. It must be noted, however, that this is not a consequence of the Carnot limitations of thermo-mechanical conversion devices, but of the electrical and chemical irreversibities outlined above.

---

[4] The same concept may be applied to the operation of the Daniel cells and all electrochemical devices.

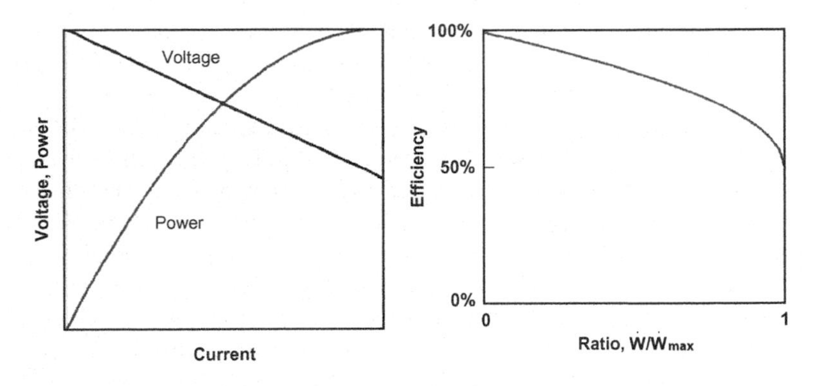

Fig. 12.12 Voltage, power and efficiency of a fuel cell or battery

Another source of irreversibilities in the operation of fuel cells stems from the fact that fuel cells and batteries produce direct voltage and direct current. The same applies to other direct energy conversion devices, such as photovoltaic solar cells. Most of the electric appliances and equipment used in households and industry are designed to use alternating current. If the power produced by the fuel cells is to be used for the supply of electricity to residential communities, the power must be converted or "inverted" to alternating current. Industrial inverters, which are placed at the point of power production, are commonly used for this purpose. A small amount of power, typically 10–20%, is lost in the inverter. This loss of power is expressed by the efficiency of the inverter, which is defined as the ratio of the electric power output to the electric power input. As with all devices, the lost power in the inverter is dissipated as heat and must be removed.

There are several figures of merit or efficiencies that describe the operation of fuel cells, as well as that of other electrochemical cells. These figures of merit are expressed in terms of the voltage or the current obtained from the cell or in terms of the total electric energy produced in the fuel cell. The following three different measures of the ability of a fuel cell to produce voltage, current and power, and which are often called efficiencies of fuel cells, have been commonly used:

$$\eta_V = \frac{V}{V_{\max}}, \quad \eta_I = \frac{I}{I_{\max}} \quad and \quad \eta = \frac{VI}{-\dot{n}\Delta H^o} = \frac{\dot{W}}{-\dot{n}\Delta H^o}, \tag{12.29}$$

where $\dot{n}$ is the rate of moles of the fuel that is fed to the fuel cell. The first two figures of merit are often called *voltage efficiency* and *current efficiency* of the fuel cell. The last expression is the *energetic efficiency* of the cell or, simply, the *efficiency* of the fuel cell. This figure of merit is the ratio of the instantaneous electric power produced to the chemical power delivered to the cell. Figure 12.12 shows the external voltage and power as a function of the current as well as the efficiency of the fuel cell as a function of the power produced. A glance at Eqs. (12.28) and (12.29) and this figure proves that the external voltage and the efficiency of the fuel cell decrease significantly with the current and power produced. Actually, at maximum power, the voltage at the electrodes is $V = 0.5V_{max}$ and the efficiency of the fuel cell is 50%. This is achieved when $R_i = R_e$. The maximum efficiency of the fuel cell is reached when the current approaches zero and the power produced also approaches zero. Therefore, in fuel cells as well as in batteries there is a trade-off between achieving maximum power and high energetic efficiency. For this reason, practical fuel cells operate at conditions below maximum power, conditions that make their efficiencies higher than 50%.

It is observed in Eq. (12.29) that the enthalpy difference of the underlying reaction is used in the expression for the energetic efficiency, instead of the Gibbs free energy ($-\Delta G^o$), which is more relevant to chemical reactions. The reason is historical and stems from the fact that common fuels, such as methane, hydrogen, or ethyl alcohol, have been routinely used for the production of heat. When these fuels are burned in a conventional way for the production of power, the maximum energy (heat) produced is $-\Delta H^o$, not $-\Delta G^o$. Hence, the former quantity has been traditionally used with these fuels and was extended to the determination of the efficiencies of fuel cells. Since the maximum electric energy that may be produced by a fuel cell is equal to ($-\Delta G^o$), and given the relationship between ΔG^o and ΔH^o, the maximum power efficiency of a fuel cell is given by the expression:

$$\eta_{\max} = \frac{-\Delta G^o}{-\Delta H^o} = 1 - T\frac{\Delta S^o}{\Delta H^o}, \tag{12.30}$$

where ΔS^o is the entropy change of the overall reaction.[5]

It must be noted that the entropy change of several fuel-oxidant reactions is negative, especially the entropy change of the combustion of liquid fuels. This implies that $\eta_{max} > 1$. For example, the combustion reactions of methanol (CH_3OH) and ethanol (C_2H_5OH) to form carbon dioxide and water at 298 K and 1 atm have $\Delta G^o = -689{,}420$ and $-1{,}299{,}640$ kJ/kmol respectively. The enthalpy change of these combustion reactions, ΔH^o, is $-676{,}270$ and $-1{,}234{,}810$ kJ/kmol respectively. The maximum theoretical efficiencies of fuel cells that would operate with methanol and ethanol would be $\eta_{max} = 1.02$ and 1.05. Of course, the maximum efficiencies are achieved at extremely low values of the flowing current and of the power produced.

Although efficiencies higher than 1 (100%) are uncommon in engineering applications, they are not impossible and do not by any means violate any physical laws. Most of the efficiencies, as well as other figures of merit, such as coefficients of performance for heat pumps and refrigerators, have been defined by engineering convention, not by natural laws. Although these figures of merit have been defined so that, in most cases, their numerical values are in the range between 0 and 100%,

[5] Most of the fuel cells operate at atmospheric pressure and for this reason the relevant properties ΔH^o and ΔG^o, which may be readily found in thermodynamic or chemical tables, are used here. In the rare case when the fuel cell operates at a different pressure, the relevant properties would be ΔH and ΔG, evaluated at the temperatures and pressures of the pertinent reactions.

in practice they may take any numerical value. Of all such figures of merit, it is only the Carnot efficiency, which stems from the second law of thermodynamics that is always constrained to be less than 100%. As it may be concluded from Eq. (12.30), all fuels that react with oxygen and have a negative entropy of reaction, would also yield $\eta_{max} > 1$ when they are used in fuel cells.

Problems

1. A small, 40 MW peak-power gas turbine operates every day from 10:00 am to 6:00 pm during the months of May, June, July and August. The gas turbine is shut down the rest of the time. Determine the availability factor and the capacity factor of this power plant.
2. A 1,000 MW nuclear power plant operates at full power during the day and at 80% from 11:00 pm to 6:00 am every day of the year. The plant is recharged with fuel and maintained every 18 months. During re-charging and maintenance, the plant does not produce any power for 20 days. Determine the availability factor and the capacity factor of this power plant.
3. A large, 800 MW nuclear power plat produces electricity at full capacity during the day. 40% of its electric capacity during the night is diverted to pump water from a river to a mountainous lake at 740 m higher elevation. The pumped water is used for the production of additional power during the daytime. The same conduits are used for the return of the water and the pump-motor assemblies double as turbine-generator pairs when the flow is reversed. It is estimated that the frictional and other losses during the pumping/returning operation are 15% of the head (that is equivalent to 111 m). The efficiency of the turbomachinery is 70% when they operate in the pump-motor mode and 75% when they operate in turbine-generator mode. What is the peak power this combination of nuclear/hydroelectric power plant produces and how much energy is lost annually by the storage-generation part of this power plant?
4. A 120 m^3 air vessel has been designed to store pressurized air. A compressor fills this vessel with 420 K, 100 atm air. During the storage period, the temperature of the air drops to 370 K. Determine:

   a) The mass of air stored in the vessel.
   b) The total exergy of the air when the filling process stops.
   c) The final pressure of the air and the total exergy of the air.
   d) The exergy loss during the storage period.

5. A 9,000 m^3 underground cavern has been constructed for the storage of pressurized air at 25 bar. An above-ground water reservoir maintains the pressure of the air constant during the filling and emptying processes. The temperature during the filling process is 178°C. What is the total mass of the stored air and what is the maximum amount of work this air–water system may produce? During the storage time, the temperature of the air drops to 83°C. As a result, the volume of the air shrinks and water seeps into the cavern to maintain the constant

pressure of 25 bar, without producing any work. What is the maximum amount of work the air–water system may produce at this state?

6. A 10,000 m^3 underground cavern has been proposed for the storage of pressurized air at 32 bar. An above-ground water reservoir maintains the pressure of the air constant at 32 bar during the filling and emptying processes. If the temperature during the filling process is 155°C, what is the total mass of the stored air and what is the maximum amount of energy this air–water system may produce? After a few months of operation, it is discovered that fissures develop in the cavern and that air leaks at an average rate of 1.2 kg/s. How much air mass is leaked during a 12 h storage period and how much volume does this correspond to? If the temperature of the air also drops to 37°C during the storage period, what is the maximum total energy that may be recovered from this storage facility?
7. A steel torsion bar is to be used for the storage of energy. The bar is cylindrical (R = 5 cm) and has a length of 1.2 m. What is the energy stored in the bar when it is rotated by 15°? What are the friction forces at the grips of two ends of the bar that would maintain this shear?
8. A flywheel is constructed with thin spokes and resembles a bicycle wheel. Assuming that the mass of the flywheel is 500 kg and that all the mass is concentrated at a radius of 2 m, what would be the rpm of the flywheel if it is to provide 0.1 MW power for 1 min?
9. Rock has often been suggested as a medium for the thermal storage of energy. A cylindrical piece of rock with diameter 10 m and height 20 m is used for the storage of thermal energy. The rock is insulated and has an average heat transfer coefficient, U = 0.28 kW/m^2 K. The density of the rock material is 2,650 kg/m^3 and its specific heat capacity is 0.72 kJ/kgK. The temperature of the rock is raised to 500°C and the ambient temperature is 25°C. Determine:

   a) The total energy stored in the rock in MWh.
   b) The temperature of the rock 12 h after the heating process stops.
   c) The heat that has escaped from the rock during this 12 h period.
   d) The loss of exergy of the rock during this 12 h period.

10. The power demand during a typical summer day in San Antonio, Texas is depicted in Fig. 12.2. This power demand is typical of urban environments in the OECD countries during the hot summer season. 60% of the demand between the hours noon to 10:00 pm is due to air-conditioning. In order to reduce peak power demand it is suggested that eutectic salts be used in conjunction with air conditioners to provide “coolness.” The eutectic salts will be frozen from midnight to 10:00 am and will provide cooling from noon to 10:00 pm. Draw a diagram of a typical power demand during a typical summer day if 20% of the consumers in San Antonio adopt this air-conditioning method. What is the percent reduction of the peak power?
11. The Gibbs free energy change of the Pb-PbO_2 reaction with sulfuric acid is ΔG^o = 286,160 kJ/kmol. Determine what is the maximum voltage an electrochemical cell based on this reaction will produce.

12. The combustion of methane produces $\Delta H^{o} = 800{,}320$ kJ/kmol of heat, which is typically used in a Rankine or Brayton cycle for the production of electricity. If the overall efficiency of the thermal cycle is 43%, what is the mass flow rate of methane that would power a 10 MW small power plant? If the same quantity of methane were to be used in a system of fuel cells with 70% efficiency, what is the power these fuel cells would produce?
13. A steel vessel of 0.5 m^3 may be used for the storage of pressurized air or pressurized hydrogen. The maximum pressure for air is 100 bar, and the maximum pressure for hydrogen 70 bar. What is the maximum electric work that may be obtained from the storage of the two substances? What is the ratio of the two and what do you conclude about the suitability of the two gases for energy storage?
14. Under normal operation, a hydrogen–oxygen fuel cell produces a current of 3.6 A and a voltage of 0.86 V. What is the internal resistance of this fuel cell? What is the energetic efficiency of this fuel cell?
15. A hydrogen–oxygen fuel cell is designed to produce 3.2 MW of electric power for a small town. The efficiency of the fuel cell is 72% and the efficiency of the inverter is 88%. For the rated power of 3.2 MW calculate:

    a) The flow rates (kmol/s) of hydrogen and oxygen consumed by the fuel cell.
    b) The waste heat generated in the fuel cell.
    c) The waste heat from the inverter.

    Suggest ways to remove the waste heat from the fuel cell and the inverter.

References

1. EL-Wakil MM (1984) Power Plant Technology, McGraw-Hill, Newyork
2. Tester JW, Drake EM, Driscoll MJ, Golay MW, Peters WA (2005) Sustainable Energy: Choosing among options, MIT Press, Cambridge

Chapter 13
Energy Conservation and Efficiency

Abstract In a strict sense, the term "energy conservation" is a misnomer. As demonstrated in Chap. 3 with the First Law of Thermodynamics, energy is conserved by nature. "Energy conservation" does not require any action by engineers or by the general population. What is usually meant by this colloquial term is the eventual conservation of natural primary energy resources, when we perform tasks that are necessary for the functioning of the human society. The colloquial "energy conservation" may be more scientifically expressed as *exergy conservation*, *minimum exergy destruction* or *minimum entropy production* while the societal tasks are being performed. The use of the exergy concept and exergetic calculations ultimately lead to the minimum consumption of primary energy resources and, finally, to the conservation of natural resources. While most of the other chapters in this book pertain to the supply side of the energy equation, conservation and improved efficiency are directly related to the demand side. By "conserving energy" and forgoing the use of a fraction of primary energy sources for the performance of societal tasks, the human society demands less primary sources that supply the global energy demand. As a result, less primary energy is demanded and consumed and the *natural resources* that supply this energy are conserved. In this chapter we distinguish between conservation and higher efficiency, we apply the concept of exergy in order to perform tasks with the minimum possible exergy destruction and we present several examples of methods that lead to the lesser consumption of natural resources in the areas of transportation and comfort in buildings. These methods include the use of fluorescent or LED devices for lighting, geothermal heat pumps for heating and air-conditioning, evaporative cooling, and electric cars.

E. E. (Stathis) Michaelides, *Alternative Energy Sources*,
Green Energy and Technology, DOI: 10.1007/978-3-642-20951-2_13,

13.1 Societal Tasks, Energy Consumption, Conservation and Higher Efficiency

Energy conservation and *energy efficiency* appear to be synonymous concepts and are often used interchangeably in the energy field. However, there are subtle differences between the two. The best way to perceive these differences is to start with the concept of a *societal task* that needs to be fulfilled with the use of energy. The task in this case is any function of individuals, groups or the entire society, which is normally accomplished by the use of energy resources. Examples of such tasks are as follows:

- Maintaining a warm and comfortable temperature for the residents of a building in Berlin, Germany, during the winter months.
- Maintaining a cool and comfortable temperature for the residents of the households of Houston, Texas during the summer months.
- Providing adequate lighting in the classrooms of a High School so that students may read their books comfortably.
- Cooking 0.4 kg of pasta to produce a meal.
- Producing 10 gallons of gasoline from crude oil.
- Producing one ton of cement.
- Transporting three hundred passengers from London to Oxford.
- Manufacturing 1,500 m of 8-gauge copper wire.

From the beginning, it is important to realize that the human society does not demand energy *per se.* Humans use energy in order to accomplish several simple or complex tasks that are similar to the ones listed in the last eight bullets. When one looks at these tasks, one realizes quickly that energy use is not an end by itself, but the means to accomplish these and other similar tasks. If the tasks are accomplished, humans are satisfied, the economy moves along and the society functions well. How much energy is consumed for the accomplishment of these tasks is entirely irrelevant to the society and to human comfort, satisfaction and happiness. Simply put, if the societal tasks are accomplished, the human society is progressing and humans are happy, regardless of the amount of energy consumed. The actual amount of energy consumed is immaterial to the human society as long as "business is done." It is only when humans are unable to fulfill such tasks, because of the lack of sufficient energy supply or because energy has become too costly, that discontent and occasionally frustration and wrath manifest themselves in the societies and the nations. This discontent was demonstrated in several examples of violence during the energy crisis of the 1970s in the USA and Britain.

Similarly, when it is said that "energy demand" increases or will increase in the future, what is actually meant is that either humans will perform a larger number of societal tasks, which will require the use of more energy, or that more humans will perform the same number of individual tasks, which will require the use of more energy. Based on the statistical information of Chap. 1, increased affluence in a society typically implies that more such tasks will be required to be accomplished. For example, when a society becomes more affluent and the "standard of living" rises, a society typically requires more air-conditioning for the summer months, which increases significantly the use of electric energy. Similarly, when the population increases, the energy use also increases because more humans desire such tasks to be accomplished. Therefore, if one writes the total consumption of energy, E, as a function of the total number of tasks performed, TSK, one obtains the following expression for the rate of increase of the energy demand as a function of time:

$$\frac{dE}{dt} = \left[\frac{\partial E}{\partial (TSK)}\right] \frac{d(TSK)}{dt} + \frac{\partial E}{\partial E_T} \frac{\partial E_T}{\partial t}, \tag{13.1}$$

where E_T is a measure of the average energy consumed per task. This simple and very general equation points to the fact that energy consumption may decrease if the average amount of energy consumed per task decreased or if the total number of tasks decreased. The first is related to what is commonly called "energy conservation" and to the increased efficiency of processes. The second is related to the total population on the planet that uses energy.

The accomplishment of the societal tasks may be done in more than one ways, methods or processes, which use different amounts of energy. Let us consider the first task in the previous list: to maintain a comfortable temperature for the residents of a building in Berlin during the winter months. This task may be accomplished in a variety of ways including the following:

1. Use the existing natural gas burner of the building and maintain the interior temperature at 24°C, as it was done in the last twenty years.
2. Use the existing natural gas burner of the building, maintain the temperature at 20°C and ask the residents to wear an extra sweater to keep warmer. Less energy will be used for the heating of the building.
3. Replace the existing natural gas burner with a heat pump, which consumes electricity instead of natural gas and maintains the temperature at 24°C. Because of the higher coefficient of performance of the heat pump, even if we account for the natural gas energy conversion to electric energy, less overall energy will be consumed for the fulfillment of the task, which is to maintain a comfortable temperature for the inhabitants of a building.

Obviously, the first alternative is the *status quo* of accomplishing the task: do nothing and continue with the same method the task was accomplished in the past. The second alternative necessitates that the residents of the building will have to change their behavior by putting on an extra sweater or another piece of clothing. This alternative implies that the residents will have to cooperate or to assist in the accomplishment of the task. In this case there is *conservation of energy* because an action of cooperation is required by the residents for less energy to be used. The third alternative is typical of an increased *energy efficiency* project. With the replacement of the hot-water burner by a heat pump, the temperature of the

Table 13.1 Several actions that lead to energy conservation and energy efficiency. All actions lead ultimately to natural resource conservation, which is the minimization of natural resource consumption

Energy conservation action	Energy efficiency action
Switch off lights when out of a room	Replace incandescent bulbs with fluorescent ones
Use carpooling for the daily commute	Replace the *BMW-381i* with a *Ford Escort* for the daily commute
Increase the thermostat temperature from 21 to 24°C in the summer. Do the opposite in the winter	Use a geothermal heat pump to replace the old air-conditioning and heating systems
Implement the daylight energy savings time	Install thermal insulation in the roof of buildings
Mandate a maximum speed on the highways	Install an additional feed-water heater in the electric power plant

building is maintained at the same level of 24°C and no cooperation is required of the residents of the building. The use of less energy is accomplished by the higher efficiency of the heat pump without any discomfort to the residents of the building.

Thus, *energy conservation* implies an action or a "sacrifice" from the society, such as: to switch off lights; to drive less miles; to use more bicycles rather than cars for transportation; to reduce the building thermostat in the winter and raise it in the summer, etc. On the other hand, *energy efficiency* is typically accomplished by the replacement of machinery, equipment or processes, with all other conditions remaining the same and does not require any "sacrifice" of comfort from humans. The lesser consumption of natural resources may be accomplished by either *energy conservation* or improved *energy efficiency* or a combination of both.

It must be noted that, occasionally, and especially when laymen are speaking, the terms *energy conservation* and *energy efficiency* are used interchangeably to denote the substitution of one energy form that consumes natural resources with an alternative energy form. For example, the substitution of a gas water heater with a solar heater, and the production of electric power from a wind turbine may be claimed as "energy conservation measures." For the correct accounting of energy resources and optimized results, the scientist or engineer must be able to differentiate between substitution, efficiency and conservation. Table 13.1 gives several examples of *energy conservation* and *energy efficiency* actions as used in everyday life.

The common characteristic of both energy conservation and energy efficiency efforts is that societal tasks are performed with a lesser use of primary energy resources. In the case of the second and third alternatives for the heating of the building in Berlin, either one uses a heat pump or raises the thermostat, the net effect will be a lesser use of heating fuel. Both alternatives and, by extent, both energy conservation measures and energy efficiency measures result in the lesser consumption of primary energy sources. When the engineer or the entire society strive to consume less energy primary sources for the accomplishment of the societal tasks it is useful to know what is the *minimum amount of energy* the accomplishment of a certain task requires. For example, what is the minimum amount of energy one will have to use for the production of one ton of cement or for cooking 0.4 kg of pasta? This minimum for a given task may become a *benchmark* for the energy consumption and the task of the engineers will be to design processes and equipment that would more closely approach the minimum. As explained in Chap. 3, this benchmark may be formulated using the concept of exergy.

13.2 The Use of the Exergy Concept to Reduce Energy Resource Consumption

The concept of exergy may be used by engineers and scientists to make decisions on the choice of processes and equipment that best utilize the energy resources. In the following sections we will demonstrate with practical examples how the concept of exergy may be used in choosing or improving the engineering processes and systems that perform a given task by utilizing the minimum amount of primary energy resources.

13.2.1 Utilization of Fossil Fuel Resources

Let us assume that we have a certain amount of a natural resource, e.g. a mass of methane, CH_4, equal to one kmol, or 16 kg, and we wish to design an energy conversion system that would maximize the amount of electric work we produce. Since methane is a hydrocarbon, a common way to convert its chemical energy into work is to combine the resource with oxygen from the atmosphere and burn it to release its chemical energy in the form of heat. Accordingly, methane and oxygen undergo a chemical reaction and produce water and carbon dioxide.

$$CH_4 + 2O_2 \rightarrow CO_2 + 2H_2O. \tag{13.2}$$

The reaction also releases a large quantity of heat. When one kmol of methane combines with oxygen in a conventional burner, it produces $\Delta H^o = 800{,}320$ kJ of heat. In a very commonly used process, methane is fed in the combustion chamber of a gas turbine, which utilizes a Brayton cycle and releases the heat of reaction to the working fluid. The turbine produces power and work as shown schematically in Fig. 3.9. In a well-designed power cycle, which has an overall thermal efficiency approximately equal to 40%, this process will produce an amount of electric work: W = 320,128 kJ/kmol. This amount of work may vary a little, depending on the efficiency of the gas cycle, which for most conventional gas power plants would be in the range 45–30%.

A moment's reflection will show that methane is a chemical substance, where the available energy is stored in the form of chemical energy. The maximum work that may be extracted from one kmol of any substance is given by the negative Gibbs free energy change during the reaction, $-\Delta G^o$, as expressed in Eq. (3.36). A glance at Thermodynamics Tables shows that for methane, $\Delta G^o = -816{,}650$ kJ/kmol and, hence, the maximum amount of work, which may be extracted from 1 kmol of methane would be $W_{max} = 816{,}650$ kJ.[1] Since this quantity of work is more than 2.5 times the work from the Brayton cycle mentioned above, it becomes apparent that there must be an alternative way to utilize methane and obtain an amount of work that is significantly higher than the amount of 320,128 kJ, which is produced by the original Brayton cycle power plant.

It is apparent that the combustion of methane in a conventional work-producing cycle would only supply a fraction of the maximum work. The combustion of all fossil fuels yields significantly less than the maximum work, which might be obtained from them, primarily because the chemical energy of the fuels is first converted to heat. The subsequent conversion of heat into work is subjected to the Carnot limitations, which were explained in Chap. 3. The discrepancy in the actual work obtained and the *ideal* or maximum work that might be obtained from the natural resource of the 1 kmol of methane, may lead the scientists and engineers to devise an alternative method, a process or a combination of processes that would potentially produce the maximum amount of work, $-\Delta G^o$, from these fuels.

When searching for a better alternative method than the combustion of fossil fuels, it becomes apparent that the maximum work from fossil fuels may be recovered by Fuel Cells, which were described in detail in Sect. 12.6. Fuel Cells convert directly the chemical energy to electricity, are not subjected to the Carnot limitations and may potentially convert the full amount of $-\Delta G^o$ into electric work. The proper use of the concept of exergy and the knowledge of the maximum work that may be obtained from this resource, points to the use of a Fuel Cell rather than a thermal power plant. Fuel cells in principle may convert the entire amount of the reaction's Gibbs free energy, $-\Delta G^o$, into electric energy. However, the thermal efficiency of fuel cells, which are now at a developing stage, is approximately 70–75%. This implies that only 70–75% of $-\Delta G^o$ will be converted to electric energy. Even this fraction is significantly higher than the thermal efficiency of a typical thermal power plant. Instead of using a burner/boiler with fossil fuels, the fossil fuel resources would produce significantly more electric work if they were used in a direct energy conversion device, such as a fuel cell.

It must be noted that, this example leads us to use an entirely different method for the conversion of the chemical energy of methane into work and it does not merely lead to small-scale improvements of the original method. While the exergy analysis determines the maximum work that may be produced from a natural resource it does not reveal directly the method, the system, or the equipment that will produce this maximum work. This is the assignment of the engineer or the scientist who is tasked with the production of the maximum work and who has to

[1] Methane combustion is one of the reactions with negative entropy change, $\Delta S < 0$. The maximum work, $-\Delta G^o$, that may be obtained from a given mass of methane is slightly higher than the heat obtained upon combustion, $-\Delta H^o$.

interpret the exergy calculations and to design the appropriate processes and equipment that would produce the maximum work. Also, the indication of a certain type of equipment, as with the fuel cell, does not necessarily imply that the maximum work will be produced, because the efficiency of all equipment will be less than 1. The invaluable contribution of the exergy concept to engineering is to furnish the amount of the maximum work that can be obtained from a resource or supply the value of the limit, which the engineers may strive to achieve. The knowledgeable engineer or scientist will then determine the method and will design the equipment and the systems to achieve a performance as close as it is feasible to obtain this optimum, within the economic, social, and environmental constraints of the design process.

13.2.2 Minimization of Energy or Power Used for a Task

In addition to helping determine the maximum work an energy resource may produce, the concept of exergy may also be used for the determination of processes or equipment when work or power is consumed. In such cases, exergy will help determine the minimum work, or power, that may accomplish a given task or a given process, where work or power must be used.

We use work or power in our everyday activities in order to accomplish certain tasks. For example, we use electric power in a refrigerator in order to keep food at a temperature below 5°C, or we use gasoline in a car in order to transport ourselves from home to work. The tasks here are the preservation of food and our transportation from home to work, not the consumption of electricity and gasoline, respectively. It must be emphasized that these tasks are not the consumption of a certain amount of energy or power. Since, power or work is necessary for the accomplishment of the tasks and since both have a cost, it is evident that rational consumers will try to accomplish all these societal tasks by using the minimum possible amount of work or power.

From the beginning we must recall that, according to thermodynamic convention, work produced by a system is positive, while work consumed by a system is negative. From the thermodynamic point of view, the work consumed by a refrigerator for the preservation of food during a day will be a negative number. Let us assume that we have three types of refrigerators, which during a year fulfill the task of keeping the food below 5°C, that consume 1,356 kWh, 1,672 kWh, and 2,198 kWh.[2] A rational consumer and environmentally conscious citizen would choose the refrigerator that consumes 1,356 kWh per year, which is the appliance that consumes the *minimum* work during a year. However, according to the thermodynamic convention, which is depicted in Fig. 3.4, the actual numerical values of the work required to run the three

[2] In many countries, refrigerators and other household appliances are sold by the manufacturer with an estimate on their annual consumption of energy.

refrigerators are respectively −1,356 kWh; −1,672 kWh; and −2,198 kWh. Of these, the value −1,356 is, actually, the *maximum*. Therefore, in order to conserve energy resources, we have chosen the *maximum* numerical value that represents the work. What is colloquially called the *minimum amount of work consumed*, is actually a *maximum* in the thermodynamic convention, which may be interpreted as the maximum work produced. Since exergy always determines the maximum absolute value of work involved with a process, the concept of exergy may be used for the determination of the best way to utilize the energy resources, not only in work-producing processes, where the value of the work is positive, but also in work-consuming processes, where the value of the work is algebraically negative. The following example of air compression will illustrate how one may use the concept of exergy in decision making processes involving the consumption of lesser work or power.

Let us assume that the task of a given process is to compress a mass of 1 kg of air at atmospheric pressure and temperature, which is 300 K,[3] to 20 atm. This process could be the compression process in a gas power cycle or in a process of energy storage via compressed air as described in Chap. 12. A typical compressor, with an 80% isentropic efficiency would consume approximately 506 kJ (actually,−506 kJ, according to thermodynamic convention). This compressor would compress the air to the required 20 atm, and by doing so, would also increase its temperature from 300 K to approximately 780 K. Since the task is solely the increase of the pressure, a moment's reflection proves that the increase of the air temperature is not necessary for the accomplishment of the given task. The significant temperature increase is a consequence of the type of equipment that was used for the accomplishment of the task, the almost isentropic compressor. Indeed, this rise in temperature is responsible for a good part of the 506 kJ of work required for this compression. Even if we had a perfectly isentropic compressor (and such devices do not exist) the amount of work spent would have been 405 kJ and the exit temperature approximately 693 K, which is, still, significantly high.

Now, let us employ the concept of exergy to determine if the task of pressurizing the air from 1 to 20 atm may be accomplished using a lesser amount of work. Since air is a compressible substance, we may determine the *maximum work* that is required by the physical principles, the Laws of Thermodynamics, for this process. The initial state of this process corresponds to the environmental state, which will be denoted by the subscript *0*, and the final state, denoted by the subscript *1*. Thus, $P_0 = 1$ atm, $T_0 = 300$ K and $P_1 = 20$ atm. Re-writing Eq. (3.31) for the compression process *0–1* and using the assumption that air is an ideal gas[4] with constant specific heats we obtain the maximum specific work for this process as follows:

[3] In the absence of other information, the ambient temperature is usually taken to be 25°C (298 K). Also, most of the data on chemical reactions pertain to a reference temperature 25°C.

[4] The ideal gas assumption is not essential in the use of exergy for better efficiency. Any other equation of state or tables for the gas may be used. In such cases a closed algebraic form for the maximum work may not be feasible and one may need to obtain a numerical solution.

$$w_{max} = e_0 - e_1 = h_0 - h_1 - T_0(s_0 - s_1) = c_P(T_0 - T_1) - c_P T_0 ln\frac{T_0}{T_1} + RT_0 ln\frac{P_0}{P_1}. \tag{13.3}$$

Of the three terms in the last part of this equation, only the term $RT_0ln(P_0/P_1)$ is directly relevant to the task of air pressurization. As expected, this term has a negative value since $P_0 < P_1$, signifying that work must be spent for the process to occur. The other two parts of Eq. (13.3), which pertain to the initial and final temperature of the system, $c_P(T_0 - T_1) - c_P T_0 ln(T_0/T_1)$, are irrelevant to the original task, which is to provide pressurized air and does not necessarily ask for the air to be at a higher temperature.

It may be easily proven by elementary calculus that, for all values of T_0 and T_1, except for $T_0 = T_1$:

$$c_P(T_0 - T_1) - c_P T_0 \ln(T_0/T_1) < 0, \tag{13.4}$$

The fact that this expression is negative implies that additional work must be spent for the air to increase its temperature during the pressurization process. Hence, the part of Eq. (13.3) that contains the temperature terms always adds to the work required for the completion of the task, even though it does not pertain directly to the given task. One may also observe that this part of the equation becomes equal to zero if $T_0 = T_1$. Therefore, if the condition $T_0 = T_1$ is satisfied in the compressor, that is we have an isothermal compression of the gas, the absolute value of the specific work required would always be less than that required by an isentropic compressor. It is seen again that the exergy analysis of the air pressurization process points to the best alternative process, the isothermal compression. The "*minimum work*" required for this isothermal compression is actually an algebraic and thermodynamic *maximum*, and in this case it is given by the expression:

$$w_{max} = RT_0 ln\frac{P_0}{P_1} = -258 \text{ kJ/kg}. \tag{13.5}$$

This may be interpreted that *at least* −258 kJ of work must be performed for the pressurization of 1 kg of gas, or that *at least* 258 kJ/kg of work must be consumed for the task to be performed.

One may observe that this amount of work is significantly lower than that required for even the idealized isentropic compression, which is 405 kJ/kg in absolute value. Ironically, although the value −258 kJ/kg is a maximum, in practice we call this "minimum work," because we typically consider only the absolute value of the work consumed, not its algebraic value. It is apparent in this example that the correct use of thermodynamic theory, and especially of the concept of exergy, provides only the numerical value of this "minimum work," and only gives scant indications on the process, equipment or the engineering system to be used. It is up to the ingenuity of the good scientist or engineer to design the engineering system that best approximates this optimum.

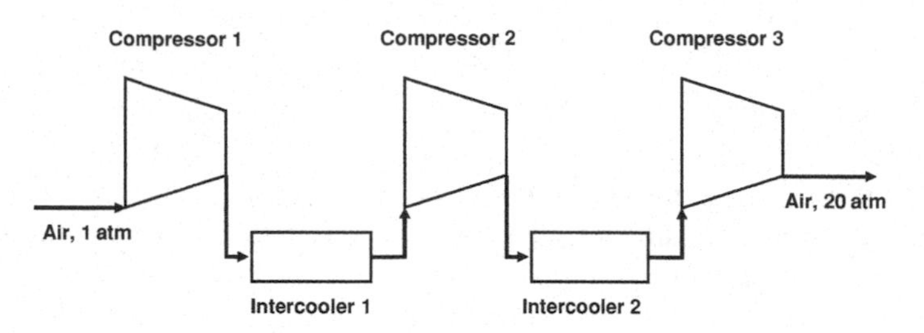

Fig. 13.1 A series of three compressors with intercoolers

Table 13.2 Compressor input work, in kJ/kg, for the compression of air from 1 to 20 atm

Isentropic compression, no intercooler, $\eta = 80\%$	506
Isentropic compression, no intercooler, $\eta = 100\%$	405
Isentropic compressions, one intercooler, $\eta = 80\%$	400
Isentropic compressions, two intercoolers, $\eta = 80\%$	371
Isentropic compressions, many intercoolers, $\eta = 80\%$	323
Isothermal compression, $\eta = 100\%$	258

It must be noted that, because isothermal processes require a significant amount of heat transfer from the compressed gas, such processes are very slow and rather difficult to achieve in practice, if a significant quantity of gas is to be compressed. In engineering practice a number of smaller, isentropic compressors with intercoolers, which were mentioned in Sect. 3.6.2, are used for the reduction of the gas exit temperature. The intercoolers are essentially heat exchangers that admit fluid from a compressor, cool the fluid to a lower temperature and supply it to the next compressor for further pressurization as it is shown in Fig. 13.1. This is repeated until the fluid is pressurized to the desired level. In the example of the pressurization of 1 kg of air, when only one intercooler is used with two 80% efficient compressors, the work required is 400 kJ; when two intercoolers are used, the work required is reduced to 371. An almost isothermal process may be achieved by using a large number of intercoolers with almost isentropic compressors. If a very large number of compressors and intercoolers are used with 80% efficient compressors, then the amount of required work is 258/0.8 = 323 kJ/kg. Table 13.2 gives a summary of the absolute value of the work required during these processes of the compression of atmospheric air to 20 atm. It is apparent that a good engineer will use the exergy method and design the compression process with a couple of intercooling stages to achieve significant "energy savings" for the performance of the given task.

13.2.3 Combination of Tasks: Cogeneration

More than one task needs to be accomplished simultaneously in several industrial applications. For example, a refinery uses a high amount of heat at moderate temperatures (110–130°C) as well as a significant amount of electric power for the operation of its turbomachinery. A typical commercial supermarket in the winter uses a significant amount of heat for space heating as well as electric power to run the refrigerators and freezers. The two tasks in this case are the production of an amount of electric power $\dot{W}$ as well as a rate of heat $\dot{Q}$. The tasks may be accomplished separately, e.g. by means of a vapor cycle, which uses the chemical energy of a fossil fuel at a rate $\dot{m}_w$, for the production of the power and a separate burner for the production of heat using a similar fossil fuel at a rate, $\dot{m}_q$. An exergy analysis of the two processes would prove immediately that the two separate systems would destroy a great deal more exergy and, thus, would consume a great deal more fuel (natural resources) than a single *cogeneration* system, which generates both power and heat. The cogeneration of electric power and heat satisfies both tasks by producing simultaneously the required amounts of heat and electric power using a single cycle, which may be a vapor or a gas cycle.

Cogeneration of heat from a vapor cycle may be achieved in one of the following two methods:

a) Using a condenser at higher temperature than the temperature at which heat is required and,
b) Extracting a fraction of the steam from the turbine (bleeding) at suitable pressure and temperature.

The schematic diagram of the first method is identical to the one depicted in the typical Rankine vapor cycle of Fig. 3.8, with the condensate extracted at higher pressure and the condenser being used as the heat exchanger that transfers the heat for the accomplishment of the heat addition task. The second method is depicted in Fig. 13.2. Steam is extracted from the turbine at state 5 and is diverted to a heat exchanger which delivers the heat of condensation for the fulfillment of the heating task. The condensate at state 6 is pumped to the boiler and, thus, returns to the original power cycle. If the mass flow rate of the extracted steam is denoted by $\dot{m}_1$, the rate of heat extracted from the cogeneration cycle is given by the following expression:

$$\dot{Q} = \dot{m}_1 (h_5 - h_6). \tag{13.6}$$

Cogeneration may also be achieved by a gas cycle, using the turbine exhaust gas.

Of course, the extraction of the steam fraction from the power cycle implies that a higher amount of steam must be heated in the cogeneration cycle for the production of the power $\dot{W}$ and that the amount of fuel used in the co-generation

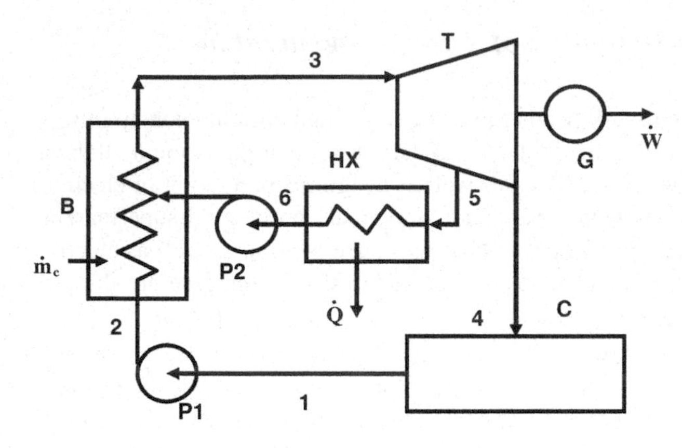

Fig. 13.2 Cogeneration of heat and electric power from a single cycle

process, $\dot{m}_c$, is higher than the original amount of fuel, $\dot{m}_w$, which produced the original power, $\dot{W}$, in the conventional power plant. However, it may be easily proven that the rate of fuel, which must be spent for the separate production of heat and power, $\dot{W}$ and $\dot{Q}$, is always greater than the rate of fuel used in the cogeneration process:

$$\dot{m}_w + \dot{m}_q < \dot{m}_c. \tag{13.7}$$

The last inequality indicates that the cogeneration method always results in the conservation of primary energy resources. The examples of methane combustion, air pressurization, and cogeneration show how the concept of exergy and *exergy analysis* may be used to determine the maximum work and power that may be extracted from a natural resource; the minimum work required for the completion of a task; and the minimum amount of energy resources to be used for the completion of two tasks. The analysis of engineering systems using the exergy concept reveals the optimum value and points to the method that should be used for optimum results. The analysis also assists the scientist or the engineer to choose equipment and processes that lead to an optimum design of the pertinent system. An exergy analysis is a valuable scientific tool that should be used in the design of energy conservation projects as well as projects that are used for the harnessing of natural energy resources.

13.2.4 Waste Heat Utilization

All power plants reject very large quantities of heat, the so-called *waste heat,* to the environment. A typical 400 MW coal-fired power plant has thermal efficiency approximately 40% and rejects approximately 600 MW to its surroundings. Even though this is an enormous amount of heat power, it is not used in practice

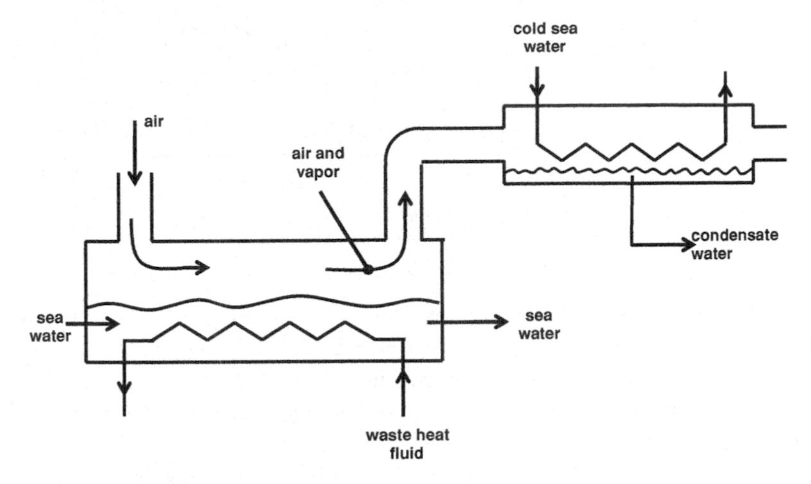

Fig. 13.3 Use of waste heat in desalination

because it is rejected from the condenser at low temperatures, which are typically close to 40°C. Very few practical applications may use such low temperature heat. Because there is a great deal of such low temperature heat available near all the steam power plants, including fossil-fueled and nuclear power plants, substantial energy savings may be generated if users of low temperature heat would collocate with power plants. Among the applications that may use such low temperature heat are the following:

1. *Sea-water desalination*, which is very important in arid regions that are close to the sea, such as the Texas part of the Gulf of Mexico, the countries of the Persian Gulf and the countries that border the Red Sea. A schematic diagram of this desalination process is shown in Fig. 13.3. The colder sea water on the left enters a vessel where it is heated by the waste heat water to a higher temperature (35–42°C). A constant stream of air from the top extracts water vapor and produces an air stream with higher humidity. This stream of air is then directed to another vessel/condenser, where the additional humidity is condensed by the ambient sea water. The condensate is fresh water, which is collected at the bottom of this vessel and sent for consumption. For more fresh water condensate, the warmer saline water output of the condenser may be mixed with the sea water input of the evaporator at the left.
2. *Soil heating for agriculture*. In general, a small increase of the soil temperature causes higher crop yield and, oftentimes, multiple crops per season. The waste heat from a power plant may be transferred by warm water in underground piping to increase the soil temperature in neighboring fields and greenhouses. In temperate climates, the crop yield often doubles when the soil temperature is maintained between 30 and 34°C.
3. *Aquaculture*. Similar to agriculture, heated ponds have a significantly higher yield of fish. The water of the ponds may be controlled and, when necessary,

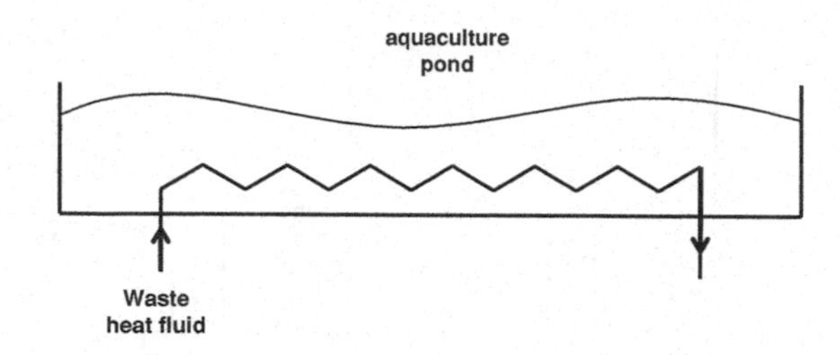

Fig. 13.4 Use of waste heat in aquaculture

heated up by the waste heat to temperatures in the range 30–35°C where not only the fish protein yield is higher per unit surface of the pond, but also where different, more desirable aquatic species may grow to be harvested. Figure 13.4 shows a schematic diagram for the use of the waste heat from a nearby power plant to aquaculture.

4. *District heating* is the heating of a large number of buildings by a single heat source. Geothermal water may be used for district heating systems as was seen in Sect. 9.5. The large quantities of heat produced by power plants may be similarly used for the heating of nearby buildings. When the electric power plant is located close to a district of a city, this district may satisfy a great deal of its space heating needs using the waste heat of the plant. Again, water may be the heat transfer medium that transports the heat from the condenser of the power plant to the building interior.[5]

It is apparent that, when the distance of the heat source to the consumption of waste heat is long, e.g. more than a few kilometers, the use of the waste heat becomes more difficult, primarily for three reasons:

a) The temperature of the heat carrying medium, usually water, drops by heat transfer to the surroundings;
b) The pumping power for the transport of the water becomes significant; and
c) The capital cost in piping and pumps required for the transfer of heat becomes high enough to render the project uneconomical.

13.3 Conservation and Efficiency Measures in Buildings

Approximately one third of the energy consumption in the OECD countries is spent in private and public buildings. The temperature of the buildings is maintained within a narrow range throughout the year, for the comfort of their

[5] Apart from waste heat utilization, other sources of hot water, such as water from hot springs and aquifers, have been used for district heating. Oftentimes such district heating systems are classified as *geothermal district heating*.

inhabitants. Since the ambient temperature varies significantly and the inside temperature remains almost constant, heat enters the buildings during the hot summer months and leaves the buildings during the cold winter months. In order to compensate for these natural heat transfer processes and maintain the almost constant temperature in the buildings, significant amounts of heat must be supplied to the buildings during winter and removed during the summer. Heating during the winter; air-conditioning during the summer; lighting throughout the year; and the supply of hot water throughout the year are tasks that consume most of the energy required in the buildings. In addition, appliances, such as refrigerators, microwave ovens, television systems, computer, and communication systems consume significant amounts of energy in the form of electric power. It is apparent that any reduction in the amount of work and heat required for the performance of these tasks would result in the conservation of energy resources. The following subsections describe some of the methods that may be implemented or are currently used for the minimization of the total primary energy use in buildings.

13.3.1 Use of Fluorescent Bulbs or Light Emitting Diodes

The typical incandescent bulb provides very low amount of light energy in comparison to the electric energy input. Typical efficiencies of incandescent bulbs—defined as visible radiation energy divided by the electric energy input—are in the range 2–4%. Typical efficiencies of florescent bulbs are in the range 10–12% and those of sodium lamps approach 20%. The efficiency of Light Emitting Diodes (LED) is close to 60%. It is apparent that the substitution of an incandescent bulb with a fluorescent bulb or, better, a LED would increase significantly the lighting efficiency of a building. For example, the amount of lighting produced by a 100 W incandescent bulb may be provided by a 25 W fluorescent bulb or by a LED that consumes only 5 W. In a typical multi-use commercial building where lighting is provided for 50% of the time during the year, that is for 4,380 hrs/yr, the mere substitution of a single incandescent bulb with its fluorescent equivalent would save 60*60*4,380*(100 − 25) J = 1,183 MJ of electricity, or 328.5 kWh per year.

This is not the final number of the energy savings. The energy used by all the lighting devices dissipates in the building and heats up the air. Since a building requires cooling during the summer and heating during the winter, the lighting energy saved does not need to be removed by the air-conditioning system during the summer and must be supplied by the heating system during the winter. In calculating the total energy savings, the location of the building makes a significant difference. For example, if the building where the 1,183 MJ of lighting energy are saved is located in Fort Worth, Texas, USA where air-conditioning is required 65% of the days during a year (2,847 hrs/yr at the 50% utilization rate of the building) and heating 15% of the year (657 hrs/yr) and if heating and cooling are accomplished with a heat pump system that has coefficients of performance 2.8

Table 13.3 Summary of annual energy savings, in kWh, in a large building where 10,000 W of incandescent light bulbs are substituted with fluorescent lights that consume 2,500 W and produce the same luminescence

Source of savings	Location: Fort Worth, TX	Location: Berlin
Electricity to lights	32,850	32,850
From air-conditioning	7,630	2,350
Additional heating	(1,300)	(5,190)
Total	39,180	30,010

for cooling and 3.8 for heating, then an additional 2,847*60*60*(100 – 75)/2.8 J = 275 MJ or 76.3 kWh of electricity is saved from the cooling requirements of the building and an additional 657*60*60*(100 – 25)/3.8 J = 46.7 MJ or 13.0 kWh must be supplied for the heating requirements of the building during winter. The total annual savings for this building would be (328.5 + 76.3 – 13.0) = 391.8 kWh.

If this building were in Berlin, Germany, where heating is required for 60% of the year (2,628 hrs/yr at the 50% utilization rate) and cooling for 20% (876 hrs/yr) the cooling savings amount to 23.5 kWh/yr and the heating supplement by the heat pump system is 51.9 kWh/yr for total savings (328.5 + 23.5 – 51.9) = 300.1 kWh/yr.

It is also apparent that, if the substitution of the 100 W incandescent bulbs were achieved by LEDs, which would consume only 5 W, the corresponding electric energy savings would be significantly higher in both locations. Based on this simple example for the substitution of a single incandescent bulb, Table 13.3 gives a summary of the energy savings in a large building resulting by the substitution of the equivalent of 10,000 W of incandescent bulbs with fluorescent ones that consume only 2,500 W and provide the same amount of luminescence. All numbers are in kWh per year. Throughout the calculations, it is assumed that lighting is required in the building for an average of 12 hours per day of the year.

Because commercial and residential buildings utilize several kW of electric power for lighting, it is apparent that significant energy savings are realized with the mere substitution of traditional lighting devices by more efficient ones, such as fluorescent bulbs or LED's. The energy savings also result in significant cost reduction for the operation of buildings. It is also apparent that energy savings are higher in buildings that are air-conditioned for longer fractions of the year. The substitution of incandescent bulbs with fluorescent lights or LED's produces higher savings in hotter climates, where the air-conditioning season is longer.

While the substitution of incandescent bulbs with more efficient lighting devices is a measure that falls under the energy efficiency category, energy conservation measures pertinent to lighting uses may be also implemented in residential or commercial buildings. Among these measures are switching off lights when they are not needed and switching off light emitting appliances, such as televisions and computer screens, when they are not in use.

13.3.2 Use of Heat Pump Cycles for Heating and Cooling

The refrigeration or heat-pump cycle, which was described in Sect. 3.6.3 and depicted schematically in Fig. 3.12, essentially creates two heat sources: the first at a low temperature, T_L, and the second at a higher temperature, T_H. When in the heat pump mode of the cycle, a quantity of heat, Q_H, is dissipated during the condensation process (process 2–3) of the cycle to the interior of a building at temperature T_H. Simultaneously, the cycle absorbs heat, Q_L, from the environment, which is at temperature T_L during the evaporation process (process 3–4) and consumes work W during the compression process (process 1–2). The coefficient of performance of the heat pump, $\beta_{hp} = Q_H/W$, is a measure of the work consumed and the heat transferred to the building.

The coefficient of performance of well-designed heat pumps may reach significantly high values, typically 4 to 5. This implies that for every unit of work that the heat pump consumes, four to five units of heat are transferred to the building. Therefore, the heat pump becomes a very efficient way for space heating, especially if it replaces electric heating. A heat pump is also a good alternative to gas heating if the coefficient of performance is sufficiently high. The following example illustrates an efficient use of a heat pump for space heating.

Consider a large building that has a heating need of $23*10^6$ kJ during a cold winter day. If the building is heated by a burner that uses natural gas with heating value 43,000 kJ/kg and the combustion efficiency of the burner is 92%, then $23*10^6/43{,}000/0.92$ kg = 581 kg of gas would be needed to heat up the building. Now let us consider an alternative heating method with a heat pump system that has a coefficient of performance 4.8. In this case, the heating requirement of $23*10^6$ kJ may be supplied to the building with the consumption of $23*10^6/4.8$ kJ = $4.79*10^6$ kJ of electric work by the heat pump.

It must be recognized in this example that the electric work must be produced in a power plant at the expense of a significantly higher amount of heat. Since typical efficiencies of electric power plants are approximately 35%, the heat requirement at the power plant would be $4.79*10^6/0.35$ kJ = $13.69*10^6$ kJ. If the power plant used natural gas for the production of electricity at the same efficiency as the burner of the building, this amount of heat would be produced from $13.69*10^6/43{,}000/0.92$ kg = 346 kg of natural gas. Therefore the substitution of the old burner with a heat pump system results in overall savings of 235 kg of natural gas per day. The savings are the consequence of the significantly higher coefficient of performance, β_{hp}, of the heat pump. In general, if the overall thermal efficiency of the power plant is η_t, and the transmission efficiency for the electric energy is η_{tr}, the substitution of the burner heating system with a heat pump system results in savings if the following inequality is satisfied:

$$\beta_{hp}\eta_t\eta_{tr} > 1. \qquad (13.8)$$

An additional advantage of utilizing a heat pump cycle for buildings is that, during the summer, the operation of the heat pump system may be reversed to become an air-conditioning system: since the refrigeration cycle creates a cold and a hot heat source, the colder heat source may be utilized in the summer to supply the same building with cooler air. In practice this is achieved by reversing the air streams in the building, which remove heat from the evaporator of the cycle (process 4–1 of Fig. 3.12) and add heat to the condenser of the cycle (process 2–3 of Fig. 3.12) respectively. Because the same equipment are utilized as heating and cooling equipment, the use of the heat pumps/air-conditioners has become widespread, both in large buildings, such as schools, hospitals, office complexes, and manufacturing plants as well as for smaller residential buildings.

A further advantage for the use of heat pump systems for cooling is that these systems have the capability to supply hot water to a building at no additional expense. A glance at the refrigeration cycle of Fig. 3.12 proves that the heat removal process from the cycle, process 2–3, occurs at relatively high temperatures, especially at the superheat part of this process. A fraction of the heat rejected during this process may be used to raise the temperature of water in a closed tank and use it as the hot water supply of the building, where the heat pump operates. Since typical upper refrigeration cycle temperatures are in the range 60–70°C and typical hot water temperatures are in the range 45–50°C, it is apparent that the refrigeration cycle may be used for the entire supply of hot water in a building, or at least a fraction of this supply. In practice this is accomplished with a heat exchanger coil that passes through the hot water tank. The immediate result of this heating process is that the hot water heater consumes less natural gas or electricity for the heating of hot water. For example, in the case of the gas heater, if the heat pump cycle provides a quantity of heat Q_{hw} to the water tank of the heater, and the lower heating value of the gas is $(\mathrm{LHV})_g$, then the mass of gas that has been saved and not burned for the production of domestic hot water is:

$$m_g = \frac{Q_{hw}}{(LHV)_g}. \tag{13.9}$$

A schematic diagram of the hot water production process is shown in Fig. 13.5, where the heat exchanger coil supplies the heat to the bulk of the water in the heater. The subscript r pertains to the refrigerant and the subscript w to the water. When the refrigeration cycle operates in the heating mode, the heat pump must operate at longer times to supply the additional heat to the water heater. When the refrigeration cycle works in the cooling mode all the heat in the condensation process 2r–3r must be dissipated and the heating of water does not add to the energy consumption of the building. Therefore, the application of this method for domestic water heating is particularly advantageous in hot climates where the refrigeration cycle operates in the cooling mode for extended periods of the year and heat needs to be dissipated somewhere. It has been documented that heat pump cycles have provided 100% of the hot water needs of households in the

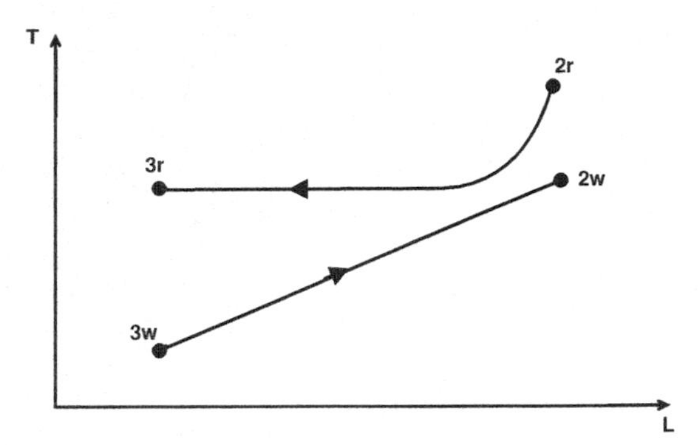

Fig. 13.5 Hot water supply from the condenser of a refrigeration cycle

southwestern part of the U.S.A. during the months April to October and from 30 to 100% in the winter months, without any additional cost of operation.

13.3.3 Geothermal Heat Pumps

One of the major drawbacks for the use of heat pumps for the heating or cooling of buildings is that they exchange heat with the atmospheric air, whose temperature is variable. During the heating season, the evaporator of the cycle removes heat, Q_L, from the atmospheric air. During the cooling season, the condenser of the cycle dissipates heat, Q_H, also to the atmosphere. However, the atmospheric temperature is highly variable and ranges in some regions from −20°C during the colder winter months to more than 40°C during the summer months. The ambient temperature variability makes the design of the refrigeration cycle difficult to optimize. Elementary calculations on the refrigeration cycle show that the coefficient of performance of a heat pump drops significantly when the outside air temperature is very low. Similarly, the coefficient of performance of the air-conditioning cycle drops significantly when the atmospheric air temperature is at its extreme highs. The c.o.p. of an air-conditioner is significantly lower when the outside temperature is 42°C than when the outside temperature is 30°C. As a consequence, the thermodynamic performance of a heat pump or an air-conditioning system deteriorates significantly when the system is needed more, that is during the coldest (as heat pump) and during the hottest (as air-conditioner) days of the year. In some cases a supplemental source or system for the cooling or heating of the building is required because of this performance deterioration.

Since the reason for this performance deterioration is the variability of the atmospheric temperature, which reaches extremes during a year, it is apparent that, if another heat reservoir with almost constant temperature were available, where heat could be absorbed during the heating season and dissipated during the cooling season, the refrigeration cycle used for the heating and cooling of the buildings would operate with higher c.o.p. throughout the year and would not be affected by seasonal temperature extremes. Moreover, this system would operate more

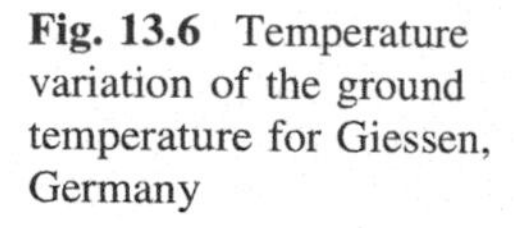

Fig. 13.6 Temperature variation of the ground temperature for Giessen, Germany

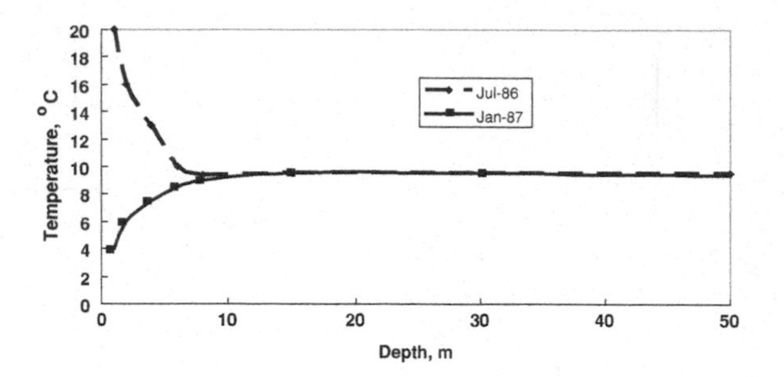

reliably, because it would not be affected by large temperature variations. The ground is such a heat reservoir: while the surface of the ground is variable and follows the atmospheric temperature, the ground temperature deeper than 6 m (20 ft) is almost constant, down to depths of 100 m (330 ft). Actually, the annual variability of the ground temperature reduces by a factor of 2 at a depth of 1.2 m (4 ft). The constant ground temperature depends on the latitude of the region and is approximately 6°C (42°F) in Scandinavia; 10°C (50°F) in Germany and England; 15°C (59°F) in Southern Italy; 18°C (65°F) in Dallas, Texas; and close to 20°C (69°F) in San Antonio, Texas. Figure 13.6 shows the ground temperature variation for the city of Giessen, Germany [1] for the months of July and January. It is observed that the ground temperature below 8 m is almost constant in both winter and summer, at 9.5°C. Also, it is observed that, even close to the surface of the ground, e.g. at 3 m, the temperature remains almost constant throughout the year.

A geothermal heat pump (GHP) utilizes a refrigeration cycle with the ground as the medium to dissipate heat from its condenser during the summer as well as to absorb heat by its evaporator during the winter months. The heat exchange to the ground is accomplished by a system of tubes that carry the cooling or heating fluid. Three main types of heat exchangers are used in the design of GHP's: two of them use a horizontal configuration and one uses a vertical configuration.

1. The horizontal ground piping system, which consists of a series of horizontal trenches with pipes at depths between 2 to 5 m. The heat exchange fluid, typically water, is forced by a pump to circulate in the system of tubes and, thus, dissipates the heat to the level of the pipes during the cooling operation and absorbs heat during the heating operation. Given that the depth of the trenches is shallow, the heat is dissipated close to the ground, where, nevertheless, the temperature is almost constant throughout the year.
2. The trench, spiral collector system, for which the trenches dug are at a similar depth, 2–5 m, but wider. The heat exchanger tubing is placed in the trenches in circular loops, in a configuration that is similar to the toy *slinky*. A large number of loops in the "slinky" facilitate the heat transfer from the circulating fluid to the ground.
3. The vertical loop system, which is essentially a long vertical U-tube heat exchanger placed in the ground. This system requires the least amount of

surface area, but typically extends to depths between 50 and 100 m. The piping of the tubes is of plastic material and a special bonding agent (bentonite or another heat-conducting gel) is used to fill the entire well and to provide structural stability to the U-tube.

The first two types of GHP systems exchange heat close to the surface of the ground and are suited for regions where more heating than cooling is required during the year. These systems are not suitable for climates where the cooling season is much more extended than the heating season, such as the southern part of USA, where significant amounts of heat power need to be dissipated in the ground. The third system, with wells extending to 120 m, is more suitable to be used in hot and arid climates. The deep wells dissipate the heat deeper into the ground and, typically, do not raise the ground temperature appreciably. However, it has been observed that, even with the deep wells, the ground temperature rises by 3–6°C. The ground temperature rise occurs when the air-conditioning use is highest during the summer and very large quantities of heat are dissipated.

A figure of merit for refrigerators, heat pumps and air-conditioners is the Seasonal Energy Efficiency Ratio (SEER) defined as the ratio of the Btu's of refrigeration, or heat removal, to the electric energy input in Watt-hours. The relationship between the SEER and the c.o.p. of a refrigeration cycle is:

$$\beta_{ref} = \frac{Q_L}{W} = (SEER)\frac{1055}{3600} = 0.293(SEER). \tag{13.10}$$

Because of the almost constant ground temperatures, the substitution of a conventional air-conditioning system with a GHP is usually accompanied by a significant increase in the SEER of the equipment. As a result of such a substitution, the air-conditioning system performs the same cooling task while consuming less electric power. This has the obvious beneficial effects of less spending for cooling by the owner of the building; and less pollution associated with the electricity consumed for the environment. It also has a rather unexpected advantage for the utility (power generation company): in many regions, where the climate is warm, the peak power demand occurs in the summer as a result of the use of high air-conditioning demand. For regions, such as the Southwest of the USA, the summer power demand by far exceeds the winter demand as it is demonstrated in Figs. 12.1 and 12.2. As a consequence, the peak power demand occurs in the summer and is highly dependent on the need for cooling. The growth of population in these regions implies significant growth of the peak electric power and significant capital expenditure for the power generation corporation in new power plants and maintenance for older ones. Oftentimes, it is financially advantageous for such a corporation to invest in energy efficiency measures, such as higher SEER for air-conditioning systems, than in new power plants. Several power generation corporations offer a rebate to consumers that install GHP's. This trend is more apparent with the publicly owned utilities, which do not derive a significant financial advantage from selling more power to consumers. For

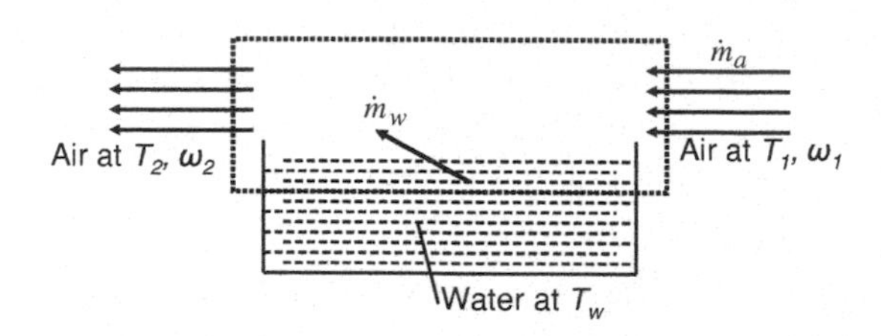

Fig. 13.7 Evaporative cooling

example, since 2009 the San Antonio, Texas, utility (CPS) has offered a rebate to its customers for the installation of GHP's with higher SEER than conventional air-conditioning systems.

13.3.4 Adiabatic Evaporation

Adiabatic evaporation in hot and arid climates may reduce significantly the cooling needs of a building. When a stream of warm and dry air passes on top of a water body, a fraction of the water evaporates. The warm air stream provides the latent heat for the vaporization of the liquid water and, hence the temperature of the air stream decreases, oftentimes significantly. The specific humidity of the air stream increases simultaneously. Figure 13.7 depicts a schematic diagram of the evaporative cooling process. The mass balance in the control volume denoted by the dashed-line rectangle is as follows:

$$\dot{m}_w = \dot{m}_a(\omega_2 - \omega_1), \tag{13.11}$$

where $\dot{m}_a$ is the mass flow rate of the air; $\dot{m}_w$ is the mass flow rate of the water that evaporates and enters the air stream; and ω is the specific humidity of the air. Similarly the energy balance—first law of thermodynamics—may be written as follows for this control volume:

$$\dot{m}_a(h_{a1} + \omega_1 h_{v1}) + \dot{m}_w h_w = \dot{m}_a(h_{a2} + \omega_2 h_{v2}). \tag{13.12}$$

The subscripts v and a denote the water vapor and air respectively. Because the temperature difference T_2–T_1 is small, the enthalpy difference of the air is equal to the product of the specific heat and the temperature difference. Hence the cooling effect, or temperature reduction, $T_2 - T_1$, is:

$$T_2 - T_1 = \frac{1}{c_P}(\omega_1 h_{v1} - \omega_2 h_{v2} + \omega_2 h_w - \omega_1 h_w) \tag{13.13}$$

or approximately, in terms of the latent heat of vaporization of water, h_{fg}:

$$T_2 - T_1 \approx \frac{h_{fg}}{c_P}(\omega_1 - \omega_2). \tag{13.14}$$

Table 13.4 Outlet temperature drop in evaporative cooling

T_1 (°C)	φ_1 (%)	φ_2 (%)	T_2 (°C)
35	10	60	21
35	20	70	23
40	20	70	26
40	20	50	30
35	30	60	23

The evaporative cooling effect can be significant in dry weather and may be used to partially or totally cool the air of a building. In addition, evaporative cooling may be achieved with rudimentary and inexpensive equipment and this makes it very attractive from the economic point of view. For example blowing air with a fan over a small open water container or injecting small water droplets in front of an air fan will result in significant evaporative cooling. Locating residences at the shores of the sea or a lake, where local, natural breezes bring cooler air has the same cooling effect. Table 13.4 shows the evaporative cooling effect, in terms of the relative humidity, φ, for several inlet and outlet conditions. It is observed that 10–18°C cooling of the ambient air may be achieved by merely increasing its relative humidity from the range 10–30 to 50–70%. In most cases, this temperature drop cools the ambient air sufficiently for additional air-conditioning not to be needed. This type of natural cooling has been used for centuries in the Mediterranean countries and especially in the islands, where artificial air-conditioning was almost unknown until the beginning of the twenty-first century.

13.3.5 District Cooling

The idea of *district cooling* is similar to that of *district heating* but is not based on the use of waste heat. District cooling is based on the higher efficiency of larger and better maintained air-conditioning installations in comparison to smaller and less efficient installations. The typical household air-conditioning plant is a smaller 2–10 kW engine. Affordability and low-price (not highest efficiency) is the primary design consideration of these smaller units. As a result, the coefficient of performance of typical small building air-conditioners is between 2.0 and 3.0. A much larger refrigeration unit, which might supply chilled water to a district of several buildings, may be designed with a significantly higher coefficient of performance. Water cooling of the system's condenser alone would increase the coefficient of performance of the unit by 1 to 1.5 units. A larger compressor, frequent maintenance and continuously controlled operation would further increase the coefficient of performance of the refrigeration system.

A schematic diagram of district cooling is shown in Fig. 13.8. A large refrigeration plant cools water and maintains it at low (8–15°C) temperature in the *chiller,* which is a large, well-insulated tank. The cold water is pumped by

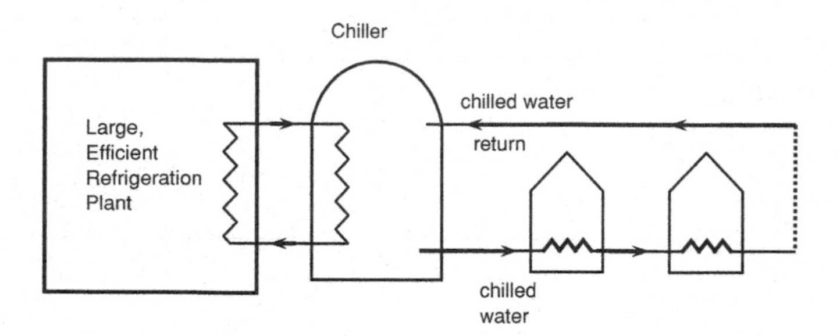

Fig. 13.8 Schematic diagram of the district cooling concept

underground ducts to nearby buildings, where it absorbs heat from the interior air and thus, supplies the cooling effect. The returned, warmer water is cooled in the chiller by the refrigeration unit.

District cooling operates at best when the buildings are close enough for the water of the chiller not to be heated significantly by heat transfer with the ambient air or the ground. For this reason, district cooling works best with larger buildings in high density areas, such as high rise buildings, commercial centers, large hotels, municipal centers or a combination of them. Several districts in city centers of the USA and southern Europe have adopted this concept of cooling with significant energy savings. Use of district cooling in suburban areas, where the building density is low, the distances long and there is significant heat transfer to the chilled water from the environment, does not result in appreciable energy savings and sometimes may even add to the energy consumption.

13.3.6 Other Energy Conservation Measures for Buildings

In addition to space heating and cooling, the hot water use in buildings consumes high amounts of thermal energy. Gas-heated hot water heaters are by far better to be used than electric hot water heaters. This happens because the electricity used in the latter emanates from an electric power plant with efficiency typically in the range 30–40%. A higher quantity of heat ($Q = W/\eta_t$) has been used for the production of the electric energy that is converted back into thermal energy in the electric water heater. In contrast, a burner has 90–98% thermal energy conversion efficiency.

Let us consider that the task at hand is the temperature rise of 200 kg of water (approximately 56 gallons) from 20 to 50°C, which requires 25.14 MJ of thermal energy [$H_2-H_1 = m*c_p(T_2-T_1)$]. The best way to supply hot water is by the use of heat pump systems, which were described in Sect. 13.3.2 and Fig. 13.5. However, the installation of heat pump systems requires a significant capital outlay, which most homeowners do not wish to spend. For this reason, there are currently two common water heater systems that supply hot water in most OECD countries, gas and electric heaters.

In order to evaluate the gas and electric heater systems used in domestic applications, we will assume that the thermal energy to the water may be supplied by one of the following two methods in the two types of heaters:

a. Burning natural gas in a burner with 95% efficiency, or
b. Using electricity from a plant with an overall efficiency 35%, which uses natural gas as its fuel.

Bearing in mind that 1 scf (standard cubic foot) of natural gas typically provides 1.072 MJ, using alternative (a) would consume 24.7 scf of natural gas. Using alternative (b) would consume 67.0 scf of natural gas. Therefore, switching from alternative (b) to alternative (a) would save 42.3 scf of the primary energy source, the natural gas. As a rule, and because electricity is generated to a large extend from thermal energy with 30–40% efficiency, the final conversion of electric energy into heat always wastes a high amount of primary energy resources. The simple conversion of the chemical energy of the primary energy resource into thermal energy, e.g. in a burner, is a much better, and most often a more economical way to produce thermal energy, whenever the latter is needed.

Even a gas-heated hot water heater dissipates a significant amount of energy to its surroundings. The heater maintains a constant water temperature in the range 45–50°C, and is typically located in an attic or a basement, where the air temperature is significantly lower. As a result, the heater looses constantly heat by conduction and convection. The heat loss is higher during the night hours when the ambient temperature is lower. Ironically, hot water is not needed during the night. Water heaters in the USA consume an average of 12% of the annual energy consumption of the average household and most of it is dissipated in the environment. Better insulation of the water heater is a good conservation measure that reduces the average heat losses of the heater.

A water heater design that supplies hot water only when it is demanded is another way to reduce the energy spent on hot water. This heater is smaller and is located close to the demand of the hot water, which is the bathroom or the kitchen sink. The heater is equipped with a sensor that ignites the gas when water flows in its pipe. When there is water flow, gas combustion in the heater supplies heat to the coils inside the heater and, thus, increases the temperature of the water that exits to the desired temperature. When water is not needed, the gas valve is closed by the flow sensor and the combustion stops. The main advantage of this self-igniting heater is that it almost does not store any hot water. It only supplies hot water on demand. When the demand ceases there are no large quantities of hot water in the heater, which cool later with a large loss of energy. The technology for these heaters is well-known and such heaters have been used in several European countries since the 1950s. Because 56% of the thermal energy supplied to the typical water heaters of the USA households is lost to their surroundings, the wider application of these smaller self-igniting heaters would reduce significantly the average household energy demand.

A natural method that reduces the heating and cooling needs of buildings is the use of awnings at windows that predominantly face the Sun, that is the southern side of buildings in the northern hemisphere or the northern side in the southern hemisphere. Well-designed awnings reduce significantly the sunlight that enters a window during the hot season, when the sun is at the highest and would allow the Sun rays to enter during the cold season when the Sun is at the lowest in the horizon. More solar energy enters the building during the heating season and less during the cooling season. This combination reduces both the heating and the cooling requirements of the building. Use of simple awnings in the south-facing windows may reduce by 15–25% the cooling needs of a household in the southern part of the USA and of most Mediterranean countries at similar latitudes.

Other conservation methods that would decrease the energy consumption in buildings are as follows:

1. Use of double-glazed windows. They increase significantly the insulation of the buildings.
2. Use of better fitting doors and windows that reduce the air draft in buildings. Similarly, in older buildings, use insulating adhesive strips to better seal doors and windows.
3. Wider use of fans for the circulation of air, which improves human comfort and reduces the need for air-conditioning.
4. Placing the heating and air-conditioning vents on the floor rather than the ceiling. This keeps the supply of hot and cold air where it is mostly needed by the occupants.
5. Use of diffuse natural lighting rather than artificial lighting. In addition to conserving the electric energy needed for the lights, this measure also reduces the cooling needs of the buildings.
6. At the time of planning and construction, optimize the orientation of the buildings and fenestration (windows) placement for less energy consumption.
7. Use retractable awnings outside the windows.
8. Use screens and thin films to reduce the insolation during the summer.
9. Better insulate attics, basements and other exposed parts of the buildings.
10. Use programmable thermostats that allow the internal temperature to be outside the human comfort zone, when the buildings are not occupied.

In all these cases, the reduction of energy consumption would also reduce the cost of maintaining the building. Older buildings, which were built in an era of cheaper energy and are very costly to maintain may be *retrofitted* to reduce their annual energy consumption. The retrofitting results in significant savings for the owners, who are able to amortize the cost of retrofitting in a short time, typically 2–6 years. Buildings that were constructed before the 1970s and especially ones where air-conditioning has been added after the original construction are prime candidates for significant monetary savings from energy retrofits. In most OECD countries local governments and local utilities provide loans and rebates to assist with the capital expenditures of building retrofitting.

13.4 Conservation and Improved Efficiency in Transportation

The transportation sector consumes approximately 30% of the net energy demand in most OECD countries and encompasses both the transportation of individuals and of goods. The wider recreational use of the automobile after the 1950s and the exponentially increasing trade among nations have contributed significantly to the consumption of energy, especially to the consumption of liquid fuels such as gasoline, diesel oil, and kerosene. In trying to devise methods for the reduction of the energy consumption in this sector of the economy, one must always keep in mind that the societal task, which needs to be fulfilled is the transport of persons or goods from one location to another and that several alternative methods are usually available. For example: 100 tons of potatoes need to be transported from Idaho to New York City; or 35 passengers need to travel from Marseille, France to Barcelona, Spain. The societal task of the transport of potatoes is fulfilled when the potatoes are transported by truck, by train, by airplane, or by a fleet of passenger cars. The societal task of the passenger transportation may be accomplished by ship, airplane, train, bus, or, by individual cars. It becomes immediately apparent that, of the available methods to accomplish the transportation task, one would use the least amount of energy or fuel. For example, transportation of goods and persons by train consumes always less energy than other land forms of transportation. Transportation by car is most often the most energy consuming method.

Given all the available means of transportation, public transportation in cities, and especially the metro/train transportation is the least energy consuming method, when it is used by a high percentage of the citizenry. However, personal convenience, societal values or status, and long-time habits play an important role in the way our society uses energy for transportation. For example, residents of most cities in the USA (with the exception of New York City) consider that it is more convenient for them to drive to their work rather than use public transport, while the vast majority of Parisians would never drive to work and would take the bus/metro combination, where they also read their daily paper or a book. In addition, for several individuals it is a status symbol to drive a larger, more expensive and, typically, higher energy consuming car than a smaller compact car. Occasionally, the short-life of some goods dictates that these goods be transported faster at the expense of higher amounts of energy. For example, tulips from Holland delivered to flower shops in Philadelphia must be transported by airplane instead of the more economical and less energy consuming ship.

A great deal may be accomplished in the reduction of the energy consumption in the transportation sector, both for the transport of goods and for individuals. At first, it must be recalled that bulk transport of goods in trains or ships is by far less energy consuming (in kJ per kg of goods transported) than the transport by trucks or by airplanes. Therefore, whenever there is no *a priori* reason to do otherwise—e.g. the goods have a short life and may be damaged—transportation by train or boat should be preferred. At the same time, it makes sense both economically and from the energy point of view, to pool resources and fill the entire transportation medium during a trip: A filled to capacity train, truck, airplane, or ship transports goods cheaper and with less specific energy consumption (kJ/kg transported) than a partially filled vehicle. For a long time, nautical companies have been using several methods, including visits to a number of ports, to fill their ships with cargo. Commercial airlines are increasingly using similar strategies to fill the aircrafts in most trips by using several methods, such as differential pricing, shared codes and offering flights at periods of high demand.

Human transportation is different than goods transportation because safety and convenience play more important roles than energy savings. While safety is paramount and should not be compromised, a certain amount of convenience may be often sacrificed for energy efficiency. For example, commuting by bus for a suburban community, while less convenient than individual cars would save a significant amount of fuel. In addition, it will contribute to the reduction of traffic congestion, to lesser emissions and to better environmental quality. Among the ways individuals and communities may reduced their energy consumption for transportation are the following:

1. Car-pooling, where several individuals from the same suburban area share a ride to the workplace. This reduces the number of cars on the road, gasoline consumption and congestion. Several municipalities encourage this practice by designating fast-moving High-Occupancy Vehicle (HOV) lanes on the highways and by assisting in the formation and coordination of car-pooling groups.
2. Establishment or wider use of bus routes and, if possible, of light rail systems.
3. Substitution of older vehicles that have low mileage with others of high mileage, especially those vehicles used for daily commute. It includes the wider use of hybrid cars, which store and re-use some of the energy that is typically wasted in stops by converting it to electric energy. This also entails the purchase of newer cars with smaller, more efficient engines.
4. Development of close-to-work communities, which are advocated by several urban planners and are designed to reduce the daily commute to work.
5. Traffic light synchronization. Frequent stops, which entail decelerations and accelerations, increase significantly the energy consumption of a vehicle. Optimized traffic patterns with average constant vehicle speeds between 35 and 45 miles per hour are desirable in urban and suburban environments for the minimization of energy consumption as well as pollution mitigation.
6. Smooth connection of highways by the use of dedicated entrances and exits without traffic lights and delays that slow down the traffic and induce higher gasoline consumption.

7. Avoidance of peak hours of traffic by the adoption of staggered work hours in large municipalities. For example, in the U.S.A. cities the working day for most businesses is rigid, starts between 8:00 am and 9:00 am and ends between 5:00 and 6:00 pm. This creates peak hour "traffic jams" in most municipalities, which slow the traffic significantly, increase the total fuel consumption and most often, contribute significantly to air pollution. The adoption of more flexible working hours (e.g. start between 7:00 am and 10:00 am and end between 4:00 and 7:00 pm) would alleviate the peak hour "traffic jams," and would reduce congestion and the commuting time of the workers.
8. Wider use of the hybrid cars, which through a system of an electric motor/generator and batteries, recover and store for later use some of the mechanical energy that is typically dissipated by the brakes.
9. Wider use of smaller cars or electric cars for commuting to work.

13.4.1 Electric Cars

When in motion, vehicles counteract ground friction and air friction forces and need power to do so. A vehicle may also need to spend additional work in order to counteract the gravity force and move up a hill.[6] This work is produced by the engine, which in the large majority of the vehicles is an internal combustion (IC) engine. It is well known that the internal combustion engine is rather inefficient with an overall thermal efficiency, η_{tv}, between 8 and 15%. This overall efficiency is for the entire system of the vehicle and covers the IC engine, incomplete combustion, emissions, transmission, etc. It is also well known that the existing electric power plants produce electric power at a significantly higher overall efficiency, η_{t}, between 32 and 38% or even higher. An electric battery may be charged with efficiency, η_{B}, approximately equal to 97%. Electric cars use motors that convert the electric energy in a battery to work with efficiencies, η_{M}, close to 90%.[7]

Typical vehicles are fitted with an IC engine, which has an average overall efficiency, $\eta_{tv} = 12.5\%$ and needs a quantity of work, W, in order to travel a distance L with specific driving patterns (instantaneous and average speed, number of stops, etc.). Traveling this distance is the required task. For the performance of this task, the vehicle will consume an amount of gasoline that would be equivalent to $Q_{IC} = W/\eta_{tv}$, or $Q_{IC} = 8\ W$.

[6] Because of friction in the engine and transmission only a small fraction of the work spent during the ascent of a hill is recovered during the descent.

[7] Internal combustion engines and thermal power plants convert thermal energy/heat to work and are subject to Carnot limitations (Eq. 3.22). The charging of a battery and the operation of a motor are forms of direct energy conversion and they are not subject to Carnot limitations.

Now let us assume that we substitute the internal combustion engine in the car with an electric motor that is powered by a system of batteries, which are charged overnight. The efficiency of the battery charging, η_B, is 97%, the efficiency of the motor, η_M is 90%, and the efficiency of the thermal power plant that produces the electric power for the batteries is $\eta_T = 36\%$. The plant-to-wheels overall efficiency of the electric car is:

$$\eta_E = \eta_T * \eta_B * \eta_M = 0.36*0.97*0.90 \text{ or } 31.4\%. \quad (13.15)$$

Therefore, when the internal combustion engine is substituted with an electric motor, the heat needed at the power plant for the production of the work W is equal to $Q_E = W/0.314 = 3.2$ W. The required heat savings for this substitution are equal to (8–3.2)W = 4.8 W or 60% of the original value $Q_{IC} = 8$ W. If a similar fuel is used in the vehicle and the power plant, this substitution causes a 60% reduction of the fuel. Actually, most of the vehicles use gasoline or diesel, while electric power plants typically use coal or nuclear fuel, which are cheaper and more abundant. Hence, the fuel savings are more meaningful to national economies, which import oil or gasoline.

While wider use of electric vehicles has advantages from the energy efficiency point of view, as well as because electric vehicles are pollution free, electric vehicles have not been widely adopted because of the following disadvantages:

1. The overall electric car engine system costs significantly more than that of the internal combustion engine.
2. The typical distance traveled with one charge—the range of the electric vehicle—is limited to approximately 80 miles (130 km).[8]
3. The typical time for the recharge of the batteries is significant, usually more than two hours. This prevents the use of electric vehicles in long trips that may require battery recharging. The few "fast-recharge" batteries that have been invented, and which claim recharge cycles of the order of 10 minutes, are costly, may create problems with the local electricity grid, and have not been convincingly tested for reliability and long-term endurance.
4. Typical battery life is currently limited to a few hundred recharges, beyond which the performance of the battery deteriorates. New materials and technological advances in this area may alleviate this problem in the near future, when it is anticipated that battery lives will extend to thousands of recharges.

[8] The daily range of several city buses is less than this figure. Several municipalities in the USA and Europe have adopted the use of electric buses. In addition to the significant energy savings, these buses do not have gas emissions and do not contribute to environmental pollution and noise pollution in the cities.

13.4.2 Fuel Cell Powered Vehicles

An electric vehicle powered by a *fuel cell* does not suffer from the last three limitations associated with the battery charging and discharging of an electric car. The fuel cell is an open thermodynamic system operating continuously and consumes hydrogen or another fuel. The fuel may be stored in a tank in a way similar to the gasoline or diesel tanks, which all current vehicles with internal combustion engines use. The fuel cell converts the stored hydrogen directly to electricity, which is used in the electric motor that provides the power for the vehicle. Figure 13.9 shows the energy conversion processes associated with an electric car powered by a fuel cell. Hydrogen is produced by electrolysis. The chemical energy in the hydrogen fuel is directly converted to direct current electricity, which is used by the motor to propel the vehicle. Typical efficiencies of fuel cells, η_{FC}, are approximately 70%. With a motor efficiency of 90% as in the example of the electric car, the hydrogen-to-wheels efficiency of a fuel cell powered vehicle would be 67.5%.

However, it must be recalled that hydrogen is not an abundant fuel and must be produced artificially, usually by the decomposition of water. Assuming that hydrogen is produced by the process of electrolysis, which has typical conversion efficiency, $\eta_{EL} = 70\%$, and that the electric energy for this process is produced at a thermal power plant with thermal efficiency $\eta_T = 36\%$, then the overall plant-to-wheels efficiency of this vehicle would be: 0.36*0.7*0.7*0.9 = 15.9%. Even this lower number is higher than the efficiency of most internal combustion cars. It is also anticipated that, with the wider use of hydrogen as a fuel, hydrogen production efficiencies and fuel cell efficiencies will improve significantly to deliver plant-to-wheels efficiencies close to 25%, or almost twice as much as those of the typical IC engines that are currently used for transportation. An additional advantage of the fuel cell powered cars is the significant reduction of pollution that is currently caused by the IC engines.

Because the technology of the electric vehicles is known, it is expected that a higher number of such vehicles will be in the roads in the next two decades, especially public buses, vans, school buses, and short-haul trucks. The substitution of internal combustion engines with electric motors will have the overall effect of lesser liquid fuel consumption, primarily of gasoline and diesel; lesser primary energy consumption; and lesser air pollution in cities.

From the point of view of the electric power supply, because the battery charging will be primarily accomplished during the night hours, when electric utilities have ample capacity to produce more power, the additional electric power consumed will not impose a strain on the peak power demand of the electric grid. On the contrary, the wider use of electric vehicles would make smoother the daily demand fluctuations (Figs. 12.1 and 12.2) of electricity and will save a significant amount of primary liquid hydrocarbon resources, mainly crude oil.

Whether or not electric cars will be widely adopted by the public as private cars depends to a large extend on the *range* of the electric cars—the maximum driving

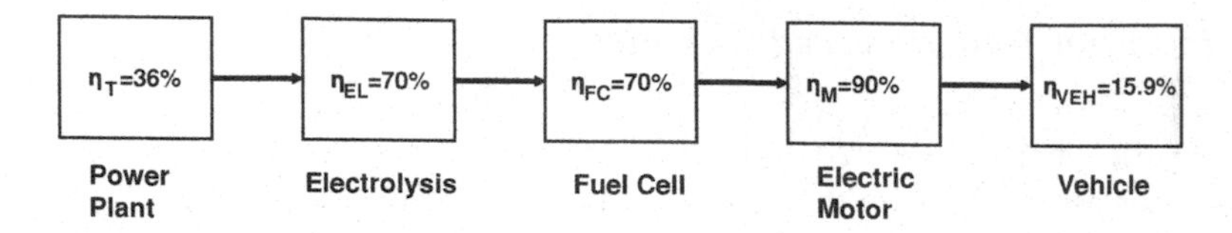

Fig. 13.9 Energy conversion diagram for a fuel cell powered vehicle. Typical efficiencies of the individual processes are shown in the boxes of the diagram

distance before charging—that will be achieved in the near future. Private cars are used for commuting as well as for vacation and other long-distance travels. Having to stop for extended periods of time for battery recharging is an inconvenience that the public is not willing to endure. It follows from the processes in Fig. 13.9, that the improvement in the efficiency of fuel cells and in the production of hydrogen, or of another suitable fuel, as well as the wider availability of hydrogen will have a profound effect in the wider adoption of fuel-cell powered cars. These cars will only need to refuel with hydrogen or another suitable fuel, a rather fast process that can be quickly accomplished in appropriate hydrogen fuel stations. The wider adoption of fuel cell powered private cars and trucks will almost eliminate the need for liquid hydrocarbons in the society and will signal the advent of the *Hydrogen Economy*, which was mentioned in Sect. 12.5.

Problems

1. For each of the following three societal tasks provide details on the methods and engineering systems they are currently met/accomplished.
   - Maintaining a comfortable temperature for the residents of the households of Houston, Texas during the summer months.
   - Provide adequate lighting in the classrooms of a High School so that students may learn comfortably.
   - Cook 0.4 kg of pasta.
     For each one of these tasks provide an alterative method to accomplish it and state if primary energy resource savings would result from the alternative method.
2. Consider the heating and cooling needs of your household. What improvements can be made to reduce the overall energy resource consumption? Separate these improvements into conservation and improved efficiency measures.
3. One of the reasons the transportation sector consumes a great deal more energy in the USA is the commute to work by individuals. The average, one-way commuting trip in Houston, Texas is 21.5 miles. Because of traffic jams the average mileage of the cars in the same area is 12 miles per gallon. Consider that four neighbors driving daily the average commuting distance decide to carpool. What is the amount of gasoline saved weekly and annually? What

would be the annual gasoline savings if 500,000 persons carpooled in a similar manner? Consider that, in the latter case, the driving conditions will get better to the point that the average mileage will improve from 12 to 16 miles per gallon. How many tons of CO_2 are not emitted to the atmosphere annually because of the carpooling activity? The week in Houston has five working days and the year 52 weeks. Also, you may consider that gasoline is composed solely of octane.

4. A tank of 0.7 m^3 is to be filled with air at 12 bar. What is the minimum electric work needed? How much electric work is used by a reciprocating compressor with an isentropic efficiency 72%?
5. 150 kg/hr of steam are needed for an industrial process. The steam is produced by an electric heater. The local power plant uses natural gas for the production of electricity with an overall efficiency 38% (this includes the transportation losses). What would be the percent reduction of the natural gas consumption if the electric heater were to be substituted with a gas heater?
6. A large oil refinery uses 40 MW of heat, which is supplied by steam at 120°C and 1 bar. The refinery also uses 18 MW of electric power. Design a Rankine cycle and a small power plant that satisfies the heating and electric needs of this refinery.
7. A large food processing plant uses 10 kg/s of steam at 140°C, 1 bar. The plant also needs 5 MW of electric power to run its electric components and sells another 10 MW to the local utility. Design a Rankine cycle to accomplish these tasks.
8. A five-ton[9] air-conditioning unit for a household is to be replaced with a GHP system. As a result, the c.o.p. of the system is expected to improve from 2.8 to 3.9. The air-conditioning system is used in this household 2,350 hours per year. What are the annual energy savings due to this improvement? If all this energy came from a coal power plant of 37% overall efficiency, how many less kg of CO_2 are emitted because of this improvement every year?
9. What does the SEER number mean for an air-conditioner? The total rated power for the local utility is 5,000 MW, of which 95% is used during peak demand hours and the power generating factor for the utility is 52%. The air-conditioning of the customers account for 60% of the peak demand and 22% of the total annual energy demand. As a result of an *energy conservation campaign,* the utility company plans to improve the air-conditioning equipment of its customers from 11.2 to 13.5 SEER. What will be the reduction of

[9] One ton of air-conditioning is 12,000 Btu per hour. Typical air-conditioning units and large refrigerators are rated in tons.

the total electricity demand for the utility and the reduction of the peak power that is needed?[10]

10. A large supermarket, which operates 24 hours every day for 361 days per year, is to substitute 950 kW of fluorescent lamps with LED's that will provide the same amount of illumination, but will only consume 150 kW. The supermarket is heated for 65 days every year and air-conditioned for 210 days. Using typical efficiency and c.o.p. values determine the annual electric energy savings resulting from this energy efficiency measure.
11. Five large hotels in Miami, USA, use their own air-conditioning units, which have a c.o.p. 3.1. It is proposed that the cooling systems of the five hotels be substituted with a larger, modern unit that would have a c.o.p. 3.7. The total installed capacity of the five hotels is 2,600 tons of air-conditioning and the current systems are in use for 3,200 hours every year. Determine: a) the annual energy savings from this substitution and b) the annual avoidance of CO_2 emissions if, currently, 72% of the electric energy in the Miami area is obtained from coal power plants and 28% from nuclear power plants.
12. The humidity in Tucson, Arizona frequently hovers near 40%, while the temperature is 38°C. Determine the temperature reduction that would occur in a building if the humidity were to increase adiabatically to 60% and to 70%.
13. In 2009 the USA consumed 17,910,000 barrels of oil per day, of which 62% was used by small cars and light trucks with an average mileage 18.5 gallons per mile. If the national standard were to increase to 22 miles per gallon, what would be the annual savings in barrels per year? What are the monetary savings, if the average price of oil is $90/bbl?
14. It is proposed that a percentage of cars be substituted by electric cars. Japan used 4,680,000 barrels of oil per day in 2009, with 69% going to small cars and trucks. If 20% of these were to be converted to electric cars and trucks, what would be the annual savings in barrels of oil? Assume typical efficiencies. If all these oil savings come from octane (C_8H_{18}) what is the annual avoidance in the total CO_2 emissions?
15. It is proposed that a percentage of cars be substituted by cars powered by hydrogen fuel cells. Germany used 2,460,000 barrels of oil per day in 2009, with approximately 70% going to small cars and trucks. If 20% of these were to be converted to fuel cell cars and trucks, what would be the annual savings in barrels of oil in the country? Assume typical efficiencies. If all these oil savings come from octane (C_8H_{18}) what is the annual avoidance in the total CO_2 emissions? All the hydrogen will be produced by electrolysis with electricity provided by wind power plants.

[10] It appears to be counterintuitive for a utility company to seek demand reduction. However, given the high capital cost and environmental restrictions on the construction of new power plants, many local utilities in areas of growing demand prefer the demand reduction strategy to constructing new and more costly power plants. Thus, they satisfy the new customers with the surplus power that comes from conservation and higher efficiency measures.

16. "The OECD countries have an enormous potential to conserve energy in comparison to the developing countries. Therefore, developing countries should be excluded from all international protocols and agreements on CO_2 reduction." Comment in a short essay of 250–300 words.
17. "If we were to only apply energy conservation measures in the USA, we would not have to build another electric power plant until 2045." Comment in a short essay of 250–300 words.
18. A compressor in a Brayton cycle raises the air pressure from 1 atm and 27°C to 25 atm and has an isentropic efficiency 78%. It is proposed to substitute this compressor with two other compressors and an intercooler. Both compressors will have a pressure ratio of 5 (that is the first produces air at 5 atm and the second at 5*5 = 25 atm) and efficiency 80%. The intercooler cools the air after the first compressor to 37°C. Determine:

    a. The specific work, in kJ/kg required for the operation of the old and the new system of compressors.
    b. If the Brayton cycle admits 2 kg/s of air and is in operation for 25% of the time, the annual power savings from this substitution.

Reference

1. Ochsner K (2008) Geothermal heat pumps—a guide for planning and installing. Earthscan, London

Chapter 14
Economics of Energy Projects

Abstract When one asks the question, "why there are no more wind or solar photovoltaic power generation units in the world?" the simple answer is: "because fossil fuels and nuclear energy were cheaper sources to produce electricity in the near past." When a new coal power plant produces a kWh at 3.5 cents and an old nuclear power plant for less than 1 cent per kWh, it is difficult for an electricity producing corporation to justify purchasing a wind turbine that would produce electricity at 12 cents per kWh or a solar farm that would produce at 16 cents per kWh. When the price of electricity produced from renewable sources of energy becomes less than the price produced from conventional energy sources, the economic considerations will favor the development of more geothermal units, solar plants, wind generators, etc. A combination of rising fossil fuel prices, the probable initiation of carbon credits, and a favorable regulatory environment that provides tax credits and accelerated depreciation may change these economic and financial circumstances for alternative energy. This chapter provides a succinct exposition of the entire decision making process that leads to the construction and operation of a power plant, from the realization of the need for more electric power, to the enumeration of the alternatives, to the choice of the optimum alternative solution that maximizes profitability for a corporation or, equivalently, provides the needed amount of electric energy at minimum cost. Central to the financial considerations are the concept of the time-value of funds and the Net Present Value method for the appraisal of an investment. Because this book is aimed for the engineering student with little or no knowledge of economics and finance, the level of this chapter is rather elementary. There are no highly sophisticated economic concepts and methods to be presented, while simple and comprehensive definitions of all the concepts used are offered in the text.

E. E. (Stathis) Michaelides, *Alternative Energy Sources*,
Green Energy and Technology, DOI: 10.1007/978-3-642-20951-2_14,

14.1 Introduction

The decision-making process on energy projects is similar to the decision-making process of other engineering projects and involves the following steps:

1. The need for a project is identified. For example: the increase of the population of the City of Westerheimer necessitates the addition of 100 MW of electric power within 5 years.
2. Alternative solutions to the problem are formulated as projects, e.g. the increased electricity demand in the City may be met by one of the following: an additional nuclear power plant, an additional gas-fired plant, a new solar-thermal plant, a wind energy farm, etc.
3. The alternative projects are evaluated. This step includes a feasibility analysis and a detailed economic analysis of all the alternatives, which were identified in step (2). In evaluating the projects, all factors must be examined and some of the proposed alternative solutions may be excluded for reasons other than economic considerations. For example, the nuclear power moratorium the citizens of Westerheimer voted upon in 1988 precludes the consideration of a nuclear power plant.
4. A decision is made on the best alternative solution and the engineering project is identified and fully specified. For example, the electric company that supplies the City of Westerheimer will build from June 2012 to May 2015 a wind energy farm with wind turbines of total rated power 180 MW (to account for the lower availability factors of the wind turbines) at the Windstrum site and the project will be completed by the end of July 2015.

It must be noted that the final decision is specific enough to allow the planning and construction of the facility, but also allows sufficient flexibility for the next stages of the project. For example, it does not specify, which wind turbines will be bought and installed, how the wind turbine towers will be built, which towers will be built first, or what will be the spatial arrangement of the towers. These decisions will be made during the planning and construction stages of the project and include several engineering and environmental considerations.

This chapter will focus on steps 2, 3, and 4 of the decision making process, with particular emphasis on the formulation and evaluation of the alternatives.

14.1.1 Fundamental Concepts and Definitions

The definitions of several concepts used in the fields of economics and management, which are helpful in the decision making process for energy projects, are given in this section in alphabetical order:

Average cost: The total of all *fixed* and *variable* costs calculated over a period of time, usually one year, divided by the total number of units produced, e.g. "the average cost of electricity is $0.059/kWh."
Average revenue: The total revenue over a period of time, usually one year, divided by the total number of units produced, e.g. "the average revenue of electricity is $0.072/kWh."
Average profit: The difference between average revenue and average cost. Using the two examples above, the average profit is $0.013/kWh.
Fixed costs: All costs that are not affected by the level of business activity or production level, such as rents, insurance, property taxes, administrative salaries, and interest on borrowed capital.
Life cycle cost: The sum of all costs, fixed and variable, associated with a project from its inception to its conclusion. The life cycle costs include the planning costs as well as any abandonment, disposal, or storage costs.
Marginal or incremental cost: The cost associated with the production of one additional unit of output.
Marginal or incremental revenue: The revenue resulting from the production of one additional unit of output.
Opportunity cost: The opportunity to use scarce resources, such as capital, to achieve monetary/financial advantage. For example, an opportunity cost to building a new power plant for a company is not to build and invest their capital in 7% interest bearing securities.
Salvage value: The price paid by a willing buyer for a plant or business after all operations have ceased.
Sunk costs: All costs associated with past activity that may not be recovered and do not affect any future costs or revenues.
Time horizon: The time from the inception to the end of the project, including any disposal or storage of equipment and products.
Variable costs: Costs associated with the level of business activity or output level, such as fuel cost, materials cost, labor cost, distribution cost, etc. The variable costs increase monotonically with the number of units produced.

14.2 The Decision Making Process

The decision making process for engineering projects follows a well-defined need and is accomplished by multi-step, structured procedures, simulations, and mathematical modeling techniques. An economic analysis is made for a few carefully selected alternative solutions to the problem and the final decision takes into consideration the economic analysis as well as previous experience, environmental and social factors as well as engineering input.

14.2.1 Developing a List of Alternatives

The first stage of this process is the reformulation of the problem or need in several ways, which may stimulate creative thinking and inspire at least one alternative solution. For example, the problem that the City of Westerheimer will need another 100 MW of electric power by 2015 may be phrased in several ways including the following:

1. "By 2015 we need to install another 100 MW of electric capacity." Alternative: Build a 100 MW power plant.
2. "By 2015 we will be short of 100 MW of electric power." Alternative: Start conservation efforts that would save 100 MW.
3. "By 2015 we need to increase the power produced by our plants by 100 MW." Alternative: Build additional reheaters and feed-water heaters in the three existing coal power plants and increase their rated capacity by a total of 100 MW.
4. "We need 100 MW of green power by 2015." Alternative: Build a wind farm or a solar plant that will produce 100 MW.
5. "Unless we have another 100 MW of power by 2015, we will not be able to sustain the City's growth." Alternative: Restrict housing building permits and impede the growth of the City.
6. "Can we buy 100 MW electric capacity by 2015?" Alternative: Buy the additional energy from the nearby City of Oesterheimer, which is expected to have a surplus of electric production capacity between 2015 and 2030.

The second stage is the development of a long list of alternative solutions to the problem/need. Some of these alternatives may be mutually exclusive, while others may be combined to yield another, usually superior, alternative. Two management processes are used for the development of the longer list of alternatives, *brainstorming* and *the nominal group technique.*

A. The **conventional brainstorming process** is based on the principles that "quantity of ideas and possible solutions evolves into quality" and "deferment of judgment and selection." The process involves the following three phases:

   1. Preparation,
   2. The brainstorming session, and
   3. The evaluation session.

During the preparation phase, the participants are selected and a loosely defined statement of the problem or need is circulated. The number of participants is limited to individuals who have the ability and experience to contribute or have demonstrated creative abilities *vis a vis* the problem to be discussed. Typically, five to twelve individuals constitutes a good group for brainstorming. During the brainstorming session, ideas are generated and duly recorded. Participants are encouraged to pose different questions to the problem at hand, which would elicit alternative solutions, as in items 1–6 above. For the unhindered expression of the participating individuals and the smooth flow of ideas it is essential that:

1. All participants perceive they have equal standing during the session. Unconventional seating arrangements (round table or theater seating), casual dressing and use of first names without titles facilitate the smooth exchange of ideas.
2. Quantity of solutions/ideas is encouraged from the beginning. The concept of quality comes at a later stage.
3. Critique or list of disadvantages is ruled out. Evaluation and critique are parts of the process to follow.
4. Improvement of contributed solutions and combination of solutions is actively encouraged.

The evaluation stage will follow at the end of the brainstorming session if all participants are expected to attend. If this is not desirable, the evaluation stage will become a separate session with a smaller number of participants. During this session the outcomes of each solution are evaluated relative to the problem at hand and a smaller list of alternatives is formulated for further evaluation.

B. The **nominal group session** is a more formal meeting with the objective to achieve consensus among the participants. The list of participants and the loosely-defined statement of the problem or need are circulated to all participants before the meeting and solutions are sought. The actual meeting has a more formal setting and includes the following parts:

   1 All individual participants are expected to present and discuss their ideas in an orderly fashion. Questions and criticism of the ideas is allowed at this stage.
   2 Clarification and modification by an individual of the ideas based on the criticisms and suggestions follows and is expected as part of this process.
   3 Group modification and clarification of each idea.
   4 Ranking of all the solutions to the problem by individual voting or other means that are previously specified.
   5 Final discussion, clarification, and, perhaps, a final modification of the short list of solutions to produce group consensus on the final list.

It must be noted that, in both processes, the alternative projects to be established may be *mutually exclusive,* that is only one of the projects may be pursued at the exclusion of all others, or *complementary,* when a group of projects may be pursued simultaneously or in series.

14.3 The Time-Value of Money

After the short list of ideas/alternatives has been selected, a detailed economic evaluation for the projects in the short list is undertaken. This evaluation takes into account only the economic aspects of the chosen alternatives and determines the *profitability* of the projects. Other issues, which may not be quantified and do not affect materially the cash flow and the profitability of the project, are called *intangible items* and are left out at this stage. Such intangible items are "the public good", "a greener planet", or "overall national security." These items must have been discussed and must have been part of the short-list selection process that resulted in the short list of the alternatives. The economic analysis that follows treats the projects strictly as investments and the final decision is based on whether or not a project would be profitable to be developed for the corporation. In a capitalist, market-oriented system of for-profit corporations money-loosing projects are not initiated and are not developed.

The concept of the time-value of money is based on the premise that $1 today is worth more than $1 a year from now, the latter is worth more than $1 two years from now, and so on. The time-value of monetary funds is intricately related to the concepts of:

(a) *return to capital*, which stipulates that an amount of capital invested must yield more capital at the end of the investment period;
(b) *interest rate* or *discount rate*, r, which is the percentage of additional funds that must be earned for the lending of capital; and
(c) current and expected future *inflation*, which increases the cost of goods in the future.

When capital investments, such as energy production or conservation investments, are appraised there is an inherent risk that the investment may not succeed and that all or part of the capital may be lost. This *investment risk* is a justification for the charging of an interest rate and the expectation of higher return on the invested capital. In general, the higher the investment risk the higher would be the expected return on the capital.

Some types of investments are considered *risk free*. The interest rate associated with them, r_{rf}, is the lowest interest charged by the capital markets. Risk-free investments are usually short-term government securities, such as 3 or 6 month governmental obligations (called *treasury paper* in the USA), which typically yield very low interest. Since the government is a very stable debtor, the investors are certain that their capital and interest will be paid in full.

The return on the capital expected for other types of debt may be significantly higher: Investors will lend funds to a corporation at rates 1–10% above the r_{rf} rate and to individuals at higher rates than these. The *premium* over the rate r_{rf} depends on the financial strength of the corporation, the nature of the investment and the duration of the loan. Loans to financially weaker corporations are riskier and carry higher interest. Longer duration loans, e.g. 30 years versus 2 years, are also riskier and in general, command higher interest rates. Certain energy-related activities, such as drilling for oil and gas or building solar photovoltaic energy farms are considered among the riskiest investments. The interest rates charged for these investments and the expected return on the capital are among the highest in the energy production industry.

It is apparent from the above, that there is not only one value for the interest rate but a whole array, ranging from the risk-free, r_{rf}, to significantly higher values. There are two attributes one must consider related to the interest rate: The first is the value and the second the duration: e.g. 6% per year, 1% per month 37% per five years, etc. Typically, interest rates are quoted per year (*per annum*) but other durations may also be contracted.

14.3.1 Simple and Compound Interest

The *simple interest* is equal to the product of the initial capital invested, which is also called *the principal, P*; the interest rate charged, r; and the number of periods the loan/investment is made, N. The formula for calculating the simple interest, I, at the end of N periods is:

$$I = PNr. \tag{14.1}$$

With simple interest the total to be paid at the end of the N periods for the whole debt is T = P + I. Thus, a sum of $10,000, which is lent at a 6% simple interest rate for 5 years will yield $10,000*5*0.06 = $3,000 interest. The total to be paid after five years is $13,000.

While the simple interest is easy to apply, it is not frequently used in contemporary commercial practice. The *compound interest* is used in most commercial transactions. According to the concept of compound interest, accrued interest at the end of each period is added to the original capital, P. Hence, during the subsequent periods, interest is earned not only on the original principal, but on the previously earned interest, which is added to the principal. The formula for calculating the total amount to be paid with compound interest is:

$$T = P(1 + r)^{N}, \tag{14.2}$$

and the interest is: I = T − P. With compound interest, the total interest on $10,000 invested at 6% after 5 years is: $3,382, which is significantly higher than the simple interest.

It is apparent from the above, that the time period of compounding is as important as the value of interest rate, r, in the calculation of the interest paid and a borrower must carefully consider this variable. For example, several credit card lenders charge interest compounded daily. On an average credit card balance of $1,000, a 15% annual rate (compounded annually) results in $150 interest paid at the end of one year. On the same balance, a daily compounded interest rate of 15/365% results in $162 interest.

14.3.2 Cash Flow, Equivalence and Present Value

The objective of the decision making process is the evaluation or appraisal of several alternative solutions to a need, as it was elaborated in Sect. 14.2.1. Every alternative solution involves a complex timetable of investments, expenditures and revenues or other receipts of funds. The concept of *cash flow* or *net cash flow* is a tool that keeps track of the net influx of funds for each year of the duration of the project. The *net cash flow* is defined as the sum of all yearly receipts minus all yearly expenses associated with the development, operation, and closure or abandonment of a project. Such projects may be as simple as the lending of principal in the first year to obtain the principal plus interest in the future, to a complex investment in the construction and operation of a solar energy field that is expected to last over a period of several decades. In the more complex cases, cash flow diagrams are often used to depict graphically the progress of a project and the yearly cash flow from the project. Net expenditures appear as negative values and net receipts as positive values.

Consider the construction of a photovoltaic generation plant where the owner and operator must invest $500,000 in the first year and $1,000,000 in the second year. The plant is completed at the end of the second year and operates continuously for ten more years. The total revenue of the plant's operation is $200,000 during the third year and increases afterwards at a rate of 7% annually. The total expenses associated with the operation of the solar project are $50,000 during the third year and increase at a rate of 5% annually in the subsequent years of operation. At the end of the project, the end of the twelfth year, the operations discontinue. At this point, the salvage value of the solar energy project is $300,000. Table 14.1 shows all the monetary expenditures and receipts from the inception to the end of the project and the net cash flow for every year.

The cash flow diagram for this solar project is depicted in Fig. 14.1. The cash flow diagram is a succinct and convenient depiction of the annual income, from which a more detailed financial analysis may be made.

The monetary amounts comprising the cash flow pertain to funds earned or spent during the corresponding year. For example, if the first year for the solar energy field project is 2010, the cash flow item 250.8 pertains to the tenth year, or the calendar year 2019. The concept of the time value of money complicates the appraisal method by inserting the concept time as an additional variable: Given that when one compares solutions or projects there must be a common basis for the comparison, how does one evaluate several potential projects that involve a multitude of monetary expenses and receipts over the lives of the projects that span several years or decades? It is apparent that for such an evaluation or appraisal of the projects, one must reduce all the monetary expenses and benefits of each

Table 14.1 Yearly expenditures, revenues and cash flow for the small solar photovoltaic power plant. In thousands of $US

Year	Investment	Sum of receipts	Sum of expenses	Salvage value	Cash flow
1	−500	0	0		−500
2	−1,000	0	0		−1,000
3	0	200	−50		150
4	0	214	−52.5		161.5
5	0	229.0	−55.1		173.9
6	0	245.0	−57.9		187.1
7	0	262.2	−60.8		201.4
8	0	280.5	−63.8		216.7
9	0	300.1	−67.0		233.1
10	0	321.2	−70.4		250.8
11	0	343.6	−73.9		269.8
12	0	367.7	−77.6	300	590.1

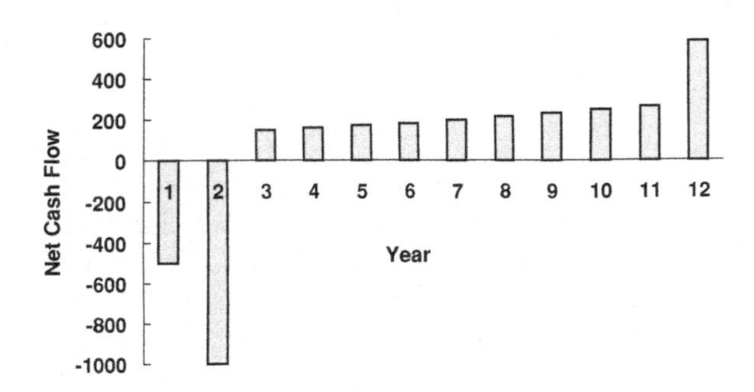

Fig. 14.1 Cash flow diagram for a small solar photovoltaic power plant

project to an *equivalent basis.* Equivalency here implies that an informed investor is indifferent between two sums of funds to be received at two different times in the future. For example, if an investor is indifferent between $100,000 to be paid in 2014 and $120,000 in 2017, then these two sums are *equivalent* to this investor. The *Present Value, PV,* of future funds or future returns offers a way of establishing this equivalent basis for the comparison of several projects. The PV establishes the currently equivalent monetary value of future funds for an investor or a corporation.

The present value, PV, of a principal amount of funds, M, to be obtained after N time periods at a discount rate, r_d, is defined as:

$$\mathrm{PV} = \frac{\mathrm{M}}{(1 + \mathrm{r_d})^N}, \tag{14.3}$$

that is an amount equal to PV today is accepted to be equivalent to a greater amount M after N periods of time. A glance at Eqs. (14.2) and (14.3) proves that the present value is directly related to the compound interest concept, with the discount rate being equivalent to the interest rate. The PV is equivalent to the principal invested today and the future amount M is equivalent to the total sum, T, which is expected after N periods of the investment. Another way to look at the present value is that an informed investor is indifferent between having a sum equal to PV today or a sum equal to M, N periods of time later. By evaluating the present value of all the cash flows for all the alternative projects, one establishes an equivalent basis for the appraisal of alternative projects that have different durations of cash flow and life cycles. Also one establishes a common basis for a rational economic comparison of the alternative projects. This will aid the final stages of the decision-making process.

Oftentimes in financial calculations there is a constant amount of funds, expenditures or incomes, which is repeated for several years. This is called an *annuity* and, since it is earned or expended over several years, it has a present value. The present value of the constant amount, A, earned or spent in every year, starting in year 1 and ending in year N is:

$$PV_A = \frac{A}{(1+r_d)} + \frac{A}{(1+r_d)^2} + \ldots + \frac{A}{(1+r_d)^N} = A\left[\frac{1/(1+r_d)^{N+1} - 1}{1/(1+r_d) - 1}\right] = \\ = \frac{A}{(1+r_d)^N}\left[\frac{(1+r_d)^{N+1} - 1}{r_d}\right]. \tag{14.4}$$

Associated to the annuity concept is the present borrowing of a principal sum of funds with the promise to repay both the principal and the interest in equal amounts over a fixed number of time periods. This is the concept most of the housing *mortgages* are based on, as well as most bank loans to small companies. In consideration of funds received at present, P, the owner promises to make a number of equal and regular payments in the future. In such cases, the term of the mortgage is counted in months and the agreed upon annual interest rate, i, takes the place of the discount rate. The periodic monetary payment, M, is given by the expression:

$$M = \frac{iP(1+i)^N}{(1+i)^{N+1} - 1}. \tag{14.5}$$

For a typical house mortgage with 30 year duration, the number of monthly periods is N = 30*12 = 360, and the interest rate, i, is the monthly interest. The latter is equal to 1/12th of the quoted annual rate. For example if the quoted annual rate is 6%, i = 0.06/12 = 0.005.

14.3.3 Cash Flow Calculations

The cash flow is a convenient concept in financial accounting that summarizes all the revenues and expenses associated with a project during a period of time, typically one calendar year. The following items are summed to make up the cash flow of a project, with the positive items denoted by (+) and negative items denoted by (−):

Revenues (+): All cash receipts from selling the product or services related to the project.
Interest earned (+): Any interest earned by funds directly associated with the project.
Tax credits (+): Credits on paid and unpaid taxes granted by local, regional, and national governments. These are typically a percentage (10–30%) of the project investment. Because of the worldwide energy challenge and climate change regulations, alternative energy projects are beneficiaries of tax credits in most OECD countries. However, for a corporation to take advantage of tax credits, it must have taxable income during the year at which credits are sought. Certain tax credits may roll in future years.
Rebates (+): All monetary receipts that are directly connected to the investment of the project and are received as a result of the investment.
Salvage value (+): The fair market price of equipment that is not used and may be sold in the market.
Capital expenditures (−): All expenses of capital equipment and construction associated with the project.
Other fixed costs (−): These include: rents of space or equipment, management and administration costs, interest paid, insurance, etc. The fixed costs do not depend on the units of energy produced and must be paid even if the business or project does not produce any units.
Variable costs (−): These are monotonically increasing with the level of production. Among the variable costs are the cost of fuel, labor costs excluding administration, and distribution costs.
Taxes (−): These are usually a percentage of the *taxable income* of a corporation. Detailed tax codes in every country specify what the *tax rate* is and how the taxable income may be calculated. While a full description of the tax code is beyond this short exposition, an item of importance for energy projects is the *depreciation schedule*. The depreciation schedule, which is different in the tax codes of different countries, specifies the way capital expenditures of a project may be *amortized* (or subtracted from the taxable income). In general, high depreciation expenses reduce the taxes paid by corporations.
Closure costs (−): Any cost associated with the termination of the project. For energy projects, this may include land clean-up and rehabilitation, well shutting, disposal of equipment, and, in the case of nuclear power plants, the decontamination of the equipment and the long-term storage of nuclear waste.

14.3.4 A Note on the Discount Rate and Interest Rates

It is apparent that the discount rate, r_d, is a fundamental concept in the calculation of the present value of any future sum, M. The choice of the right value for the appropriate discount rate is of paramount importance in any engineering project.

As with the interest rates, there is not a single discount rate used by corporations, governmental agencies and individuals, but a range of discount rates that depend upon the following parameters:

1. *The type of entity/corporation that performs the project (public or private).* In general, public corporations do not aim for profits and may borrow funds at lower rates than private corporations and, since their cost of capital is lower, their discount rates are lower too.
2. *The risk of the project.* The more risky the project (e.g. because of new technology utilization) the higher is the return sought and the higher would be the discount rate.
3. *The overall economy and ease of borrowing capital.* In general, when the overall economy is healthy it is easier to borrow funds. Interest rates are lower and so are the discount rates of corporations.
4. *Inflation.* Inflation or expected future inflation is directly and monotonically correlated with discount rates.
5. *The length of the project.* Longer projects are associated with higher economic uncertainty and risk and the discount rates used may be significantly higher. For that matter, the long term interest rate charged by financial institutions is in general higher than short term interest rate.

The discount rate, r_d, used by corporations and public utilities is significantly higher than the interest rate, r, charged by financial institutions. It is usually tied to an interest rate by a simple formula such as: $r_d = r+\delta r$, where δr is in the range 3–15%. Corporations have a range of discount rates, which depend on the perceived risk and the duration of the project (short, intermediate, or long term). The interest rate, to which the discount rate may be tied, is usually one of the following:

1. The *inter-bank discount rate*, which is the rate banks charge each other for borrowing funds and (in the USA) is set by the Federal Reserve Board.
2. The *short-term interest rate* of government securities. This is typically the interest rate on the 6 month or 12 month Treasury bonds.
3. The *intermediate interest rate,* which is the interest rate of 10 year Treasury bonds.
4. The *long-term interest rate* of government securities. This is typically the interest rate on 20 or 30 year Treasury bonds.
5. The *prime rate* charged by financial institutions. This rate is usually offered to the best and most risk-free customers of the institutions.
6. The *cost of borrowed* capital. This represents the interest rate paid by the corporation when it issues long-term bonds. Because every corporation and public utility has different financial ratings and ability to attract capital by issuing bonds, the cost of borrowed capital varies significantly between the several corporations and public utilities.
7. The *expected return on equity.* This rate is used for internally financed projects and is equal to the return expected by the shareholders of the corporation.

This rate also varies significantly between the different corporations and public utilities.

Typically, large energy projects are financed by a combination of public borrowing and equity. The pertinent discount rate is between the values of the last two rates. If any part of the project is financed by banks, one of the first four rates may come in the formula that determines the discount rate. It must be noted that the interest rates are not constant and vary daily (usually by a small amount). A glance at the financial pages of most newspapers shows the previous-daily values of the first five interest rates, which were mentioned above. The last two pertain to individual corporations and are not publicized.

14.4 Investment Appraisal Methods

When the list of the possible alternative projects has been established and a way has been determined for the equivalency of the future cash flow, one must use a method to evaluate the alternatives and to make a final decision on which project to develop. Every good method of project evaluation must have the following characteristics:

1. It takes into account the time horizon of the project.
2. It takes into account all cash flows pertinent to the project for the entire time horizon.
3. It defines an acceptable way to discount future cash flows and uses an *equivalent basis* of future cash flows.

In the next subsections, several methods that are commonly used for the evaluation of investments in energy projects are presented and compared.

14.4.1 The Net Present Value (NPV)

This method is sometimes called *present worth or net present worth.* The NPV of a project is the sum of the discounted annual cash flows from the inception to the disposal of the project. Consider a project that commences at the present year, which is usually denoted as year 0, and finishes at year n, that is n + 1 years later. The pertinent cash flow streams are: CF_0, CF_1, $CF_2 \ldots$, CF_n and the discount rate is r_d. The equation that defines the NPV of this project is:

$$NPV = CF_0 + \frac{CF_1}{(1 + r_d)} + \frac{CF_2}{(1 + r_d)^2} + \ldots + \frac{CF_n}{(1 + r_d)^n} = \sum_{i=0}^{n} \frac{CF_i}{(1 + r_d)^i}. \qquad (14.6)$$

If the NPV of a project is positive, the entity that considers it (a public utility or a for-profit corporation) incurs a net present benefit or profit from its development and execution. If the NPV is negative, there is no advantage to the entity from pursuing the project, because it is equivalent to a present loss. The latter project is not undertaken. It is apparent that the discount rate, r_d, plays a very important role in the calculation of the NPV and this is one of the most difficult parameters to determine precisely. The r_d value for projects is usually established by the management after taking into account the factors listed in the previous section.

During the decision making process, the NPV's of all alternatives are calculated. For a project to be acceptable, its NPV must be positive. Therefore, all alternative projects with negative NPV must be rejected and only the ones with positive NPV are considered. If these alternatives are exclusive, that is only one must be selected, the alternative project with the highest positive NPV is selected to be developed. If the projects are not exclusive, they are ranked in order of decreasing NPV, with the first few projects having the priority for further development. In the latter case, the decision of how many projects to pursue is dictated by resource availability considerations, such as: a) the availability of capital (do we have enough funds available for the investments needed?); and b) the capacity of the entity to pursue simultaneously these projects (can we design all the power plants? Do we have enough supervising engineers for all the projects?).

The NPV is a straightforward way of calculating the present equivalent of all the benefits and costs associated with the energy projects and provides an excellent method to assess the benefits of all energy production and conservation projects. It also helps to prioritize and rank projects. For these reasons, the NPV method is recommended for the appraisal of all energy projects.

14.4.2 Average Return on Book (ARB)

The "Book" in this method refers to the *Book Value* of an investment, which is an accounting concept. The Book Value is defined as the initial monetary value of the investment minus the accumulated depreciation. According to this method, a project is undertaken if the ARB is greater than an acceptable rate, which is usually determined by management.

Let us consider a project that starts in year 0 with a $12,000 investment, straight line depreciation is allowed during the years 1–3. The yearly net income of this project is $2,700. Table 14.2 shows the Book Value of this project during the four years of the life of the project:

The Average Book Value is (12,000 + 8,000 + 4,000 + 0)/4 = 6,000 and the ARB is 2,700/6,000 = 0.45 or 45%. This project would be undertaken if the acceptable ARB rate to the management is less than 45%.

The ARB method suffers from several deficiencies: First, there is no allowance that immediate income is more valuable than future income. Thus, the *equivalent basis* of revenues and costs is not an inherent attribute of this method. Secondly, it

Table 14.2 Calculation of book value

	Year 0	Year 1	Year 2	Year 3
Investment	12,000			
Net income	2,700	2,700	2,700	2,700
Accumulated depreciation	0	4,000	8,000	12,000
Book value	12,000	8,000	4,000	0

does not use the cash flows, which are real and tangible flows, but the accounting income, which is an accounting concept. The accounting net income does not take into account all capital expenses for a project during the year they occur, but only takes into account the allowed depreciation of these costs. Thirdly, the method uses the Average Book Value, another accounting concept, which does not differentiate between recent and distant capital expenditures. And, fourthly, the acceptable ARB is usually based on historical facts and is more difficult to determine based on the current economic environment and the current interest rates. For all these reasons, the ARB is not an investment appraisal method that is recommended to be used with energy projects, which are typically long-term projects.

14.4.3 The Pay-Back Period (PBP)

According to this method, the initial investment on the project must be recovered from net receipts/income within a specified period of time, which ranges from two to six years. This is the *payback period* and is calculated by counting the number of years it takes for the cumulative cash flows to equal the initial investment. Thus, a project that requires an initial investment of $1,000,000 and pays net cash flows of $100,000 in the first two years and $210,000 thereafter would have a payback period of six years: During the first five years the cumulative cash flow is $830,000, which is less than the initial investment, while during the first six years the cumulative cash flow is $1,040,000 which surpasses the initial investment of $1,000,000.

The payback period method makes use of the cash flows, which is the correct parameter for the calculation of the annual costs and benefits of the project. However, it does not differentiate between recent and distant cash flows or between capital expenditures that occur at the beginning or at another time during the life of the project. Therefore, the *equivalence basis* is not inherent in the PBP method. Another disadvantage of the method is that the recovery of the initial capital may be achieved in a fraction of years, while the rule for the acceptance of a project is always expressed in whole years. Table 14.3 shows the cash flows, the payback period and the NPV, calculated at a discount rate of 10%, for three projects, which have the same initial investment of $2,000,000. It is apparent that, although all projects have the same payback period of two years, one of the

Table 14.3 Payback period and net present value for three hypothetical projects. Numbers in $1,000

Project	CF_0	CF_1	CF_2	CF_3	PBP (yrs)	NPV
A	−2,000	1,000	1,000	5,000	2	3,492
B	−2,000	0	2,000	5,000	2	3,409
C	−2,000	1,000	1,000	100,000	2	74,857

projects (C) is superior to the first two by a wide margin, because of the large income that occurs after the initial payback period. Even when one examines only projects A and B, the first project is superior because it provides positive cash flow during the first year. The NPV, which is listed in the last column, is a better way to make a rational choice among the three projects.

The *Discounted Payback Period* (*DPBP*) method has been proposed, where the cash flows are discounted by a rate as in the NPV method. While this modification takes into account the time value of funds, it still does not take into account any funds that are obtained after the cutoff period. In addition, the value of the Payback Period (in years) is arbitrary and does not take into account the current economic environment and the ability of the corporation or utility to borrow and use funds. In the projects of Table 14.3 the large payoff in year 3 of project C does not receive the emphasis it deserves.

14.4.4 Internal Rate of Return (IRR)

The IRR method is connected directly and, in most cases, is equivalent to the NPV method: The IRR of a project is the value of rate of return, r_{ir}, which makes the NPV of the project equal to zero. The expression that defines the rate r_{ir} is:

$$CF_0 + \frac{CF_1}{(1+r_{ir})} + \frac{CF_2}{(1+r_{ir})^2} + \ldots + \frac{CF_n}{(1+r_{ir})^n} = \sum_{i=0}^{n} \frac{CF_i}{(1+r_{ir})^i} = 0. \quad (14.7)$$

For the application of the IRR method, the nonlinear Eq. 14.7 is solved to obtain the value of r_{ir}. For the decision making process, if $r_{ir} > r_d$, then the project should be undertaken and if $r_{ir} < r_d$ the project should be abandoned.

It is apparent that calculating the IRR is more cumbersome than calculating the NPV. The IRR is based on the NPV and in most cased the two methods are equivalent and may be used interchangeably. However, there are cases when the IRR method is not equivalent to the NPV. This stems from the non-linearity in the computation of the IRR and applies to mutually exclusive projects and projects that include investments at different timeframes. Table 14.4 below, shows three mutually exclusive projects where the NPV and IRR methods yield different results. While projects A and B have the same initial investment, project A has a higher IRR because the returns occur early in the life of the project. In this case,

Table 14.4 IRR and NPV for three mutually exclusive projects. Numbers are in \$1,000

	CF_0	CF_1	CF_2	CF_3	CF_4	CF_5	CF_6 to CF_∞	IRR, %	NPV at 10%
A	−9,000	6,000	5,000	4,000	0	0	0	33	3,592
B	−9,000	1,800	1,800	1,800	1,800	1,800	1,800	20	9,000
C	0	−6,000	1,200	1,200	1,200	1,200	1,200	20	6,000

the NPV and the IRR will yield different rankings for the two projects as long as the value of r_d used in the NPV method is lower than 15.6%. If $r_d > 15.6$ the two methods yield the same rankings. Projects B and C have the same IRR (20%), but the NPV of project B is higher, primarily because of the higher initial investment.

While the NPV and the IRR methods are equivalent in most cases, the NPV is simpler and leads to better decisions, especially when one deals with mutually exclusive projects with complex payoff schedules.

14.4.5 Profitability Index (PI)

The *profitability index* is sometimes referred to as *benefit-cost ratio* and is defined as the sum of all discounted future positive cash flows divided by the initial investment. If all initial investment occurs in year 0 and is denoted by I, and the cash flows in years 1 to n are denoted as CF_1, CF_2, CF_3,..., and CF_n, respectively, the expression for the PI is as follows:

$$PI = \left(\frac{CF_1}{(1+r_d)} + \frac{CF_2}{(1+r_d)^2} + \ldots + \frac{CF_n}{(1+r_d)^n} \right) / I = \frac{1}{I} \sum_{i=1}^{n} \frac{CF_i}{(1+r_d)^i}. \tag{14.8}$$

For the decision-making process, if $PI > 1$, the project is profitable and should be undertaken.

The PI method has several common features with the NPV method. For example, the time-value of funds is inherent in both methods and the present value of all positive cash flows is used in the PI method. Both methods may also use the same discount rate. However, when the investment occurs over a period of several years and when the investment and revenue streams follow a complex pattern and are mixed during several years, as it often occurs in the early years of an energy project, the PI method becomes more difficult to use and may lead to an erroneous decision between two mutually exclusive projects.

It is apparent, that the NPV method is a relatively simple method to use, especially with long-term and complex projects. This method is recommended to be used for energy generation and improved efficiency projects because it is well defined, it is relatively simpler to use than the other methods, and it is widely accepted by the business and engineering communities.

14.5 Use of the NPV Method for Electricity Generation Projects

Having established the advantages of using the NPV method as a tool for energy project investment appraisal, we may now study how to use this method in a practical situation:

Let us consider the building of a wind farm for the production of electricity in a location in Western Texas. The nominal power rating of this wind farm is 50 MW. The investment for this project is estimated to be $40,000,000 ($40 M) to be spent as follows: $10 M in year 0 for the preparation of the field and the design of the farm; $25 M in year 1 to be spent mostly on the wind turbines/generators; and $5 M in year 2 for the installation of the turbines/generators and connection to the grid. The corporation plans to put $20 M of investment into the project and finance the other $20 M in year 1 of the project by issuing 12 year bonds at an interest rate of 7%. Based on the wind conditions in the area, it is determined that the plant will produce on average 300,000 kWh per day for 365 days per year, or an average of $1.095*10^8$ kWh/yr. No electricity will be produced in years 0 and 1, while the plant is in construction, 50% of the total capacity will be produced in year 2 and 100% thereafter. It is estimated that the useful life of this power plant will be 12 years after the completion of the installation. At that point, the year 14 of the project, the whole wind farm may be sold to another corporation for $8 M. The fixed costs of the plant are estimated to be $100,000 per year, starting in year 0 and increasing at an annual rate of 3%. The operating costs are estimated to be $250,000/yr, starting in year 2 and increasing at 7% annually for the duration of the project. The average price of electricity is currently $0.047/kWh and it is expected to increase at an annual rate of 3.5%. The corporation is taxed at 28% rate and a straight-line, 10 year depreciation schedule is allowed. Given that wind farm technology is still considered risky, the corporate discount rate, r_d, for this type of investment is 15%.

A clarification is needed on the financing of this project: It is common practice for energy producing corporations to issue interest-bearing bonds to finance part of their projects, in this case 50% of the total investment. For simplicity, it will be assumed that the total revenue of the bonds will be collected at the end of year 0, interest will be paid for the bonds during years 1 through 12 and the whole principal of $20 M for the bonds will be repaid to the bondholders at the beginning of year 13. Thus, in year 1 the actual investment by the corporation is $5 M with the other $20 M coming from the issuing of bonds that will be repaid in year 13.

Regarding the cash flow, the revenue of this project comes entirely from the production of electricity, which is: 0 during years 0 and 1; $0.5475*10^8$ kWh during year 2; and $1.095*10^8$ kWh during years 3–14; and from the salvage value at the end of the project. The costs of the project consist of: investment of capital, annual fixed and variable costs; closing costs of $100,000 at the end of the project; interest expense on the bonds; and taxes paid by the corporation. The taxes are 28% of the *taxable income* of the corporation, which is for simplicity defined as the total annual revenue minus annual fixed and variable expenses, minus interest, minus the allowable depreciation. Given the schedule of the investment for this project, the depreciation of all the expenses is allowed over a period of 10 years. Thus, the allowable depreciation is as follows:

- for the year 0 the allowable depreciation is $1 M (= $10 M/10)
- for the year 1 the allowable depreciation is $3.5 M ($10 M/10 + $25 M/10)
- for years 2 through 9 the allowable depreciation is $4.0 M ($10 M/10 + $25 M/10 + $5 M/10)
- for year 10 the allowable depreciation is $3 M ($25 M/10 + $5 M/10)
- for year 11 the allowable depreciation is $0.5 M ($5 M/10).

Thus, at the end of year 11 the entire investment has been depreciated and there is no depreciation allowed in the subsequent years. It will be assumed that the corporation has other profitable activities to take advantage of the accrued depreciation during years 0 and 1, when there is no revenue from the project. In these two years, the project will incur a "negative tax," which is equivalent to a tax benefit for the corporation by offsetting taxes from other operations. Table 14.5 shows the steps for the calculation of the NPV for this project. Under the assumptions stated above, the NPV of the project is a negative $1,011,078, which dictates that the corporation should not undertake the development of this wind farm.

The question arises then, "under what conditions would this *green energy* project be profitable?" And the obvious answer is "if some parameters changed to favor the NPV." For example, if the current price of electricity were $0.05/kWh and everything else remained the same, the NPV would be a positive $148,094 and the project would be worth undertaking. This implies that, if consumers are willing to pay half a cent per kWh more, this renewable energy project would materialize. Similarly, if the project were determined to be less risky by the corporation and the discount rate were 13% instead of 15%, the NPV would be a positive $426,635 and, again, this project would be worth undertaking by the utility. Projects become less risky, when corporations accumulate a great deal of expertise in operating them and when certain financial guarantees are received from local or state governments, which is the subject of the next section.

The formulae used for the computation of the cash flow are given here for reference purposes. It must be pointed out that these formulae and the Tables in this chapter are for illustration purposes only. Accounting and the tax laws differ from country to country and, hence, the *taxable income* is differently defined in different countries.

Total Revenue = (Revenue from Electricity) + (Salvage Value).
Total Costs = (Capital Investment) + (Closing Costs) + (Fixed Costs) + (Variable Costs) + (Interest on Bonds) + (Bond Repayment).
Pre-Tax Income = (Total Revenue)–(Closing Costs)–(Fixed Costs)–(Variable Costs)–(Interest on Bonds)–(Bond Repayment).

Table 14.5 NPV calculation of the wind farm—basic scenario

NPV of a 50 MW wind farm,basic scenario

Years	0	1	2	3	4	5	6	7
Revenue								
Price, per kwh	0.0470	0.0486	0.0503	0.0521	0.0539	0.0558	0.0578	0.0598
kWh produced	0	0	54,750,000	109,500,000	109,500,000	109,500,000	109,500,000	109,500,000
Revenue from electricity	0	0	2,756,530	5,706,017	5,905,727	6,112,428	6,326,363	6,547,785
Salvage value	0	0	0	0	0	0	0	0
Total revenue	0	0	2,756,530	5,706,017	5,905,727	6,112,428	6,326,363	6,547,785
Costs								
Capital investment	10,000,000	5,000,000	5,000,000	0	0	0	0	0
Closing costs	0	0	0	0	0	0	0	0
Fixed costs	100,000	103,000	106,090	109,273	112,551	115,927	119,405	122,987
Variable costs	0	0	250,000	267,500	286,225	306,261	327,699	350,638
Interest on bonds	0	1,400,000	1,400,000	1,400,000	1,400,000	1,400,000	1,400,000	1,400,000
Bond repayment	0	0	0	0	0	0	0	0
Total costs	10,100,000	6,503,000	6,756,090	1,776,773	1,798,776	1,822,188	1,847,104	1,873,625
Tax calculation								
Pre-tax income	−1,00,000	1,503,000	1,000,440	3,929,244	4,106,951	4,290,239	4,479,258	4,674,160
Depreciation	1,000,000	3,500,000	4,000,000	4,000,000	4,000,000	4,000,000	4,000,000	4,000,000
Tax credit	0	0	0	0	0	0	0	0
Taxable income	−1,100,000	5,003,000	−2,999,560	−70,756	106,951	290,239	479,258	674,160
Tax	−308,000	1,400,840	−839,877	−19,812	29,946	81,267	134,192	188,765
Cash flow	−9,792,000	5,102,160	−3,159,683	3,949,056	4,077,005	4,208,972	4,345,066	4,485,395
Discount factor	0.8696	0.7561	0.6575	0.5718	0.4972	0.4323	0.3759	0.3269
Discount cash flow	−8,514,783	3,857,966	−2,077,543	2,257,885	2,026,992	1,819,655	1,633,471	1,466,284
Net present value	−1,011,078							

Table 14.5 (continued)

NPV of a 50 MW wind farm,basic scenario

Years	8	9	10	11	12	13	14
Revenue							
Price, per kwh	0.0619	0.0641	0.0663	0.0686	0.0710	0.0735	0.0761
kWh produced	109,500,000	109,500,000	109,500,000	109,500,000	109,500,000	109,500,000	109,500,000
Revenue from electricity	6,776,958	7,014,151	7,259,647	7,513,734	7,776,715	8,048,900	8,330,611
Salvage value	0	0	0	0	0	0	8,000,000
Total revenue	6,776,958	7,014,151	7,259,647	7,513,734	7,776,715	8,048,900	16,330,611
Costs							
Capital investment	0	0	0	0	0	0	0
Closing costs	0	0	0	0	0	0	100,000
Fixed costs	126,677	130,477	134,392	138,423	142,576	146,853	151,259
Variable costs	375,183	401,445	429,547	459,615	491,788	526,213	563,048
Interest on bonds	1,400,000	1,400,000	1,400,000	1,400,000	1,400,000	0	0
Bond repayment	0	0	0	0	0	20,000,000	0
Total costs	1,901,860	1,931,923	1,963,938	1,993,038	2,034,364	20,673,066	814,307
Tax calculation							
Pre-tax income	4,875,098	5,082,229	5,295,708	5,515,696	5,742,351	7,375,834	15,516,304
Depreciation	4,000,000	4,000,000	3,000,000	500,000	0	0	0
Tax credit	0	0	0	0	0	0	0
Taxable income	875,098	1,082,229	2,295,708	5,015,696	5,742,351	7,375,834	15,516,304
Tax	245,027	303,024	642,798	1,404,395	1,607,858	2,065,233	4,344,565
Cash flow	4,630,071	4,779,205	4,652,910	4,111,301	4,134,493	−14,689,400	11,171,739
Discount factor	0.2843	0.2472	0.2149	0.1869	0.1625	0.1413	0.1229
Discount cash flow	1,316,155	1,181,346	1,000,111	768,432	671,971	−2,076,033	1,372,945
Net present value							

Production 109,500,000 kWh/yr; Price growth rate = 0.035% annually; Internal Rate of Return = 15%; Fixed cost growth rate = 3%; Variable cost growth rate = 7%; Corporate tax rate = 28%

Taxable Income = (Pre-Tax Income)–(Depreciation).
Tax = (Taxable Income)*(Tax Rate)–(Tax Credit).
Cash Flow = (Total Revenue)–(Total Costs)–(Tax).
Discount factor for year i: $1/(i + r_d)^i$.
Discounted Cash Flow = (Cash Flow)*(Discount factor for year i).
NPV = Sum of all Discounted Cash Flows.

14.5.1 NPV and Governmental Incentives or Disincentives

It is apparent from the calculations of the last section that one may find several combinations of prices, variable and fixed costs, growth rates, etc. that render a given project worth undertaking to a corporation. The use of a spreadsheet helps significantly with the calculations. It must be noted, however, that these parameters are not within the control of a corporation. The determination of the future values of most of these parameters represents mere assumptions that are made by accountants and engineers, using past experiences on similar projects. However, recent experience dictates that many of these parameters (e.g. fuel costs in the decade 2001–2010) may vary widely and unpredictably. This adds to the risk of a project and contributes to corporations using higher discount rates.

There are several parameters that determine the NPV of a project, which are controlled by the local or central governments. Among these parameters are:

(a) The tax rate (28% in the example of Table 14.5).
(b) The allowable schedule of depreciation.
(c) Offering tax credits for certain *green energy* activities and projects.
(d) Guaranteed price for the energy produced or yearly rate increases.

Items b) and c) are offered currently as incentives for the renewable energy projects in several OECD countries. Recently, renewable energy and conservation projects have attracted significant credits from 5 to 30% in most OECD countries. Such incentives play a significant role in reducing the risk of renewable energy projects and, also, in making the NPV or the project positive and the project worth developing. We will consider here two such cases, both of which affect the taxation of the corporation:

1. *Allow the capital costs of the project to be depreciated faster*: Instead of a 10 year depreciation schedule, let us assume that the capital equipment may be depreciated with the straight-line method within 5 years. Thus, the allowable depreciation in year 0 is \$2 M; in year 1 it is \$7 M; in year 2, \$8 M; in year 3, \$8 M; in year 4, \$8 M; in year 5, \$6 M; and in year 6, \$1 M. This is called *accelerated depreciation schedule.* The effect of only this schedule of accelerated depreciation is to make the NPV of the project from a negative \$1,011,078 to a positive \$665,287. The positive NPV will

swing the decision-making process in favor of the construction of the plant. The calculations for this example are shown in Table 14.6. All parameters are the same as in the previous case except for the allowable depreciation. It is apparent that only this, seemingly minor modification, would swing the decision-making process from negative to positive. An accelerated schedule of depreciation allows a corporation to receive certain tax benefits earlier rather than later on the funds it has already spend. Since the early benefits of a project affect the NPV the most, an accelerated depreciation schedule always has a positive effect on the NPV.

2. *Tax credits on the investment*: Tax credits work in a similar way, with the corporation receiving back in the form of reduced taxes a percentage of its investment in equipment and facilities. Let us assume that there is a governmental 10% tax credit on the investment for renewable energy projects, such as the wind farm under consideration. Although tax credits vary from country to country, 10% is on the lower end of the range of rates that have been offered for such projects in the U.S.A., France, Germany, Greece, the United Kingdom, and Italy during the decade 2001–2010. With a 10% tax credit on renewable energy investments, the corporation saves from other taxable activities $1 M during year 0, $2.5 M in year 1 and $0.5 M in year 2. In this case, the tax credits are sufficient to change the negative NPV of the basic scenario to a positive value of $2,077,605 as shown in Table 14.7, which has been calculated with the same parameters as those for Table 14.5, but with a 10% tax credit on the investment for the wind farm.

It must be emphasized that, for a corporation to receive the tax benefits of accelerated depreciation and the tax credits, it must have other profitable taxable operations, whose taxes would be offset by the depreciation benefit. If a corporation does not have other profitable operations, claiming the depreciation and credits may be delayed for later years in some countries. However, this delay would have a negative impact on the NPV. Individuals, smaller corporations or small start-up companies typically do not benefit from accelerated depreciation schedules. Direct subsidies and rebates may be more effective incentives for smaller corporations and individuals.

3. *A regulatory disincentive:* The regulatory environment, which is largely controlled by central, regional and local governments, may also impose incentives and disincentives to energy projects. An obvious disincentive is the taxation of alternative energy activities or by-products, such as the imposition of a disposal fee on nuclear waste, which may be justified on the grounds that it takes funds to process and store the waste produced by nuclear power plants. Another disincentive that rarely comes to the attention of the public is a prolonged delay of the commencement of the energy project because of a local judicial decision, usually on a perceived environmental or ecological effect of the energy project. Consider, for example, the case of the wind farm with accelerated depreciation. As shown in Table 14.6, the project has a NPV of $665,287 and the corporation

Table 14.6 NPV calculation of the wind farm—basic scenario with accelerated depreciation

NPV of a 50 MW wind farm, accelerated depreciation

Years	0	1	2	3	4	5	6	7
Revenue								
Price, per kwh	0.0470	0.0486	0.0503	0.0521	0.0539	0.0558	0.0578	0.0598
kWh produced	0	0	54,750,000	109,500,000	109,500,000	109,500,000	109,500,000	109,500,000
Revenue from electricity	0	0	2,756,530	5,706,017	5,905,727	6,112,428	6,326,363	6,547,785
Salvage value	0	0	0	0	0	0	0	0
Total revenue	0	0	2,756,530	5,706,017	5,905,727	6,112,428	6,326,363	6,547,785
Costs								
Capital investment	10,000,000	5,000,000	5,000,000	0	0	0	0	0
Closing costs	0	0	0	0	0	0	0	0
Fixed costs	100,000	103,000	106,090	109,273	112,551	115,927	119,405	122,987
Variable costs	0	0	250,000	267,500	286,225	306,261	327,699	350,638
Interest on bonds	0	1,400,000	1,400,000	1,400,000	1,400,000	1,400,000	1,400,000	1,400,000
Bond repayment	0	0	0	0	0	0	0	0
Total costs	10,100,000	6,503,000	6,756,090	1,776,773	1,798,776	1,822,188	1,847,104	1,873,625
Tax calculation								
Pre-tax income	−100,000	−1,503,000	1,000,440	3,929,244	4,106,951	4,290,239	4,479,258	4,674,160
Depreciation	2,000,000	7,000,000	8,000,000	8,000,000	8,000,000	6,000,000	1,000,000	0
Tax credit	0	0	0	0	0	0	0	0
Taxable income	−2,100,000	−8,503,000	−6,999,560	−4,070,756	−3,893,049	−1,709,761	3,479,258	4,674,160
Tax	−588,000	−2,380,340	−1,959,877	−1,139,812	−1,090,054	−478,733	974,192	1,308,765
Cash Flow	−9,512,000	−4,122,160	−2,039,683	5,069,056	5,197,005	4,768,972	3,505,066	3,365,395
Discount factor	0.8696	0.7561	0.6575	0.5718	0.4972	0.4323	0.3759	0.3269
Discount cash flow	−8,271,304	−3,116,945	−1,341,125	−2,898,249	2,583,830	2,061,758	1,317,684	1,100,154
Net present value	665,287							

Table 14.6 (continued)

NPV of a 50 MW wind farm, accelerated depreciation

Years	8	9	10	11	12	13	14
Revenue							
Price, per kwh	0.0619	0.0641	0.0663	0.0686	0.0710	0.0735	0.0761
kWh produced	109,500,000	109,500,000	109,500,000	109,500,000	109,500,000	109,500,000	109,500,000
Revenue from electricity	6,776,958	7,014,151	7,259,647	7,513,734	7,776,715	8,048,900	8,330,611
Salvage value	0	0	0	0	0	0	8,000,000
Total revenue	6,776,958	7,014,151	7,259,647	7,513,734	7,776,715	8,048,900	16,330,611
Costs							
Capital investment	0	0	0	0	0	0	0
Closing costs	0	0	0	0	0	0	100,000
Fixed costs	126,677	130,477	134,392	138,423	142,576	146,853	151,259
Variable costs	375,183	401,445	429,547	459,615	491,788	526,213	563,048
Interest on bonds	1,400,000	1,400,000	1,400,000	1,400,000	1,400,000	0	0
Bond repayment	0	0	0	0	0	20,000,000	0
Total costs	1,901,860	1,931,923	1,963,938	1,998,038	2,034,364	20,673,066	814,307
Tax calculation							
Pre-tax income	4,875,098	5,082,229	5,295,708	5,515,696	5,742,351	7,375,834	15,516,304
Depreciation	0	0	0	0	0	0	0
Tax credit	0	0	0	0	0	0	0
Taxable income	4,875,098	5,082,229	5,295,708	5,515,696	5,742,351	7,375,834	15,516,304
Tax	1,365,027	1,423,024	1,432,798	1,544,395	1,607,858	2,065,233	4,344,565
Cash Flow	3,510,071	3,659,205	3,812,910	3,971,301	4,134,493	−14,689,400	11,171,739
Discount factor	0.2843	0.2472	0.2149	0.1869	0.1625	0.1413	0.1229
Discount cash flow	997,781	904,499	819,559	742,265	671,971	−2,076,033	1,372,945
Net present value							

Production 109,500,000 kWh/yr; Price growth rate = 0.035% annually; Internal Rate of Return = 15%; Fixed cost growth rate = 3%; Variable cost growth rate = 7%; Corporate tax rate = 28%

Table 14.7 NPV calculation of the wind farm—basic scenario with 10% tax credit

NPV of a 50 MW wind farm,basic scenario with 10% tax credit

Years	0	1	2	3	4	5	6
Revenue							
Price, per kwh	0.0470	0.0486	0.0503	0.0521	0.0539	0.0558	0.0578
kWh produced	0	0	54,750,000	109,500,000	109,500,000	109,500,000	109,500,000
Revenue from electricity	0	0	2,756,530	5,706,017	5,905,727	6,112,428	6,326,363
Salvage value	0	0	0	0	0	0	0
Total revenue	0	0	2,756,530	5,706,017	5,905,727	6,112,428	6,326,363
Costs							
Capital investment	10,000,000	5,000,000	5,000,000	0	0	0	0
Closing costs	0	0	0	0	0	0	0
Fixed costs	100,000	103,000	106,090	109,273	112,551	115,927	119,405
Variable costs	0	0	250,000	267,500	286,225	306,261	327,699
Interest on bonds	0	1,400,000	1,400,000	1,400,000	1,400,000	1,400,000	1,400,000
Bond repayment	0	0	0	0	0	0	0
Total costs	10,100,000	6,503,000	6,756,090	1,776,773	1,798,776	1,822,188	1,847,104
Tax calculation							
Pre-tax income	−100,000	−1,503,000	1,000,440	3,929,244	4,106,951	4,290,239	4,479,258
Depreciation	0	1,000,000	3,500,000	4,000,000	4,000,000	4,000,000	4,000,000
Tax credit	1,000,000	2,500,000	500,000	0	0	0	0
Taxable income	−100,000	−2,503,000	−2,499,560	−70,758	106,951	290,239	479,258
Tax	−1,028,000	−3,200,840	−11,998,777	−19,812	29,946	81,267	134,192
Cash flow	−9,072,000	−3,302,160	−2,799,683	3,949,056	4,077,005	4,208,972	4,345,066
Discount factor	0.8696	0.7561	0.6575	0.5718	0.4972	0.4323	0.3759
Discount cash flow	−7,888,695	−2,496,907	−1,840,837	2,257,885	2,026,992	18,196,555	1,633,471
Net present value	1,212,774						

Table 14.7 (continued)

NPV of a 50 MW wind farm,basic scenario with 10% tax credit

Years	7	8	9	10	11	12	13	14
Revenue								
Price, per kwh	0.0598	0.0619	0.0641	0.0663	0.0686	0.0710	0.0735	0.0761
kWh produced	109,500,000	109,500,000	109,500,000	109,500,000	109,500,000	109,500,000	109,500,000	109,500,000
Revenue from electricity	6,547,785	6,776,958	7,014,151	7,259,647	7,513,734	7,776,715	8,048,900	8,330,611
Salvage value	0	0	0	0	0	0	0	8,000,000
Total revenue	6,547,785	6,776,958	7,014,151	7,259,647	7,513,734	7,776,715	8,048,900	16,330,611
Costs								
Capital investment	0	0	0	0	0	0	0	0
Closing costs	0	0	0	0	0	0	0	100,000
Fixed costs	122,987	126,677	130,477	134,392	138,423	142,576	146,853	151,259
Variable costs	350,638	375,183	401,445	429,547	459,615	491,788	526,213	563,048
Interest on bonds	1,400,000	1,400,000	1,400,000	1,400,000	1,400,000	1,400,000	0	0
Bond repayment	0	0	0	0	0	0	20,000,000	0
Total costs	1,873,625	1,901,860	1,931,923	1,963,938	1,993,038	2,034,364	20,673,066	814,307
Tax calculation								
Pre-tax income	4,674,160	4,875,098	5,082,229	5,295,708	551,566	5,742,351	7,375,834	15,516,304
Depreciation	4,000,000	4,000,000	4,000,000	3,000,000	500,000	0	0	0
Tax credit	0	0	0	0	0	0	0	0
Taxable income	674,160	875,098	1,082,229	2,295,708	5,015,696	5,742,351	7,375,834	15,516,304
Tax	188,765	245,027	303,024	642,798	1,404,395	1,607,858	2,065,233	4,344,565
Cash flow	4,485,395	4,630,071	4,779,205	4,652,910	4,111,301	4,134,493	−14,689,400	11,171,739
Discount factor	0.3269	0.2843	0.2472	0.2149	0.1869	0.1625	0.1413	0.1229
Discount cash flow	1,466,284	1,316,155	1,181,346	100,111	768,432	671,971	−2,076,033	1,372,945
Net present value								

Production 109,500,000 kWh/yr; Price growth rate = 0.035% annually; Internal Rate of Return = 15%; Fixed cost growth rate = 3%; Variable cost growth rate = 7%; Corporate tax rate = 28%

goes ahead with the construction and operation of the plant. However, a local environmental group determines that the wind turbines will be harmful to a migratory species of Canadian geese that happen to pass near the wind farm site. The environmental group persuades a local judge to issue an injunction for the construction and operation of the wind farm, pending a "...complete and thorough environmental impact of the proposed plant." The corporation appeals this decision to a higher court and, eventually, prevails in the court system of the country. However, the effect of this judicial process has been to delay the operation of the plant for 12 months (a very short time for most judicial systems). This causes revenue to start being generated in year 3 instead of year 2. The results are shown in Table 14.8. It is observed that the effect of this simple twelve-month delay is to reduce the NPV of the project from a positive $665,287 to a negative $71,839. This delay and the negative NPV would classify this project as unprofitable for the corporation. Longer delays in the commencement of the power production may become disastrous and even threaten the viability of smaller corporations, thus, adding significantly to the risk of projects. This risk is reflected by an increase in the discount rate used by corporations and has a detrimental effect on the NPV and the commercial viability of such energy projects.

14.5.2 Use of the NPV Method for Improved Efficiency Projects

The NPV method is a general and sound financial accounting method that may be used in all the projects and not in power plant construction projects alone. Because it is a general method, it may be used in energy conservation as well as in improved energy efficiency projects. These projects are very similar to the electricity generation projects, but usually of shorter duration. The main difference between the electricity generation and the conservation/efficiency projects is that the "revenue" is actually generated from energy savings, which need to be quantified. Also, it is very common for conservation and energy efficiency projects to attract tax credits from central or local governments.[1] Oftentimes, these incentives are augmented by electricity production companies, who perceive these projects as a means to avoid or to delay new generation projects that are more costly and more capital intensive. In addition, because these involve small engineering projects that may be completed in a few weeks, the savings or revenues typically start accruing immediately (during year zero).

[1] In the wake of the debate on environmental change and global warming, the governments of several countries offer generous incentives for energy conservation/efficiency projects in the form of tax credits and accelerated depreciation.

Table 14.8 NPV of the wind farm—basic scenario with accelerated depreciation as in Table 14.6, but with a 12 month delay in the production of power

NPV of a 50 MW wind farm, accelerated depreciation, 12 month delay in construction

years	0	1	2	3	4	5	6
Revenue							
Price, per kwin	0.0,470	0.0,486	0.0,503	0.0521	0.0539	0.0558	0.0578
kWh produced	0	0	0	54,750,000	109,500,000	109,500,000	109,500,000
Revenue from electricity	0	0	2,756,530	5,706,017	5,905,727	6,112,428	6,326,363
Salvage Value	0	0	0	0	0	0	0
Total revenue	0	0	2,756,530	5,706,017	5,905,727	6,112,428	6,326,363
Costs							
Capital investment	10,000,000	5,000,000	5,000,000	0	0	0	0
Closing Cost	0	0	0	0	0	0	0
Fixed costs	100,000	103,000	106,090	109,273	112,551	115,927	119,405
Variable cost	0	0	250,000	267,500	286,225	306,261	327,699
Interest on bonds	0	1,400,000	1,400,000	1,400,000	1,400,000	1,400,000	1,400,000
Bond repayment	0	0	0	0	0	0	0
Total costs	10,100,000	6,503,000	6,756,090	1,776,773	1,798,776	1,822,188	1,847,104
Tax calculations							
Pre-tax income	−100,000	−1,503,000	1,000,440	3,929,244	4,106,951	4,290,239	4,479,258
Depreciation	2,000,000	7,000,000	8,000,000	8,000,000	8,000,000	6,000,000	1,000,000
Tax credit	0	0	0	0	0	0	0
Taxable income	−2,100,000	−8,503,000	−6,999,560	−4,070,756	−3,893,049	−1,709,761	3,479,258
Tax	−588,000	−2,380,840	−1,959,877	−1,139,812	−1,090,054	−478,733	974,192
Cash flow	−9,512,000	−4,122,160	−2,039,683	5,069,056	5,197,005	4,768,972	3,505,066
Discount factor	0.8696	0.7561	0.6575	0.5718	0.4972	0.4323	0.3759
Discount cash flow	−8,271,304	−3,116,945	−1,341,125	2,898,249	2,583,830	2,061,758	1,317,684
Net present value	−71,839						

Table 14.8 (continued)

NPV of a 50 MW wind farm, accelerated depreciation, 12 month delay in construction

years	7	8	9	10	11	12	13	14
Revenue								
Price, per kwin	0.0598	0.0619	0.0641	0.0663	0.0686	0.0710	0.0735	0.0761
kWh produced	109,500,000	109,500,000	109,500,000	109,500,000	109,500,000	109,500,000	109,500,000	109,500,000
Revenue from electricity	6,547,785	6,776,958	7,014,151	7259,647	7,513,734	7,776,715	8,048,900	0
Salvage Value	0	0	0	0	0	0	0	800,0000
Total revenue	6,547,785	6,776,958	7,014,151	7259,647	7,513,734	7,776,715	8,048,900	800,0000
Costs								
Capital investment	0	0	0	0	0	0	0	0
Closing Cost	0	0	0	0	0	0	0	100,000
Fixed costs	122,987	126,677	130,477	13,492	138,423	142,576	146,853	151,259
Variable cost	350,638	375,183	401,445	429,547	459,515	491,788	526,213	563,048
Interest on bonds	1,400,000	1,400,000	1,400,000	1,400,000	1,400,000	1,400,000	0	0
Bond repayment	0	0	0	0	0	0	20,000,000	0
Total costs	1,873,625	1,901,860	1,931,923	1,963,938	1,998,038	2,034,364	20,673,066	814,307
Tax calculations								
Pre-tax income	4,674,160	4,875,098	5,082,229	5,295,708	5,515,696	5,742,351	7,375,834	7,185,693
Depreciation	0	0	0	0	0	0	0	0
Tax credit	0	0	0	0	0	0	0	0
Taxable income	4,674,160	4,875,098	5,082,229	5,295,708	5,515,696	5,742,351	7,375,834	7,185,693
Tax	1,308,765	1,365,027	1,423,024	1,482,798	1,544,395	1,607,858	2,065,233	2,011,994
Cash flow	3,365,395	3,510,071	3,659,205	3,812,910	3,971,301	4,134,493	−14,689,400	5,173,699
Discount factor	0.3269	0.2843	0.2472	0.2149	0.1869	0.1625	0.1413	0.1229
Discount cash flow	1,100,154	997,781	904,499	819,559	742,265	671,971	−2,076,033	535,819
Net present value								

Production 109,500,000 kWh/yr; Price growth rate = 0.035% annually; Internal Rate of Return = 15%; Fixed cost growth rate = 3%; Variable cost growth rate = 7%; Corporate tax rate = 28%

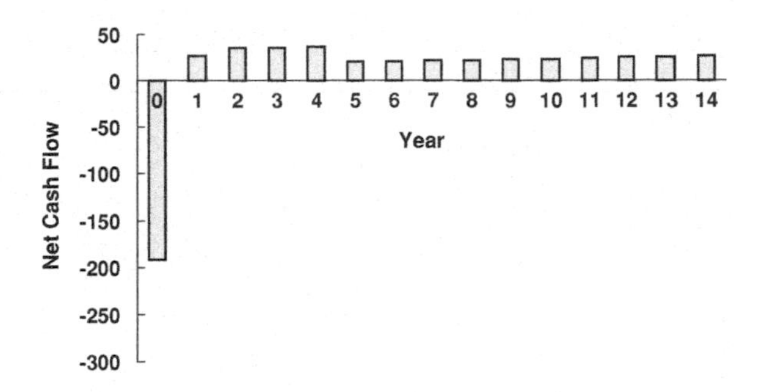

Fig. 14.2 Cash flow diagram for the installation of a GHP system

Let us look at the substitution of a simple, conventional air-conditioning unit in a small commercial building with a Geothermal Heat Pump (GHP). The building is located in Fort Worth, Texas, USA and is occupied and used throughout the year. It is proposed that the old air-conditioning and heating systems of the building be replaced with a GHP. While the mechanical systems will be replaced, the duct-work and air distribution systems in the building will be kept intact to reduce the overall cost and inconvenience of the replacement. The entire replacement of the system will cost $300,000. It is expected that the more efficient GHP will result in savings of 230,000 kWh per year from the air-conditioning load of the building. At $0.095 /kWh[2] this amounts to a savings of $21,850 per year for the owner of the building. It will also save the equivalent of $4,450 per year from the heating costs during the winter and the hot water supply during the entire year (see Sect. 13.3.3 for a complete analysis). The total savings for the project amount to $26,300 per year, start accruing in year zero and are expected to increase at a rate of 3%, because of higher energy prices in the future. In addition, the federal U.S.A. tax code allows for a 30% tax credit for the owner of the building and depreciation of the investment in five years. The GHP system is expected to have a life-time of 15 years and does not need significant maintenance. It may be assumed that the discount rate for the investment is 12% and the tax rate for the owner of the building is 28%.

Figure 14.2 shows the cash flow diagram for this project. The cash flows in the first five years, shown as years 0–4, include the tax benefits from the depreciation of the investment. The NPV of the project will be given by the following formula:

$$NPV = CF_0 + \frac{CF_1}{(1.12)} + \frac{CF_2}{(1.12)^2} + \ldots + \frac{CF_{14}}{(1.12)^{14}} = \sum_{i=0}^{14} \frac{CF_i}{(1.12)^i}, \tag{14.9}$$

In this case it is observed that, after the first five years when the tax credits and depreciation are accounted for, the cash flow consists of the energy savings only. At year 4 the energy savings amount to $29,601 and, for all the subsequent years,

[2] Because air-conditioners use at peak demand periods, the cost of electricity to the owner of the building is more expensive during the hours of their operation.

Table 14.9 NPV calculation for the substitution of an older air-conditioning system by a GHP

Internal rate of return	12%					
Tax rate	28%					
NPV for the replacement of the air-conditioning unit with a GHP						
Years	0	1	2	3	4	5–14
Revenue						
Revenue from energy savings	26,300	27,089	27,902	28,739	29,601	203,875
Costs						
Capital investment	300,000			0	0	0
Total costs	300,000	0	0	0	0	0
Tax calculation						
Pre-tax income	26,300	27,089	27,902	28,739	29,601	203,875
Depreciation	42,000	42,000	42,000	42,000	42,000	0
Tax credit	90,000	0	0	0	0	0
Taxable income	−15,700	−14,911	−14,098	−13,261	−12,399	203,875
Tax	−94,396	−4,175	−3,948	−3,713	−3,472	57,085
Cash flow calculation						
Cash flow	−179,304	31,264	31,849	32,452	33,073	146,790
Discount factor	1.0000	0.8929	0.7972	0.7118	0.6355	0.5674
Discounted cash flow	−179,304	27,914	25,390	23,099	21,018	83,293
Net present value	1,410					

the savings increase at a constant rate of 3% and are discounted by 12%. One may use this simple feature of this problem and the general property of the series

$$1 + x + x^2 + x^3 + x^4 + \ldots + x^n = \frac{1 - x^n}{1 - x}, \tag{14.10}$$

to calculate the discounted cash flows from year 5 to year 14 as follows:

$$\frac{CF_5}{(1.12)^5} + \frac{CF_6}{(1.12)^6} + \frac{CF_7}{(1.12)^7} + \ldots + \frac{CF_{14}}{(1.12)^{14}} =$$
$$\frac{CF_5}{(1.12)^5}\left[1 + \frac{1.03}{1.12} + \left(\frac{1.03}{1.12}\right)^2 + \left(\frac{1.03}{1.12}\right)^3 + \ldots + \left(\frac{1.03}{1.12}\right)^9\right] = 6.589\frac{CF_5}{(1.12)^5}. \tag{14.11}$$

Equation (14.11) shows that the discounted cash flow from the energy savings between years 5–14 amount to $83,293. These computations and the pretax income for the building owner are shown in Table 14.9. It may be seen in the Table that the NPV of this project is a positive $1,410, which implies that it is worth substituting the old air-conditioning system with the Geothermal Heat Pump.

The positive NPV in this case justifies completely the investment for the energy efficiency project. It is evident, however, that the tax credit of 30% plays a very important role in the positive NPV and the justification of this project. Without the tax credit the NPV of the project would have been strongly negative. This is a common characteristic of most alternative energy and improved efficiency projects: because of the price structure of the fossil fuels, the maturity of fossil fuel technology and the long-term engineering experience in fossil fuel projects, switching to alternative energy sources typically becomes uneconomical without economic and financial externalities, such as regulations or governmental incentives. The latter are justified by the following reasons:

- The potential adverse effects of carbon dioxide build-up.
- The declining supply of fossil fuels and the certainty that they will have to be replaced in the future at considerable expense to the society.
- The creation of new professions and new jobs that may support the economy in the future.
- Energy security for the countries that import fossil fuels.

In addition, it has been observed that several electricity generation companies (utilities) are willing to subsidize energy efficiency projects, such as the one described in this section. This trend is more evident with the publicly owned utilities sector. For example, since 2009, the San Antonio, Texas utility (CPS) offers a $400 rebate per ton of air-conditioning that is substituted by a more efficient GHP. The underlining reason for this rebate is that more efficient air-conditioning units cause lesser peak power demand during the summer months. This has three beneficial effects for the power producing corporations that justify the small investment of the rebates:

1. The lesser peak power demand implies that the most inefficient and expensive power producing units will not need to be operated for the production of power.
2. Any growth in the demand for electricity, e.g. because of population growth, will not need to be met immediately, thus deferring the high capital cost associated with the building of new power plants further in the future. It must be noted here that the average cost of adding 1 MW of electric producing capacity by constructing a new power plant in 2009 was approximately $5,000,000.
3. There are lesser atmospheric emissions of pollutants from the power plants of the corporation. This is often a mandate of the citizens the electricity generating corporation serves.

14.6 Project Financing for Alternative Energy Technology

It is apparent from the examples of the previous sections that the building of new alternative energy power plants and the completion of energy efficiency projects requires significant injection of capital. Energy projects are capital intensive and, especially, new technology for the production of electric power from renewable sources requires significant capital investment. In the beginning of the Twenty First Century the world economy and the international financial markets are very much influenced by the 1985–2000 era when the new inventions in electronics, computers and internet created the "new world economy" with several very successful information, high-tech companies, such as Microsoft, Apple, Google, Yahoo, etc. The creation of these, now gigantic, corporations involved an excellent idea and a relatively very small amount of capital. The capital was typically provided by the teams of investors that in the U.S.A. are called "angels" (a few hundred thousand dollars) and "venture capitalists" (a few million dollars). The financial rewards to these investors from successful projects were momentous, typically 10–1,000 times return on the invested capital. The rewards were also very quick to materialize, typically within 3–5 years, when the original company made its first Initial Public Offering (IPO) of stock.

The timeframe of the "high technology" projects and their financial models are not applicable to energy production investment models for the following reasons:

1. Oftentimes, the technology involved in alternative energy projects is not new. This implies that these projects are not very risky and, as a consequence, they are not as profitable as certain high technology projects.
2. Energy production and conservation projects are easier to be duplicated and patents that may ensure high profitability are more difficult to obtain.
3. All energy projects are long-term projects with payoffs that extend from 20 to 60 years. The original investors that are risking their capital, typically, do not recoup their investment within 2–5 years in an Initial Public Offering of the company's stock.
4. The investment required for even a small alternative energy power plant is very high in comparison to investments in internet-related corporations.

Because of these reasons, it is not rational to expect that the free market will provide the means to finance all the alternative energy projects that have reasonable potential to become successful and profitable. The continuation of regulatory intervention and governmental subsidies may be necessary for the initial financing and the long-term success of alternative energy and energy efficiency projects, at least until the expected or perceived depletion of the fossil fuel resources brings in a more favorable and realistic price regime for alternative energy. An additional reason for the continuation of governmental subsidies and intervention is the reduction of atmospheric pollution, which is a significant public health benefit.

Problems

1. You are considering the construction and operation of a 5 MW solar photovoltaic power plant. Enumerate: three of your expected fixed costs and three of your expected variable costs.
2. List all the costs you will expect to have from the construction and operation of a geothermal power plant, which is to produce energy for 40 years.

3. Because of population growth, it is expected that the city you live in will need an additional 300 MW of electric power. List all the alternatives to satisfy the increased demand.
4. The expected average rate of return in the next decade is 4%. What is the value of $60,000 ten years from now?
5. A financial company uses a 5% rate of return today and expects that this rate of return will increase by one percentage point in each of the next nine years. How much is today's sum of $100,000 worth to this company ten years from now?
6. You borrow $250,000 for a business venture to be repaid three years from now. You have the option to take a simple interest rate of 40 or a 1% rate compounded monthly interest. Which option will you choose?
7. What is the present value of $1,000,000 twenty years from now? The discount rate is 7%.
8. A small geothermal power plant is expected to generate a net income of $300,000 in each of the next thirty years. What is the present value of this monetary stream if the discount rate is 6.5%?
9. From the point of view of an investment company, explain what the difference between the following is:
(a) The inter-bank discount rate.
(b) The prime rate.
(c) The cost of borrowed capital.
(d) The return on equity.
10. For the small photovoltaic power plant of Table 14.1 determine:

    (a) The net present value for a discount rate 7%.
    (b) The payback period.
    (c) The average return on book if the straight-line depreciation is taken.
    (d) The internal rate of return.
    (e) The profitability index.

11. For the small photovoltaic power plant of Table 14.1, what would be the net present value if there were a 25% investment credit in addition to the cash flow shown in the years when the investment is made?
12. [3]The initial investment for a geothermal power plant is $60,000,000 and is to be paid in two years (year 0 and year 1) equally. The electricity generated by the plant is expected to bring revenue of 10,000,000 per year increasing at a rate of 3% annually. The total annual expenses of the plant will be 1,500,000 increasing at a rate 4% annually. The company that owns the plant plans to borrow 50% of the capital expenditure at a rate 6%. The discount rate for this company is 12% and its taxation rate is 35%. A 10 year depreciation period is allowed for this power plant. The plant is expected to generate electricity for

[3] Construct a spreadsheet to facilitate the solution of the next four problems.

30 years after which it will have zero value. What is the present value of this investment? Should the company undertake the investment?

13. What would be the answer to the plant of problem 12 if the company were allowed to borrow 80% of the cost of the geothermal power plant at an annual rate of 5%.
14. In order to promote the development of alternative energy sources the government allows 5 year depreciation for the project of problem 12. What is the net present value of the investment?
15. In order to accelerate the use of geothermal power, the government allows a 20% tax credit for investment in geothermal energy (in the years the investment is made) as well as 5 year depreciation for the project of problem 12. What is the net present value of the investment now?
16. "The government does not need to subsidize energy projects. In a free market economy we should allow the market forces to determine if a project is going to succeed or fail." Comment.

Index

E. E. (Stathis) Michaelides, *Alternative Energy Sources*,
Green Energy and Technology, DOI: 10.1007/978-3-642-20951-2,

CPSIA information can be obtained
at www.ICGtesting.com
Printed in the USA
LVOW13*0037170717
541563LV00005B/249/P